Food Pac

Food Packaging

Advanced Materials, Technologies, and Innovations

Edited by
Sanjay Mavinkere Rangappa
Jyotishkumar Parameswaranpillai
Senthil Muthu Kumar Thiagamani
Senthilkumar Krishnasamy
Suchart Siengchin

CRC Press is an imprint of the
Taylor & Francis Group, an **informa** business

First edition published 2021
by CRC Press
6000 Broken Sound Parkway NW, Suite 300, Boca Raton, FL 33487-2742

and by CRC Press
2 Park Square, Milton Park, Abingdon, Oxon, OX14 4RN

CRC Press is an imprint of Taylor & Francis Group, LLC

Library of Congress Cataloging-in-Publication Data

Names: Rangappa, Sanjay Mavinkere, editor. | Parameswaranpillai, Jyotishkumar, editor. | Thiagamani, Senthil Muthu Kumar, editor. | Krishnasamy, Senthilkumar, editor. | Siengchin, Suchart, editor.
Title: Food packaging : advanced materials, technologies, and innovations / edited by Sanjay Mavinkere Rangappa, Jyotishkumar Parameswaranpillai, Senthil Muthu Kumar Thiagamani, Senthilkumar Krishnasamy, Suchart Siengchin.
Other titles: Food packaging (CRC Press)
Description: First edition. | Boca Raton, FL : CRC Press, 2020. | Includes bibliographical references and index. | Summary: "This one-stop reference is aimed at packaging materials researchers working across various industries. With chapters written by leading international researchers, it offers a broad view of important developments in food packaging. The book presents an extensive survey of food packaging materials and modern technologies and demonstrates the potential of various materials for use in demanding applications. It discusses use of polymers, composites, nanotechnology, hybrid materials, coatings, wood-based, and other materials in packaging. It also describes biodegradable packaging, antimicrobial studies, and environmental issues related to packaging materials"-- Provided by publisher.
Identifiers: LCCN 2020030268 (print) | LCCN 2020030269 (ebook) | ISBN 9780367335380 (hardback) | ISBN 9780429322129 (ebook)
Subjects: LCSH: Food--Packaging--Technological innovations.
Classification: LCC TP374 .F6532 2020 (print) | LCC TP374 (ebook) | DDC 664/.09--dc23
LC record available at https://lccn.loc.gov/2020030268
LC ebook record available at https://lccn.loc.gov/2020030269

ISBN: 978-0-367-33538-0 (hbk)
ISBN: 978-0-367-55600-6 (pbk)
ISBN: 978-0-429-32212-9 (ebk)

Typeset in Times
by Deanta Global Publishing Services, Chennai, India

Contents

Preface

The storage, protection, and transportation of food products is an important area of research in the 21st century. Even today, approximately 30% of the food around the world is wasted. Extending shelf life and safe packing are essential to preventing the loss of food materials. Food packaging of synthetic nonbiodegradable plastics is often used to extend the shelf life of the food due to its versatility in performance because of its good thermomechanical properties, water barrier properties, low cost, and because it is lightweight. Approximately 35% of the plastics produced are used in the food packaging industry and this covers ca. 60% of the total plastic waste generated by volume. However, the plastic waste generated after usage is not properly recycled. This causes dumping of plastic waste in landfill or the natural environment. Therefore, because of the environmental issues of nonbiodegradable plastics, biodegradable plastics are preferred for food packaging. The demand for high-quality food has led to the introduction of active agents in food packaging. The active agents, such as enzymes, essential oils, nanoparticles, etc., can impart antimicrobial properties. Thus, the active agents can enhance the protection and shelf life of the food.

The high acceptance of active food packaging systems based on biodegradable polymers is expected to increase the production of biodegradable polymers from both synthetic and natural resources. Over the last decade, research in the field of active food packaging has been booming. This has led to an upsurge in the number of related papers, reviews, and patents. Therefore, we believe it is important to edit a book on *Food Packaging: Advanced Materials, Technologies, and Innovations.* We hope that this book will benefit researchers, engineers, faculty members, and students working in the area of food packaging.

This book consists of 14 chapters that describe the recent advances in food packaging. Chapter 1 provides an overview of bio-based materials for active food packaging. Chapter 2 reviews the different biodegradable polymers and their applications in the food packaging industry. Chapter 3 discusses the recent developments in multilayer biodegradable polymer films for food packaging applications. Chapter 4 provides an overview of the environmental impacts of food packaging. Special emphasis has been given to the life cycle assessment of the packaging materials. Chapter 5 provides an overview of several approaches (antimicrobial sachet, antimicrobial surface coating, antimicrobial polymers as packaging material, etc.) adopted for antimicrobial food packaging applications. Chapter 6 reviews the utilization of bioactive substances such as plant extracts, peptides, essential oils, and bacteriocins for the protection and extension of shelf life of food materials. Chapter 7 discusses the importance of biodegradable polymer matrices such as cellulose and polylactic acid for food packing applications. Chapter 8 addresses the emerging electrospun nanofiber technology for the development of active and intelligent food packaging systems. Chapter 9 discusses the role of nanoparticles in improving shell life and the protection of food material. Chapter 10 focuses on the recent developments in intelligent food packaging that enables effective monitoring of the freshness and quality of food. Chapter 11 highlights the recent advances in chitin/chitosan films for food packaging applications. The different processing methods, strategies to develop active food packaging systems with extended shell life, intelligent food packing, etc., are discussed in detail. Chapter 12 discusses the characteristic properties of chitosan-based hybrid nanocomposites and their possible applications in food packaging. Chapter 13 discusses the manufacturing, characteristics, and safety evaluation of medium-density fiberboard for its applications as a packaging material of fresh fruits and vegetables. Chapter 14 provides an overview of the recent progress of nanocomposites in active and intelligent food packaging systems

The editors thank all the contributors for their chapters and the CRC Press publishing team for their guidance.

Editors

Sanjay Mavinkere Rangappa is a Research Scientist at the Natural Composites Research Group Lab, Academic Enhancement Department, King Mongkut's University of Technology North Bangkok, Thailand.

Dr. Sanjay Mavinkere Rangappa received a B. E. (Mechanical Engineering) from Visvesvaraya Technological University, Belagavi, India, in 2010; an M. Tech. (Computational Analysis in Mechanical Sciences) from VTU Extension Center, GEC, Hassan, in 2013; a Ph.D. (Faculty of Mechanical Engineering Science) from Visvesvaraya Technological University, Belagavi, India, in 2017; and Postdoctorate from King Mongkut's University of Technology North Bangkok, Thailand, in 2019. He is a Life Member of Indian Society for Technical Education (ISTE) and an Associate Member of Institute of Engineers (India). He is a reviewer for more than 50 international journals (for Elsevier, Springer, Sage, Taylor & Francis Group, Wiley), book proposals, and international conferences. In addition, he has published more than 85 articles in high-quality international peer-reviewed journals, three editorial corner, more than 20 book chapters, one book as an author, 15 books as an editor, and also presented research papers at national/international conferences. His current research areas include natural fiber composites, polymer composites, and Advanced Material Technology. He is a recipient of the DAAD Academic exchange Project-Related Personnel Exchange Program (PPP) between Thailand and Germany to the Institute of Composite Materials, University of Kaiserslautern, Germany. He has received a Top Peer Reviewer 2019 award, a Global Peer Review Award powered by Publons and the Web of Science Group.

Jyotishkumar Parameswaranpillai is a Research Professor at the Center of Innovation in Design and Engineering for Manufacturing, King Mongkut's University of Technology North Bangkok, Thailand.

Dr. Jyotishkumar Parameswaranpillai received his Ph.D. in Polymer Science and Technology (Chemistry) from Mahatma Gandhi University. He has research experience in various international laboratories such as Leibniz Institute of Polymer Research Dresden (IPF) Germany, Catholic University of Leuven, Belgium, and University of Potsdam, Germany. He has published more than 100 papers in high-quality international peer-reviewed journals on polymer nanocomposites, polymer blends and alloys, and biopolymer, and has edited five books. He received numerous awards and recognitions including the prestigious Kerala State Award for the Best Young Scientist 2016, the INSPIRE Faculty Award 2011, and the Best Researcher Award 2019 from King Mongkut's University of Technology North Bangkok.

Senthil Muthu Kumar Thiagamani is an Associate Professor at the Department of Mechanical Engineering, Kalasalingam Academy of Research and Education, Tamil Nadu, India.

Dr. Senthil Muthu Kumar Thiagamani received his Diploma in Mechanical Engineering from the Directorate of Technical Education, Tamil Nadu in 2004, obtained his Bachelor of Engineering in Mechanical Engineering from Anna University Chennai in 2007, and Masters of Technology in Automotive Engineering from Vellore Institute of Technology, Vellore in 2009. He received his Doctor of Philosophy in Mechanical Engineering (specializing in bio-composites) from Kalasalingam Academy of Research and Education (KARE) in 2018. He has also completed his postdoctoral research (natural fiber composites) in the Department of Mechanical and Process Engineering at The Sirindhorn International Thai-German Graduate School of Engineering (TGGS), King Mongkut's University of Technology (KMUTNB), Thailand. He is a member of international bodies such as the Society of Automotive Engineers and the International Association of Advanced Materials. He started his academic career as Lecturer in Mechanical Engineering at KARE in 2010. He has around ten years of teaching and research experience. He is also a visiting researcher at KMUTNB. His research interests include biodegradable polymer composites and characterization. He has authored more than 25 peer-reviewed journal articles and book chapters. He is currently editing three books on the theme of bio-composites.

Senthilkumar Krishnasamy is a Research Scientist at the Center of Innovation in Design and Engineering for Manufacturing (CoI-DEM), King Mongkut's University of Technology North Bangkok, Thailand.

Dr. Senthilkumar Krishnasamy graduated as a Bachelor in Mechanical Engineering from Anna University, Chennai, India, in 2005. He then chose to continue his master's studies and was graduated with a master's degree in CAD/CAM from Anna University, Tirunelveli, in 2009. He has obtained his Ph.D. from the Department of Mechanical Engineering—Kalasalingam University (2016). He had been working in the Department of Mechanical Engineering, Kalasalingam Academy of Research and Education (KARE), India, from 2010 to 2018 (October). He has completed his postdoctoral fellowship at Universiti Putra Malaysia, Serdang, Selangor, Malaysia and King Mongkut's University of Technology (KMUTNB) North Bangkok under the research topics of "Experimental investigations on mechanical, morphological, thermal and structural properties of kenaf fiber/mat epoxy composites and Sisal composites" and "Fabrication of Eco-friendly hybrid green composites on tribological properties in a medium-scale application," respectively. His area of research interests includes the modification and treatment of natural fibers, nanocomposites, and hybrid-reinforced polymer composites. He has been published in more than 25 International journal papers, three book chapters, spoken at 12 international conferences and six national conferences in the field of natural fiber composites. He is one of the editors for three approved books.

Suchart Siengchin is President of King Mongkut's University of Technology North Bangkok, Thailand.

He is affiliated with the Department of Materials and Production Engineering (MPE), The Sirindhorn International Thai-German Graduate School of Engineering (TGGS), King Mongkut's University of Technology North Bangkok, Thailand.

Prof. Dr.-Ing. habil. Suchart Siengchin received his Dipl.-Ing. in Mechanical Engineering from University of Applied Sciences Giessen/Friedberg, Hessen, Germany, in 1999, M.Sc. in Polymer Technology from University of Applied Sciences Aalen, Baden-Wuerttemberg, Germany, in 2002, M.Sc. in Material Science at the Erlangen-Nürnberg University, Bayern, Germany, in 2004, Doctor of Philosophy in Engineering (Dr.-Ing.) from Institute for Composite Materials, University of Kaiserslautern, Rheinland-Pfalz, Germany, in 2008, and carried out postdoctoral research in Kaiserslautern University and School of Materials Engineering, Purdue University, the United States. In 2016, he received the habilitation at the Chemnitz University in Sachen, Germany. He worked as a Lecturer for the Production and Material Engineering Department at The Sirindhorn International Thai-German Graduate School of Engineering (TGGS), KMUTNB. He has been full Professor at KMUTNB and became the President of KMUTNB. He won the Outstanding Researcher Award in 2010, 2012, and 2013 at KMUTNB. His research interests are in Polymer Processing and Composite Material. He is Editor-in-Chief of *KMUTNB International Journal of Applied Science and Technology* and the author of more than 150 peer-reviewed journal articles. He has participated with presentations in more than 39 international and national conferences with respect to Materials Science and Engineering topics.

Contributors

B. Ashok
Department of Physics
Osmania University College of Engineering
Hyderabad, India

Anna Rafaela Cavalcante Braga
Department of Chemical Engineering
Universidade Federal de São Paulo (UNIFESP)
São Paulo, Brazil

M. Chandrasekar
School of Aeronautical Sciences
Hindustan Institute of Technology and Science
Chennai, India

Yu-Shen Cheng
Department of Chemical and Materials Engineering
National Yunlin University of Science and Technology
Douliu, Taiwan

Jorge Alberto Vieira Costa
Laboratory of Biochemical Engineering
College of Chemistry and Food Engineering
Federal University of Rio Grande (FURG)
Rio Grande, Brazil

Fernanda Domingues
Centro de Investigação em Ciências da Saúde (CICS-UBI)
Universidade da Beira Interior
Covilhã, Portugal

and

Departamento de Química
Faculdade de Ciências
Universidade da Beira Interior
Covilhã, Portugal

Mariana Buranelo Egea
Goiano Federal Institute of Education
Science and Technology
Campus Rio Verde
Rio Verde, Brazil

Gabriel da Silva Filipini
Laboratory of Food Technology
School of Chemistry and Food
Federal University of Rio Grande
Rio Grande, Brazil

R. A. Ilyas
Laboratory of Biocomposite Technology
Institute of Tropical Forestry and Forest Products (INTROP)
Advanced Engineering Materials and Composites Research Centre (AEMC)
Department of Mechanical and Manufacturing Engineering
Universiti Putra Malaysia
Serdang, Malaysia

Aswathy Jayakumar
School of Biosciences
Mahatma Gandhi University
Kottayam, India

Radhakrishnan E. K.
School of Biosciences
Mahatma Gandhi University
Kottayam, India.

Jasila Karayil
Department of Chemistry
Government Polytechnic Women's College
Calicut, India

Suelen Goettems Kuntzler
Laboratory of Microbiology and Biochemistry
College of Chemistry and Food Engineering
Federal University of Rio Grande (FURG)
Rio Grande, Brazil

Ailton Cesar Lemes
School of Chemistry
Department of Biochemical Engineering
Federal University of Rio de Janeiro (UFRJ)
Rio de Janeiro, Brazil

Ângelo Luís
Centro de Investigação em Ciências da Saúde (CICS-UBI)
Universidade da Beira Interior
Covilhã, Portugal

and

Laboratório de Fármaco-Toxicologia
UBIMedical
Universidade da Beira Interior
Covilhã, Portugal

Vilásia Guimarães Martins
Laboratory of Food Technology
School of Chemistry and Food
Federal University of Rio Grande
Rio Grande, Brazil

Liana Noor Megashah
Department of Bioprocess Technology
Faculty of Biotechnology and Biomolecular Sciences
Universiti Putra Malaysia
Serdang, Malaysia

Michele Greque de Morais
Laboratory of Microbiology and Biochemistry
College of Chemistry and Food Engineering
Federal University of Rio Grande (FURG)
Rio Grande, Brazil

Juliana Botelho Moreira
Laboratory of Microbiology and Biochemistry
College of Chemistry and Food Engineering
Federal University of Rio Grande (FURG)
Rio Grande, Brazil

Lawrence Yee Foong Ng
Laboratory of Biopolymer and Derivatives
Institute of Tropical Forestry and Forest Products
Universiti Putra Malaysia
Serdang, Malaysia

Daiane Nogueira
Laboratory of Food Technology
School of Chemistry and Food
Federal University of Rio Grande
Rio Grande, Brazil

Mohd Nor Faiz Norrahim
Research Centre for Chemical Defence (CHEMDEF)
Universiti Pertahanan Nasional Malaysia
Kuala Lumpur, Malaysia

Josemar Gonçalves de Oliveira-Filho
São Paulo State University (UNESP)
School of Pharmaceutical Sciences
Araraquara, Brazil

Farah Nadia Mohammad Padzil
Laboratory of Biopolymer and Derivatives
Institute of Tropical Forestry and Forest Products
Universiti Putra Malaysia
Serdang, Malaysia

Jyotishkumar Parameswaranpillai
Center of Innovation in Design and Engineering for Manufacturing
King Mongkut's University of Technology North Bangkok
Bangkok, Thailand

A. Rodríguez Bernaldo de Quirós
Department of Analytical Chemistry
Nutrition and Food Science
Faculty of Pharmacy
University of Santiago de Compostela
Santiago de Compostela, Spain

Sanjay M. R.
Natural Composites Research Group Lab
Academic Enhancement Department
King Mongkut's University of Technology North Bangkok
Bangkok, Thailand

Sabareesh Radoor
Department of Mechanical and Process Engineering
The Sirindhorn International Thai—German Graduate School of Engineering (TGGS)
King Mongkut's University of Technology North Bangkok
Bangkok, Thailand

Noor Farisha Abd. Rahim
Department of Quality Assurance
Putra Business School
Serdang, Malaysia

A. Varada Rajulu
International Research Centre
Kalasalingam University
Krishnankoil, India

Maria Râpă
Processing of Metallic and Ecometallurgical Materials Department
University Politehnica of Bucharest
Bucharest, Romania

Haniyeh Rostamzad
Fisheries Department
Faculty of Natural Resources
University of Guilan
Sowmeh Sara, Iran

F. Salgado
SP Consulting
Poio-Pontevedra, Spain

S. M. Sapuan
Laboratory of Biocomposite Technology
Institute of Tropical Forestry and Forest Products (INTROP)
Advanced Engineering Materials and Composites Research Centre (AEMC)
Department of Mechanical and Manufacturing Engineering
Universiti Putra Malaysia
Serdang, Malaysia

R. Sendón
Department of Analytical Chemistry
Nutrition and Food Science
Faculty of Pharmacy
University of Santiago de Compostela
Santiago de Compostela, Spain

K. Senthilkumar
Center of Innovation in Design and Engineering for Manufacturing
King Mongkut's University of Technology North Bangkok
Bangkok, Thailand

Nur Sharmila Sharip
Laboratory of Biopolymer and Derivatives
Institute of Tropical Forestry and Forest Products (INTROP)
Universiti Putra Malaysia
Selangor, Malaysia

Siti Shazra Shazleen
Laboratory of Biopolymer and Derivatives
Institute of Tropical Forestry and Forest Products (INTROP)
Universiti Putra Malaysia
Serdang, Malaysia

Suchart Siengchin
Department of Materials and Production Engineering (MPE)
The Sirindhorn International Thai—German Graduate School of Engineering (TGGS)
King Mongkut's University of Technology North Bangkok
Bangkok, Thailand

Filomena Silva
Agencia Aragonesa para la Investigación y el Desarollo (ARAID)
Zaragoza, Spain

and

Faculty of Veterinary Medicine
University of Zaragoza
Zaragoza, Spain

Malinee Sriariyanun
Department of Mechanical and Process Engineering
The Sirindhorn International Thai-German Graduate School of Engineering
King Mongkut's University of Technology North Bangkok
Bangkok, Thailand

Karthikeyan Subramaniam
Department of Automobile Engineering
Kalasalingam Academy of Research and Education
Anand Nagar, Krishnankoil, Tamilnadu, India

R. Syafiq
Laboratory of Biocomposite Technology
Institute of Tropical Forestry and Forest Products (INTROP)
Advanced Engineering Materials and Composites Research Centre (AEMC)
Department of Mechanical and Manufacturing Engineering
Universiti Putra Malaysia
Serdang, Malaysia

Atthasit Tawai
Department of Mechanical and Process Engineering
The Sirindhorn International Thai-German Graduate School of Engineering
King Mongkut's University of Technology North Bangkok
Bangkok, Thailand

Ana Luiza Machado Terra
Laboratory of Microbiology and Biochemistry
College of Chemistry and Food Engineering
Federal University of Rio Grande (FURG)
Rio Grande, Brazil

Senthil Muthu Kumar Thiagamani
Department of Automobile Engineering
School of Mechanical and Automotive Engineering
Kalasalingam Academy of Research and Education
Anand Nagar, India

Theivasanthi Thirugnanasambandan
International Research Centre
Kalasalingam University
Krishnankoil, India

Sandhya Alice Varghese
Department of Mechanical and Process Engineering
The Sirindhorn International Thai-German Graduate School of Engineering (TGGS)
King Mongkut's University of Technology
Bangkok, Thailand

Cornelia Vasile
"Petru Poni" Institute of Macromolecular Chemistry
Physical Chemistry of Polymers Department
Iasi, Romania

P. Vazquez-Loureiro
Department of Analytical Chemistry
Nutrition and Food Science
Faculty of Pharmacy
University of Santiago de Compostela
Santiago de Compostela, Spain

Tengku Arisyah Tengku Yasim-Anuar
Department of Bioprocess Technology
Faculty of Biotechnology and Biomolecular Sciences
Universiti Putra Malaysia
Serdang, Malaysia

1 Bio-Based Materials for Active Food Packaging
Dream or Reality

Ângelo Luís, Fernanda Domingues, and Filomena Silva

CONTENTS

1.1 INTRODUCTION

The use of plastic packaging is on the rise, explained by the need to reduce food waste and the increased demand due to population growth and market expansion (Groh et al., 2019; Narancic and O'Connor, 2019). However, there are also increasing concerns about the damage caused to the environment and human health (Groh et al., 2019; Narancic and O'Connor, 2019). These concerns include littering and accumulation of nondegradable plastics in the environment (Andrady, 2011), generation of secondary micro- and nano-plastics (Hernandez et al., 2019), and release of hazardous chemicals during manufacturing and use, as well as following landfilling, incineration, or improper disposal leading to pollution of the environment (Groh et al., 2019; Narancic and O'Connor, 2019).

Advances in food packaging have played a vital role across the world (Marsh and Bugusu, 2007). Simply stated, packaging maintains the benefits of food processing after the process is completed, enabling foods to travel safely for long distances from their point of origin and still be wholesome at the time of consumption (Lee et al., 2015; Sohail, Sun, and Zhu, 2018). However, packaging technology must balance food protection with other issues, including energy and material costs, heightened social and environmental consciousness, and strict regulations on pollutants and disposal of solid waste (Marsh and Bugusu, 2007).

Among developed countries, the European Union (EU) has the strictest arsenal of rules. "EU Framework Regulation 1935/2004/EC" makes risk assessment and risk management compulsory for the introduction of any new substance, material, or industrial practice (active/intelligent packaging system, recycling) regardless of whether the material is plastic or not (Zhu, Guillemat, and Vitrac, 2019). While not mentioning design, "EU Regulation 2023/2006/EC" encourages the developments of good manufacturing practices and quality assurance systems at all stages of production and handling for each of the 17 groups of materials and their combinations accepted for food contact (Zhu, Guillemat, and Vitrac, 2019). The reduction of the environmental impact of packaging and its wastes has been enforced in the EU for the past 25 years via the "Directive 94/62/EC" (Zhu, Guillemat, and Vitrac, 2019).

New packaging systems are an effective way to extend and/or maintain the shelf life of foods and to preserve the quality of a range of fresh products such as vegetables, fruits, meat, and fish (Lee et al., 2015; Janjarasskul and Suppakul, 2018). Active packaging (AP) systems have been developed by the food industry, and their effectiveness in extending the shelf life of fresh foods has been reported in several studies (Lee et al., 2015; Janjarasskul and Suppakul, 2018; Al-Tayyar, Youssef, and Al-hindi, 2020). For example, active packaging inhibits microbial contamination and spoilage in fresh foods by employing antimicrobial packaging materials or gaseous agents or antimicrobial inserts in the package headspace (Lee et al., 2015; Janjarasskul and Suppakul, 2018; Pellerito et al., 2018).

Polymer biocomposites are novel materials that have gained increased interest from researchers in polymer science and engineering, since they are novel, high-performance, lightweight, and eco-friendly materials that can replace traditional nonbiodegradable plastic packaging (Al-Tayyar, Youssef, and Al-hindi, 2020).

Therefore, the major aim of this chapter is to make an overview of bio-based materials for active food packaging keeping also in mind the circular economy and the degradation path of these materials.

1.2 ACTIVE FOOD PACKAGING: GENERAL DEFINITIONS AND SOME EXAMPLES

Food packaging, in a broad sense, refers to the barrier that protects food products from environmental disturbances such as oxygen, moisture, light, dust, chemical, and microbiological contamination. At its beginnings, packaging materials were inert polymers made of cardboard, glass, and, ultimately, plastic that only served this protection purpose; but over the years, packaging has evolved to incorporate technologies such as vacuum (VP) and modified atmosphere packaging (MAP). These technologies have long been used to maintain product freshness and increase its shelf life by controlling the gas atmosphere of the packaged food, delaying oxygen-dependent processes such as food oxidation and microbial growth. Despite the success attained by such technologies, the food industry still requires newer and better performing technologies to decrease food losses and extend even further the shelf life of packaged products, which resulted in the introduction of active packaging. According to "European Regulation (EC) No 450/2009," active packaging systems "deliberately incorporate components that would release or absorb substances into or from the packaged food or the environment surrounding the food" (European Commission, 2009). In general, active packaging systems can be divided into two main groups: absorbers (active scavenging systems) and emitters (active-releasing systems) (Yildirim et al., 2018) (see Figure 1.1).

Furthermore, AP can be divided according to the main problem they are supposed to tackle in the food product, such as oxidation (antioxidant packaging), microbial growth (antimicrobial packaging), increased moisture (moisture absorber) and ripening control (ethylene absorber) (see Table 1.1).

1.2.1 ANTIOXIDANT PACKAGING

Antioxidant compounds can be divided into two groups according to their function: primary or chain breaking antioxidants and secondary or preventive antioxidants (Mishra and Bisht, 2011). Primary antioxidants can accept free radicals (e.g., lipid

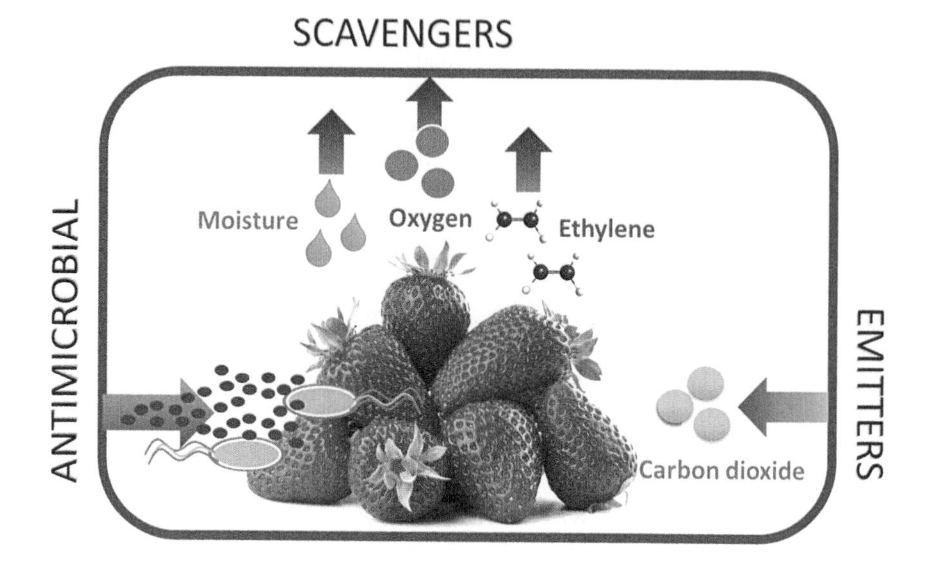

FIGURE 1.1 Types of active food packaging.

or peroxyl radicals) leading to the delay of autooxidation initiation or the interruption of its propagation by converting free radicals to stable radicals or non-radical products usually through the donation of hydrogen atoms (Mishra and Bisht, 2011). In contrast, secondary antioxidants act by preventing oxidation processes by a variety of routes such as binding transition metal ions, scavenging oxygen, replenishing hydrogen atoms to primary antioxidants, absorbing UV light and inhibiting oxidizing enzymes (Silva, Becerril, and Nerin, 2019). For instance, phenolic compounds and hydroxylamines act as multi-functional antioxidants (Ambrogi et al., 2017) as they can function either as primary or secondary antioxidants through different mechanisms (Mishra and Bisht, 2011). In terms of antioxidant packaging (AOXP) technologies, the two main types available are oxygen scavengers/absorbers and free radical scavengers. Oxygen absorbers react with the oxygen in the packaging atmosphere, ceasing its action on the food product. The most widely used oxygen scavenger is ferrous oxide although there are other compounds available such as catechols, ascorbic acid, sulfites, and enzymes such as glucose oxidase or catalase (Nerin, Vera, and Canellas, 2018; Kerry, O'Grady, and Hogan, 2006). Free radical scavengers such as synthetic ones like selenium nanoparticles, butylated hydroxytoluene (BHT), butylated hydroxyanisole (BHA) and tert-butylhydroquinone (TBHQ) or natural ones such as plant extracts and essential oils react with free radicals so that they are unable to engage in further oxidative chain reactions. Over the years, AOXP has evolved from using synthetic antioxidants incorporated in plastic materials such as polyethylene (PE), polypropylene (PP), polyethylene terephthalate (PET) or polyvinyl alcohol (PVA) to using natural antioxidants incorporated in the same plastic materials and, finally, the most recent tendency is the incorporation of natural antioxidants (green tea, olive leaf extracts, resveratrol, alpha-tocopherol, essential oils, etc.) in more eco-friendly and sustainable materials such as bioplastics

TABLE 1.1
Types of Active Packaging and Examples of Strategies Used

Issue	Effects on food quality	Type of packaging	Strategy	References
Oxidation	Changes in organoleptic properties (taste, color, odor, etc.). Changes in nutritional value (oxidation of vitamins, proteins, fatty acids, etc.). Formation of oxidation products potentially harmful to humans.	Antioxidant	Use of oxygen scavengers (ferrous oxide, catechols, ascorbic acid, sulfites, photosensitive dyes, unsaturated hydrocarbons, and enzymes such as glucose oxidase and catalase) to control oxygen levels in the packaging atmosphere. Use of free radical scavengers (butylated hydroxytoluene (BHT), butylated hydroxyanisole (BHA), and tert-butylhydroquinone (TBHQ), plant extracts, essential oils, selenium nanoparticles) to inhibit secondary oxidation chain reactions caused by free radicals.	(Silva, Becerril, and Nerin, 2019; Borzi et al., 2019; Vera, Canellas, and Nerín, 2018; Otoni et al., 2016)
Moisture	Texture and appearance changes. Increase susceptibility to microbial damage.	Moisture absorbers/ scavengers	Desiccants (clay, silica, and zeolites), bentonite, poly-acrylic acid sodium salts	(Dey and Neogi, 2019)
Ethylene production	Ripening acceleration in fruits and greens. Chlorophyll degradation. Shelf-life decrease. Increased microbial contamination.	Ethylene absorbers/ inhibitors	1-Methylcyclopropene (1-MCP) and silver inhibit ethylene action by competing with its binding site in fruits and vegetables. Potassium permanganate removes ethylene through its oxidation. Biofilters with microorganisms remove ethylene by microbial degradation. Zeolites and clays absorb ethylene Palladium-based absorbers scavenge ethylene.	(Smith et al., 2009; Keller et al., 2013)

(Continued)

TABLE 1.1 (CONTINUED)
Types of Active Packaging and Examples of Strategies Used

Issue	Effects on food quality	Type of packaging	Strategy	References
Microbial growth	Changes in organoleptic properties due to the growth of spoilage microorganisms. Development of foodborne diseases due to the consumption of pathogen-contaminated food. Production of toxic secondary microbial metabolites (mycotoxins).	Antimicrobial	Carbon dioxide emitters increase CO_2 concentration and decrease pH, decreasing the growth of aerobic microorganisms and slowing down enzymatic reactions. Antimicrobial-releasing materials contain synthetic (organic acids, ethyl lauryl arginate) or natural (essential oils, bacteriophages, bacteriocins, enzymes, etc.) antimicrobials that cause microbial (bacteria, fungi, yeast) growth inhibition.	(Silva, Becerril, and Nerin, 2019; Silva et al., 2019; Silva, Domingues, and Nerín, 2018; Otoni et al., 2016; Nguyen Van Long, Joly, and Dantigny, 2016)

(e.g., poly(lactic acid)) and biopolymers (starch, protein, chitosan, etc.). These packages with antioxidant properties have been applied to a vast array of food products that are susceptible to oxidation such as fatty foods (peanuts, cereals, butter, oils), meat (beef, pork, chicken), fish and vegetables (mushrooms) (for a more detailed review, please refer to Silva, Becerril, and Nerin (2019).

1.2.2 Water Absorbers/Scavengers

Although moisture control can be achieved with modified atmosphere and vacuum packaging and the use of low barrier polymers (micro-perforated films, enhanced permeability films), only the use of desiccants or absorbers can be designated as active moisture packaging. Other strategies are known as passive moisture packaging (Yildirim et al., 2018). Water scavengers/absorbers can be divided into two main types depending on whether they are able to control relative humidity or moisture (see Figure 1.2).

Relative humidity controllers absorb humidity in the headspace of the package as is the case of desiccants. Some examples of compounds used as desiccants are silica gel, clays, molecular sieves (zeolite, bentonite, alumina silicate salts, etc.), humectant salts (sodium, calcium and magnesium chloride, calcium sulfate) and other humectant compounds such as sorbitol and calcium oxide (Yildirim et al., 2018). These compounds are usually incorporated into packages as sachets, bags, or integrated into pads.

Moisture absorbers are typically placed underneath the food product in the form of pads, sheets, blankets, or humidity-regulating trays so that they can absorb the liquid released by the product (Yildirim et al., 2018). They are usually used for foods with high water activity such as fish, poultry meat, and fruits and vegetables, and contain superabsorbent substances such as polyacrylate salts, humectant salts, silicates,

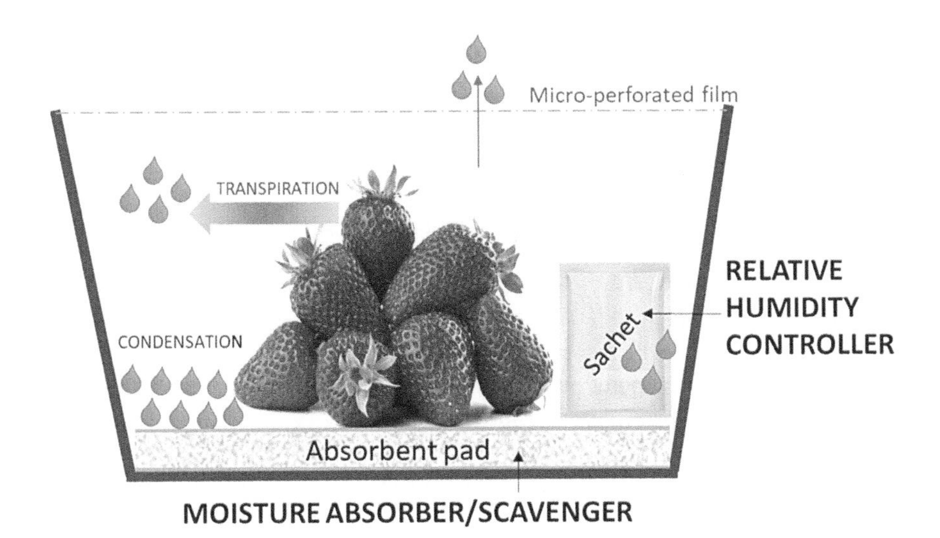

FIGURE 1.2 Examples of moisture active packaging technologies.

and starch copolymers, like galactoxyloglucan, added to the perforated polymer (Yildirim et al., 2018; Gaikwad, Singh, and Ajji, 2018). Additionally, these compounds can also be divided into moisture absorbers, as silica gel for instance, which can physically absorb and hold water molecules, and moisture scavengers, such as calcium chloride, which absorbs water through chemical reactions (Gaikwad, Singh, and Ajji, 2018).

1.2.3 Ethylene Scavengers/Absorbers

Ethylene (C_2H_4) is a natural plant growth regulator and is known to play a role in many physiological processes from seed germination to fruit ripening and senescence (Smith et al., 2009). The fact that ethylene is responsible for plant ripening means that this gas has an impact on the storage life of many fruits, vegetables, and ornamental crops (Keller et al., 2013). Ethylene production depends on the type of fruit or plant and we can distinguish between two classes of fresh produces: climateric and nonclimateric. Climateric fruits/plants are those that ripen after harvest (banana, avocado, tomato, etc.) and that ripen faster in response to an ethylene trigger ("climateric" period). In contrast, nonclimateric fruits (leafy vegetables, strawberries, oranges, etc.) do not ripen after harvest and ethylene exposure will not trigger ripening but will result in decreased shelf life. Furthermore, the ethylene production rate, fresh produce sensitivity to ethylene, and the resultant reactions to the gas differ from produce to produce. For instance, in the case of fruits, we can distinguish between very low ethylene producers (<0.1 μL/kg fruit/h) such as cut flowers or cherries, moderate (banana, tomato), high (avocado, apricot), and very high ethylene producers (>100 μL ethylene/kg fruit/h) such as apples or passion fruits (Keller et al., 2013; Smith et al., 2009). In the case of ethylene sensitivity, it is not directly connected with ethylene production since, for instance, very low ethylene producers such as cherries have a low sensitivity (3–5 ppm) to ethylene, while cut flowers, also very low ethylene producers, have a high sensitivity (0.01 0.5 ppm) to ethylene (Keller et al., 2013; Smith et al., 2009). As a result, ethylene can be physiologically active at very low concentrations within the ppm-ppb range. Therefore, even small amounts of ethylene during shipping and storage of fruits, vegetables, and ornamental plants can have deleterious effects on its shelf life leading to faster ripening and, eventually, deterioration of these fresh products and consequent product losses, reaching as high as 80% in some cases (Keller et al., 2013). Taking all this into consideration, over the past decades, the food packaging industry started to develop new strategies for ethylene control through two major routes: ethylene blocking by inhibiting of its action in foods or ethylene removal by its adsorption or oxidation (see Table 1.2). Ethylene blockers or removers can be incorporated into the packaging systems through several routes such as their incorporation in the packaging material (film, paper, etc.) or in the form of sachets or pads placed inside the package.

1.2.4 Antimicrobial Packaging

Antimicrobial packaging has the main goal of reducing, retarding, or even stopping microbial growth by interacting with the packaged food (direct contact) or the

TABLE 1.2
Strategies Used to Control Ethylene

Ethylene control route	Class of agent	Technology	Effects	References
Blocking		1-methylcyclopropene (1-MCP)	Volatile compound that blocks ethylene binding sites both in fruits and flowers.	(Watkins, 2006)
		Silver thiosulfate	Silver ions (Ag^+) bind to ethylene receptors and block its signal transduction. To improve its uptake and transport, Ag^+ is applied in the complex silver thiosulfate (STS). Exclusive for flowers and plants.	(Hyde et al., 2019; Smith et al., 2009)
		Aminoethoxyvinylglycine	Plant growth regulator by blocking the production of ethylene in the plant fruit.	(Saltveit, 2005)
		Calcium chloride	Calcium inhibits enzymes (1-aminocyclopropanecarboxylic acid synthase and 1-aminocyclopropanecarboxylic oxidase) responsible for endogenous ethylene production activity of both, and the production of endogenous ethylene.	(Zhang et al., 2019)
		Salicylic acid	Reduction of ethylene formation from 1-aminocyclopropanecarboxylic acid through enzymatic inhibition	(Leslie and Romani, 1988)

(*Continued*)

TABLE 1.2 (CONTINUED)
Strategies Used to Control Ethylene

Ethylene control route	Class of agent	Technology	Effects	References
Removal	Catalysts	Platinum/alumina	Catalytic oxidation of ethylene to water and carbon dioxide at 100–120°C.	(Conte et al., 1992)
		Titanium dioxide	Photocatalytic oxidation of ethylene at 3°C.	(Hussain et al., 2011)
	Oxidizing agents	Potassium permanganate	Incomplete oxidation of ethylene to carbon dioxide and water with the generation of other partially oxidized species.	(Mansourbahmani et al., 2018)
	Sorbents	Activated carbon	High ethylene adsorption capacity and selectivity due to its well-developed pore structure, unique surface (presence of functional groups), and large pore volume.	(Ye et al., 2010; Zhou et al., 2016)
		Zeolites	Due to their large surface area and cation exchange capacity, zeolites act as molecular sieves for ethylene. Additionally, zeolites can be modified with divalent ions such as Cu^{2+} or Zn^{2+} which increases its affinity for ethylene.	(Coloma et al., 2014; Pierucci et al., 2019)
	Catalyst/sorbent	Palladium-based material	Palladium-impregnated zeolites can act both as catalyst and absorber of ethylene under dry conditions or at 100% relative humidity.	(Tzeng et al., 2019; Smith et al., 2009)

package headspace (indirect contact) (Otoni, Espitia, Avena-Bustillos, and McHugh, 2016b). By controlling both spoilage and pathogenic microbial populations, antimicrobial packaging ensures consumer safety, while increasing food shelf life (Silva, Becerril, and Nerin, 2019). Nowadays, antimicrobial packaging can come in several forms such as sachets or pads containing volatile antimicrobials, polymer films with direct incorporation of antimicrobial substances (extrusion, casting) and coating or adsorption of antimicrobials onto the surface of the polymer (Radusin et al., 2013). Given the current interest in the development of antimicrobial packaging materials, the list of antimicrobials being studied for this purpose continues to grow taking into consideration the new trends for green consumerism. This led to a change in the type of antimicrobials applied with a preference toward natural antimicrobials (e.g., bacteriocins, enzymes, phages, biopolymers, natural extracts and compounds, essential oils and their components and metal nanoparticles) over synthetic ones (e.g., organic acids, triclosan, antibiotics, chlorine dioxide, nitrites, and ammonium salts). Although there is a vast array of antimicrobials used, not each one of them is suitable for every application. Firstly, the antimicrobial spectrum of the compound should be considered, meaning that the specific antimicrobial used should be active against the microbial flora of the food product to be packaged. Secondly, the regulatory status of the antimicrobial compound by the food safety agencies such as the US Food and Drug Administration and the European Food Safety Authority should also be considered. Last but not least, the compatibility between the packaging material used and the selected antimicrobial selected should also be examined (Dobrucka, 2015). Table 1.3 summarizes some of the antimicrobial compounds used in the development of antimicrobial food packaging.

1.3 EUROPEAN DIRECTIVE ON CIRCULAR ECONOMY: THE CHANGE FROM PLASTICS TO BIO-BASED MATERIALS

Traditional plastics derived from fossil fuels are a family of hundreds of materials with a wide range of properties (transparency, strength, flexibility, light weight, thermal performance, easy and low-cost production, and simple sterilization methods), making them ideal materials to be applied in many industrial and consumer products (Narancic and O'Connor, 2019; Luzi et al., 2019; Fortunati, Mazzaglia, and Balestra, 2019). In 2016, 335 million tons of plastic were produced globally, reflecting their popularity and widespread application (Narancic and O'Connor, 2019). Plastics help society to reduce food waste by, for example, providing better barrier properties and thus delaying food spoilage. In addition, the plastics industry in the EU employed 1.5 million people and generated a turnover of €340 billion in 2015. Approximately 40% of the produced plastic goes to the food packaging sector. In the EU, 70% of the collected plastic waste ends in landfills or is incinerated (Narancic and O'Connor, 2019). A serious alarm is the field of food packaging, which every year generates enormous quantities of petroleum-based wastes that are stored in particular areas around the planet, causing enormous negative effects and high recycling costs (Luzi et al., 2019; Russell, 2014). In future years, especially in 2030 and 2050, it is projected that the quantity of plastic waste due to the packaging sector will have grown by two-fold and three-fold, respectively (Luzi et al., 2019). It is estimated that 8 million tons of

TABLE 1.3
Mode of Action of Commonly Used Antimicrobials Used in Active Food Packaging

Antimicrobial class	Microbial action	Mode of action	References
Organic acids	*Bacillus subtilis* *Listeria monocytogenes* *Staphylococcus aureus* *Escherichia coli* *Candida albicans*	Salts of lactic acid, such as sodium lactate and potassium lactate, exert greater inhibitory effects against Gram-positive bacteria than against Gram-negative bacteria and offer antifungal activity against certain *Aspergillus species*. Potassium sorbate inhibits the germination of bacterial spores.	(Akhter et al., 2019; Birck et al., 2016; Liang et al., 2019)
Bacteriocins	*Listeria monocytogenes* *Clostridium perfringens* *Staphylococcus aureus* *Pseudomonas aeruginosa* *Aeromonas hydrophila* *Escherichia coli* *Salmonella* Typhimurium	These positively charged compounds can interact electrostatically with the negative charges of the phosphate groups on the microbial cell membranes, resulting in the generation of pores in the membrane and subsequent cell death. The broad spectrum bacteriocins from Gram-positive bacteria have a more suitable use for food applications than the ones from Gram-negative bacteria as the latter mainly contain lipopolysaccharides (LPS) or other endotoxins that can pose toxicity issues.	(Mapelli et al., 2019; Meira et al., 2017; Zimet et al., 2019; Woraprayote et al., 2013)
Enzymes	Pseudomonads *Listeria innocua* *Brochothrix thermosphacta* *Escherichia coli* *Enterococcus faecalis* *Shewanella putrefaciens* *Pseudomonas fluorescens* *Psychrotrophs* Total viable counts	Lysozyme is most effective against Gram-positive bacteria due to its ability to break down the glycosidic bonds of peptidoglycan in the cell wall of these bacteria. Lactoferrin, a whey protein that binds ferric ions, exerts its antimicrobial activity by depriving microbial cells of iron and by altering the permeability of Gram-negative bacteria due to its interaction with LPS components. The glucose oxidase (GO) enzyme, a flavoprotein purified from different types of fungi exerts its antimicrobial activity by catalyzing the formation of hydrogen peroxide and gluconic acid through the oxidation of β-D-glucose. Lactoperoxidase catalyzes the oxidation of thiocyanate ion (SCN-) which generates oxidized products such as hypothiocyanite (OSCN-) and hypothiocyanous acid (HOSCN) that act as antimicrobial agents by causing the irreversible oxidation of sulfhydryl (SH) groups present in microbial enzymes and other proteins.	(Bugatti et al., 2018; Jasour et al., 2015; Murillo-Martínez et al., 2013)

(*Continued*)

TABLE 1.3 (CONTINUED)
Mode of Action of Commonly Used Antimicrobials Used in Active Food Packaging

Antimicrobial class	Microbial action	Mode of action	References
Bacteriophages	*Pseudomonas fluorescens* *Escherichia coli* O157:H7 *Listeria monocytogenes*	Lytic bacteriophages are viruses able to infect and lyse bacterial cells and, because of microbial cell lysis, release large number of progeny phages, which can then continue the infection cascade. As they are specific for a host cell, they do not interact with other microorganisms or eukaryotic cells in the environment and so, they do not cause illness neither in animals nor humans.	(Alves et al., 2019; Lone et al., 2016; Amarillas et al., 2018)
Metal nanoparticles	*Listeria monocytogenes* *Escherichia coli* *Staphylococcus aureus*	Metal-based NPs present low toxicity to eukaryotic (mammalian cells) as they can differentiate prokaryotic (bacterial cells) from mammalian cells through bacteria's metal transport system and metalloproteins. When acting selectively with the bacterial cell, these NPs trigger antimicrobial action through three main routes: interaction with the lipid by-layer, interaction with cytosolic proteins and through oxidative stress due to the generation of reactive oxygen species (ROS).	(Kim and Song, 2018; Marcous, Rasouli, and Ardestani, 2017; Lotfi, Ahari, and Sahraeyan, 2019)
Essential oils	*Staphylococcus* spp. *Listeria monocytogenes* *Botrytis cinerea* Total viable counts Pseudomonads Lactic acid bacteria *Aspergillus flavus* *Penicillium citrinum*	EOs antimicrobial activity is mainly a consequence of their hydrophobicity that enables them to partition the lipid layer of cell membranes and mitochondria, increasing their permeability, leading to the ion and small molecule leakage and, to a greater extent, cell lysis and death; They also disturb the cytoplasmic membrane by disrupting proton motive force, electron flow, active transport, and efflux. The lipophilic character of EOs makes them accumulate in lipid bilayers and disturb protein-lipid interactions.	(Campos-Requena et al., 2017; Artiga-Artigas, Acevedo-Fani, and Martín-Belloso, 2017; Hager et al., 2019; Niu et al., 2018; Chein, Sadiq, and Anal, 2019)

(Continued)

TABLE 1.3 (CONTINUED)
Mode of Action of Commonly Used Antimicrobials Used in Active Food Packaging

Antimicrobial class	Microbial action	Mode of action	References
Natural compounds	*Brochotrix thermosphacta* *Pseudomonas fluorescens* *Campylobacter jejuni* *Campylobacter coli* Total viable counts Pseudomonads Psychrotrophs Lactic acid bacteria *Escherichia coli* *Staphylococcus aureus* *Pseudomonas aeruginosa* *Listeria monocytogenes* *Salmonella* Typhimurium *Bacillus subtilis* *Bacillus cereus*	Multi-target action on microbial cells being able to disrupt membrane function and structure, interrupt DNA/RNA synthesis/function, interfere with intermediary metabolism, induce coagulation of cytoplasmic constituents, and interfere with cell-to-cell communication. This wide action on the microbial cell subsequently results in a broad spectrum of antimicrobial activity of these compounds and to a decreased risk in the arise of microbial resistance mechanisms.	(Alzagameem et al., 2019; Silva, Domingues, and Nerín, 2018; Balti et al., 2017)

plastic leak into the ocean each year, being equivalent to dumping the contents of one garbage truck into the ocean every minute, which is expected to increase to four trucks per minute by 2050 if no action is taken (Guillard et al., 2018).

The first reports of plastic litter in the oceans in the early 1970s drew minimal attention from the scientific community (Andrady, 2011). In the following decades, with accumulating data on the ecological consequences of such debris, the topic received increasingly sustained research interest (Andrady, 2011). Most studies have focused on the entanglement of marine mammals, cetaceans, and other species in net fragment litter and on "ghost fishing" (Andrady, 2011). Ingestion of plastics by birds and turtles is extensively documented worldwide and at least 44% of marine bird species are known to ingest plastics (Andrady, 2011). Plastics are now ubiquitous in the marine environment, and urgent action is required to mitigate this worsening trend. A 2014 study (from six years of research by the 5 Gyres Institute) estimated that 5.25 trillion plastic particles (weighing 269,000 tons) are floating in the sea. Due to its durability, the lifespan of plastic is estimated to be hundreds to thousands of years (Xanthos and Walker, 2017).

The aspiration of a circular economy is to shift material flows toward a zero waste and pollution production system (Tecchio et al., 2017). The sustainable use of resources is considered a keystone of the European roadmap to 2050, and to target this transition, the European Commission (EC) proposed an EU action plan for the circular economy in 2015 (Tecchio et al., 2017; Bartl, 2018). The EU hosts some of the world's most developed waste management systems and an ambitious policy commitment to the circular economy (Scheinberg et al., 2016). A circular economy would turn goods that are at the end of their service life into resources for others, closing loops in industrial ecosystems and minimizing waste (Stahel, 2016). It would change the economic logic because it replaces production with sufficiency: reuse what you can, recycle what cannot be reused, repair what is broken, remanufacture what cannot be repaired (Stahel, 2016). The most widely used definition of circular economy is the one formulated by the Ellen MacArthur Foundation in the early 2010s,

> [A]n industrial system that is restorative or regenerative by intention and design. It replaces the end-of-life concept with restoration, shifts towards the use of renewable energy, eliminates the use of toxic chemicals, which impairs reuse, and aims at the elimination of waste through the superior design of materials, products, systems, and, within this, business models.
>
> **(Milios, 2018)**

This is in contrast with linear economic activities (i.e., where resources are rapidly consumed, and production processes do not account for their unsustainable exploitation nor their recovery) that rely exclusively on the shrinking pool of Earth's natural resources and impose potential risks in the long run to the society (Milios, 2018). If linear production and consumption practices are complemented—and gradually substituted—by circular material flows, substantial resource efficiency improvements can be achieved (Milios, 2018).

In 2008, the waste hierarchy principle was included in the "Waste Framework Directive 2008/98/EC" (WFD), which was subsequently transposed into the national law of EU Member States. The European WFD defines the waste hierarchy as the priority order of operations to be followed in the management of waste: prevention, preparing for reuse, recycling, other recovery (including energy recovery), and disposal (Pires and Martinho, 2019; Bartl, 2018). In 2015, the "Circular Economy Strategy from EU COM/2015/0614" defended the role of waste management based on a waste hierarchy as the way to lead to the best overall environmental outcome and to get valuable materials back into the economy, as mentioned above (Pires and Martinho, 2019; Bartl, 2018). The circular economy is a concept promoted both to policymakers and to the overall business community, especially in the EU but also worldwide (Iaquaniello et al., 2018). Differently from a standard recycling practice, the circular economy emphasizes, among other aspects, waste utilization to close the cycle and minimize the use of resources. The recent EU action plan for the circular economy (European Commission 2018b, 2018a) requires revised legislative proposals on waste to include "an ambitious and credible long-term path for waste management and recycling" (Iaquaniello et al., 2018; Bartl, 2018).

The EC plan for a circular economy includes a specific component for the use of plastics more effectively and with the production of less waste. The section of the document entitled "A Vision for Europe's New Plastics Economy" offers the following: "A smart, innovative and sustainable plastics industry, where design and production fully respect the needs of reuse, repair, and recycling, brings growth and jobs to Europe and helps cut EU's greenhouse gas emissions and dependence on imported fossil fuels" (Rhodes, 2018). With this in mind, several specific measures were proposed:

- Innovative materials and alternative feedstocks for plastic production are developed and used where evidence clearly shows that they are more sustainable compared to the non-renewable alternatives. This supports efforts on decarbonization and creating additional opportunities for growth (European Commission, 2018b, 2018a).
- The EU is taking a leading role in a global dynamic, with countries engaging and cooperating to halt the flow of plastics into the oceans and taking remedial action against plastic waste already accumulated. Best practices are disseminated widely, scientific knowledge improves, citizens mobilize, and innovators and scientists develop solutions that can be applied worldwide (European Commission, 2018b, 2018a).

The "EU Strategy for Plastics in a Circular Economy" and a "Staff Working Document" were published on January 16, 2018 in the context of the Circular Economy Package, along with a report on oxo-degradable plastics (European Commission, 2019b, 2019c). The "EU Strategy for Plastics" builds upon four pillars:

- Improving the economics and quality of recycling plastics, with actions related to improving product design, boosting recycled content and improving separate collection of plastic waste (European Commission, 2019b, 2019c)

- Curbing plastic waste and littering, with actions to reduce single-use plastics, tackle sea-based sources of marine litter, monitor and curb marine litter more effectively, actions on compostable and biodegradable plastics and actions to curb microplastics pollution (European Commission, 2019b, 2019c)
- Driving investment and innovation toward circular solutions, with actions to promote investment and innovation in the value chain (European Commission, 2019b, 2019c)
- Harnessing global action, with actions at bilateral and multilateral level as well as actions related to international trade (European Commission, 2019b, 2019c)

Encouraged by a favorable European regulation, recent innovative research has focused on developing bioplastics from organic waste streams (crop residues, agro-food by-products, sewage sludge, etc.) seeking to enter a circular economy concept that does not compete with food usage and that is fully biodegradable to respond to the overwhelming negative externalities of our plastic packaging (Guillard et al., 2018; Fortunati, Mazzaglia, and Balestra, 2019; Russell, 2014).

Guidance on cascading use of biomass was published on November 16, 2018 to promote efficient use of bio-based resources through dissemination of best practices and support for innovation in the bioeconomy (European Commission, 2019b). The guidance explains cascading and provides some principles and practices to inspire stakeholders when applying it (European Commission, 2019b). The updated "Bioeconomy Strategy and Action Plan" propose 14 concrete actions in three priority areas (European Commission, 2019b):

- Strengthen and scale-up the bio-based sectors, unlock investments, and markets.
- Deploy bioeconomies rapidly across the whole of Europe.
- Understand the ecological boundaries of the bioeconomy.

The promotion of bio-based materials and products, whenever possible and relevant, will be ensured during the development of the EU Ecolabel and Green Public Procurement (GPP) criteria for new or existing product groups, according to environmental footprint results, and in line with available EU standards and technical reports, as well as with the strategic approach for EU Ecolabel and GPP (European Commission, 2019b).

Bio-based materials, such as wood, crops, or fibers, can be used to manufacture a wide range of products, namely new food packaging materials, as well as for biofuel and other energy uses (European Commission, 2016, 2019a). Apart from providing an alternative to fossil-based products, bio-based materials are also renewable, biodegradable, and compostable. At the same time, these materials require special attention due to the need to minimize their lifecycle environmental impact, making sustainable sourcing an important priority (European Commission, 2016, 2019a). In a circular economy, a cascading use of renewable resources should be promoted together with innovation in new materials, chemicals, and processes. The EC is promoting the efficient use of bio-based resources through a series of measures, such as the promotion of the cascading use of biomass and support to innovation in the bioeconomy, and has included a target for recycling wood packaging in the revised legislative proposal on waste, as well as a provision to ensure that the bio-waste is collected separately (European Commission, 2016, 2019a).

1.4 BIO-BASED PACKAGING MATERIALS: OBTENTION OF MATERIALS COMMERCIALLY AVAILABLE AND THEIR DEGRADATION PATH

Bio-based polymers are considered a valid replacement of petroleum synthetic polymeric matrices (Luzi et al., 2019); they are obtained by the processing of renewable resources (vegetable and animal wastes) and offer several positive aspects, such as environmental advantages, disintegratability and degradability, improved possibility to recycle the polymeric wastes, no presence of toxic components, and high biocompatibility, when compared to the conventional petrochemical polymeric matrices (Luzi et al., 2019).

According to the American Society for Testing and Materials (ASTM), a biodegradable material is a material that degrades because of the enzymatic action of the naturally occurring microorganisms such as bacteria, fungi, and algae (Kale et al., 2007; Song et al., 2009); while a compostable material is the one that undergoes degradation by biological processes during composting to yield carbon dioxide, water, inorganic compounds, and biomass at a rate consistent with other known compostable materials and leaves no visually distinguishable or toxic residues (Kale et al., 2007; Song et al., 2009). Therefore, all compostable materials are biodegradable, but the reverse is not true (Kale et al., 2007; Song et al., 2009).

What follows is a general overview of the commercially available bio-based materials most employed in the development of novel and innovative food packaging systems with special focus on their obtention and degradation path.

1.4.1 Polysaccharides

Polysaccharides can be used as interesting substitutes for synthetic polymers in film production because they present excellent mechanical and structural properties, but they have a poor barrier against water vapor (Domínguez et al., 2018).

1.4.1.1 Starch

Starch is perhaps the most popular plant polysaccharide for bioplastic formation because of its abundance, cost-effectiveness, and excellent film-forming abilities (Thakur et al., 2019; Al-Tayyar, Youssef, and Al-hindi, 2020). Several starches from previously unexplored sources, either alone or in combination with other biopolymers, have been assessed as biodegradable packaging agents or as bio-based coatings to extend the shelf life of fresh products (Menzel et al., 2019; Tanetrungroj and Prachayawarakorn, 2018; Basiak, Lenart, and Debeaufort, 2017; Farrag et al., 2018; A. Ali et al., 2019).

Starch is composed of amylose and amylopectin. The amylose is a linear and long molecule built up of 1,4-linked β-D-glucose, although in the amylopectin chains the glucose monomers are linked through α-1,6-linkages, determining an extremely branched arrangement (Luzi et al., 2019; Thakur et al., 2019). Therefore, the molecular structure of amylose is simpler than amylopectin, showing a linear structure with few α-1,6-branches. Amylopectin generally is the main component of starch, composed of short chains and a high number of α-1,6-branches (5% of the molecule), while amylose, in general, is randomly arranged among the amylopectin molecules in the amorphous regions (Luzi et al., 2019; Thakur et al., 2019).

In Europe, starch was until the 18th century made primarily from wheat. From then on, however, potato became an increasingly important raw material for European-produced starch. Fifty years later, the first maize starch was also produced in Europe (The European Starch Industry, 2013, 2018). Today, Europe remains the world's leading producer of potato starch, with European potato starch being exported to all corners of the globe. Over the years European potato starch farmers grouped together to form agricultural cooperatives and most of the EU potato starch-producing companies that exist today are still agricultural cooperatives (The European Starch Industry, 2013, 2018). Nevertheless, the EU starch industry faces challenges internationally, namely the huge economies of scale of the US starch industry, the protected developing starch sector in China, and the tax breaks supporting the Thai tapioca starch industry (The European Starch Industry, 2013, 2018).

As mentioned earlier, microorganisms can directly consume natural polymers like starch, since enzymatic reactions can reduce their molecular weight in extracellular environments, i.e., outside the cells of the microorganisms (Kale et al., 2007). The polymer chains are enzymatically cleaved, and the portions that are small enough are transferred into the cells and consumed. The resulting smaller molecules are much more susceptible to enzymatic attack, making the polymer degrade much faster (Kale et al., 2007).

1.4.1.2 Agar

The inner matrix of red algal cell walls from the family *Gracilariaceae*, *Gelidiaceae*, *Pterocladiaceae*, and *Gelidiellaceae* contains agar as the most abundant cell wall polysaccharide (Lee et al., 2017). Agar and other algal cell wall polysaccharides have a biological function analogous to that of hemicellulose in terrestrial plants, but with higher flexibility that can withstand strong ocean waves and currents (Lee et al., 2017). Agar is a collective term used to describe a mixture of gelling polysaccharides made up of D- and L-galactose (Lee et al., 2017).

Agar have diverse applications as colloids in food packaging coatings and films (Malagurski et al., 2017; Wang et al., 2018; Sousa and Gonçalves, 2015), pharmaceuticals, cosmetic, medical and biotechnology industries. Global production of agar has escalated from 6,800 tons ($82.2 million) in 2002 to 9,600 tons ($173 million) in 2009, with *Gracilaria* (80%) and *Gelidium* (20%) as the largest agar industrial sources (Lee et al., 2017). Agar is also a biodegradable polysaccharide allowing the harnessing of power of microorganisms present in the selected environment to completely remove bioplastic products designed for biodegradability from the environmental compartment via the microbial food time chain in a timely, safe and efficacious manner (Rujnić-Sokele and Pilipović, 2017).

1.4.1.3 Alginate

Alginates are naturally occurring, indigestible polysaccharides, commonly produced by and refined from various genera of brown algae (mainly, *Laminaria hyperborean, Macrocystis pyrifera, Ascophyllum nodosum*; and to a lesser extent, *Laminaria digitate, Laminaria japonica, Eclonia maxima, Lesonia negrescens, Sargassum* sp.) (Parreidt, Müller, and Schmid, 2018). Some bacteria such as *Azotobacter vinelandii* or mucoid strains of *Pseudomonas aeruginosa* also synthesize alginate like polymers as exopolysaccharide. The molecular structure of alginates is composed of unbranched, linear binary copolymers of β-D-mannuronic acid and α-L-guluronic acid residues linked by 1-4 glycosidic bonds (Parreidt, Müller, and Schmid, 2018).

Alginate is an appealing film-forming compound because of its nontoxicity, biodegradability, biocompatibility, and low price (Tavassoli-Kafrani, Shekarchizadeh, and Masoudpour-Behabadi, 2016). Its functional properties, thickening, stabilizing, suspending, film-forming for active food packaging (Yadav, Liu, and Chiu, 2019; Tang et al., 2018), gel-producing and emulsion-stabilizing have all been well studied (Tavassoli-Kafrani, Shekarchizadeh, and Masoudpour-Behabadi, 2016).

To extract the alginate, seaweed is broken into pieces and stirred with a hot alkali solution, usually sodium carbonate. After about 2 h, the alginate dissolves as sodium alginate originating a very thick slurry that contains undissolved parts of seaweed, mainly cellulose. The solution is diluted with very large amounts of water. Then, the solution is forced through a filter cloth in a filter press along with a filter aid such as diatomaceous earth. The last step is the precipitation of the alginate from the filtered solution, either as alginic acid or calcium alginate (Tavassoli-Kafrani, Shekarchizadeh, and Masoudpour-Behabadi, 2016). Pretreatment (before alkaline extraction) of the seaweed with acid leads to a more efficient extraction, a less colored product, and reduced loss of viscosity during extraction because lower amounts of phenolic compounds are present (Tavassoli-Kafrani, Shekarchizadeh, and Masoudpour-Behabadi, 2016).

1.4.1.4 Chitin/Chitosan

Chitin or poly(β-(1→4)-N-acetyl-D-glucosamine) is a natural polysaccharide of major importance, first identified in 1884. This biopolymer is synthesized by an enormous number of living organisms and it is the most abundant natural polymer, after cellulose (Younes and Rinaudo, 2015). In its native state, chitin occurs as ordered crystalline microfibrils, which form structural components in the exoskeleton of arthropods or in the cell walls of fungi and yeast (Younes and Rinaudo, 2015). So far, the main commercial sources of chitin are crab and shrimp shells. In industrial processing, chitin is extracted by acid treatment to dissolve the calcium carbonate followed by alkaline solution to dissolve proteins (Younes and Rinaudo, 2015).

Chitin has more applications when transformed into chitosan (by partial deacetylation under alkaline conditions) (Younes and Rinaudo, 2015). Chitosan is a random copolymer with a molar fraction degree of acetylation (DA) of β-(1→4)-N-acetyl-D-glucosamine and a fraction (1-DA) of β-(1→4)-D-glucosamine. The degree of acetylation of chitosan is characterized by the molar fraction of N-acetylated units (Younes and Rinaudo, 2015).

The chitin materials possess many instinctive characteristics, such as biocompatibility and biodegradability (Zhu et al., 2019). It is noted that the construction of chitin-based materials via green technology is a sustainable pathway, and the novel chitin materials fabricated from chitin nanocrystals as well as via direct dissolution/regeneration have been shown to have a wide range of applications (Zhu et al., 2019), namely in the food industry (Kaya et al., 2018).

Chitosan films have been researched for years because of their excellent performance, especially in the food industry, in relation with the reduction of environmental impacts, since they are biodegradable (Wang, Qian, and Ding, 2018; Gomes et al., 2019; Homez-Jara et al., 2018). Chitosan-based films can be used as food packaging materials and extend the shelf life of food, in the form of pure chitosan films, chitosan/biopolymer films, chitosan/synthetic polymer films, chitosan derivative films, etc. (Wang, Qian, and Ding, 2018; Gomes et al., 2019; Homez-Jara et al., 2018).

1.4.1.5 Xanthan Gum

Xanthan gum was discovered in the 1950s at the National Center for Agricultural Utilization of the United States Department of Agriculture (USDA). The first industrial production of xanthan gum was performed in 1960 followed by the availability of commercial product in 1964 (Kumar, Rao, and Han, 2018). This is the second microbial polysaccharide, after dextran (early 1940s), which was industrially commercialized. Xanthan gum is biodegradable, non-toxic, non-sensitizing, and does not cause any eye or skin irritation, and has also been approved by the US FDA in 1969 (Kumar, Rao, and Han, 2018).

Xanthan gum is a complex exopolysaccharide produced by the plant-pathogenic bacterium *Xanthomonas campestris* pv. campestris. It consists of D-glucosyl, D-mannosyl, and D-glucuronyl acid residues in a molar ratio of 2:2:1 and variable proportions of O-acetyl and pyruvyl residues (Becker et al., 1998). Because of its physical properties, it is widely used as a thickener or viscosifier in both food and non-food industries. Xanthan gum is also used as a stabilizer for a wide variety of suspensions, emulsions, and foams (Becker et al., 1998).

1.4.1.6 Xylans

There are three main sub-groups of hemicelluloses that have a backbone of 1→4 linked β-D- pyranose (hemiacetal sugars) residues; namely mannans, xylans, and xyloglucans. The most abundant hemicellulose type is xylan, although its composition varies among species. It consists of β-D-xylopyranosyl (xylose) residues that are linked via β-(1→4) glycosidic bonds (Naidu, Hlangothi, and John, 2018). This polysaccharide is generally available in large quantities in wood, forest, and by-products of the pulp and paper mills as well as agro-industries (Queirós et al., 2017).

Hydrogen bonding between individual hemicellulose cells, as well as ester and ether linkages of hemicellulose with lignin, inhibit the extraction of hemicellulose from the plant material (Naidu, Hlangothi, and John, 2018). Recently, xylans and its derivatives, namely carboxymethyl xylans, have been studied as promising renewable and biodegradable raw materials for the development of edible films for active food packaging (Sousa et al., 2016; Luís et al., 2019; Queirós et al., 2017).

1.4.1.7 Pullulan

Pullulan is one of the polymers obtained from the fermentation broth of the black yeast, as *Aureobasidium pullulans*. Pullulan was first reported in 1958 and its structure was detailed in 1959 (Prajapati, Jani, and Khanda, 2013; Singh, Kaur, and Kennedy, 2019). The prerequisite for the fermentative production of pullulan is carbon source, nitrogen source, and other essential nutrients for adequate growth of *A. pullulans*. The nutrients used to produce pullulan are expensive, which increases their production cost. However, the waste generated by many agro-based industries is very rich in organic/inorganic compounds required for the growth of *A. pullulans*. These wastes can be used as an alternative substrate for pullulan production by submerged or solid-state fermentation (Prajapati, Jani, and Khanda, 2013; Singh, Kaur, and Kennedy, 2019). Pullulan is a linear glucan and its structure consists of maltotriose as repeating units. Each maltotriose unit constitutes 2 α-(1→4) bonded glucopyranose rings interlinked by α-(1→6) linkage (Prajapati, Jani, and Khanda, 2013; Singh, Kaur, and Kennedy, 2019). Pullulan possesses unique physicochemical properties: non-ionic polymer without any toxicity, mutagenicity, carcinogenicity, and its viscosity is comparatively lower than other polymers (Prajapati, Jani, and Khanda, 2013; Singh, Kaur, and Kennedy, 2019). Owing to these distinctive properties, pullulan has potential applications in food, pharmaceutical, and biomedical fields. Pullulan is an edible biodegradable polymer and it has been certified harmless for usage in food products by food safety regulations in many countries (Prajapati, Jani, and Khanda, 2013; Singh, Kaur, and Kennedy, 2019). It can be used as a stabilizer, binder, intensifier, beverage filler, dietary fiber, thickener, texture improver, food packaging material (Chu et al., 2019; Morsy et al., 2014; Xiao et al., 2017; Tabasum et al., 2018).

1.4.1.8 Carrageenan

Carrageenans are natural hydrophilic polymers with a linear chain of partially sulfated galactans, which present high potential for film-forming. These sulfated polysaccharides are extracted from the cell walls of various red seaweeds (Li et al., 2014; Tavassoli-Kafrani, Shekarchizadeh, and Masoudpour-Behabadi, 2016). The most usual seaweeds for extraction of carrageenans are *Kappaphycus alvarezii* and *Eucheuma denticulatum*. However, some scientists extract carrageenans from *Hypnea musciformis* (Wulfen) *Lamoroux* and *Solieria filiformis* seaweeds (Li et al., 2014; Tavassoli-Kafrani, Shekarchizadeh, and Masoudpour-Behabadi, 2016). The main source of commercial carrageenan is *Chondrus crispus* species of seaweed known as carrageen moss or Irish moss in England, and carraigin in Ireland. The name carraigin was first used by Stanford in 1862 for the extract of *C. crispus*. It has been used in Ireland since 400 AD as a gelation agent and as a home remedy to cure coughs and colds. The term "carrageenan" is more recent and has been used by several authors after 1950 (Li et al., 2014; Tavassoli-Kafrani, Shekarchizadeh, and Masoudpour-Behabadi, 2016).

Carrageenans are mainly composed of D-galactose residues linked alternately in 3-linked-β-D-galactopyranose and 4-linked-α-D-galactopyranose units, being classified according to the degree of substitution that occurs on their free hydroxyl groups. Substitutions are generally either the addition of ester sulfate or the presence of the 3,6-anhydride on the 4-linked residue (Li et al., 2014; Tavassoli-Kafrani, Shekarchizadeh, and Masoudpour-Behabadi, 2016). In addition to D-galactose and 3,6-anhydro-D-galactose as the main sugar residues and sulfate as the main substituent, other carbohydrate residues may be present in carrageenans preparations, such as glucose, xylose, and uronic acids, as well as some substituents, such as methyl ethers and pyruvate group (Li et al., 2014; Tavassoli-Kafrani, Shekarchizadeh, and Masoudpour-Behabadi, 2016).

Carrageenan is water-soluble, biocompatible and biodegradable, and it has been extensively studied in food, dairy and pharmaceutical industries as gelling, stabilizing and emulsifying agent (Kanmani and Rhim, 2014; Liu et al., 2019).

1.4.1.9 Pectin

Pectins represent a group of structurally heterogeneous polysaccharides widely distributed in primary cell walls and the middle lamella of higher plants. Pectic polysaccharides are vital structural components of plant cell walls, and they are often associated with other cell wall polysaccharides such as cellulose and hemicelluloses (Dranca and Oroian, 2018; Lara-Espinoza et al., 2018; Naqash et al., 2017). The diverse structural and macromolecular properties of pectins, such as galacturonan methoxylation, galacturonic acid content, the composition of neutral sugars, and molecular weight, are dependent on the pectin source and set the basis for multiple food and non-food applications of this complex polysaccharide (Dranca and Oroian, 2018; Lara-Espinoza et al., 2018; Naqash et al.. 2017).

The exact chemical composition and structure of pectin is still under discussion due to the high complexity of this molecule. Elucidation of pectin structure is important to understand its role in plant growth and development, during ripening of fruits, in food processing and as a nutritional fiber (Dranca and Oroian, 2018; Lara-Espinoza et al., 2018; Naqash et al., 2017). The structure of pectins is very difficult to determine because their composition varies with the source and conditions of extraction, location, and other environmental factors. Pectin can also change during its isolation from plants, storage, maturity degree, and processing of raw plant material (Dranca and Oroian, 2018; Lara-Espinoza et al., 2018; Naqash et al., 2017).

Pectin can be found in almost all plants, but commercially most pectins are obtained from citrus fruits like orange, lemons, grapefruit, and apples. These materials contain a high amount of pectic substances and can be found available as residues from juice production. Although the color may be different depending on the source of pectin, it will not significantly affect its technical use (Dranca and

Oroian, 2018; Lara-Espinoza et al., 2018; Naqash et al., 2017). Fruits like quince, plums, and gooseberries contain much more pectin compared to soft fruits like cherries, grapes, and strawberries. Dried apple pulp generally contains 15 to 20% pectin and dried citrus peel range between 30 to 35% of pectin. Other typical levels of pectin in fruit like apricot, cherries, orange, and carrots are 1%, 0.4%, 0.5–3.5%, and 1.4%, respectively, based on fresh weight (Dranca and Oroian, 2018; Lara-Espinoza et al., 2018; Naqash et al., 2017).

In the food sector, pectin's traditional usage as a gelling agent, thickening agent, and stabilizer is being complemented by the emerging utilization of pectin as a fat replacer and health-promoting functional ingredient (Manrich et al., 2017). The capacity of this biodegradable material to improve the water resistant properties of edible films has been of relevance to numerous research studies (Manrich et al., 2017). In the domain of food preservation, pectin as a biopolymer finds application in packaging and as a carrier molecule for antimicrobials, antioxidants, and other related compounds (Naqash et al., 2017).

1.4.1.10 Cellulose-Derived

Cellulose is the most abundant polymer on the planet with plants producing 180 billion tons per year by photosynthesis from the conversion of CO_2 and water. Cellulose has widely been used for two centuries in commodity and industrial products such as pulp and paper, textiles, and coatings. Available on all the inhabited continents, cellulose is biodegradable, non-toxic, sustainable, easy to functionalize, wettable, and flexible (Raghuwanshi and Garnier, 2019). These properties make cellulose an attractive material for engineering advanced applications such as low-cost, biodegradable and disposable biodiagnostics, point of care analytical devices, and membranes for separation (dialysis). Currently, the main biomedical applications of cellulose include biodiagnostics, such as blood typing, and pregnancy tests, tissue engineering, scaffolds, eye care solutions, coatings, active food packaging, and sensors (Raghuwanshi and Garnier, 2019).

The majority of cell walls in plants consist of cellulose, hemicelluloses, and lignin, where lignin presents at about 10–25% by dry weight and acts as a binder between cellulose and hemicelluloses components (Sharma et al., 2018). It is the lignin, which confers the stiffness and strength with its binding function and gives protection to the cell wall. The other two main components of the plant cell wall i.e., cellulose and hemicelluloses represent about 35–50% and 20–35% of dry weight of lignocellulosic biomass, respectively (Sharma et al., 2018).

Cellulose is a linear polysaccharide with repeating units of cellobiose (disaccharide D-glucose) units linked by $\beta\rightarrow1,4$ linkage and there are strong intramolecular or intermolecular hydrogen bonding between adjacent glucose units in the same chain or different chain through the open hydroxyl groups present in glucose monomer units (Sharma et al., 2018).

Cellulose films of nanoscale thickness (1–100 nm) are transparent, smooth (roughness <1 nm), and provide a large surface area interface for biomolecules immobilization and interactions. These attractive film properties create many possibilities for both fundamental studies and applications, namely in food packaging (Kontturi and Spirk, 2019; Raghuwanshi and Garnier, 2019).

Several cellulose derivatives have also been exploited as new bio-based materials for active food packaging, namely: carboxymethylcellulose (CMC), methyl cellulose (MC), cellulose acetate (CA), ethyl cellulose (EC), cellulose nanocrystals, cellulose nanofibers, microfibrillated cellulose, and bacterial cellulose.

CMC, a typical anionic polysaccharide produced by the alkali-catalyzed reaction of cellulose with chloroacetic acid, is one of the fundamental derivatives of cellulose. It has been used as thickener or stabilizer in different foods since it has been approved as generally recognized as safe (GRAS). Due to the presence of various hydroxyl and carboxylic groups, it has good moisture sorption and water binding characteristics (Akhtar et al., 2018). It exhibits thermal gelation, making CMC-based films be extensively used in food and pharmaceutical industries. CMC is being applied widely in different edible film formulations, because of its good film-forming ability, biodegradability, biocompatibility, odorlessness, tastelessness, nontoxicity, and hydrophilicity (Akhtar et al., 2018).

The simplest cellulose derivative is MC, in which the hydroxyl residues of cellulose are replaced by methyl groups. MC has excellent film-making properties, with high solubility, low oxygen, and lipid permeability; and it is a promising biopolymer for active food packaging (Matta, Tavera-Quiroz, and Bertola, 2019).

CA is one of the most important organic esters obtained from natural sources. Its applications include fibers, membranes, films (Silva et al., 2016) and plastics. CA can be widely used in the food packaging industry because of its versatility in processing method that can be used to process it, including extrusion, injection, and compression, as well as its comparatively low cost and performance ratio compared to other polymers. CA is also lightweight, microwavable, and has good optical properties. In addition, CA biocomposites have good barrier properties which prolong the shelf life of food, protecting it from attack by microorganisms and oxidation (Melo et al., 2018).

EC is a kind of water-insoluble cellulose ether with favorable mechanical properties, relatively low cost, and good film-forming performance. To protect core substances and achieve sustainable drug release, EC is widely used as a coating and encapsulating agent for pharmaceutical products (Yang et al., 2014). EC films have been widely used in drug-controlled release as a coating polymer on solid pharmaceutical dosage forms due to its inherent properties, such as good mechanical strength and chemical resistance, excellent durability and solubility in various kinds of organic solvents, wide accessibility, nontoxicity and low cost (Feng et al., 2015).

Recently, highly crystalline nanoscale material, namely cellulose nanocrystals, has garnered a tremendous level of attention from many research communities. Cellulose nanocrystals are broadly needle-shaped nanometric or rod-like particles, that have at least one dimension <100 nm and exhibit a highly crystalline nature (Trache et al., 2017). They can be produced from diverse starting materials that include algal cellulose, bacterial cellulose, bast fibers, cotton linters, microcrystalline cellulose, tunicin, and wood pulp. These nanocrystals impart attractive combinations of biophysicochemical characteristics such as biocompatibility, biodegradability, light weight, nontoxicity, stiffness, renewability, sustainability, optical transparency, low thermal expansion, gas impermeability, adaptable surface chemistry, and improved mechanical properties (Trache et al., 2017).

Cellulose nanofibers, which can be extracted from cellulose fibers have excellent mechanical properties like glass and carbon fibers and have been studied as good film-forming materials (Silva et al., 2019). There are various kinds of treatments to extract cellulose nanofibers from cellulose fibers such as chemical, mechanical, or a combination of these treatments (Fazeli, Keley, and Biazar, 2018; Liu et al., 2020). During the last few years, the cellulose nanofiber extracted from hemp, *Helicteres isora* plant, banana peels, wheat straw, sisal, *Aloe vera* rind, bamboo, and cassava bagasse have been used to improve the mechanical properties and water sensitivity of thermoplastic starch matrix (Fazeli, Keley, and Biazar, 2018; Liu et al., 2020). Nanofibrillation and pretreatment are required procedures for the extraction of cellulose nanofibers from natural fibers. Pretreatment processes such as acid or alkali treatment, treatment with ionic liquids, enzymatic pretreatment, and steam explosion treatment, were first used to remove non-cellulosic materials in plants. Some nanofibrillation technologies such as high-pressure homogenizers, grinders, and ultrasonic method were used to generate high shear forces to separate the fibrils from the purified cellulose fibers (Fazeli, Keley, and Biazar, 2018; Liu et al., 2020).

Microfibrillated cellulose prepared from wood pulp is a kind of cellulose nanofiber of high aspect ratio and network structure, containing both amorphous and crystalline regions. Compared to other nanofibers, microfibrillated cellulose is low cost, which makes its application a real economic benefit (Deng et al., 2016; Oinonen et al., 2016). A great number of hydroxyl groups on the surface of microfibrillated cellulose not only strengthen the intermolecular interaction between the fibers, but also result in severe aggregation of microfibrillated cellulose. However, too many hydroxyl groups make them sensitive to moisture when used to form free-standing microfibrillated cellulose films (Deng et al., 2016; Oinonen et al., 2016).

Bacterial cellulose consists of a translucent and gelatinous film, formed by interwoven indefinite-length cellulose microfibrils, distributed in random directions. Bacterial cellulose is produced extracellularly by the Gram-negative bacterial cultures of *Gluconacetobacter*, *Acetobacter*, *Agrobacterium*, *Achromobacter*, *Aerobacter*, *Sarcina*, *Azobacter*, *Rhizobium*, *Pseudomonas*, *Salmonella*, and *Alcaligenes*. Among them, the most efficient bacterial cellulose producer belongs to the *Gluconacetobacter* genus (Picheth et al., 2017). During the biosynthesis, bacterial cellulose forms a pellicle constituted of a random microfibrillar network of cellulose chains aligned in parallel, interspersed among amorphous regions that occupy 90% of the material total volume. The bacterial cellulose structure is composed of (1→4)-D-anhydroglucopyranose chains bounded through β-glycosidic linkages (Picheth et al., 2017).

Microorganisms can directly attack the molecules of cellulose and its derivatives since they can produce enzymes to cleave, or depolymerize, the natural polymer backbones, and consequently molecular weight reduction can happen outside the microbial cells (Kale et al., 2007; Ahmed et al., 2018). For other biodegradable polymers, before being utilized by microorganisms, the molecular weights must be reduced to a point at which the molecules can enter the microbial cells by other means of degradation, such as hydrolysis or photodegradation (Kale et al., 2007; Ahmed et al., 2018).

1.4.2 Proteins

Proteins are promising materials for the preparation of bio-based food packaging. The use of proteins presents several important advantages, such as good mechanical, physical (resistance and flexibility) and optical (transparency) properties, while presenting a great barrier to aromas, oxygen, and organic vapors, together with a selective permeability to other gases (CO_2, O_2) (Domínguez et al., 2018). In fact, some authors concluded that protein-based films have physical, mechanical, barrier, and thermal characteristics like some plastic films, such as polyvinyl chloride (Domínguez et al., 2018). Furthermore, proteins are generally superior to polysaccharides in their ability to form films with high mechanical and barrier properties, providing also a higher nutritional value. Moreover, they can be used for controlled release of additives and bioactive compounds (Zink et al., 2016; Ramos et al., 2012).

1.4.2.1 Whey

Milk possesses a protein system formed by two major families: caseins (which are insoluble) and whey proteins (which are soluble). The former account for ca. 80% (w/w) of the whole protein concentration and can easily be recovered from skim milk via isoelectric precipitation, through addition, or *in situ* production of acid, or via rennet-driven coagulation, both of which release whey as a by-product (Ramos et al., 2012; Chen et al., 2019). Whey proteins may in turn be recovered from whey via ultrafiltration, or else via centrifugation or regular filtration following thermal precipitation (Ramos et al., 2012; Chen et al., 2019). Besides their intrinsically nutritive value, whey proteins exhibit several functional properties that are essential for the formation of edible films, being also biodegradables (Ramos et al., 2012; Chen et al., 2019).

1.4.2.2 Casein

Casein, one of the milk proteins, as mentioned above, contains four main subunits: κ-casein, β-casein, α s1-casein, and α s2-casein, which make up 13%, 36%, 38%, and 10% of the casein composition, respectively. The unique properties of these protein fractions affect the film-forming ability of casein (Chen et al., 2019). Their molecular weights are between 19 and 25 kDa and their average isoelectric point is between 4.6 and 4.8. All the four caseins are amphiphilic and have ill-defined structures (Elzoghby, El-Fotoh, and Elgindy, 2011). Casein may be used in pharmaceutical products either in the form of acid casein, which has a low aqueous solubility, or sodium caseinate which is freely soluble except near its isoelectric point (Elzoghby, El-Fotoh, and Elgindy, 2011).

Casein can easily form films from aqueous solutions without further processing because of the strong interchain cohesion caused by their random-coil nature together with the great number of formed intermolecular hydrogen, hydrophobic and electrostatic bonds (Chen et al., 2019). Several properties of casein, such as biodegradability, high thermal stability, nontoxicity, the ability to bind small molecules and ions, and micelle formation capability, make it a good material for biodegradable films. Due to ready availability, water solubility, emulsification capability, and their high nutritional value, caseins are desirable biomaterials for the preparation of edible films (Chen et al., 2019).

1.4.2.3 Zein

Zein is a class of alcohol-soluble prolamine storage proteins in corn, which was first isolated from whole white maize. Based on solubility and sequence homology, zein can be separated into four classes: α-zein (19 and 22 kDa), β- zein (14 kDa), γ-zein (16 and 27 kDa), and δ-zein (10 kDa). Among these, α-zein comprises 70–85% of the total fraction of zein mass, and γ-zein accounts for 10–20% as the second most abundant fraction (Shukla and Cheryan, 2001; Shi and Dumont, 2014; Y. Zhang et al., 2015; Arcan, Boyacı, and Yemenicioğlu, 2017). All zein groups are abundant in hydrophobic and neutral amino acids (e.g., leucine, proline, and alanine) and contain some polar amino acid residues, such as glutamine (Shukla and Cheryan, 2001; Shi and Dumont, 2014; Y. Zhang et al., 2015; Arcan, Boyacı, and Yemenicioğlu, 2017).

The first commercial zein plant began operation in 1938. Notably, zein-based fiber textiles were sold for clothing production during the 1950s. Unfortunately, with the development of synthetic fibers, the reproducible fiber gradually disappeared. However, using zein as a coating material for candies, enriched rice, dried fruits, and nuts began in the middle of the 20th century and has continued until now (Shukla and Cheryan, 2001; Shi and Dumont, 2014; Zhang et al., 2015; Arcan, Boyacı, and Yemenicioğlu, 2017). Based on the long history of zein usage in food and related evaluation data, zein was approved in 1985 by the US FDA as a GRAS excipient for film coating of pharmaceuticals, mainly involving tablets (Shukla and Cheryan, 2001; Shi and Dumont, 2014; Zhang et al., 2015; Arcan, Boyacı, and Yemenicioğlu, 2017).

Zein has been recognized for its film-forming properties for a long time (Luís, Domingues, and Ramos, 2019). Biodegradable zein films can be easily produced by solvent casting, extrusion, and spin casting. Solvent casting, like the tablet-coating process, requires that zein be dissolved in a hydrated organic solvent, followed by drying at room temperature or under specified conditions, which generally involves the development of hydrophobic, hydrogen and limited disulfide bonds between zein chains (Shukla and Cheryan, 2001; Shi and Dumont, 2014; Zhang et al., 2015; Arcan, Boyacı, and Yemenicioğlu, 2017).

1.4.2.4 Gelatin

Gelatin is composed of 50.5% carbon, 6.8% hydrogen, 17% nitrogen and 25.2% oxygen. Gelatin is derived from the fibrous insoluble protein called collagen (by chemical denaturation) and is obtained from bones, skin and connective tissue generated as waste during animal slaughtering and processing (Hanani, Roos, and Kerry, 2014; Chiellini et al., 2004). Collagen consists of rigid bar-like molecules that is organized in fibers interconnected by covalent bonds. These molecules have three polypeptide chains arranged in a triple helix that is stabilized by hydrogen and hydrophobic bonds (Hanani, Roos, and Kerry, 2014; Chiellini et al., 2004).

The most used commercial gelatins are derived from bones and hides taken from cattle and pigs. Currently, rapid progress on development of gelatin sources from skin and bones derived from fish and poultry is underway and may serve as important commercial gelatin sources into the future. This is because proteins derived from these sources have never had any association with bovine spongiform encephalopathy (BSE), may be more ethnically and religiously acceptable, but more importantly, may possess novel and unique functional properties (Hanani, Roos, and Kerry, 2014; Chiellini et al., 2004). Gelatin from fish and poultry have received considerable attention in recent years. Fish gelatin has similar functional characteristics to mammalian gelatin. It is either derived from skin, bones, fins, scales, or swim bladder (Hanani, Roos, and Kerry, 2014; Chiellini et al., 2004).

Gelatin offers good processability properties both in aqueous media and in the melt, exhibiting also the ability to form biodegradable films (Hanani, Roos, and Kerry, 2014; Chiellini et al., 2004).

1.4.3 Flour-Based

Flours are natural blends of starch, protein, lipids, and fibers. Recently, there has been an increasing interest in developing edible films using agriculture crop flours because of their low cost, being easy to obtain and wide availability compared to pure components such as polysaccharides and proteins (Drakos, Pelava, and Evageliou, 2018; Gutiérrez et al., 2016; Nouraddini, Esmaiili, and Mohtarami, 2018; Díaz et al., 2019). The properties of a flour film depend on the various natural and intrinsic molecular interactions taking place between starch, protein, lipid, and fiber components as well as the conditions during the drying and the production process. Another advantage of using flours in film development is that they come from various botanical sources and thus, their compositions vary (Drakos, Pelava, and Evageliou, 2018; Gutiérrez et al., 2016; Nouraddini, Esmaiili, and Mohtarami, 2018; Díaz et al., 2019).

In recent years, many studies have focused on the evaluation of biodegradable films from a variety of flours such as banana, chia, quinoa, achira, triticale, semolina, and rice. According to the obtained results, most of the films made from the flours had a heterogeneous structure with poor mechanical properties and water vapor permeability compared to those from starches. Thus, it has been revealed that the combination of flour with starch improves the mentioned characterization of the films (Drakos, Pelava, and Evageliou, 2018; Gutiérrez et al., 2016; Nouraddini, Esmaiili, and Mohtarami, 2018; Díaz et al., 2019). Legume flours are also a good source of material for film formation due to their high content of starch and protein; some of them also contain a significant amount of lipids. Lentil, defatted soy and grass pea flours have been studied as film-forming materials (Drakos, Pelava, and Evageliou, 2018; Gutiérrez et al., 2016; Nouraddini, Esmaiili, and Mohtarami, 2018; Díaz et al., 2019).

1.5 BIO-BASED ACTIVE PACKAGING MATERIALS: FOCUS ON EXAMPLES TESTED *IN VIVO*

Since 2015, there has been a growing trend in the use of natural-based, eco-friendlier packaging materials for the incorporation of active agents such as antioxidants, ethylene absorbers, and antimicrobials. For the purpose of this revision and bearing in mind that all packaging materials should be subjected to proof-of-concept testing, only research studies containing *in vivo/in situ* testing (e.g., in the food product) will be analyzed. When investigating the use of these bio-based polymers in published and indexed studies (see Figure 1.3), chitosan is, by far, the most used one (28%), followed, in descending order, by cellulose-based materials (carboxymethyl cellulose, methylcellulose, cellulose acetate, cellulose nanocrystals, cellulose nanofibers or microfibrillated cellulose) (16%), gelatin (mainly from fish such as puffer fish skin, starfish, tilapia fish skin, as well as from nutraceutical capsules waste) (12%), starch (from cassava, corn, sweet potato, rye) (10%), protein-based (from fish, quinoa, distiller dried grains, grain sorghum, whey, casein, seaweed, soy) (9%) and alginate (7%). Other less-used polymers include agar (1%), carrageenan (1%), flour-based (from banana, triticale, semolina, rice, pomelo peel, and lentils) (4%), gum (Arabic, Guam, from *Acacia* and basil seeds) (3%), xanthan (1%), pectin (3%), zein (4%), and others such as polyurethane made from Mahua oil components (Indumathi, and Rajarajeswari, 2019), kenaf fibers (Tawakkal, Cran, and Bigger, 2017) and *Aloe vera* gel (Ali et al., 2016).

To design these new bio-based materials, natural-based polymers derived from polysaccharides and proteins such as chitosan, gelatin, alginate, zein, and cellulose have been used alone or in combination in the form of blends, composites or multilayer films. Although some combinations containing three natural-based polymers have been described: casein protein/alginate/pectin (Gautam and Mishra, 2017) and alginate/CMC/carrageenan (Shankar and Rhim, 2018) (see Table 1.4); usually the most preferred combinations are the binary ones such as chitosan/protein (Robledo et al., 2018), gelatin/starch (Malherbi et al., 2019), cellulose/chitosan (Hu, Wang, and Wang, 2016; Wang et al., 2019), alginate/cellulose (Rezaei and Shahbazi, 2018) and Mahua

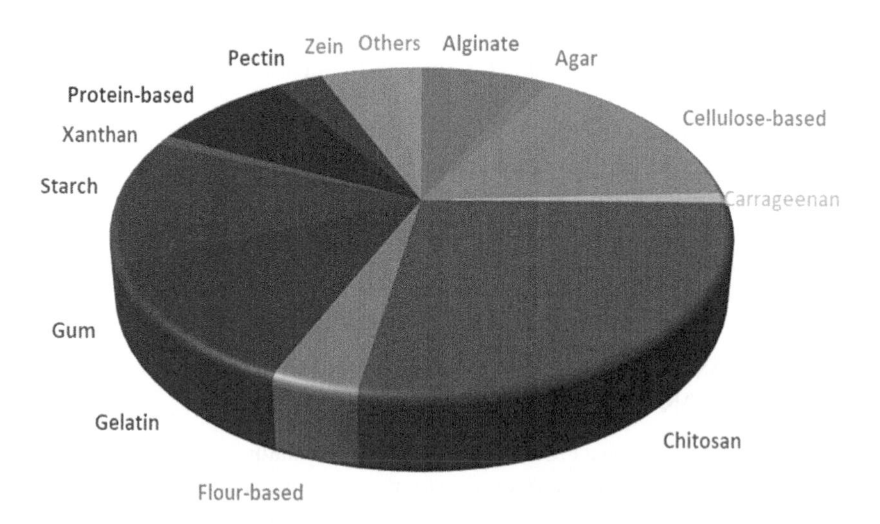

FIGURE 1.3 Distribution of the published and indexed (Web of Knowledge®) research studies containing bio-based polymers.

TABLE 1.4
Combination of Natural Bio-Based Polymers Used for Active Food Packaging.

Polymer combination	Active packaging	Type of packaging	Food product	References
Chitosan enriched with cellulose nanoparticles	Antimicrobial and antioxidant	Film	Minced beef	(Shahbazi and Shavisi, 2018)
Quaternized chitosan/carboxymethyl cellulose blend	Antimicrobial	Coating with a film-forming solution	Bananas	(Hu, Wang, and Wang, 2016)
Alginate/carboxymethylcellulose composite	Antimicrobial	Edible coating	Silver carp fillets	(Rezaei and Shahbazi, 2018)
Carboxymethyl cellulose/chitosan composite	Antimicrobial	Film	Veal “carpaccio”	(Rollini et al., 2017)
Microfibrillated cellulose/chitosan composite				
Alginate/carboxymethyl cellulose/carrageenan blend	Antimicrobial	Coating of paper pouches	Fish paste	(Shankar and Rhim, 2018)
Carboxymethyl cellulose/chitosan blends	Antimicrobial	Edible coating	Strawberries	(Shahbazi, 2018)
Quinoa protein/chitosan	Antimicrobial	Edible coating	Strawberries	(Robledo et al., 2018)
Starch/gelatin	Antioxidant	Sachets	Olive oil	(Malherbi et al., 2019)
Starch/chitosan	Antimicrobial	Wrapping film	Strawberries	(Wang et al., 2019)
Chitosan/methylcellulose composite	Antimicrobial	Wrapping film	Freshcut wax gourd wax	(H. Wang et al., 2019)
Cellulose nanocrystal/chitosan composite	Antimicrobial	Film	Fresh pork loin	(Khan et al., 2016)
Pectin/gelatin multilayer film	Antimicrobial	Label	Mung bean seeds	(Saade et al., 2018)
Polyurethane/chitosan	Antimicrobial	Wrapping film	Carrots	(Indumathi and Rajarajeswari, 2019)
Alginate/chitosan	Antimicrobial and antioxidant	Coating	Pork pieces	(Kulig et al., 2017)

(Continued)

TABLE 1.4 (CONTINUED)
Combination of Natural Bio-Based Polymers Used for Active Food Packaging

Polymer combination	Active packaging	Type of packaging	Food product	References
Chitosan/cellulose acetate phthalate	Antimicrobial	Film	Grapes	(Indumathi, Saral Sarojini, and Rajarajeswari, 2019)
Agar/fish protein	Antimicrobial	Wrapping film	Flounder fillets	(Rocha et al., 2018)
Undaria pinnatifida (seaweed) protein/gelatin	Antimicrobial	Wrapping film	Smoked chicken breast	(Rocha et al., 2018)
Chitosan/alginate multilayer	Antioxidant	Edible coating	Mangoes	(Yin et al., 2019)
Gelatin/chitosan nanofibers	Antimicrobial	Wrapping film	Chicken fillets Cheese	(Amjadi et al., 2019)
Gelatin/microcrystalline cellulose	Antimicrobial	Wrapping film	Bread	(Figueroa-Lopez, Andrade-Mahecha, and Torres-Vargas, 2018)

oil-derived thermoplastic polyurethane/chitosan (Indumathi and Rajarajeswari, 2019). Additionally, besides being used on its own, some of these combinations can also be used for the coating of other commonly used natural-based packaging materials such as paper to increase its barrier properties (Shankar and Rhim, 2018).

Other times, these polymers are used in combination with synthetic, biodegradable ones such as poly(lactic acid) (PLA), polycaprolactone (PCL), polyethylene oxide and poly(vinyl alcohol) to take advantage of the improved barrier and processing properties of these materials, while ensuring that the final polymers would be biodegradable (see Table 1.5). Furthermore, the addition of bio-based polymers to synthetic ones such as ethylene vinyl alcohol (EVOH) with very low biodegradability, leads to the formation of biodegradable composites.

For instance, polyethylene oxide has been used in combination with zein and soy protein isolate for the development of antimicrobial packaging materials incorporating ginger essential oil as active agent (da Silva et al., 2018). Also, PLA has been used in combination with some bio-based polymers as, for example, chitosan (Fathima et al., 2018) and zein (Altan, Aytac, and Uyar, 2018). However, the most preferred combination is with PVA that has been combined with starch (Liu et al., 2017; Zhijun Wu et al., 2017; Mustafa et al., 2019), Arabic gum (Muppalla and Chawla, 2018), pectin (Yang et al., 2019), and gelatin (Kanatt and Chawla, 2018). Furthermore, natural-based polymers have also been used as coatings for plastic materials such as low density polyethylene (Caro et al., 2016; Huang et al., 2018), polyethylene (Tas et al., 2019; Santonicola et al., 2017), polypropylene (Navikaite-Snipaitiene et al., 2018; Mahboobeh Kashiri et al., 2019; Mahboobeb Kashiri et al., 2017), or even as a component coating/film of common plastic-based packages (Carvalho et al., 2019; Moreno et al., 2018). However, for the purpose of this chapter and keeping in mind the EU circular economy guidelines for plastic materials, these studies will not be further discussed.

TABLE 1.5
Biodegradable Films Resulting from the Combination of Natural-Bio-Based Polymers with Synthetic Polymers

Polymer combination	Active packaging	Type of packaging	Food product	References
Starch/PVA blend	Antimicrobial	Wrapping film	Figs	(Wu et al., 2017)
Chitosan/PCL multilayer	Antimicrobial and antioxidant	Film	Chicken breast fillets	(Sogut and Seydim, 2019)
Zein/PLA	Antimicrobial	Fiber	Bread	(Altan, Aytac, and Uyar, 2018)
Nanochitosan/PLA	Antimicrobial	Pouches	Prawns	(Fathima et al., 2018)
Starch/PVA	Antimicrobial	Film	Milk	(Mustafa et al., 2019)
Gelatin/PVA	Antimicrobial	Pouches	Minced chicken	(Kanatt and Chawla, 2018)
Xanthan/PLA	Antimicrobial	Film	Meat	(Radford et al., 2017)

In the next section we will discuss, in a more detailed manner, the development of active packaging using one single bio-based polymer, having selected the six most used ones (chitosan, cellulose-based polymers, gelatin, starch, protein-based polymers, and alginate) over the past five years for this purpose.

1.5.1 Chitosan-Based Active Packaging Materials

As mentioned before, over recent years, chitosan has been the preferred bio-based polymer for the development of active food packaging, mainly with antimicrobial properties (see Table 1.6) due to the intrinsic antimicrobial activity of chitosan itself. Although many of the antimicrobial active packaging strategies described used an active agent, there are also some examples in which chitosan is used alone as active agent for the development of coatings with antioxidant or antimicrobial activities. The most preferred active agents incorporated into the chitosan matrix are of natural origin, such as essential oils and their major components (ginger EO, rosemary EO, cinnamon EO, and oregano EO, among others), natural compounds and extracts (gallic acid, tocopherol, olive pomace flour, Kombucha tea, sumac extract, etc.) and phages (*E. coli* O15:H7 phage) with only a few examples using inorganic or synthetic agents (citric acid, potassium permanganate, and ε-polylysine). These active agents can be incorporated directly into the chitosan matrix or sometimes incorporated in encapsulation systems such as nanoclays (montmorillonite, laponite, etc.) or liposomes. Furthermore, nanoclays can also be added to the chitosan matrix as reinforcement agents to improve the structural and barrier properties of chitosan films. Chitosan, in the form of coatings, wrapping films, and pouches, has been used to package a variety of food products such as fruits, bread, fresh meat, fish fillets, and nuts using an air atmosphere or in combination with modified atmosphere packaging or vacuum packaging.

1.5.2 Cellulose and Cellulose-Based Packaging Materials

Cellulose, both plant-derived or bacterial-derived, carboxymethyl cellulose, cellulose acetate, ethyl cellulose or cellulose nanofibers have been used to produce films, membranes, pads, sheets, and pouches, to package produces such as green tea, fresh vegetables and fruits, cheese, fresh meat, and smoked salmon (see Table 1.7). Similarly to chitosan, the active agents incorporated are mainly of natural origin such as extracts and EOs (green tea extract, rosemary extract, eugenol, ginger EO, cinnamon EO, etc.), bacteriocins (sakacin-A) and nanoparticles (silver and titanium oxide) that led to the development of several antioxidant and mostly antimicrobial packaging materials.

1.5.3 Gelatin-Based Packaging Materials

From all protein-based materials, gelatin from several sources (porcine skin, fish skin and bone, duck feet, and starfish) has been the preferred one to create packaging films and fibers (see Table 1.8). The resulting packaging materials with antimicrobial and antioxidant activities have been tested *in vivo* in fresh meat, fruits and

TABLE 1.6
In Vivo Testing of Active Food Packaging Based on Chitosan. VP—Vacuum Packaging, MAP—Modified Atmosphere Packaging, EO—Essential Oil, MMT—Montmorillonite, TBARS—Thiobarbituric Acid Reactive Species, CFU—Colony Forming Units

Active agent	Additives/ Techniques	Active packaging	Type of packaging	Food product	Outcome	References
NA	NA	Antimicrobial	Coating for paperboard	Bilberries Red currants	An increased shelf life can be achieved between 2 and 4 days, using 2% chitosan-coated food contact materials due to a decrease in *Botrytis cinerea* infection.	(Hefft, 2018)
Olive pomace flour	NA	Antioxidant	Bag	Walnut	Chitosan films combined with olive residues presented significant protective effect against oxidation (peroxide and trienes formation) in the nuts during the 31 days of storage. Chitosan-packaged samples maintained their omega 3 and omega 6 content during storage.	(Crizel, Rios, Alves, et al., 2018)
α-tocopherol	MMT	Antioxidant	Film	Salmon	Films with 15% tocopherol and 10% tocopherol + MMT led to lower oxidation values (TBARS) until the sixth day of packaging, with higher increase at the eighth day due to the pro-oxidant effect of tocopherol at high concentrations.	(Dias et al., 2019)

(*Continued*)

TABLE 1.6 (CONTINUED)
In Vivo Testing of Active Food Packaging Based on Chitosan. VP—Vacuum Packaging, MAP—Modified Atmosphere Packaging, EO—Essential Oil, MMT—Montmorillonite, TBARS—Thiobarbituric Acid Reactive Species, CFU—Colony Forming Units

Active agent	Additives/ Techniques	Active packaging	Type of packaging	Food product	Outcome	References
Kombucha tea	NA	Antimicrobial	Film	Minced beef	Chitosan/Kombucha tea film significantly retards microbial growth (psychrotrophs, total viable counts, coagulase-positive *Staphylococcus*). Shelf life extension of minced beef meat up to 3 days.	(Ashrafi, Jokar, and Nafchi, 2018)
Rosemary EO	MMT	Antimicrobial and antioxidant	Wrapping film	Minced poultry	Meat wrapped with biofilm showed a reduction of 1.2–2.1 log CFU/g on the total viable counts when compared to control films. Active films also succeeded on retarding poultry lipid peroxidation and discoloration.	(Souza et al., 2019)
Citric acid	NA	Antioxidant	Pouch	Green chilies	Citric acid films showed antioxidant properties as compared to neat chitosan films and were found to enhance green chili shelf-life.	(Priyadarshi, Sauraj, Kumar, and Negi, 2018)

(*Continued*)

TABLE 1.6 (CONTINUED)
In Vivo Testing of Active Food Packaging Based on Chitosan. VP—Vacuum Packaging, MAP—Modified Atmosphere Packaging, EO—Essential Oil, MMT—Montmorillonite, TBARS—Thiobarbituric Acid Reactive Species, CFU—Colony Forming Units

Active agent	Additives/ Techniques	Active packaging	Type of packaging	Food product	Outcome	References
Escherichia coli O157:H7 phage	NA	Antimicrobial	Edible coating	Tomatoes	Chitosan-based edible coating can stabilize the phage without significant loss in lytic activity of phage over a period of one week. Approximately 3 log differences in *E. coli* levels between the control and the treated samples.	(Amarillas et al., 2018)
Gallic acid	NA	Antioxidant and antimicrobial	Edible coating	Pork steak	Gallic acid increased chitosan antimicrobial activity (total viable counts). Films with 0.2% and 0.4% of gallic acid had lower lipid oxidation and myoglobin oxidation. Films with 0.4% gallic acid also exhibited a pro-protein oxidation effect.	(Fang et al., 2018)
Apricot kernel EO	NA	Antimicrobial	Pouch	Bread	Films showed better antimicrobial and antioxidant properties as compared to neat chitosan films and successfully inhibited the fungal growth on packaged bread slices.	(Priyadarshi, Sauraj, Kumar, Deeba, et al., 2018)

(Continued)

TABLE 1.6 (CONTINUED)

In Vivo Testing of Active Food Packaging Based on Chitosan. VP—Vacuum Packaging, MAP—Modified Atmosphere Packaging, EO—Essential Oil, MMT—Montmorillonite, TBARS—Thiobarbituric Acid Reactive Species, CFU—Colony Forming Units

Active agent	Additives/ Techniques	Active packaging	Type of packaging	Food product	Outcome	References
Eucalyptus globulus EO	NA	Antimicrobial	Wrapping film	Sliced sausage	The highest antibacterial activity against *L. monocytogenes* in sliced sausage was obtained by chitosan films containing 1.5% EO (1.01 log reduction).	(Azadbakht et al., 2018)
Trachyspermum ammi EO	NA	Antimicrobial	Edible film	Chicken meat	Chitosan film incorporated with 2% *T. ammi* EO had the highest inhibitory effects on total viable counts, psychrotrophs and coliform bacteria.	(Karimnezhad et al., 2017)
Ag nanoparticles	Laponite	Antimicrobial	Wrapping film	Litchi	Ag films exhibited good antifungal activity and effectively extended the storage life of litchi by decreasing mold growth and fruit deterioration.	(Wu et al., 2018)
Cinnamon EO	NA	Antimicrobial	Film	Peanut kernels	Combination of CS and cinnamon EO coating restricted *A. flavus* and *P. citrinum* contamination to 9.8% and 13.4%, respectively, in artificially inoculated peanut kernels at 28°C for 14 days of storage.	(Chein, Sadiq, and Anal, 2019)

(*Continued*)

TABLE 1.6 (CONTINUED)
In Vivo Testing of Active Food Packaging Based on Chitosan. VP—Vacuum Packaging, MAP—Modified Atmosphere Packaging, EO—Essential Oil, MMT—Montmorillonite, TBARS—Thiobarbituric Acid Reactive Species, CFU—Colony Forming Units

Active agent	Additives/ Techniques	Active packaging	Type of packaging	Food product	Outcome	References
Escherichia coli O157:H7 phage	Liposomes	Antimicrobial	Wrapping film	Beef meat	Chitosan film embedded with liposome-encapsulated phage exhibited high antibacterial activity against *Escherichia coli* O157:H7, without impacting the organoleptic properties of beef.	(Cui, Yuan, and Lin, 2017)
NA	Liquid smoke	Antimicrobial	Coating	Tilapia fish fillets	Oxidation parameters such as TBARS were reduced in the tilapia fillets with liquid smoking. Chitosan coating was effective in the control of the microorganisms (total viable counts and psychrotrophs) during storage.	(Santos et al., 2017)
Sumac hydro-alcoholic extract *Zataria multiflora* EO	NA	Antimicrobial and antioxidant	Coating	Beef meat	Active films exhibited antimicrobial activity with reductions in total viable counts, lactic acid bacteria, Pseudomonads, *Enterobacteriaceae* and yeasts-mold counts. Active films showed antioxidant activity with reduced TBARS and peroxide values. The highest antimicrobial effect was achieved in chitosan/*Zataria multiflora* films with 4% sumac extract.	(Langroodi, Tajik, and Mehdizadeh, 2018)

(*Continued*)

TABLE 1.6 (CONTINUED)
In Vivo Testing of Active Food Packaging Based on Chitosan. VP—Vacuum Packaging, MAP—Modified Atmosphere Packaging, EO—Essential Oil, MMT—Montmorillonite, TBARS—Thiobarbituric Acid Reactive Species, CFU—Colony Forming Units

Active agent	Additives/ Techniques	Active packaging	Type of packaging	Food product	Outcome	References
ε-polylysine	Nanofibers via electrospinning	Antimicrobial	Film	Chicken meat	ε-polylysine /chitosan nanofiber films were successful in inhibiting *Salmonella typhimurium* and *Salmonella enterica* (3 log reduction) on chicken.	(Lin et al., 2018)
Chrysanthemum EO	Nanofibers via electrospinning	Antimicrobial and antioxidant	Film	Beef meat	Active films inhibited *L. monocytogenes* growth on beef with an inhibition rate of 99.91%, 99.97%, and 99.95% at the temperature of 4°C, 12°C, and 25°C, respectively, after 7 days of storage. Active films also revealed antioxidant effects as shown by the reduction in TBARS values after 12 days of storage at 4°C.	(Lin et al., 2019)
Pink pepper residue extract	NA	Antimicrobial and antioxidant	In combination with MAP (100% CO_2)	Salmon fillets	The results showed that films with extract significantly reduced lipid oxidation. Bacterial counts (total viable counts, psychrotrophs, and lactic acid bacteria) were significantly lower in films with pink pepper extract.	(Merlo et al., 2019)

(Continued)

TABLE 1.6 (CONTINUED)
In Vivo Testing of Active Food Packaging Based on Chitosan. VP—Vacuum Packaging, MAP—Modified Atmosphere Packaging, EO—Essential Oil, MMT—Montmorillonite, TBARS—Thiobarbituric Acid Reactive Species, CFU—Colony Forming Units

Active agent	Additives/ Techniques	Active packaging	Type of packaging	Food product	Outcome	References
Oregano EO	NA	Antimicrobial	Coating Film	Pomegranate fruit	Both packaging types (coating and film) inhibited *Botrytis* spp. Growth. Inhibitory effect was higher for coating (direct contact) as opposed to samples exposed to the films (vapor phase-indirect contact).	(Munhuweyi et al., 2017)
Rosemary EO Ginger EO	Montmorillonite (MMT)	Antimicrobial and antioxidant	Wrapping film	Chicken meat	Films with MMT or EOs showed antioxidant and antimicrobial (total viable counts and coliforms) effects during storage, with no significant difference when compared with films with MMT+EOs.	(Souza, and Fernando, 2018)
Ginger EO	NA	Antimicrobial	Wrapping film	Barracuda fish fillets	Films with 0.3 % ginger EO film inhibited the growth of H_2S-producing bacteria and pseudomonads during storage at 2°C for 20 days.	(Remya et al., 2016)

(Continued)

TABLE 1.6 (CONTINUED)

In Vivo Testing of Active Food Packaging Based on Chitosan. VP—Vacuum Packaging, MAP—Modified Atmosphere Packaging, EO—Essential Oil, MMT—Montmorillonite, TBARS—Thiobarbituric Acid Reactive Species, CFU—Colony Forming Units

Active agent	Additives/ Techniques	Active packaging	Type of packaging	Food product	Outcome	References
NA	NA	Antimicrobial	Coating in combination with VP	Pork meat	1% and 2% chitosan coatings were able to significantly reduce *L. monocytogenes* counts by more than 1.5 log CFU/g with respect to control. Active films also inhibited the growth of mesophilic bacteria and were particularly effective on lactic acid bacteria and yeasts.	(Serio et al., 2018)
Ginger EO	MMT	Antimicrobial and antioxidant	Wrapping films	Chicken meat	The incorporation of ginger EO or MMT enhanced the biopolymer activity, by reducing lipid oxidation and microbiological (total viable counts and coliforms) growth of the poultry meat. Films with ginger EO+MMT did not show any significant differences with films only with EO or MMT.	(Souza et al., 2018)
Ginger+cinnamon EO mixtures	NA	Antimicrobial and antioxidant	Wrapping films	Pork meat	Chitosan films with cinnamon+ginger EOs were effective in retarding total viable counts as well as lipid oxidation (TBARS). The highest antioxidant and antimicrobial activities were observed in chitosan films incorporated with 1% of each EO.	(Wang et al., 2017)

(Continued)

TABLE 1.6 (CONTINUED)
In Vivo Testing of Active Food Packaging Based on Chitosan. VP—Vacuum Packaging, MAP—Modified Atmosphere Packaging, EO—Essential Oil, MMT—Montmorillonite, TBARS—Thiobarbituric Acid Reactive Species, CFU—Colony Forming Units

Active agent	Additives/ Techniques	Active packaging	Type of packaging	Food product	Outcome	References
Potassium permanganate	NA	Ethylene absorber	Wrapping film	Tomatoes	Fruits packaged with films with 7 g of potassium permanganate had a higher hardness value than the control sample at room temperature and now visible change in color (hue value). At refrigerated temperature, the packaged sample had a hardness value lower than the control.	(Warsiki, 2018)
Gallic acid	NA	Antimicrobial and antioxidant	Coating	Mackerel fillets	Films with gallic acid were more effective than films just with chitosan alone in inhibiting microbial (total viable counts) growth and lipid oxidation (TBARS). Gallic acid films extended the shelf life of mackerel by 6 days compared with the control group.	(Wu et al., 2016)
Garcinia atroviridis	NA	NA	Wrapping film	Indian mackerel	With the edible film of 5% (v/v) *Garcinia atroviridis* applied, the Indian mackerel was found to have a longer shelf life (3.5 days) as compared to the control (2.5 days) with a significant reduction in total viable counts.	(Zaman, Lin, and Phing, 2018)

TABLE 1.7

In Vivo Testing of Active Food Packaging Based on Cellulose and Cellulose Derivatives. EO—Essential Oil, TBARS—Thiobarbituric Acid Reactive Species, CFU—Colony Forming Units, HDPE—High Density Polyethylene), NP—Nanoparticle, HNTs—Halloysite nanotubes

Cellulose type	Active agent	Active packaging	Type of packaging	Food product	Outcome	References
Carboxymethyl cellulose (CMC)	Microencapsulated rice grass extract	Antioxidant	Pouch	Green tea	CMC films with rice grass effectively prevented green tea lipid oxidation. Total phenolic content of green tea packed in CMC films with rice extract was equivalent to that packed in HDPE with similar shelf life.	(Rodsamran and Sothornvit, 2018)
Cellulose	Silver NPs	Antimicrobial	Paper packet	Tomatoes Cut cabbage	Packets impregnated with nanoparticles exhibited significant antimicrobial properties. Periodic evaluation of stored vegetables in these packets demonstrated enhanced shelf-life with no significant changes in nutritional values.	(Singh and Sahareen, 2017)
Bacterial cellulose	Pomegranate peel extract Green tea extract Rosemary extract	Antimicrobial and antioxidant	Membrane	Mushrooms	Rosemary films showed the highest antioxidant property, but the antimicrobial activity of PPE was more than other extracts. Films with pomegranate peel had the highest effect on preserving total phenol and ascorbic acid contents. Films with pomegranate peel extract (50%) showed the lowest total viable counts after 15 days (6.5 log CFU/g).	(Moradian, Almasi, and Moini, 2018)

(*Continued*)

TABLE 1.7 (CONTINUED)
In Vivo Testing of Active Food Packaging Based on Cellulose and Cellulose Derivatives. EO—Essential Oil, TBARS—Thiobarbituric Acid Reactive Species, CFU—Colony Forming Units, HDPE—High Density Polyethylene), NP—Nanoparticle, HNTs—Halloysite nanotubes

Cellulose type	Active agent	Active packaging	Type of packaging	Food product	Outcome	References
Cellulose acetate	*Cymbopogon citratus*	Antioxidant	Wrapping film	Coalho cheese	Cheese samples wrapped with active films showed no change in texture and were able to maintain cheese yellow color during the 25 days.	(Oliveira et al., 2017)
Carboxymethyl cellulose	Zinc oxide (ZnO) NPs	Antimicrobial	Film	Pork meat	ZnO nanoparticle-coated film inhibits microbial spoilage of pork samples by decreasing the number of total viable counts during refrigerated storage for 14 days.	(Suo et al., 2016)
Paper	Eugenol	Insecticidal	Grafted paper	Flour	Active paper showed significantly enhanced bioactive properties (insecticide/insectifuge) against *T. castaneum* weevils.	(Muratore, Barbosa, and Martini, 2019)
Cellulose acetate	Pink pepper EO	Antimicrobial	Film	Sliced cheese	In contaminated cheese with *L. monocytogenes*, *S. aureus*, *E. coli*, or *S.* Typhimurium, all films were shown to be active against Gram-positive pathogens, while only films with 6% EO have shown to be active also against Gram-negative pathogens.	(Dannenberg et al., 2017)

(Continued)

TABLE 1.7 (CONTINUED)
In Vivo Testing of Active Food Packaging Based on Cellulose and Cellulose Derivatives. EO—Essential Oil, TBARS—Thiobarbituric Acid Reactive Species, CFU—Colony Forming Units, HDPE—High Density Polyethylene), NP—Nanoparticle, HNTs—Halloysite nanotubes

Cellulose type	Active agent	Active packaging	Type of packaging	Food product	Outcome	References
Paper	Silver NPs Titanium oxide NPs Zeolites	Antimicrobial	Sheet	Bread	Paper with silver and titanium was the most effective in the preservation of nutritional compounds in the bread. Paper with silver, titanium, and zeolites prolonged the microbiological safety of bread in terms of yeasts and molds content for 10 days at 20°C and 12 days at 4°C.	(Mihaly-Cozmuta et al., 2017)
Carboxymethyl cellulose	Ginger EO	Antimicrobial	Edible coating	Cheese	Cheese samples with active coatings showed a decrease in mold and yeast growth during storage time, with a linear decrease during 24 days with a complete inhibition at day 18.	(Tapiero Cuellar, 2017)
Ethyl cellulose	Cinnamon EO encapsulated into Ag^{+}/ Zn_{2}^{+}-permutite	Antimicrobial	Pad	Chinese bayberry	Treatment with the permutite/cinnamon EO antimicrobial pads resulted in a significantly lower decay incidence as shown by the visual observation in the decrease of fungal growth.	(Niu et al., 2018)

(Continued)

TABLE 1.7 (CONTINUED)
In Vivo Testing of Active Food Packaging Based on Cellulose and Cellulose Derivatives. EO—Essential Oil, TBARS—Thiobarbituric Acid Reactive Species, CFU—Colony Forming Units, HDPE—High Density Polyethylene), NP—Nanoparticle, HNTs—Halloysite nanotubes

Cellulose type	Active agent	Active packaging	Type of packaging	Food product	Outcome	References
Carboxymethyl cellulose	Okra mucilage ZnO NPs	Antimicrobial and antioxidant	Wrapping film	Chicken breast	Microbial (total viable counts, coagulase-positive *Staphylococcus*, and lactic acid bacteria) growth was inhibited by active films containing okra mucilage and ZnO nanoparticles. The increase of lipid oxidation was also significantly restricted by incorporation of okra mucilage and ZnO NPs CMC films with 50% okra mucilage and ZnO had the best inhibitory effects on the microbial growth and chemical changes in the chicken breast meat.	(Mohammadi et al., 2019)
Cellulose nanofibers	Sakacin-A	Antimicrobial	Film	Smoked salmon	Active films reduced *Listeria innocua* counts by about 2.5–3 log cycles after 28 days of storage at 6°C	(Mapelli et al., 2019)

TABLE 1.8
In Vivo Testing of Active Food Packaging Based on Gelatin. VP—Vacuum Packaging, EO—Essential Oil, TBARS—Thiobarbituric Acid Reactive Species

Gelatin source	Active agent	Active packaging	Type of packaging	Food product	Outcome	References
Porcine skin	Thyme EO/β-cyclodextrin ε-polylysine nanoparticles	Antimicrobial	Fibers	Chicken meat	Active gelatin nanofibers exhibited excellent antimicrobial activity against *C. jejuni* on chicken. Active fiber-packaged chicken showed lower total viable counts and oxidation values (TBARS).	(Lin, Zhu, and Cui, 2018)
Fish	Carvacrol	Antimicrobial and antioxidant	Wrapping film	Breaded hake medallions	Films with 0.6% carvacrol caused a significant reduction in psychrotroph counts. Active films also showed higher antioxidant properties as demonstrated by lower total oxidation index (TOI), peroxide value (PV), and *p*-anisidine value (AV).	(Neira et al., 2019)
Tilapia fish skin	Ginger EO ZnO NPs	Antimicrobial and antioxidant	Film in combination with VP	Pork meat	Active films showed strong antibacterial activity against food spoilage bacteria (psychrotrophs, total viable counts, and lactic acid bacteria. Films also exhibited antioxidant activity toward lipid oxidation during storage.	(Zhang et al., 2017)
Olive Flounder bone	ZnO NPs	Antimicrobial	Wrapping film	Spinach	Active films showed antimicrobial activity against *L. monocytogenes* inoculated on spinach without affecting the quality of spinach, such as vitamin C content and color.	(Beak, Kim, and Song, 2017)

(*Continued*)

TABLE 1.8 (CONTINUED)
In Vivo Testing of Active Food Packaging Based on Gelatin. VP—Vacuum Packaging, EO—Essential Oil, TBARS—Thiobarbituric Acid Reactive Species

Gelatin source	Active agent	Active packaging	Type of packaging	Food product	Outcome	References
Porcine skin	Butylated hydroxyanisole	Antimicrobial	Mat	Strawberries	The shelf-life of strawberry can be prolonged effectively in the presence of active mats during storage due to the broad antifungal spectrum of the mats.	(Li et al., 2018)
Skate skin	Thyme EO	Antimicrobial	Wrapping film	Chicken meat	Chicken meat packaged in active films with 1% thyme EO showed lower *L. monocytogenes* and *E. coli* O157:H7 counts when compared to the control during storage.	(Lee et al., 2016a)
Porcine skin	ε-Polylysine	Antimicrobial	Wrapping fibers	Beef meat	Active films exhibited outstanding antibacterial activity against *Listeria monocytogenes* during 10-days storage at 4°C and 7-days storage at 12°C, without impact on the surface color and sensory properties.	(Lin, Gu, and Cui, 2018)
Puffer fish skin	*Moringa oleifera* leaf extract	Antimicrobial and antioxidant	Wrapping film	Gouda cheese	During storage, active films effectively inhibited the microbial growth of *L. monocytogenes* and retarded lipid oxidation in packaged cheese samples when compared with the control sample.	(Lee, Yang, and Song, 2016)
Nutraceutical capsules wastes	Papaya peel	Antioxidant	Wrapping film	Lard	Films with 7.5% papaya peel microparticles were the most efficient as active barriers (higher antioxidant activity) with lower content of peroxides quantified after 22 days.	(Crizel, Rios, Alves, et al., 2018)

(Continued)

TABLE 1.8 (CONTINUED)
In Vivo Testing of Active Food Packaging Based on Gelatin. VP—Vacuum Packaging, EO—Essential Oil, TBARS—Thiobarbituric Acid Reactive Species

Gelatin source	Active agent	Active packaging	Type of packaging	Food product	Outcome	References
Duck feet	Cinnamon EO	Antimicrobial	Wrapping film	Tomatoes	Active film containing cinnamon EO effectively inhibited the growth of *S. typhimurium* inoculated on cherry tomatoes during storage compared to the control. Active films also reduced color change on cherry tomatoes.	(Yang et al., 2017)
Starfish	Vanillin	Antimicrobial	Wrapping film	Crab sticks	During storage, the populations of *L. monocytogenes* inoculated on crab sticks wrapped with films containing vanillin (0.05%) were lower than those on the control sample.	(Lee et al., 2016b)
Silver carp	β-cyclodextrin/ curcumin complexes	Antioxidant	Bag	Apple juice	The slowest decline in total polyphenol content was found in gelatin films with cyclodextrin complexes containing 2.5% curcumin. The incorporation of β-cyclodextrin/curcumin complex significantly enhanced the antioxidant activity of gelatin films.	(Rezaei and Shahbazi, 2018)
Porcine skin	Moringa oil-loaded chitosan nanoparticles	Antimicrobial	Wrapping fibers	Cheese	On cheese, active nanofibers possessed high antibacterial activity against *L. monocytogenes* and *S. aureus* at 4°C and 25°C for 10 days, without any effect on the sensory quality of cheese.	(Lin, Gu, and Cui, 2019)

(*Continued*)

TABLE 1.8 (CONTINUED)
In Vivo Testing of Active Food Packaging Based on Gelatin. VP—Vacuum Packaging, EO—Essential Oil, TBARS—Thiobarbituric Acid Reactive Species

Gelatin source	Active agent	Active packaging	Type of packaging	Food product	Outcome	References
Porcine skin	*Caesalpinia decapetala* and *Caesalpinia spinosa* extracts	Antioxidant	Film	Ground beef	Active films with *Caesalpinia decapetala* and *Caesalpinia spinosa* extracts were effective in delaying lipid oxidation and deterioration of beef patty quality during storage.	(Gallego et al., 2016)
Bovine	Anthocyanins/chitosan nanocomplexes	Antioxidant	Film	Olive oil	Films containing anthocyanin/chitosan nanocomplexes delayed olive oil oxidative deterioration (peroxide value) when compared with the films composed of gelatin only after 56 days of storage.	(Wang et al., 2019)

vegetables, cheese, lard, and fruit juice. In some of these active packaging materials, active agents have been loaded in encapsulating compounds as cyclodextrins or chitosan nanoparticles to improve their stability and provide a more controlled release of such compounds from the gelatin matrix.

1.5.4 Starch-Based Packaging Materials

Starch is also one of the bio-based polymers mostly used in the development of new packaging materials, with several starch-based materials being already available at an industrial level. At the bench-top level, active packaging strategies using starch polymers have been used to produce both antioxidant or antimicrobial packaging, by the incorporation of active agents such as rosehip extract, cinnamaldehyde, *Zataria multiflora* EO, lycopene, carvacrol, thymol and nisin (bacteriocin) (see Table 1.9). To facilitate the inclusion of some of these compounds in the starch matrix, several strategies such as microencapsulation in maltodextrin, nanoemulsions, formation of nanocapsules with Tween 80 as well as the addition of nanoclays (bentonite, montmorillonite, and halloysite) have been described. These packages have been tested to inhibit oxidation of fresh meat and oils along with the improvement of the microbial safety of fresh meat, bread, fresh fruit and vegetables, and cheese.

1.5.5 Other Protein-Based Packaging Materials

Besides gelatin, other proteins derived from biological sources such as sorghum, whey, distiller dried grains, and silk have been utilized to create new active packaging materials with demonstrated antimicrobial and/or antioxidant properties in fresh food products such as meat and fish and also some processed food as well, like smoked salmon and salami (see Table 1.10) packaged in films, pouches or edible coatings. From a sustainability point of view, as these protein sources are all by-products of the food industry, the entire array of active agents used are of natural origin, ranging from enzymes (lysozyme and lactoperoxidase system), ginger, chamomile and clove EOs, tea and rosemary extracts, quercetin, and citral and lytic phages.

1.5.6 Alginate-Based Packaging Materials

Another not so common polysaccharide used for the development of active food packaging is the seaweed-derived alginate that has been described as the matrix for the study of several active edible coatings and some films that were applied to food products such as sausages, fruits, beef and chicken meat, and fish fillets (see Table 1.11). These edible coatings containing EOs, bacteriophages, or silver nanoparticles (NPs) have been effective in reducing lipid oxidation and in preventing bacterial or fungal growth.

TABLE 1.9
In Vivo Testing of Active Food Packaging Based on Starch. EO—Essential Oil, MMT – Montmorillonite, TBARS—Thiobarbituric Acid Reactive Species, CFU—Colony Forming Units, NP—Nanoparticle, NA—Not Applicable

Active agent	Additives/ Techniques	Active packaging	Type of packaging	Food product	Outcome	References
NA	Inverted sugar Sucrose	Antimicrobial	Edible coating	Mini panettone	Active coatings inhibited the growth of yeasts and molds in panettone at 25 and 45°C during 40 days of storage.	(Saraiva et al., 2016)
Oregano EO Pumpkin residue extract	NA	Antimicrobial and antioxidant	Film	Ground beef	Active films with pumpkin residue and EO demonstrated antimicrobial activity against total viable counts, coliforms and *Salmonella* Enteritidis. Active films showed significant antioxidant activity (lower TBARS values).	(Caetano et al., 2017)
Rosehip extract	NA	Antioxidant	Wrapping film	Chicken meat	Chicken breast samples wrapped with the active films containing 1.0% rosehip extract showed lower lipid oxidation (peroxide value and TBARS) than control samples.	(Go and Song, 2019)
Wine grape pomace	Microencapsulation in maltodextrin	Antioxidant	Pouch	Olive oil	Active films were able to maintain olive oil quality during storage at 40°C with exposition to fluorescent light for 12 days; while the oil packed in polypropylene pouches was degraded before the 4th day of storage.	(Stoll et al., 2017)

(Continued)

TABLE 1.9 (CONTINUED)
In Vivo Testing of Active Food Packaging Based on Starch. EO—Essential Oil, MMT—Montmorillonite, TBARS—Thiobarbituric Acid Reactive Species, CFU—Colony Forming Units, NP—Nanoparticle, NA—Not Applicable

Active agent	Additives/ Techniques	Active packaging	Type of packaging	Food product	Outcome	References
Zataria multiflora EO Cinnamaldehyde	Nanoemulsions	Antioxidant	Film	Ground beef	The best antioxidant effects (TBARS and protein oxidation) were obtained for films with *Zataria multiflora* nanoemulsions. Cinnamaldehyde addition did not improve *Zataria* active films' antioxidant activity.	(Amiri et al., 2019)
Thyme EO	MMT	Antimicrobial	Wrapping film	Spinach	The incorporation of thyme EO in the film significantly reduced *E. coli* and *S.* Typhimurium populations on fresh baby spinach leaves to below detectable levels within 5 days.	(Issa, Ibrahim, and Tahergorabi, 2017)
ε-polylysine	NA	Antimicrobial	Film	Bread	The shelf life of bread inoculated with *A. parasiticus* was increased by 1 day with the use of active films (1.6–6.5 mg polylysine/cm^2). The shelf life of bread contaminated with *P. expansum* was increased by 3 days with active films containing 6.5 mg polylysine/cm^2. Aflatoxin production was greatly inhibited by active films (93–100%).	(Luz et al., 2018)
Lycopene	Nanocapsule (Tween 80)	Antioxidant	Pouch	Sunflower oil	Films with lycopene nanocapsules provided greater protection to oxidation (peroxide value, conjugated dienes and trienes) of stored sunflower oil under accelerated oxidation conditions.	(Assis et al., 2017)

(Continued)

TABLE 1.9 (CONTINUED)

In Vivo Testing of Active Food Packaging Based on Starch. EO—Essential Oil, MMT—Montmorillonite, TBARS—Thiobarbituric Acid Reactive Species, CFU—Colony Forming Units, NP—Nanoparticle, NA—Not Applicable

Active agent	Additives/ Techniques	Active packaging	Type of packaging	Food product	Outcome	References
Cinnamon EO	Bentonite	Antimicrobial	Pouch	Meatballs	Starch films with 0.75% (w/w) sodium bentonite and 2.5% (w/w) cinnamon EO significantly inhibited microbial (total viable counts) growth up to 96 h below the legal limits (10^6 CFU/g) when compared to control films that exceeded the limit within 48 h.	(Lameerat et al., 2018)
Carvacrol Thymol	MMT	Antimicrobial	Label	Strawberries	Films containing 50:50 carvacrol:thymol were more effective in controlling *Botrytis cinerea* growth on inoculated strawberries than films containing only carvacrol by indirect contact with films.	(Campos-Requena et al., 2017)
Nisin	Halloysite	Antimicrobial	Film	Cheese	After 4 days of storage, films with 2% nisin significantly reduced the initial counts of *L. monocytogenes*. Films with 6% nisin completely inhibited *L. monocytogenes*;	(Meira et al., 2016)
Bixin	NA	Antioxidant	Pouch	Sunflower oil	Sunflower oil packaged in films with bixin, under accelerated oxidation conditions, exhibited lower oxidation rates, thus maintaining its freshness for longer.	(Pagno et al., 2016)

TABLE 1.10
In Vivo Testing of Active Food Packaging Using Protein-Based Polymers. NA—Not Applicable, EO—Essential Oil, MAP—Modified Atmosphere Packaging, VP—Vacuum Packaging, MMT—Montmorillonite, TBARS—Thiobarbituric Acid Reactive Species, CFU—Colony Forming Unit

Protein source	Active agent	Active packaging	Type of packaging	Food product	Outcome	References
Sorghum (kafirin)	Citral Quercetin	Antimicrobial and antioxidant	Wrapping film	Chicken meat	Citral-containing films showed significant higher antimicrobial (total viable counts) activity on chicken fillets stored at 2±0.5°C for 96 h. Quercetin containing films possessed the ability to inhibit the development of lipid oxidative products (TBARS).	(Giteru et al., 2017)
Whey	Lysozyme	Antimicrobial	Film	Smoked salmon	Smoked salmon slices coated by activated films containing lysozyme and oleic acid showed significantly lower *L. innocua* counts than controls after 1 week at 4°C.	(Boyaci, Korel, and Yemenicioğlu, 2016)
Whey	Ginger EO Chamomile EO	Antimicrobial and antioxidant	Edible coating	Trout fillets	Bacterial growth was inhibited in samples with 0.4% combined ginger and chamomile Eos. The addition of ginger and chamomile EOs to protein coatings showed a protective oxidation effect (TBARS and peroxide value).	(Oğuzhan Yıldız and Yangılar, 2017)
Distiller dried grains	Green tea extract Oolong tea extract Black tea extract	Antioxidant	Wrapping film	Pork meat	Pork meat wrapped with films containing the 3 extracts had less lipid oxidation (TBARS and peroxide value) than did the control. Films with green tea extract had the greatest antioxidant activity.	(Yang et al., 2016)

(*Continued*)

TABLE 1.10 (CONTINUED)
In Vivo Testing of Active Food Packaging Using Protein-Based Polymers. NA—Not Applicable, EO—Essential Oil, MAP—Modified Atmosphere Packaging, VP—Vacuum Packaging, MMT—Montmorillonite, TBARS—Thiobarbituric Acid Reactive Species, CFU—Colony Forming Unit

Protein source	Active agent	Active packaging	Type of packaging	Food product	Outcome	References
Whey	Lytic phages	Antimicrobial	Film	Meat	Phage-added films yielded a complete inactivation of DH5α and O157:H7 *E. coli* strains on meat.	(Tomat et al., 2019)
Whey	Lactoperoxidase system	Antimicrobial	Edible coating in combination with MAP	Perch fillets	Combination of lactoperoxidase films and MAP could significantly inhibit bacterial (total viable counts, psychrotrophs, *Shewanella putrefaciens*, *Pseudomonas fluorescens*) and fat oxidation (TBARS).	(Rostami, Abbaszadeh, and Shokri, 2017)
Whey	NA	Antimicrobial and antioxidant	Edible coating	Trout fillets	Glycerol addition to the film-forming solution yielded films with increased antimicrobial (total viable counts, psychrotrophs, lactic acid bacteria, *Enterobacteriaceae*) and antioxidant (TBARS, peroxide value) activities.	(Yıldız and Yangılar, 2016)
Whey	Green tea	Antioxidant	Pouch in combination with VP	Salmon	Films with green tea were effective in delaying lipid oxidation (peroxide value, *p*-anisidine value, TBARS, and hexanal) of fresh salmon samples until the 14th day of storage.	(Castro et al., 2019)
Whey	Rosemary extract	Antioxidant	Film in combination with VP	Salami	Active film was able to delay the salami's lipid (TBARS and hexanal monitoring) oxidation for, at least, 30 days.	(Andrade et al., 2019)

(*Continued*)

TABLE 1.10 (CONTINUED)
In Vivo Testing of Active Food Packaging Using Protein-Based Polymers. NA—Not Applicable, EO—Essential Oil, MAP—Modified Atmosphere Packaging, VP—Vacuum Packaging, MMT—Montmorillonite, TBARS—Thiobarbituric Acid Reactive Species, CFU—Colony Forming Unit

Protein source	Active agent	Active packaging	Type of packaging	Food product	Outcome	References
Whey	Clove EO MMT	Antimicrobial and antioxidant	Wrapping film	Tuna fillets	Films with MMT and EO were able to decrease microbial growth (especially pseudomonads) and lipid autooxidation (evaluated according to TBARS values) of tuna fillets during the storage period.	(Echeverría et al., 2018)
Silk (fibroin)	Thyme EO	Antimicrobial	Wrapping nanofibers in combination with cold plasma	Poultry meat	*Salmonella* Typhimurium counts in chicken meat and duck meat wrapped in active films decreased by 6.1 and 6.06 log CFU/g compared with control group at 25°C, respectively.	(Tabassum and Khan, 2020)

TABLE 1.11
In Vivo Testing of Active Food Packaging Based on Alginate. NA—Not Applicable, EO – Essential Oil, NP—Nanoparticle, MAP—Modified Atmosphere Packaging, CFU—Colony Forming Units, MMT—Montmorillonite, TBARS—Thiobarbituric Acid Reactive Species

Active agent	Active packaging	Type of packaging	Food product	Outcome	References
Tinospora cordifolia	Antimicrobial and antioxidant	Edible coating	Goat meat sausage	Significant antioxidant activity (lower TBARS and % oleic acid values) and microbial counts (total viable counts, psychrotrophs, yeast, and mold) for sausages packaged in films with *T. cordifolia*.	(Kalem et al., 2018)
Silver NPs	Antimicrobial	Edible coating	Strawberries Loquats	Active coatings (with silver) improved the quality parameters of strawberries and loquats, including its microbial analysis, leading to an increased shelf life of coated fruits.	(Hanif et al., 2019)
Ethanol	Antimicrobial	Edible film	Ham frankfurters	Ethanol-based edible films inhibited *Listeria monocytogenes* growth on frankfurters and ham.	(Kapetanakou, Karyotis, and Skandamis, 2016)
Rosemary EO Oregano EO	Antioxidant	Edible coating	Beef	EO coatings exhibited antioxidant activity when compared with control coatings, with oregano coatings being the most effective (46.81% decrease in lipid oxidation).	(Vital et al., 2016)
Thyme EO Oregano EO	Antimicrobial	Edible coating in combination with MAP	Fresh-cut papaya	Samples coated with 1.0% thyme EO and oregano EO were the most effective in improving microbiological food safety for papaya stored for 12 days at 4°C.	(Tabassum and Khan, 2020)
Bacteriophage	Antimicrobial	Film	Chicken fillets	Active films decreased *P. fluorescens* counts, with a 2 log CFU/g reduction up to day 2, with reductions being maintained up to 5 days of exposure (1 log CFU/g) on chicken breast fillets artificially inoculated.	(Tomat et al., 2019)
Marjoram EO	Antimicrobial	Film	Trout fillets	Alginate-MMT films enriched with 1% marjoram EO significantly delayed the growth of *L. monocytogenes* during the 15-day storage. Active films also reduced total viable counts and psychrotrophs.	(Alboofetileh et al., 2016)

1.5.7 Other Bio-Based Active Packaging

In addition to all the above mentioned bio-based polymers, other not so conventional polymers such as zein, flours from several sources (banana, triticale and pomelo peel), gums (Arabic, Guam, from basil seed and Acacia), agar, pectin, quinoa or even *Aloe vera* have also been tested for their capacity to generate active packaging materials that were able to prevent oxidation or microbial spoilage in a vast array of foods (see Table 1.12), being the preferred form as edible coatings.

TABLE 1.12
Examples of Other Bio-Based Polymers Used for Active Food Packaging That Have Been Evaluated in Food Products

Polymer	Active packaging	Type of packaging	Food product	References
Zein	Antioxidant	Coating	Melons	(Boyacı et al., 2019)
Zein	Antimicrobial	Edible coating	Fish fillets	(Hager et al., 2019)
Zein	Antioxidant	Wrapping film	Apple slices	(Vimala Bharathi et al., 2019)
Semolina flour	Antimicrobial	Film	Mozzarella cheese	(Jafarzadeh et al., 2019)
Agar	Antimicrobial	Wrapping film	Grapes	(Kumar et al., 2019)
Triticale flour	Antimicrobial	Film	Cheese	(Romero et al., 2016)
Guar gum	Antioxidant	Mat	Flaxseed oil	(Yang et al., 2019)
Arabic gum	Antimicrobial	Edible coating	Strawberries	(Tahir et al., 2018)
Pectin	Antimicrobial	Edible coating	Apricots	(Gorrasi and Bugatti, 2016)
Quinoa	Antimicrobial and antioxidant	Edible film	Trout fillets	(Alak et al., 2019)
Pectin	Antimicrobial	Edible coating	Fresh-cut persimmon	(Sanchís et al., 2017)
Basil seed gum	Antimicrobial	Edible coating	Shrimp	(Khazaei, Esmaiili, and Emam-Djomeh, 2017)
Banana flour	Antioxidant	Sachet	Roasted peanuts	(Orsuwan and Sothornvit, 2018)
Triticale flour	Antimicrobial	Pouch	Cheese slices	(Salvucci et al., 2019)
Agar	Antimicrobial	Edible coating	Cheese	(Guitián et al., 2019)
Acacia gum	Antimicrobial	Edible film	Cheese	(Goulart et al., 2019)
Zein	Antimicrobial	Wrapping nanofibers	Beef meat	(Aytac et al., 2017)
Guar gum	Antimicrobial	Film	Pomegranate arils	(Saurabh, Gupta, and Variyar, 2018)
Pomelo peel flour	Antioxidant	Film	Soybean oil	(Wu et al., 2019)
Aloe vera gel	Antimicrobial	Edible coating	Grapes	(Ali et al., 2016)

1.6 CONCLUSIONS

In the actual context of climate change coupled with the global awareness of ocean pollution mostly by plastic debris, and keeping in mind that the food packaging materials are generally produced by traditional plastics derived from fossil oil, which are nonbiodegradable, scientific research has focused on the development of new bio-based materials for active food packaging, obtained from renewable resources. Polysaccharides and proteins appear to be the most promising candidates to produce biodegradable and edible films that can also be complemented with the incorporation of bioactive compounds (plant extracts, essential oils, etc.), and thus contributing to extending the shelf life of foods.

ACKNOWLEDGMENTS

Ângelo Luís acknowledges the contract of Scientific Employment in the scientific area of Microbiology financed by Fundação para a Ciência e a Tecnologia (FCT). This work was partially supported by CICS-UBI, which is financed by National Funds from FCT and Community Funds (UIDB/00709/2020). The authors would also like to acknowledge the funding provided by the Gobierno de Aragón (Spain) and FEDER (LMP49_18).

REFERENCES

Ahmed, Temoor, Muhammad Shahid, Farrukh Azeem, Ijaz Rasul, Asad Ali Shah, Muhammad Noman, Amir Hameed, Natasha Manzoor, Irfan Manzoor, and Sher Muhammad. 2018. "Biodegradation of Plastics: Current Scenario and Future Prospects for Environmental Safety." *Environmental Science and Pollution Research International* 25(8): 7287–98. doi:10.1007/s11356-018-1234-9.

Akhtar, Hafiz Muhammad Saleem, Asad Riaz, Yahya Saud Hamed, Mohamed Abdin, Guijie Chen, Peng Wan, and Xiaoxiong Zeng. 2018. "Production and Characterization of CMC-Based Antioxidant and Antimicrobial Films Enriched with Chickpea Hull Polysaccharides." *International Journal of Biological Macromolecules* 118(A): 469–77. doi:10.1016/j.ijbiomac.2018.06.090.

Akhter, Rehana, F. A. Masoodi, Touseef Ahmed Wani, and Sajad Ahmad Rather. 2019. "Functional Characterization of Biopolymer Based Composite Film: Incorporation of Natural Essential Oils and Antimicrobial Agents." *International Journal of Biological Macromolecules* 137: 1245–55. doi:10.1016/j.ijbiomac.2019.06.214.

Al-Tayyar, Nasser A., Ahmed M. Youssef, and Rashad Al-hindi. 2020. "Antimicrobial Food Packaging Based on Sustainable Bio-Based Materials for Reducing Foodborne Pathogens: A Review." *Food Chemistry* 310: 125915. doi:10.1016/j.foodchem.2019.125915.

Alak, Gonca, Kubra Guler, Arzu Ucar, Veysel Parlak, Esat Mahmut Kocaman, Telat Yanık, and Muhammed Atamanalp. 2019. "Quinoa as Polymer in Edible Films with Essential Oil: Effects on Rainbow Trout Fillets Shelf Life." *Journal of Food Processing and Preservation* 43(12). doi:10.1111/jfpp.14268.

Alboofetileh, Mehdi, Masoud Rezaei, Hedayat Hosseini, and Mehdi Abdollahi. 2016. "Efficacy of Activated Alginate-Based Nanocomposite Films to Control *Listeria monocytogenes* and Spoilage Flora in Rainbow Trout Slice." *Journal of Food Science and Technology* 53(1): 521–30. doi:10.1007/s13197-015-2015-9.

Ali, Amjad, Ying Chen, Hongsheng Liu, Long Yu, Zulqarnain Baloch, Saud Khalid, Jian Zhu, and Ling Chen. 2019. "Starch-Based Antimicrobial Films Functionalized by Pomegranate Peel." *International Journal of Biological Macromolecules* 129: 1120–26. doi:10.1016/j.ijbiomac.2018.09.068.

Ali, Javed, Suyash Pandey, Vaishali Singh, and Prerna Joshi. 2016. "Effect of Coating of *Aloe vera* Gel on Shelf Life of Grapes." *Current Research in Nutrition and Food Science Journal* 4(1): 58–68. doi:10.12944/CRNFSJ.4.1.08.

Alkan Tas, Buket, Ekin Sehit, C. Erdinc Tas, Serkan Unal, Fevzi C. Cebeci, Yusuf Z. Menceloglu, and Hayriye Unal. 2019. "Carvacrol Loaded Halloysite Coatings for Antimicrobial Food Packaging Applications." *Food Packaging and Shelf Life* 20. doi:10.1016/j.fpsl.2019.01.004.

Altan, Aylin, Zeynep Aytac, and Tamer Uyar. 2018. "Carvacrol Loaded Electrospun Fibrous Films from Zein and Poly(Lactic Acid) for Active Food Packaging." *Food Hydrocolloids* 81: 48–59. doi:10.1016/j.foodhyd.2018.02.028.

Alves, Diana, Arlete Marques, Catarina Milho, Maria José Costa, Lorenzo M. Pastrana, Miguel A. Cerqueira, and Sanna Maria Sillankorva. 2019. "Bacteriophage ΦIBB-PF7A Loaded on Sodium Alginate-Based Films to Prevent Microbial Meat Spoilage." *International Journal of Food Microbiology* 291: 121–27. doi:10.1016/j.ijfoodmicro.2018.11.026.

Alzagameem, Abla, Stephanie Elisabeth Klein, Michel Bergs, Xuan Tung Do, Imke Korte, Sophia Dohlen, Carina Hüwe, et al. 2019. "Antimicrobial Activity of Lignin and Lignin-Derived Cellulose and Chitosan Composites against Selected Pathogenic and Spoilage Microorganisms." *Polymers* 11(4). doi:10.3390/polym11040670.

Amarillas, Luis, Luis Lightbourn-Rojas, Ana K. Angulo-Gaxiola, J. Basilio Heredia, Arturo González-Robles, and Josefina León-Félix. 2018. "The Antibacterial Effect of Chitosan-Based Edible Coating Incorporated with a Lytic Bacteriophage against *Escherichia coli* O157:H7 on the Surface of Tomatoes." *Journal of Food Safety* 38(6). doi:10.1111/jfs.12571.

Ambrogi, V., C. Carfagna, P. Cerruti, and V. Marturano. 2017. "Additives in Polymers." In: *Modification of Polymer Properties*. Elsevier Inc., 87–108. doi:10.1016/B978-0-323-44353-1.00004-X.

Amiri, Elham, Majid Aminzare, Hassan Hassanzad Azar, and Mohammad Reza Mehrasbi. 2019. "Combined Antioxidant and Sensory Effects of Corn Starch Films with Nanoemulsion of *Zataria multiflora* Essential Oil Fortified with Cinnamaldehyde on Fresh Ground Beef Patties." *Meat Science* 153: 66–74. doi:10.1016/j.meatsci.2019.03.004.

Amjadi, Sajed, Sana Emaminia, Maryam Nazari, Shabnam Heyat Davudian, Leila Roufegarinejad, and Hamed Hamishehkar. 2019. "Application of Reinforced ZnO Nanoparticle-Incorporated Gelatin Bionanocomposite Film with Chitosan Nanofiber for Packaging of Chicken Fillet and Cheese as Food Models." *Food and Bioprocess Technology* 12(7): 1205–19. doi:10.1007/s11947-019-02286-y.

Andrade, Mariana A., Regiane Ribeiro-Santos, Manuela Guerra, and Ana Sanches-Silva. 2019. "Evaluation of the Oxidative Status of Salami Packaged with an Active Whey Protein Film." *Foods* 8(9). doi:10.3390/foods8090387.

Andrady, Anthony L. 2011. "Microplastics in the Marine Environment." *Marine Pollution Bulletin* 62(8): 1596–605. doi:10.1016/j.marpolbul.2011.05.030.

Arcan, İskender, Derya Boyacı, and Ahmet Yemenicioğlu. 2017. "The Use of Zein and Its Edible Films for the Development of Food Packaging Materials." *Reference Module in Food Science*. doi:10.1016/B978-0-08-100596-5.21126-8.

Artiga-Artigas, María, Alejandra Acevedo-Fani, and Olga Martín-Belloso. 2017. "Improving the Shelf Life of Low-Fat Cut Cheese Using Nanoemulsion-Based Edible Coatings Containing Oregano Essential Oil and Mandarin Fiber." *Food Control* 76: 1–12. doi:10.1016/J.FOODCONT.2017.01.001.

Ashrafi, Azam, Maryam Jokar, and Abdorreza Mohammadi Nafchi. 2018. "Preparation and Characterization of Biocomposite Film Based on Chitosan and Kombucha Tea as Active Food Packaging." *International Journal of Biological Macromolecules* 108: 444–54. doi:10.1016/j.ijbiomac.2017.12.028.

Assis, Renato Queiroz, Stefani Machado Lopes, Tania Maria Haas Costa, Simone Hickmann Flôres, and Alessandro de Oliveira Rios. 2017. "Active Biodegradable Cassava Starch Films Incorporated Lycopene Nanocapsules." *Industrial Crops and Products* 109: 818–27. doi:10.1016/j.indcrop.2017.09.043.

Aytac, Zeynep, Semran Ipek, Engin Durgun, Turgay Tekinay, and Tamer Uyar. 2017. "Antibacterial Electrospun Zein Nanofibrous Web Encapsulating Thymol/Cyclodextrin-Inclusion Complex for Food Packaging." *Food Chemistry* 233: 117–24. doi:10.1016/j.foodchem.2017.04.095.

Azadbakht, Ehsan, Yahya Maghsoudlou, Morteza Khomiri, and Mahboobeh Kashiri. 2018. "Development and Structural Characterization of Chitosan Films Containing *Eucalyptus globulus* Essential Oil: Potential as an Antimicrobial Carrier for Packaging of Sliced Sausage." *Food Packaging and Shelf Life* 17: 65–72. doi:10.1016/j.fpsl.2018.03.007.

Balti, Rafik, Mohamed Ben Mansour, Nadhem Sayari, Lamia Yacoubi, Lotfi Rabaoui, Nicolas Brodu, and Anthony Massé. 2017. "Development and Characterization of Bioactive Edible Films from Spider Crab (*Maja crispata*) Chitosan Incorporated with Spirulina Extract." *International Journal of Biological Macromolecules* 105(2): 1464–72. doi:10.1016/j.ijbiomac.2017.07.046.

Bartl, Andreas. 2018. "The EU Circular Economy Package: A Genius Programme or an Old Hat?" *Waste Management and Research* 36(4): 309–10. doi:10.1177/0734242X18755022.

Basiak, Ewelina, Andrzej Lenart, and Frédéric Debeaufort. 2017. "Effect of Starch Type on the Physico-Chemical Properties of Edible Films." *International Journal of Biological Macromolecules* 98: 348–56. doi:10.1016/j.ijbiomac.2017.01.122.

Beak, Songee, Hyeri Kim, and Kyung Bin Song. 2017. "Characterization of an Olive Flounder Bone Gelatin-Zinc Oxide Nanocomposite Film and Evaluation of Its Potential Application in Spinach Packaging." *Journal of Food Science* 82(11): 2643–49. doi:10.1111/1750-3841.13949.

Becker, A., F. Katzen, A. Pühler, and L. Ielpi. 1998. "Xanthan Gum Biosynthesis and Application: A Biochemical/Genetic Perspective." *Applied Microbiology and Biotechnology* 50(2): 145–52. doi:10.1007/s002530051269.

Birck, C., S. Degoutin, M. Maton, C. Neut, M. Bria, M. Moreau, F. Fricoteaux, V. Miri, and M. Bacquet. 2016. "Antimicrobial Citric Acid/Poly(Vinyl Alcohol) Crosslinked Films: Effect of Cyclodextrin and Sodium Benzoate on the Antimicrobial Activity." *LWT-Food Science and Technology* 68: 27–35. doi:10.1016/j.lwt.2015.12.009.

Borzi, Fabrizio, E. Torrieri, Magdalena Wrona, and Cristina Nerín. 2019. "Polyamide Modified with Green Tea Extract for Fresh Minced Meat Active Packaging Applications." *Food Chemistry* 300. doi:10.1016/j.foodchem.2019.125242.

Boyaci, Derya, Gianmarco Iorio, Gozde Seval Sozbilen, Derya Alkan, Silvia Trabattoni, Flavia Pucillo, Stefano Farris, and Ahmet Yemenicioğlu. 2019. "Development of Flexible Antimicrobial Zein Coatings with Essential Oils for the Inhibition of Critical Pathogens on the Surface of Whole Fruits: Test of Coatings on Inoculated Melons." *Food Packaging and Shelf Life* 20. doi:10.1016/j.fpsl.2019.100316.

Boyaci, Derya, Figen Korel, and Ahmet Yemenicioğlu. 2016. "Development of Activate-At-Home-Type Edible Antimicrobial Films: An Example pH-Triggering Mechanism Formed for Smoked Salmon Slices Using Lysozyme in Whey Protein Films." *Food Hydrocolloids* 60: 170–78. doi:10.1016/j.foodhyd.2016.03.032.

Bugatti, V., L. Vertuccio, G. Viscusi, and G. Gorrasi. 2018. "Antimicrobial Membranes of Bio-Based PA 11 and HNTs Filled with Lysozyme Obtained by an Electrospinning Process." *Nanomaterials* 8(3). doi:10.3390/nano8030139.

Caetano, Karine dos Santos, Claudia Titze Hessel, Eduardo Cesar Tondo, Simone Hickmann Flôres, and Florencia Cladera-Olivera. 2017. "Application of Active Cassava Starch Films Incorporated with Oregano Essential Oil and Pumpkin Residue Extract on Ground Beef." *Journal of Food Safety* 37(4). doi:10.1111/jfs.12355.

Campos-Requena, Víctor H., Bernabé L. Rivas, Mónica A. Pérez, Carlos R. Figueroa, Nicolás E. Figueroa, and Eugenio A. Sanfuentes. 2017. "Thermoplastic Starch/Clay Nanocomposites Loaded with Essential Oil Constituents as Packaging for Strawberries—*In Vivo* Antimicrobial Synergy over *Botrytis cinerea*." *Postharvest Biology and Technology* 129: 29–36. doi:10.1016/j.postharvbio.2017.03.005.

Caro, Nelson, Estefanía Medina, Mario Díaz-Dosque, Luis López, Lilian Abugoch, and Cristian Tapia. 2016. "Novel Active Packaging Based on Films of Chitosan and Chitosan/Quinoa Protein Printed with Chitosan-Tripolyphosphate-Thymol Nanoparticles via Thermal Ink-Jet Printing." *Food Hydrocolloids* 52: 520–32. doi:10.1016/j.foodhyd.2015.07.028.

Carvalho, Raissa Alvarenga, Taline Amorim Santos, Ana Carolina Salgado Oliveira, Viviane Machado Azevedo, Marali Vilela Dias, Eduardo Mendes Ramos, and Soraia Vilela Borges. 2019. "Biopolymers of WPI/CNF/TEO in Preventing Oxidation of Ground Meat." *Journal of Food Processing and Preservation* 43(12). doi:10.1111/jfpp.14269.

Castro, Frederico V. R., Mariana A. Andrade, Ana Sanches Silva, Maria Fátima Vaz, and Fernanda Vilarinho. 2019. "The Contribution of a Whey Protein Film Incorporated with Green Tea Extract to Minimize the Lipid Oxidation of Salmon (*Salmo salar* L.)." *Foods* 8(8). doi:10.3390/foods8080327.

Chein, Su Hlaing, Muhammad Bilal Sadiq, and Anil Kumar Anal. 2019. "Antifungal Effects of Chitosan Films Incorporated with Essential Oils and Control of Fungal Contamination in Peanut Kernels." *Journal of Food Processing and Preservation* 43(12). doi:10.1111/jfpp.14235.

Chen, Hongbo, Jingjing Wang, Yaohua Cheng, Chuansheng Wang, Haichao Liu, Huiguang Bian, Yiren Pan, Jingyao Sun, and Wenwen Han. 2019. "Application of Protein-Based Films and Coatings for Food Packaging : A Review." *Polymers* 11(12). doi:10.3390/polym11122039.

Chiellini, Emo, Patrizia Cinelli, Federica Chiellini, and Syed H. Imam. 2004. "Environmentally Degradable Bio-Based Polymeric Blends and Composites." *Macromolecular Bioscience* 4(3): 218–31. doi:10.1002/mabi.200300126.

Chu, Yifu, Tian Xu, Cheng Cheng Gao, Xiaoya Liu, Ni Zhang, Xiao Feng, Xingxun Liu, Xinchun Shen, and Xiaozhi Tang. 2019. "Evaluations of Physicochemical and Biological Properties of Pullulan-Based Films Incorporated with Cinnamon Essential Oil and Tween 80." *International Journal of Biological Macromolecules* 122: 388–94. doi:10.1016/j.ijbiomac.2018.10.194.

Coloma, A., F. J. Rodríguez, J. E. Bruna, A. Guarda, and M. J. Galotto. 2014. "Development of an Active Film with Natural Zeolite as Ethylene Scavenger." *Journal of the Chilean Chemical Society* 59(2): 2409–14. doi:10.4067/S0717-97072014000200003.

Conte, J., A. El Blidi, L. Rigal, and L. Torres. 1992. "Ethylene Removal in Fruit Storage Rooms: A Catalytic Oxidation Reactor at Low Temperature." *Journal of Food Engineering* 15(4): 313–29. doi:10.1016/0260-8774(92)90012-U.

Cui, Haiying, Lu Yuan, and Lin Lin. 2017. "Novel Chitosan Film Embedded with Liposome-Encapsulated Phage for Biocontrol of *Escherichia coli* O157:H7 in Beef." *Carbohydrate Polymers* 177: 156–64. doi:10.1016/j.carbpol.2017.08.137.

Dannenberg, Guilherme da Silva, Graciele Daiana Funck, Claudio Eduardo dos Santos Cruxen, Juliana de Lima Marques, Wladimir Padilha da Silva, and Ângela Maria Fiorentini. 2017. "Essential Oil from Pink Pepper as an Antimicrobial Component in Cellulose Acetate Film: Potential for Application as Active Packaging for Sliced Cheese." *LWT—Food Science and Technology* 81: 314–18. doi:10.1016/j.lwt.2017.04.002.

Deng, Sha, Rui Huang, Mi Zhou, Feng Chen, and Qiang Fu. 2016. "Hydrophobic Cellulose Films with Excellent Strength and Toughness via Ball Milling Activated Acylation of Microfibrillated Cellulose." *Carbohydrate Polymers* 154: 129–38. doi:10.1016/j.carbpol.2016.07.101.

Dey, Aishee, and Sudarsan Neogi. 2019. "Oxygen Scavengers for Food Packaging Applications: A Review." *Trends in Food Science and Technology*. doi:10.1016/j.tifs.2019.05.013.

Dias, Marali Vilela, Viviane Machado Azevedo, Taline Amorim Santos, Cícero Cardoso Pola, Bruna Rage Baldone Lara, Soraia Vilela Borges, Nildade Fatima Ferreira Soares, Éber Antônio Alves Medeiros, and Claire Sarantópoulous. 2019. "Effect of Active Films Incorporated with Montmorillonite Clay and α-Tocopherol: Potential of Nanoparticle Migration and Reduction of Lipid Oxidation in Salmon." *Packaging Technology and Science* 32(1): 39–47. doi:10.1002/pts.2415.

Díaz, Olga, Tania Ferreiro, José Luis Rodríguez-Otero, and Ángel Cobos. 2019. "Characterization of Chickpea (*Cicer arietinum* L.) Flour Films: Effects of pH and Plasticizer Concentration." *International Journal of Molecular Sciences* 20(5). doi:10.3390/ijms20051246.

Domínguez, Rubén, Francisco J. Barba, Belén Gómez, Predrag Putnik, Danijela Bursać Kovačević, Mirian Pateiro, Eva M. Santos, and Jose M. Lorenzo. 2018. "Active Packaging Films with Natural Antioxidants to Be Used in Meat Industry: A Review." *Food Research International* 113: 93–101. doi:10.1016/j.foodres.2018.06.073.

Drakos, Antonios, Elias Pelava, and Vasiliki Evageliou. 2018. "Properties of Flour Films as Affected by the Flour's Source and Particle Size." *Food Research International* 107: 551–58. doi:10.1016/j.foodres.2018.03.005.

Dranca, Florina, and Mircea Oroian. 2018. "Extraction, Purification and Characterization of Pectin from Alternative Sources with Potential Technological Applications." *Food Research International* 113: 327–50. doi:10.1016/j.foodres.2018.06.065.

Echeverría, Ignacio, María Elvira López-Caballero, María Carmen Gómez-Guillén, Adriana Noemi Mauri, and María Pilar Montero. 2018. "Active Nanocomposite Films Based on Soy Proteins-Montmorillonite-Clove Essential Oil for the Preservation of Refrigerated Bluefin Tuna (*Thunnus thynnus*) Fillets." *International Journal of Food Microbiology* 266: 142–49. doi:10.1016/j.ijfoodmicro.2017.10.003.

Elzoghby, Ahmed O., Wael S. Abo El-Fotoh, and Nazik A. Elgindy. 2011. "Casein-Based Formulations as Promising Controlled Release Drug Delivery Systems." *Journal of Controlled Release* 153(3): 206–16. doi:10.1016/j.jconrel.2011.02.010.

European Commission. 2009. "Commission Regulation (EC) No 450/2009 of 29 May 2009 on Active and Intelligent Materials and Articles Intended to Come into Contact with Food."

European Commission. 2016. "Study on the Optimised Cascading Use of Wood." https://ec.europa.eu/growth/content/study-optimised-cascading-use-wood-0_en. Accessed: January 17, 2020.

European Commission. 2018a. "Communication from the Commission to the European Parliament, the Council, the European Economic and Social Committee and the Committee of the Regions: A European Strategy for Plastics in a Circular Economy."

European Commission. 2018b. "Proposal for a Directive of the European Parliament and of the Council on the Reduction of the Impact of Certain Plastic Products on the Environment."

European Commission. 2019a. "Circular Economy| Internal Market Industry, Entrepreneurship and SMEs." https://ec.europa.eu/growth/industry/sustainability/circular-economy_en. Accessed: January 17, 2020.

European Commission. 2019b. "Report from the Commission to the European Parliament, the Council, the European Economic and Social Committee and the Committee of the Regions on the Implementation of the Circular Economy Action Plan."

European Commission. 2019c. "Sustainable Products in a Circular Economy—Towards an EU Product Policy Framework Contributing to the Circular Economy."

The European Starch Industry. 2013. "AAF Position on Trade and Competitiveness." https://starch.eu/blog/2013/05/15/aaf-position-on-trade-and-competitiveness/. Accessed: January 18, 2020.

The European Starch Industry. 2018. "What Is Starch? What Is Used for? Why Do We Need It?" https://starch.eu/the-european-starch-industry/. Accessed: January 18, 2020.

Fang, Zhongxiang, Daniel Lin, Robyn Dorothy Warner, and Minh Ha. 2018. "Effect of Gallic Acid/Chitosan Coating on Fresh Pork Quality in Modified Atmosphere Packaging." *Food Chemistry* 260: 90–6. doi:10.1016/j.foodchem.2018.04.005.

Farrag, Yousof, Walther Ide, Belén Montero, Maite Rico, Saddys Rodríguez-Llamazares, Luis Barral, and Rebeca Bouza. 2018. "Starch Films Loaded with Donut-Shaped Starch-Quercetin Microparticles: Characterization and Release Kinetics." *International Journal of Biological Macromolecules* 118(B): 2201–7. doi:10.1016/j.ijbiomac.2018.07.087.

Fathima, P. E., Satyen Kumar Panda, P. Muhamed Ashraf, T. O. Varghese, and J. Bindu. 2018. "Polylactic Acid/Chitosan Films for Packaging of Indian White Prawn (*Fenneropenaeus indicus*)." *International Journal of Biological Macromolecules* 117: 1002–10. doi:10.1016/j.ijbiomac.2018.05.214.

Fazeli, Mahyar, Meysam Keley, and Esmaeil Biazar. 2018. "Preparation and Characterization of Starch-Based Composite Films Reinforced by Cellulose Nanofibers." *International Journal of Biological Macromolecules* 116: 272–80. doi:10.1016/j.ijbiomac.2018.04.186.

Feng, Kai, Lei Hou, Cody A. Schoener, Peiyi Wu, and Hao Gao. 2015. "Exploring the Drug Migration Process through Ethyl Cellulose-Based Films from Infrared-Spectral Insights." *European Journal of Pharmaceutics and Biopharmaceutics* 93: 46–51. doi:10.1016/j.ejpb.2015.03.011.

Ferreira, Saraiva, Luciana Emanuela, Luana de Oliveira Melo Naponucena, Verônica da Silva Santos, Rejane Pina Dantas Silva, Carolina Oliveira de Souza, Ingrid Evelyn Gomes Lima Souza, Maria Eugênia de Oliveira Mamede, and Janice Izabel Druzian. 2016. "Development and Application of Edible Film of Active Potato Starch to Extend Mini Panettone Shelf Life." *LWT—Food Science and Technology* 73: 311–19. doi:10.1016/j.lwt.2016.05.047.

Figueroa-Lopez, Kelly J., Margarita María Andrade-Mahecha, and Olga Lucía Torres-Vargas. 2018. "Development of Antimicrobial Biocomposite Films to Preserve the Quality of Bread." *Molecules* 23(1). doi:10.3390/molecules23010212.

Fortunati, Elena, Angelo Mazzaglia, and Giorgio M. Balestra. 2019. "Sustainable Control Strategies for Plant Protection and Food Packaging Sectors by Natural Substances and Novel Nanotechnological Approaches." *Journal of the Science of Food and Agriculture* 99(3): 986–1000. doi:10.1002/jsfa.9341.

Gaikwad, Kirtiraj K., Suman Singh, and Abdellah Ajji. 2018. "Moisture Absorbers for Food Packaging Applications." *Environmental Chemistry Letters*. doi:10.1007/s10311-018-0810-z.

Gallego, María Gabriela, Michael H. Gordon, Francisco Segovia, and María Pilar Almajano Pablos. 2016. "Gelatine-Based Antioxidant Packaging Containing *Caesalpinia decapetala* and Tara as a Coating for Ground Beef Patties." *Antioxidants* 5(2). doi:10.3390/antiox5020010.

Gautam, Gitanjali, and Poonam Mishra. 2017. "Development and Characterization of Copper Nanocomposite Containing Bilayer Film for Coconut Oil Packaging." *Journal of Food Processing and Preservation* 41(6). doi:10.1111/jfpp.13243.

Giteru, Stephen Gitonga, M. Azam Ali Indrawati Oey, Stuart K. Johnson, and Zhongxiang Fang. 2017. "Effect of Kafirin-Based Films Incorporating Citral and Quercetin on Storage of Fresh Chicken Fillets." *Food Control* 80: 37–44. doi:10.1016/j.foodcont.2017.04.029.

Go, Eun-Jeong, and Kyung Bin Song. 2019. "Antioxidant Properties of Rye Starch Films Containing Rosehip Extract and Their Application in Packaging of Chicken Breast." *Starch: Stärke* 71(11–12): 1900116. doi:10.1002/star.201900116.

Gomes, Laidson P., Hiléia K. S. Souza, José M. Campiña, Cristina T. Andrade, António F. Silva, Maria P. Gonçalves, and Vania M.Flosi Paschoalin. 2019. "Edible Chitosan Films and Their Nanosized Counterparts Exhibit Antimicrobial Activity and Enhanced Mechanical and Barrier Properties." *Molecules* 247. doi:10.3390/molecules24010127.

Gorrasi, Giuliana, and Valeria Bugatti. 2016. "Edible Bio-Nano-Hybrid Coatings for Food Protection Based on Pectins and LDH-Salicylate: Preparation and Analysis of Physical Properties." *LWT—Food Science and Technology* 69: 139–45. doi:10.1016/j.lwt.2016.01.038.

Goulart, Erica Monize, Paula Martins Olivo, Bruna Moura Rodrigues, Grasiele Scamaral Madrona, Paulo Cesar Pozza, and Magali Soares dos Santos Pozza. 2019. "Application of Functional Edible Films in Ricotta Cheese." *Acta Scientiarum—Technology* 41(1). doi:10.4025/actascitechnol.v41i1.36464.

Groh, Ksenia J., Thomas Backhaus, Bethanie Carney-Almroth, Birgit Geueke, Pedro A. Inostroza, Anna Lennquist, Heather A. Leslie, et al. 2019. "Overview of Known Plastic Packaging-Associated Chemicals and Their Hazards." *The Science of the Total Environment* 651(2): 3253–68. doi:10.1016/j.scitotenv.2018.10.015.

Guillard, Valérie, Sébastien Gaucel, Claudio Fornaciari, Hélène Angellier-Coussy, Patrice Buche, and Nathalie Gontard. 2018. "The Next Generation of Sustainable Food Packaging to Preserve Our Environment in a Circular Economy Context." *Frontiers in Nutrition* 5. doi:10.3389/fnut.2018.00121.

Guitián, M. Virginia, Carolina Ibarguren, M. Cecilia Soria, P. Hovanyecz, Claudia Banchio, and M. Carina Audisio. 2019. "Anti-*Listeria monocytogenes* Effect of Bacteriocin-Incorporated Agar Edible Coatings Applied on Cheese." *International Dairy Journal* 97: 92–8. doi:10.1016/j.idairyj.2019.05.016.

Gutiérrez, Tomy J., Romel Guzmán, Carolina Medina Jaramillo, and Lucía Famá. 2016. "Effect of Beet Flour on Films Made from Biological Macromolecules: Native and Modified Plantain Flour." *International Journal of Biological Macromolecules* 82: 395–403. doi:10.1016/j.ijbiomac.2015.10.020.

Hager, Janelle V., Steven D. Rawles, Youling L. Xiong, Melissa C. Newman, and Carl D. Webster. 2019. "Edible Corn-Zein-Based Coating Incorporated with Nisin or Lemongrass Essential Oil Inhibits *Listeria monocytogenes* on Cultured Hybrid Striped Bass, *Morone chrysops* × *Morone saxatilis*, Fillets During Refrigerated and Frozen Storage." *Journal of the World Aquaculture Society* 50(1): 204–18. doi:10.1111/jwas.12523.

Hanif, Javaria, Nauman Khalid, Rao Sanaullah Khan, Muhammad Faraz Bhatti, Mohammad Qasim Hayat, Muhammad Ismail, Saadia Andleeb, et al. 2019. "Formulation of Active Packaging System Using *Artemisia scoparia* for Enhancing Shelf Life of Fresh Fruits." *Materials Science and Engineering. Part C* 100: 82–93. doi:10.1016/j.msec.2019.02.101.

Hefft, Daniel Ingo. 2018. "Effects of Chitosan-Coated Paperboard Trays on the *Botrytis cinerea* Formation and Sensory Quality of Bilberries (*Vaccinium myrtillus* L.) and Redcurrants (*Ribes rubrum* L.)." *International Journal of Fruit Science* 18(3): 300–6. doi:10.1080/15538362.2017.1411863.

Hernandez, Laura M., Elvis Genbo Xu, Hans C. E. Larsson, Rui Tahara, Vimal B. Maisuria, and Nathalie Tufenkji. 2019. "Plastic Teabags Release Billions of Microparticles and Nanoparticles into Tea." *Environmental Science and Technology*. doi:10.1021/acs.est.9b02540.

Homez-Jara, Angie, Luis Daniel Daza, Diana Marcela Aguirre, José Aldemar Muñoz, José Fernando Solanilla, and Henry Alexander Váquiro. 2018. "Characterization of Chitosan Edible Films Obtained with Various Polymer Concentrations and Drying Temperatures." *International Journal of Biological Macromolecules* 113: 1233–40. doi:10.1016/j.ijbiomac.2018.03.057.

Hu, Dongying, Haixia Wang, and Lijuan Wang. 2016. "Physical Properties and Antibacterial Activity of Quaternized Chitosan/Carboxymethyl Cellulose Blend Films." *LWT—Food Science and Technology* 65: 398–405. doi:10.1016/j.lwt.2015.08.033.

Huang, Xuefeng Zeng Yun'an, Qiujin Zhu, Kuan Lu, Qian Xu, and Chun Ye. 2018. "Development of an Active Packaging with Molecularly Imprinted Polymers for Beef Preservation." *Packaging Technology and Science* 31(4): 213–20. doi:10.1002/pts.2368.

Hussain, Murid, Samir Bensaid, Francesco Geobaldo, Guido Saracco, and Nunzio Russo. 2011. "Photocatalytic Degradation of Ethylene Emitted by Fruits with TiO_2 Nanoparticles." *Industrial and Engineering Chemistry Research* 50(5): 2536–43. doi:10.1021/ie1005756.

Hyde, Peter T., Xian Guan, Viviane Abreu, and Tim L. Setter. 2019. "The Anti-Ethylene Growth Regulator Silver Thiosulfate (STS) Increases Flower Production and Longevity in Cassava (*Manihot esculenta* Crantz)." *Plant Growth Regulation*. doi:10.1007/s10725-019-00542-x.

Iamareerat, Butsadee, Manisha Singh, Muhammad Bilal Sadiq, and Anil Kumar Anal. 2018. "Reinforced Cassava Starch Based Edible Film Incorporated with Essential Oil and Sodium Bentonite Nanoclay as Food Packaging Material." *Journal of Food Science and Technology* 55(5): 1953–59. doi:10.1007/s13197-018-3100-7.

Iaquaniello, Gaetano, Gabriele Centi, Annarita Salladini, Emma Palo, and Siglinda Perathoner. 2018. "Waste to Chemicals for a Circular Economy." *Chemistry—A European Journal* 24(46): 11831–39. doi:10.1002/chem.201802903.

Indumathi, M. P., K. Saral Sarojini, and G. R. Rajarajeswari. 2019. "Antimicrobial and Biodegradable Chitosan/Cellulose Acetate Phthalate/ZnO Nano Composite Films with Optimal Oxygen Permeability and Hydrophobicity for Extending the Shelf Life of Black Grape Fruits." *International Journal of Biological Macromolecules* 132: 1112–20. doi:10.1016/j.ijbiomac.2019.03.171.

Issa, Aseel, Salam A. Ibrahim, and Reza Tahergorabi. 2017. "Impact of Sweet Potato Starch-Based Nanocomposite Films Activated With Thyme Essential Oil on the Shelf-Life of Baby Spinach Leaves." *Foods* 6(6). doi:10.3390/foods6060043.

Jafarzadeh, Shima, Jong Whan Rhim, Abd Karim Alias, Fazilah Ariffin, and Shahrom Mahmud. 2019. "Application of Antimicrobial Active Packaging Film Made of Semolina Flour, Nano Zinc Oxide and Nano-Kaolin to Maintain the Quality of Low-Moisture Mozzarella Cheese during Low-Temperature Storage." *Journal of the Science of Food and Agriculture* 99(6): 2716–25. doi:10.1002/jsfa.9439.

Janjarasskul, Theeranun, and Panuwat Suppakul. 2018. "Active and Intelligent Packaging: The Indication of Quality and Safety." *Critical Reviews in Food Science and Nutrition* 58(5): 808–31. doi:10.1080/10408398.2016.1225278.

Jasour, Mohammad Sedigh, Laleh Mehryar Ali Ehsani, and Seyedeh Samaneh Naghibi. 2015. "Chitosan Coating Incorporated with the Lactoperoxidase System: An Active Edible Coating for Fish Preservation." *Journal of the Science of Food and Agriculture* 95: 1373–78. doi:10.1002/jsfa.6838.

Kale, Gaurav, Thitisilp Kijchavengkul, Rafael Auras, Maria Rubino, Susan E. Selke, and Sher Paul Singh. 2007. "Compostability of Bioplastic Packaging Materials: An Overview." *Macromolecular Bioscience* 7(3): 255–77. doi:10.1002/mabi.200600168.

Kalem, Insha K., Z. F. Bhat, Sunil Kumar, Liwen Wang, Reshan J. Mudiyanselage, and Hina F. Bhat. 2018. "*Tinospora cordifolia*: A Novel Bioactive Ingredient for Edible Films for Improved Lipid Oxidative and Microbial Stability of Meat Products." *Journal of Food Processing and Preservation* 42(11). doi:10.1111/jfpp.13774.

Kanatt, Sweetie R., and S. P. Chawla. 2018. "Shelf Life Extension of Chicken Packed in Active Film Developed with Mango Peel Extract." *Journal of Food Safety* 38(1). doi:10.1111/jfs.12385.

Kanmani, Paulraj, and Jong Whan Rhim. 2014. "Development and Characterization of Carrageenan/Grapefruit Seed Extract Composite Films for Active Packaging." *International Journal of Biological Macromolecules* 68: 258–66. doi:10.1016/j.ijbiomac.2014.05.011.

Kapetanakou, Anastasia E., Dimitrios Karyotis, and Panagiotis N. Skandamis. 2016. "Control of *Listeria monocytogenes* by Applying Ethanol-Based Antimicrobial Edible Films on Ham Slices and Microwave-Reheated Frankfurters." *Food Microbiology* 54: 80–90. doi:10.1016/j.fm.2015.10.013.

Karimnezhad, Fatemeh, Vadood Razavilar, Amir Ali Anvar, and Soheyl Eskandari. 2017. "Study the Antimicrobial Effects of Chitosan-Based Edible Film Containing the *Trachyspermum ammi* Essential Oil on Shelf-Life of Chicken Meat." *Microbiology Research* 8(2). doi:10.4081/mr.2017.7226.

Kashiri, Mahboobeb, Josep P. Cerisuelo, Irene Domínguez, Gracia López-Carballo, Virginia Muriel-Gallet, Rafael Gavara, and Pilar Hernández-Muñoz. 2017. "Zein Films and Coatings as Carriers and Release Systems of *Zataria multiflora* Boiss. Essential Oil for Antimicrobial Food Packaging." *Food Hydrocolloids* 70: 260–68. doi:10.1016/j.foodhyd.2017.02.021.

Kashiri, Mahboobeh, Gracia López-Carballo, Pilar Hernández-Muñoz, and Rafael Gavara. 2019. "Antimicrobial Packaging Based on a LAE Containing Zein Coating to Control Foodborne Pathogens in Chicken Soup." *International Journal of Food Microbiology* 306. doi:10.1016/j.ijfoodmicro.2019.108272.

Kaya, Murat, Asier M. Salaberria, Muhammad Mujtaba, Jalel Labidi, Talat Baran, Povilas Mulercikas, and Fatih Duman. 2018. "An Inclusive Physicochemical Comparison of Natural and Synthetic Chitin Films." *International Journal of Biological Macromolecules* 106: 1062–70. doi:10.1016/j.ijbiomac.2017.08.108.

Keller, Nicolas, Marie Noëlle Ducamp, Didier Robert, and Valérie Keller. 2013. "Ethylene Removal and Fresh Product Storage: A Challenge at the Frontiers of Chemistry. Toward an Approach by Photocatalytic Oxidation." *Chemical Reviews*. doi:10.1021/cr900398v.

Kerry, J. P., M. N. O'Grady, and S. A. Hogan. 2006. "Past, Current and Potential Utilisation of Active and Intelligent Packaging Systems for Meat and Muscle-Based Products: A Review." *Meat Science* 74(1): 113–30. doi:10.1016/j.meatsci.2006.04.024.

Khan, Avik, Hejer Gallah, Bernard Riedl, Jean Bouchard, Agnes Safrany, and Monique Lacroix. 2016. "Genipin Cross-Linked Antimicrobial Nanocomposite Films and Gamma Irradiation to Prevent the Surface Growth of Bacteria in Fresh Meats." *Innovative Food Science and Emerging Technologies* 35: 96–102. doi:10.1016/j.ifset.2016.03.011.

Khazaei, Naimeh, Mohsen Esmaiili, and Zahra Emam-Djomeh. 2017. "Application of Active Edible Coatings Made from Basil Seed Gum and Thymol for Quality Maintenance of Shrimp during Cold Storage." *Journal of the Science of Food and Agriculture* 97(6): 1837–45. doi:10.1002/jsfa.7984.

Kim, Sujin, and Kyung Bin Song. 2018. "Antimicrobial Activity of Buckwheat Starch Films Containing Zinc Oxide Nanoparticles against *Listeria monocytogenes* on Mushrooms." *International Journal of Food Science and Technology* 53(6): 1549–57. doi:10.1111/ijfs.13737.

Kontturi, Eero, and Stefan Spirk. 2019. "Ultrathin Films of Cellulose: A Materials Perspective." *Frontiers in Chemistry* 7: 488. doi:10.3389/fchem.2019.00488.

Kulig, Dominika, Anna Zimoch-Korzycka, Zaneta Kró, Maciej Oziembłowski, and Andrzej Jarmoluk. 2017. "Effect of Film-Forming Alginate/Chitosan Polyelectrolyte Complex on the Storage Quality of Pork." *Molecules* 22(1). doi:10.3390/molecules22010098.

Kumar, Anuj, Kummara Madhusudana Rao, and Sung Soo Han. 2018. "Application of Xanthan Gum as Polysaccharide in Tissue Engineering: A Review." *Carbohydrate Polymers* 180: 128–44. doi:10.1016/j.carbpol.2017.10.009.

Kumar, Santosh, Jyotish Chandra Boro, Dharitri Ray, Avik Mukherjee, and Joydeep Dutta. 2019. "Bionanocomposite Films of Agar Incorporated with ZnO Nanoparticles as an Active Packaging Material for Shelf Life Extension of Green Grape." *Heliyon* 5(6). doi:10.1016/j.heliyon.2019.e01867.

Langroodi, Ali Mojaddar, Hossein Tajik, and Tooraj Mehdizadeh. 2018. "Preservative Effects of Sumac Hydro-Alcoholic Extract and Chitosan Coating Enriched along with *Zataria multiflora* Boiss Essential Oil on the Quality of Beef during Storage." *Veterinary Research Forum* 9(2): 153–61. doi:10.30466/VRF.2018.30831.

Lara-Espinoza, Claudia, Elizabeth Carvajal-Millán, René Balandrán-Quintana, Yolanda López-Franco, and Agustín Rascón-Chu. 2018. "Pectin and Pectin-Based Composite Materials: Beyond Food Texture." *Molecules* 23(4): 942. doi:10.3390/molecules23040942.

Lee, Ka Yeon, Ji Hyeon Lee, Hyun Ju Yang, and Kyung Bin Song. 2016a. "Production and Characterisation of Skate Skin Gelatin Films Incorporated with Thyme Essential Oil and Their Application in Chicken Tenderloin Packaging." *International Journal of Food Science and Technology* 51(6): 1465–72. doi:10.1111/ijfs.13119.

Lee, Ka Yeon, Ji Hyeon Lee, Hyun Ju Yang, and Kyung Bin Song. 2016b. "Characterization of a Starfish Gelatin Film Containing Vanillin and Its Application in the Packaging of Crab Stick." *Food Science and Biotechnology* 25(4): 1023–28. doi:10.1007/s10068-016-0165-9.

Lee, Ka Yeon, Hyun Ju Yang, and Kyung Bin Song. 2016. "Application of a Puffer Fish Skin Gelatin Film Containing Moringa oleifera Lam. Leaf Extract to the Packaging of Gouda Cheese." *Journal of Food Science and Technology* 53(11): 3876–83. doi:10.1007/s13197-016-2367-9.

Lee, Seung Yuan, Seung Jae Lee, Dong Soo Choi, and Sun Jin Hur. 2015. "Current Topics in Active and Intelligent Food Packaging for Preservation of Fresh Foods." *Journal of the Science of Food and Agriculture* 95(14): 2799–810. doi:10.1002/jsfa.7218.

Lee, Wei-Kang, Yi -Yi Lim, Adam Thean-Chor Leow, Parameswari Namasivayam, Janna Ong Abdullah, and Chai-Ling Ho. 2017. "Biosynthesis of Agar in Red Seaweeds: A Review." *Carbohydrate Polymers* 164: 23–30. doi:10.1016/j.carbpol.2017.01.078.

Leslie, Charles A., and Roger J. Romani. 1988. "Inhibition of Ethylene Biosynthesis by Salicylic Acid." *Plant Physiology* 88(3): 833–37. doi:10.1104/pp.88.3.833.

Li, Liang, Rui Ni, Yang Shao, and Shirui Mao. 2014. "Carrageenan and Its Applications in Drug Delivery." *Carbohydrate Polymers* 103: 1–11. doi:10.1016/j.carbpol.2013.12.008.

Li, Linlin, Hualin Wang, Minmin Chen, Suwei Jiang, Shaotong Jiang, Xingjiang Li, and Qiaoyun Wang. 2018. "Butylated Hydroxyanisole Encapsulated in Gelatin Fiber Mats: Volatile Release Kinetics, Functional Effectiveness and Application to Strawberry Preservation." *Food Chemistry* 269: 142–49. doi:10.1016/j.foodchem.2018.06.150.

Liang, Xue, Shiyi Feng, Saeed Ahmed, Wen Qin, and Yaowen Liu. 2019. "Effect of Potassium Sorbate and Ultrasonic Treatment on the Properties of Fish Scale Collagen/Polyvinyl Alcohol Composite Film." *Molecules* 24(13). doi:10.3390/molecules24132363.

Lin, Lin, Yulei Gu, and Haiying Cui. 2018. "Novel Electrospun Gelatin-Glycerin-ε-Poly-Lysine Nanofibers for Controlling *Listeria monocytogenes* on Beef." *Food Packaging and Shelf Life* 18: 21–30. doi:10.1016/j.fpsl.2018.08.004.

Lin, Lin, Yulei Gu, and Haiying Cui. 2019. "Moringa Oil/Chitosan Nanoparticles Embedded Gelatin Nanofibers for Food Packaging against *Listeria monocytogenes* and *Staphylococcus aureus* on Cheese." *Food Packaging and Shelf Life* 19: 86–93. doi:10.1016/j.fpsl.2018.12.005.

Lin, Lin, Xuefang Mao, Yanhui Sun, Govindan Rajivgandhi, and Haiying Cui. 2019. "Antibacterial Properties of Nanofibers Containing Chrysanthemum Essential Oil and Their Application as Beef Packaging." *International Journal of Food Microbiology* 292: 21–30. doi:10.1016/j.ijfoodmicro.2018.12.007.

Lin, Lin, Liao Xue, Surendhiran Duraiarasan, and Cui Haiying. 2018. "Preparation of ε-Polylysine/Chitosan Nanofibers for Food Packaging against *Salmonella* on Chicken." *Food Packaging and Shelf Life* 17: 134–41. doi:10.1016/j.fpsl.2018.06.013.

Lin, Lin, Yulin Zhu, and Haiying Cui. 2018. "Electrospun Thyme Essential Oil/Gelatin Nanofibers for Active Packaging against *Campylobacter jejuni* in Chicken." *LWT* 97: 711–18. doi:10.1016/j.lwt.2018.08.015.

Liu, Bin, Han Xu, Huiying Zhao, Wei Liu, Liyun Zhao, and Yuan Li. 2017. "Preparation and Characterization of Intelligent Starch/PVA Films for Simultaneous Colorimetric Indication and Antimicrobial Activity for Food Packaging Applications." *Carbohydrate Polymers* 157: 842–49. doi:10.1016/j.carbpol.2016.10.067.

Liu, Yunpeng, Yan Qin, Ruyu Bai, Xin Zhang, Limin Yuan, and Jun Liu. 2019. "Preparation of pH-Sensitive and Antioxidant Packaging Films Based on κ-Carrageenan and Mulberry Polyphenolic Extract." *International Journal of Biological Macromolecules* 134: 993–1001. doi:10.1016/j.ijbiomac.2019.05.175.

Liu, Zhe, Dehui Lin, Patricia Lopez-Sanchez, and Xingbin Yang. 2020. "Characterizations of Bacterial Cellulose Nanofibers Reinforced Edible Films Based on Konjac Glucomannan." *International Journal of Biological Macromolecules* 145: 634–45. doi:10.1016/j.ijbiomac.2019.12.109.

Lone, Ayesha, Hany Anany, Mohammed Hakeem, Louise Aguis, Anne Claire Avdjian, Marina Bouget, Arash Atashi, Luba Brovko, Dominic Rochefort, and Mansel W. Griffiths. 2016. "Development of Prototypes of Bioactive Packaging Materials Based on Immobilized Bacteriophages for Control of Growth of Bacterial Pathogens in Foods." *International Journal of Food Microbiology* 217: 49–58. doi:10.1016/j.ijfoodmicro.2015.10.011.

Lotfi, Somayeh, Hamed Ahari, and Razi Sahraeyan. 2019. "The Effect of Silver Nanocomposite Packaging Based on Melt Mixing and Sol–Gel Methods on Shelf Life Extension of Fresh Chicken Stored at 4°C." *Journal of Food Safety* 39(3). doi:10.1111/jfs.12625.

Luís, Ângelo, Fernanda Domingues, and Ana Ramos. 2019. "Production of Hydrophobic Zein-Based Films Bioinspired by the Lotus Leaf Surface: Characterization and Bioactive Properties." *Microorganisms* 7(8). doi:10.3390/microorganisms7080267.

Luís, Ângelo, Luísa Pereira, Fernanda Domingues, and Ana Ramos. 2019. "Development of a Carboxymethyl Xylan Film Containing Licorice Essential Oil with Antioxidant Properties to Inhibit the Growth of Foodborne Pathogens." *LWT-Food Science and Technology* 111: 218–25. doi:10.1016/j.lwt.2019.05.040.

Luz, C., J. Calpe, F. Saladino, Fernando B. Luciano, M. Fernandez-Franzón, J. Mañes, and G. Meca. 2018. "Antimicrobial Packaging Based on Ɛ-Polylysine Bioactive Film for the Control of Mycotoxigenic Fungi *in vitro* and in Bread." *Journal of Food Processing and Preservation* 42(1). doi:10.1111/jfpp.13370.

Luzi, Francesca, Luigi Torre, José Maria Kenny, and Debora Puglia. 2019. "Bio- and Fossil-Based Polymeric Blends and Nanocomposites for Packaging: Structure-Property Relationship." *Materials* 12(3): 471. doi:10.3390/ma12030471.

Malagurski, Ivana, Steva Levic, Aleksandra Nesic, Miodrag Mitric, Vladimir Pavlovic, and Suzana Dimitrijevic-Brankovic. 2017. "Mineralized Agar-Based Nanocomposite Films: Potential Food Packaging Materials with Antimicrobial Properties." *Carbohydrate Polymers* 175: 55–62. doi:10.1016/j.carbpol.2017.07.064.

Malherbi, Naiane Miriam, Ana Camila Schmitz, Remili Cristiani Grando, Ana Paula Bilck, Fábio Yamashita, Luciano Tormen, Farayde Matta Fakhouri, José Ignacio Velasco, and Larissa Canhadas Bertan. 2019. "Corn Starch and Gelatin-Based Films Added with Guabiroba Pulp for Application in Food Packaging." *Food Packaging and Shelf Life* 19: 140–46. doi:10.1016/j.fpsl.2018.12.008.

Manrich, Anny, Francys K. V. Moreira, Caio G. Otoni, Marcos V. Lorevice, Maria A. Martins, and Luiz H. C. Mattoso. 2017. "Hydrophobic Edible Films Made up of Tomato Cutin and Pectin." *Carbohydrate Polymers* 164: 83–91. doi:10.1016/j.carbpol.2017.01.075.

Mansourbahmani, Saeideh, Behzad Ghareyazie, Vahid Zarinnia, Sepideh Kalatejari, and Reza Salehi Mohammadi. 2018. "Study on the Efficiency of Ethylene Scavengers on the Maintenance of Postharvest Quality of Tomato Fruit." *Journal of Food Measurement and Characterization* 12(2): 691–701. doi:10.1007/s11694-017-9682-3.

Mapelli, Chiara, Alida Musatti, Alberto Barbiroli, Seema Saini, Julien Bras, Daniele Cavicchioli, and Manuela Rollini. 2019. "Cellulose Nanofiber (CNF)–Sakacin-A Active Material: Production, Characterization and Application in Storage Trials of Smoked Salmon." *Journal of the Science of Food and Agriculture* 99(10): 4731–38. doi:10.1002/jsfa.9715.

Marcous, A., S. Rasouli, and F. Ardestani. 2017. "Low-Density Polyethylene Films Loaded by Titanium Dioxide and Zinc Oxide Nanoparticles as a New Active Packaging System against *Escherichia Coli* O157:H7 in Fresh Calf Minced Meat." *Packaging Technology and Science* 30(11): 693–701. doi:10.1002/pts.2312.

Marsh, Kenneth, and Betty Bugusu. 2007. "Food Packaging—Roles, Materials, and Environmental Issues." *Journal of Food Science* 72(3): 39–55. doi:10.1111/j.1750-3841.2007.00301.x.

Matta, Eliana, María José Tavera-Quiroz, and Nora Bertola. 2019. "Active Edible Films of Methylcellulose with Extracts of Green Apple (*Granny Smith*) Skin." *International Journal of Biological Macromolecules* 124: 1292–98. doi:10.1016/j.ijbiomac.2018.12.114.

Meira, Stela Maris Meister, Gislene Zehetmeyer, Jóice Maria Scheibel, Júlia Orlandini Werner, and Adriano Brandelli. 2016. "Starch-Halloysite Nanocomposites Containing Nisin: Characterization and Inhibition of *Listeria monocytogenes* in Soft Cheese." *LWT—Food Science and Technology* 68: 226–34. doi:10.1016/j.lwt.2015.12.006.

Meira, Stela Maris Meister, Gislene Zehetmeyer, Júlia Orlandini Werner, and Adriano Brandelli. 2017. "A Novel Active Packaging Material Based on Starch-Halloysite Nanocomposites Incorporating Antimicrobial Peptides." *Food Hydrocolloids* 63: 561–70. doi:10.1016/j.foodhyd.2016.10.013.

Melo, Patricia Gontijo, Mariana Fornazier Borges, Jéssica Afonso Ferreira, Matheus Vicente Barbosa Silva, and Reinaldo Ruggiero. 2018. "Bio-Based Cellulose Acetate Films Reinforced with Lignin and Glycerol." *International Journal of Molecular Sciences* 19(4): 1143. doi:10.3390/ijms19041143.

Menzel, Carolin, Chelo González-Martínez, Amparo Chiralt, and Francisco Vilaplana. 2019. "Antioxidant Starch Films Containing Sunflower Hull Extracts." *Carbohydrate Polymers* 214: 142–51. doi:10.1016/j.carbpol.2019.03.022.

Merlo, Thais Cardoso, Carmen J. Contreras-Castillo, Erick Saldaña, Giovana Verginia Barancelli, Mariana Damiames Baccarin Dargelio, Cristiana Maria Pedroso Yoshida, Eduardo E. Ribeiro Junior, Adna Massarioli, and Anna Cecilia Venturini. 2019. "Incorporation of Pink Pepper Residue Extract into Chitosan Film Combined with a Modified Atmosphere Packaging: Effects on the Shelf Life of Salmon Fillets." *Food Research International* 125. doi:10.1016/j.foodres.2019.108633.

Mihaly-Cozmuta, Anca, Anca Peter, Grigore Craciun, Anca Falup, Leonard Mihaly-Cozmuta, Camelia Nicula, Adriana Vulpoi, and Monica Baia. 2017. "Preparation and Characterization of Active Cellulose-Based Papers Modified with TiO_2, Ag and Zeolite Nanocomposites for Bread Packaging Application." *Cellulose* 24(9): 3911–28. doi:10.1007/s10570-017-1383-x.

Milios, Leonidas. 2018. "Advancing to a Circular Economy: Three Essential Ingredients for a Comprehensive Policy Mix." *Sustainability Science* 13(3): 861–78. doi:10.1007/s11625-017-0502-9.

Mishra, Rojita, and Satpal Singh Bisht. 2011. "Antioxidants and Their Characterization." *Journal of Pharmacy Research* 4(8): 2744–46.

Mohammadi, Hamid, Abolfazl Kamkar, Ali Misaghi, Marija Zunabovic-Pichler, and Seyran Fatehi. 2019. "Nanocomposite Films with CMC, Okra Mucilage, and ZnO Nanoparticles: Extending the Shelf-Life of Chicken Breast Meat." *Food Packaging and Shelf Life* 21. doi:10.1016/j.fpsl.2019.100330.

Moradian, Sahel, Hadi Almasi, and Sohrab Moini. 2018. "Development of Bacterial Cellulose-Based Active Membranes Containing Herbal Extracts for Shelf Life Extension of Button Mushrooms (*Agaricus bisporus*)." *Journal of Food Processing and Preservation* 42(3). doi:10.1111/jfpp.13537.

Moraes Crizel, Tainara de, Alessandro de Oliveira Rios, Vítor D. Alves, Narcisa Bandarra, Margarida Moldão-Martins, and Simone Hickmann Flôres. 2018. "Active Food Packaging Prepared with Chitosan and Olive Pomace." *Food Hydrocolloids* 74: 139–50. doi:10.1016/j.foodhyd.2017.08.007.

Moraes Crizel, Tainara de, Alessandro de Oliveira Rios, Vítor D. Alves, Narcisa Bandarra, Margarida Moldão-Martins, and Simone Hickmann Flôres. 2018. "Biodegradable Films Based on Gelatin and Papaya Peel Microparticles with Antioxidant Properties." *Food and Bioprocess Technology* 11(3): 536–50. doi:10.1007/s11947-017-2030-0.

Moreno, Olga, Lorena Atarés, Amparo Chiralt, Malco C. Cruz-Romero, and Joseph Kerry. 2018. "Starch-Gelatin Antimicrobial Packaging Materials to Extend the Shelf Life of Chicken Breast Fillets." *LWT* 97: 483–90. doi:10.1016/j.lwt.2018.07.005.

Morsy, Mohamed K., Hassan H. Khalaf, Ashraf M. Sharoba, H. Hassan El-tanahi, and Catherine N. Cutter. 2014. "Incorporation of Essential Oils and Nanoparticles in Pullulan Films to Control Foodborne Pathogens on Meat and Poultry Products." *Journal of Food Science* 79(4): M675–84. doi:10.1111/1750-3841.12400.

Munhuweyi, Karen, Oluwafemi J. Caleb, Cheryl L. Lennox, Albert J. van Reenen, and Umezuruike Linus Opara. 2017. "*In Vitro* and *In Vivo* Antifungal Activity of Chitosan-Essential Oils against Pomegranate Fruit Pathogens." *Postharvest Biology and Technology* 129: 9–22. doi:10.1016/j.postharvbio.2017.03.002.

Muppalla, Shobita R., and S. P. Chawla. 2018. "Effect of Gum Arabic-Polyvinyl Alcohol Films Containing Seed Cover Extract of *Zanthoxylum rhetsa* on Shelf Life of Refrigerated Ground Chicken Meat." *Journal of Food Safety* 38(4). doi:10.1111/jfs.12460.

Muratore, Florencia, Silvia E. Barbosa, and Raquel E. Martini. 2019. "Development of Bioactive Paper Packaging for Grain-Based Food Products." *Food Packaging and Shelf Life* 20. doi:10.1016/j.fpsl.2019.100317.

Murillo-Martínez, María M., Salvador R. Tello-Solís, Miguel A. García-Sánchez, and Edith Ponce-Alquicira. 2013. "Antimicrobial Activity and Hydrophobicity of Edible Whey Protein Isolate Films Formulated with Nisin and/or Glucose Oxidase." *Journal of Food Science* 78(4). doi:10.1111/1750-3841.12078.

Mustafa, Pakeeza, Muhammad B. K. Niazi, Zaib Jahan, Ghufrana Samin, Arshad Hussain, Tahir Ahmed, and Salman R. Naqvi. 2019. "PVA/Starch/Propolis/Anthocyanins Rosemary Extract Composite Films as Active and Intelligent Food Packaging Materials." *Journal of Food Safety*. doi:10.1111/jfs.12725.

Naidu, Darrel Sarvesh, Shanganyane Percy Hlangothi, and Maya Jacob John. 2018. "Bio-Based Products from Xylan: A Review." *Carbohydrate Polymers* 179: 28–41. doi:10.1016/j.carbpol.2017.09.064.

Naqash, Farah, F. A. Masoodi, Sajad Ahmad Rather, S. M. Wani, and Adil Gani. 2017. "Emerging Concepts in the Nutraceutical and Functional Properties of Pectin—A Review." *Carbohydrate Polymers* 168: 227–39. doi:10.1016/j.carbpol.2017.03.058.

Narancic, Tanja, and Kevin E. O'Connor. 2019. "Plastic Waste as a Global Challenge: Are Biodegradable Plastics the Answer to the Plastic Waste Problem?." *Microbiology* 165(2): 129–37. doi:10.1099/mic.0.000749.

Navikaite-Snipaitiene, Vesta, Liudas Ivanauskas, Valdas Jakstas, Nadine Rüegg, Ramune Rutkaite, Evelyn Wolfram, and Selçuk Yildirim. 2018. "Development of Antioxidant Food Packaging Materials Containing Eugenol for Extending Display Life of Fresh Beef." *Meat Science* 145: 9–15. doi:10.1016/j.meatsci.2018.05.015.

Neira, Laura M., Silvina P. Agustinelli, Roxana A. Ruseckaite, and Josefa F. Martucci. 2019. "Shelf Life Extension of Refrigerated Breaded Hake Medallions Packed into Active Edible Fish Gelatin Films." *Packaging Technology and Science* 32(9): 471–80. doi:10.1002/pts.2450.

Nerin, Cristina, Vera Paula, and Canellas Elena. 2018. "Active and Intelligent Food Packaging." In: *Food Safety and Protection*. CRC Press, 459–91. doi:10.1201/9781315153414-14.

Nguyen Van Long, N., Catherine Joly, and Philippe Dantigny. 2016. "Active Packaging with Antifungal Activities." *International Journal of Food Microbiology*. doi:10.1016/j.ijfoodmicro.2016.01.001.

Niu, Ben, Zhipeng Yan, Ping Shao, Ji Kang, and Hangjun Chen. 2018. "Encapsulation of Cinnamon Essential Oil for Active Food Packaging Film with Synergistic Antimicrobial Activity." *Nanomaterials* 8(8). doi:10.3390/nano8080598.

Nouraddini, Mahsa, Mohsen Esmaiili, and Forogh Mohtarami. 2018. "Development and Characterization of Edible Films Based on Eggplant Flour and Corn Starch." *International Journal of Biological Macromolecules* 120(B): 1639–45. doi:10.1016/j.ijbiomac.2018.09.126.

Nur Hanani, Z. A., Y. H. Roos, and J. P. Kerry. 2014. "Use and Application of Gelatin as Potential Biodegradable Packaging Materials for Food Products." *International Journal of Biological Macromolecules* 71: 94–102. doi:10.1016/j.ijbiomac.2014.04.027.

Oğuzhan Yıldız, Pınar, and Filiz Yangılar. 2017. "Effects of Whey Protein Isolate Based Coating Enriched with *Zingiber officinale* and *Matricaria recutita* Essential Oils on the Quality of Refrigerated Rainbow Trout." *Journal of Food Safety* 37(4). doi:10.1111/jfs.12341.

Oinonen, Petri, Holger Krawczyk, Monica Ek, Gunnar Henriksson, and Rosana Moriana. 2016. "Bioinspired Composites from Cross-Linked Galactoglucomannan and Microfibrillated Cellulose: Thermal, Mechanical and Oxygen Barrier Properties." *Carbohydrate Polymers* 136: 146–53. doi:10.1016/j.carbpol.2015.09.038.

Oliveira, Marília A., Maria S. R. Bastos, Hilton C. R. Magalhães, Deborah S. Garruti, Selene D. Benevides, Roselayne F. Furtado, and Antônio S. Egito. 2017. "α, β-Citral from *Cymbopogon citratus* on Cellulosic Film: Release Potential and Quality of Coalho Cheese." *LWT—Food Science and Technology* 85: 246–51. doi:10.1016/j.lwt.2017.07.029.

Orsuwan, Aungkana, and Rungsinee Sothornvit. 2018. "Active Banana Flour Nanocomposite Films Incorporated with Garlic Essential Oil as Multifunctional Packaging Material for Food Application." *Food and Bioprocess Technology* 11(6): 1199–210. doi:10.1007/s11947-018-2089-2.

Otoni, Caio G., Paula J. P. Espitia, Roberto J. Avena-Bustillos, and Tara H. McHugh. 2016. "Trends in Antimicrobial Food Packaging Systems: Emitting Sachets and Absorbent Pads." *Food Research International*. doi:10.1016/j.foodres.2016.02.018.

Pagno, Carlos Henrique, Yuri Buratto de Farias, Tania Maria Haas Costa, Alessandro de Oliveira Rios, and Simone Hickmann Flôres. 2016. "Synthesis of Biodegradable Films with Antioxidant Properties Based on Cassava Starch Containing Bixin Nanocapsules." *Journal of Food Science and Technology* 53(8): 3197–205. doi:10.1007/s13197-016-2294-9.

Parreidt, Tugce Senturk, Kajetan Müller, and Markus Schmid. 2018. "Alginate-Based Edible Films and Coatings for Food Packaging Applications." *Foods* 7(10): 1–38. doi:10.3390/foods7100170.

Pellerito, Alessandra, Sara M. Ameen, Maria Micali, and Giorgia Caruso. 2018. "Antimicrobial Substances for Food Packaging Products: The Current Situation." *Journal of AOAC International* 101(4): 942–47. doi:10.5740/jaoacint.17-0448.

Picheth, Guilherme Fadel, Cleverton Luiz Pirich, Maria Rita Sierakowski, Marco Aurélio Woehl, Caroline Novak Sakakibara, Clayton Fernandes de Souza, Andressa Amado Martin, Renata da Silva, and Rilton Alves de Freitas. 2017. "Bacterial Cellulose in Biomedical Applications: A Review." *International Journal of Biological Macromolecules* 104(A): 97–106. doi:10.1016/j.ijbiomac.2017.05.171.

Pierucci, Sauro, Laura Piazza, Johannes De Bruijn, Ambar E. Gómez, Pedro Melín, Cristina Loyola, Víctor A. Solar, and Héctor Valdés. 2019. "Effect of Doping Natural Zeolite with Copper and Zinc Cations on Ethylene Removal and Postharvest Tomato Fruit Quality." In: *Chemical Engineering Transactions*, Vol. 75. doi:10.3303/CET1975045.

Pires, Ana, and Graça Martinho. 2019. "Waste Hierarchy Index for Circular Economy in Waste Management." *Waste Management* 95: 298–305. doi:10.1016/j.wasman.2019.06.014.

Pires, João Ricardo Afonso, Victor Gomes Lauriano de Souza, and Ana Luísa Fernando. 2018. "Chitosan/Montmorillonite Bionanocomposites Incorporated with Rosemary and Ginger Essential Oil as Packaging for Fresh Poultry Meat." *Food Packaging and Shelf Life* 17: 142–49. doi:10.1016/j.fpsl.2018.06.011.

Prajapati, Vipul D., Girish K. Jani, and Simin M. Khanda. 2013. "Pullulan: An Exopolysaccharide and Its Various Applications." *Carbohydrate Polymers* 95(1): 540–49. doi:10.1016/j.carbpol.2013.02.082.

Priyadarshi, Ruchir, Bijender Kumar Sauraj, Farha Deeba, Anurag Kulshreshtha, and Yuvraj Singh Negi. 2018. "Chitosan Films Incorporated with Apricot (*Prunus armeniaca*) Kernel Essential Oil as Active Food Packaging Material." *Food Hydrocolloids* 85: 158–66. doi:10.1016/j.foodhyd.2018.07.003.

Priyadarshi, Ruchir, Bijender Kumar Sauraj, and Yuvraj Singh Negi. 2018. "Chitosan Film Incorporated with Citric Acid and Glycerol as an Active Packaging Material for Extension of *Green Chilli* Shelf Life." *Carbohydrate Polymers* 195: 329–38. doi:10.1016/j.carbpol.2018.04.089.

Queirós, Lúcia C. C., Sónia C. L. Sousa, Andreia F. S. Duarte, Fernanda C. Domingues, and Ana M. M. Ramos. 2017. "Development of Carboxymethyl Xylan Films with Functional Properties." *Journal of Food Science and Technology* 54(1): 9–17. doi:10.1007/s13197-016-2389-3.

Radford, Devon, Brandon Guild, Philip Strange, Rafath Ahmed, Loong Tak Lim, and S. Balamurugan. 2017. "Characterization of Antimicrobial Properties of *Salmonella* Phage Felix O1 and *Listeria* Phage A511 Embedded in Xanthan Coatings on Poly(Lactic Acid) Films." *Food Microbiology* 66: 117–28. doi:10.1016/j.fm.2017.04.015.

Raghuwanshi, Vikram Singh, and Gil Garnier. 2019. "Cellulose Nano-Films as Bio-Interfaces." *Frontiers in Chemistry* 7: 1–17. doi:10.3389/fchem.2019.00535.

Ramos, Óscar L., João C. Fernandes, Sara I. Silva, Manuela E. Pintado, and F. Xavier Malcata. 2012. "Edible Films and Coatings from Whey Proteins: A Review on Formulation, and on Mechanical and Bioactive Properties." *Critical Reviews in Food Science and Nutrition* 52(6): 533–52. doi:10.1080/10408398.2010.500528.

Remya, S., C. O. Mohan, J. Bindu, G. K. Sivaraman, G. Venkateshwarlu, and C. N. Ravishankar. 2016. "Effect of Chitosan Based Active Packaging Film on the Keeping Quality of Chilled Stored Barracuda Fish." *Journal of Food Science and Technology* 53(1): 685–93. doi:10.1007/s13197-015-2018-6.

Rezaei, Fatemeh, and Yasser Shahbazi. 2018. "Shelf-Life Extension and Quality Attributes of Sauced Silver Carp Fillet: A Comparison among Direct Addition, Edible Coating and Biodegradable Film." *LWT—Food Science and Technology* 87: 122–33. doi:10.1016/j.lwt.2017.08.068.

Rhodes, Christopher J. 2018. "Plastic Pollution and Potential Solutions." *Science Progress* 101(3): 207–60. doi:10.3184/003685018X15294876706211.

Robledo, Nancy, Luis López, Andrea Bunger, Cristian Tapia, and Lilian Abugoch. 2018. "Effects of Antimicrobial Edible Coating of Thymol Nanoemulsion/Quinoa Protein/Chitosan on the Safety, Sensorial Properties, and Quality of Refrigerated Strawberries (*Fragaria* × *Ananassa*) under Commercial Storage Environment." *Food and Bioprocess Technology* 11(8): 1566–74. doi:10.1007/s11947-018-2124-3.

Rocha, Meritaine da Alemán, Viviane Patrícia Romani, M. Elvira López-Caballero, M. Carmen Gómez-Guillén, Pilar Montero, and Carlos Prentice. 2018. "Effects of Agar Films Incorporated with Fish Protein Hydrolysate or Clove Essential Oil on Flounder (*Paralichthys orbignyanus*) Fillets Shelf-Life." *Food Hydrocolloids* 81: 351–63. doi:10.1016/j.foodhyd.2018.03.017.

Rodsamran, Pattrathip, and Rungsinee Sothornvit. 2018. "Carboxymethyl Cellulose from Renewable Rice Stubble Incorporated with Thai Rice Grass Extract as a Bioactive Packaging Film for Green Tea." *Journal of Food Processing and Preservation* 42(9). doi:10.1111/jfpp.13762.

Rollini, Manuela, Erika Mascheroni, Giorgio Capretti, Veronique Coma, Alida Musatti, and Luciano Piergiovanni. 2017. "Propolis and Chitosan as Antimicrobial and Polyphenols Retainer for the Development of Paper Based Active Packaging Materials." *Food Packaging and Shelf Life* 14: 75–82. doi:10.1016/j.fpsl.2017.08.011.

Romero, Viviana, Rafael Borneo, Nancy Passalacqua, and Alicia Aguirre. 2016. "Biodegradable Films Obtained from Triticale (x *Triticosecale wittmack*) Flour Activated with Natamycin for Cheese Packaging." *Food Packaging and Shelf Life* 10: 54–9. doi:10.1016/j.fpsl.2016.09.003.

Rostami, Hosein, Sepideh Abbaszadeh, and Sajad Shokri. 2017. "Combined Effects of Lactoperoxidase System-Whey Protein Coating and Modified Atmosphere Packaging on the Microbiological, Chemical and Sensory Attributes of Pike-Perch Fillets." *Journal of Food Science and Technology* 54(10): 3243–50. doi:10.1007/s13197-017-2767-5.

Rujnić-Sokele, Maja, and Ana Pilipović. 2017. "Challenges and Opportunities of Biodegradable Plastics: A Mini Review." *Waste Management and Research* 35(2): 132–40. doi:10.1177/0734242X16683272.

Russell, David A. M. 2014. "Sustainable (Food) Packaging—An Overview." *Food Additives and Contaminants—Part A* 31(3): 396–401. doi:10.1080/19440049.2013.856521.

Saade, Carol, Bassam A. Annous, Anthony J. Gualtieri, Karen M. Schaich, Lin Shu Liu, and Kit L. Yam. 2018. "System Feasibility: Designing a Chlorine Dioxide Self-Generating Package Label to Improve Fresh Produce Safety Part II: Solution Casting Approach." *Innovative Food Science and Emerging Technologies* 47: 110–19. doi:10.1016/j.ifset.2018.02.003.

Saltveit, Mikal E. 2005. "Aminoethoxyvinylglycine (AVG) Reduces Ethylene and Protein Biosynthesis in Excised Discs of Mature-Green Tomato Pericarp Tissue." *Postharvest Biology and Technology* 35(2): 183–90. doi:10.1016/j.postharvbio.2004.07.002.

Salvucci, Emiliano, Mariana Rossi, Andrés Colombo, Gabriela Pérez, Rafael Borneo, and Alicia Aguirre. 2019. "Triticale Flour Films Added with Bacteriocin-Like Substance (BLIS) for Active Food Packaging Applications." *Food Packaging and Shelf Life* 19: 193–99. doi:10.1016/j.fpsl.2018.05.007.

Sanchís, Elena, Christian Ghidelli, Chirag C. Sheth, Milagros Mateos, Lluís Palou, and María B. Pérez-Gago. 2017. "Integration of Antimicrobial Pectin-Based Edible Coating and Active Modified Atmosphere Packaging to Preserve the Quality and Microbial Safety of Fresh-Cut Persimmon (*Diospyros kaki* Thunb. Cv. Rojo Brillante)." *Journal of the Science of Food and Agriculture* 97(1): 252–60. doi:10.1002/jsfa.7722.

Santonicola, Serena, Verónica García Ibarra, Raquel Sendón, Raffaelina Mercogliano, and Ana Rodríguez Bernaldo de Quirós. 2017. "Antimicrobial Films Based on Chitosan and Methylcellulose Containing Natamycin for Active Packaging Applications." *Coatings* 7(10). doi:10.3390/coatings7100177.

Saral Sarojini, K., M. P. Indumathi, and G. R. Rajarajeswari. 2019. "Mahua Oil-Based Polyurethane/Chitosan/Nano ZnO Composite Films for Biodegradable Food Packaging Applications." *International Journal of Biological Macromolecules* 124: 163–74. doi:10.1016/j.ijbiomac.2018.11.195.

Saurabh, Chaturbhuj K., Sumit Gupta, and Prasad S. Variyar. 2018. "Development of Guar Gum Based Active Packaging Films Using Grape Pomace." *Journal of Food Science and Technology* 55(6): 1982–92. doi:10.1007/s13197-018-3112-3.

Scheinberg, Anne, Jelena Nesic, Rachel Savain, Pietro Luppi, Portia Sinnott, Flaviu Petean, and Flaviu Pop. 2016. "From Collision to Collaboration—Integrating Informal Recyclers and Re-Use Operators in Europe: A Review." *Waste Management and Research* 34(9): 820–39. doi:10.1177/0734242X16657608.

Serio, Annalisa, Clemencia Chaves-López, Giampiero Sacchetti, Chiara Rossi, and Antonello Paparella. 2018. "Chitosan Coating Inhibits the Growth of *Listeria monocytogenes* and Extends the Shelf Life of Vacuum-Packed Pork Loins at 4°C." *Foods* 7(10). doi:10.3390/foods7100155.

Shahbazi, Yasser. 2018. "Application of Carboxymethyl Cellulose and Chitosan Coatings Containing *Mentha spicata* Essential Oil in Fresh Strawberries." *International Journal of Biological Macromolecules* 112: 264–72. doi:10.1016/j.ijbiomac.2018.01.186.

Shahbazi, Yasser, and Nassim Shavisi. 2018. "A Novel Active Food Packaging Film for Shelf-Life Extension of Minced Beef Meat." *Journal of Food Safety* 38(6). doi:10.1111/jfs.12569.

Shankar, Shiv, and Jong Whan Rhim. 2018. "Antimicrobial Wrapping Paper Coated with a Ternary Blend of Carbohydrates (Alginate, Carboxymethyl Cellulose, Carrageenan) and Grapefruit Seed Extract." *Carbohydrate Polymers* 196: 92–101. doi:10.1016/j.carbpol.2018.04.128.

Sharma, Amita, Manisha Thakur, Munna Bhattacharya, Tamal Mandal, and Saswata Goswami. 2018. "Commercial Application of Cellulose Nano-Composites—A Review." *Biotechnology Reports*: e00316. doi:10.1016/j.btre.2019.e00316.

Shi, Weida, and Marie Josée Dumont. 2014. "Review: Bio-Based Films from Zein, Keratin, Pea, and Rapeseed Protein Feedstocks." *Journal of Materials Science* 49(5): 1915–30. doi:10.1007/s10853-013-7933-1.

Shukla, Rishi, and Munir Cheryan. 2001. "Zein: The Industrial Protein from Corn." *Industrial Crops and Products* 13(3): 171–92. doi:10.1016/S0926-6690(00)00064-9.

Silva, Ângela, Andreia Duarte, Sónia Sousa, Ana Ramos, and Fernanda C. Domingues. 2016. "Characterization and Antimicrobial Activity of Cellulose Derivatives Films Incorporated with a Resveratrol Inclusion Complex." *LWT—Food Science and Technology* 73: 481–89. doi:10.1016/j.lwt.2016.06.043.

Silva, Filomena, Raquel Becerril, and Cristina Nerin. 2019. "Safety Assessment of Active Food Packaging: Role of Known and Unknown Substances." In: Belén Gómara and María Luisa Marina, *Advances in the Determination of Xenobiotics in Foods*, Bentham Science, Sharjah, UAE, 1–41. DOI: 10.2174/9789811421587119010004.

Silva, Filomena, Fernanda C. Domingues, and Cristina Nerín. 2018. "Control Microbial Growth on Fresh Chicken Meat Using Pinosylvin Inclusion Complexes Based Packaging Absorbent Pads." *LWT* 89: 148–54. doi:https. doi:10.1016/j.lwt.2017.10.043.

Silva, Filomena, Nicolás Gracia, Birgitte H. McDonagh, Fernanda C. Domingues, Cristina Nerín, and Gary Chinga-Carrasco. 2019a. "Antimicrobial Activity of Biocomposite Films Containing Cellulose Nanofibrils and Ethyl Lauroyl Arginate." *Journal of Materials Science*. doi:10.1007/s10853-019-03759-3.

Silva, Francine Tavares da, Kamila Furtado da Cunha, Laura Martins Fonseca, Mariana Dias Antunes, Shanise Lisie Mello El Halal, Ângela Maria Fiorentini, Elessandra da Rosa Zavareze, and Alvaro Renato Guerra Dias. 2018. "Action of Ginger Essential Oil (*Zingiber officinale*) Encapsulated in Proteins Ultrafine Fibers on the Antimicrobial Control *In Situ*." *International Journal of Biological Macromolecules* 118(A): 107–15. doi:10.1016/j.ijbiomac.2018.06.079.

Santos, Silva, Fábio Marcel da, Ana Irene Martins da Silva, Cláudia Brandão Vieira, Mayra Horácio de Araújo, André Luis Coelho da Silva, Maria das Graças Carneiro-da-Cunha, Bartolomeu Warlene Silva de Souza, and Ranilson de Souza Bezerra. 2017. "Use of Chitosan Coating in Increasing the Shelf Life of Liquid Smoked Nile Tilapia (*Oreochromis niloticus*) Fillet." *Journal of Food Science and Technology* 54(5): 1304–11. doi:10.1007/s13197-017-2570-3.

Singh, M., and T. Sahareen. 2017. "Investigation of Cellulosic Packets Impregnated with Silver Nanoparticles for Enhancing Shelf-Life of Vegetables." *LWT—Food Science and Technology* 86: 116–22. doi:10.1016/j.lwt.2017.07.056.

Singh, Ram Sarup, Navpreet Kaur, and John F. Kennedy. 2019. "Pullulan Production from Agro-Industrial Waste and Its Applications in Food Industry: A Review." *Carbohydrate Polymers* 217: 46–57. doi:10.1016/j.carbpol.2019.04.050.

Smith, Andrew W. J., Stephen Poulston, Liz Rowsell, Leon A. Terry, and James A. Anderson. 2009. "A New Palladium-Based Ethylene Scavenger to Control Ethylene-Induced Ripening of Climacteric Fruit." *Platinum Metals Review* 53(3): 112–22. doi:10.1595/147106709X462742.

Sogut, Ece, and Atif Can Seydim. 2019. "The Effects of Chitosan- and Polycaprolactone-Based Bilayer Films Incorporated with Grape Seed Extract and Nanocellulose on the Quality of Chicken Breast Fillets." *LWT* 101: 799–805. doi:10.1016/j.lwt.2018.11.097.

Sohail, Muhammad, Da Wen Sun, and Zhiwei Zhu. 2018. "Recent Developments in Intelligent Packaging for Enhancing Food Quality and Safety." *Critical Reviews in Food Science and Nutrition* 58(15): 2650–62. doi:10.1080/10408398.2018.1449731.

Song, J. H., R. J. Murphy, R. Narayan, and G. B. H. Davies. 2009. "Biodegradable and Compostable Alternatives to Conventional Plastics." *Philosophical Transactions of the Royal Society B* 364(1526): 2127–39. doi:10.1098/rstb.2008.0289.

Sousa, Ana M. M., and Maria P. Gonçalves. 2015. "Strategies to Improve the Mechanical Strength and Water Resistance of Agar Films for Food Packaging Applications." *Carbohydrate Polymers* 132: 196–204. doi:10.1016/j.carbpol.2015.06.022.

Sousa, Sónia C. L., Ana M. M. Ramos, Dmitry V. Evtuguin, and José A. F. Gamelas. 2016. "Xylan and Xylan Derivatives—Their Performance in Bio-Based Films and Effect of Glycerol Addition." *Industrial Crops and Products* 94: 682–89. doi:10.1016/j.tifs.2012.06.012.

Souza, Victor G. L., João R. A. Pires, Érica T. Vieira, Isabel M. Coelhoso, Maria P. Duarte, and Ana L. Fernando. 2018. "Shelf Life Assessment of Fresh Poultry Meat Packaged in Novel Bionanocomposite of Chitosan/Montmorillonite Incorporated with Ginger Essential Oil." *Coatings* 8(5). doi:10.3390/coatings8050177.

Souza, Victor Gomes Lauriano, João R. A. Pires, Érica Torrico Vieira, Isabel M. Coelhoso, Maria Paula Duarte, and Ana Luisa Fernando. 2019. "Activity of Chitosan-Montmorillonite Bionanocomposites Incorporated with Rosemary Essential Oil: From *in vitro* Assays to Application in Fresh Poultry Meat." *Food Hydrocolloids* 89: 241–52. doi:10.1016/j.foodhyd.2018.10.049.

Stahel, Wlater R. 2016. "The Circular Economy." *Nature* 531(7595): 435–38. doi:10.4324/9781315270326-38.

Stoll, Liana, Alexandre Martins da Silva, Aline Oliveira e.Silva Iahnke, Tania Maria Haas Costa, Simone Hickmann Flôres, and Alessandro de Oliveira Rios. 2017. "Active Biodegradable Film with Encapsulated Anthocyanins: Effect on the Quality Attributes of Extra-Virgin Olive Oil during Storage." *Journal of Food Processing and Preservation* 41(6). doi:10.1111/jfpp.13218.

Suo, Biao, Huarong Li, Yuexia Wang, Zhen Li, Zhili Pan, and Zhilu Ai. 2016. "Effects of ZnO Nanoparticle-Coated Packaging Film on Pork Meat Quality during Cold Storage." *Journal of the Science of Food and Agriculture* 97(7): 2023–29. doi:10.1002/jsfa.8003.

Tabassum, Nazia, and Mohammad Ali Khan. 2020. "Modified Atmosphere Packaging of Fresh-Cut Papaya Using Alginate Based Edible Coating: Quality Evaluation and Shelf Life Study." *Scientia Horticulturae* 259. doi:10.1016/j.scienta.2019.108853.

Tabasum, Shazia, Aqdas Noreen, Muhammad Farzam, Hijab Umar, Nadia Akram, Z. I. Nazli, S. A. S. Chatha, and K. M. Zia. 2018. "A Review on Versatile Applications of Blends and Composites of Pullulan with Natural and Synthetic Polymers." *International Journal of Biological Macromolecules* 120: 603–32. doi:10.1016/j.ijbiomac.2018.07.154.

Tahir, Haroon Elrasheid, Zou Xiaobo, Shi Jiyong, Gustav Komla Mahunu, Xiaodong Zhai, and Abdalbasit Adam Mariod. 2018. "Quality and Postharvest-Shelf Life of Cold-Stored Strawberry Fruit as Affected by Gum Arabic (*Acacia senegal*) Edible Coating." *Journal of Food Biochemistry* 42. doi:10.1111/jfbc.12527.

Tanetrungroj, Yossathorn, and Jutarat Prachayawarakorn. 2018. "Effect of Dual Modification on Properties of Biodegradable Crosslinked-Oxidized Starch and Oxidized-Crosslinked Starch Films." *International Journal of Biological Macromolecules* 120(A): 1240–46. doi:10.1016/j.ijbiomac.2018.08.137.

Tang, Siying, Zhe Wang, Penghui Li, Wan Li, Chengyong Li, Yi Wang, and Paul K. Chu. 2018. "Degradable and Photocatalytic Antibacterial Au-TiO_2/Sodium Alginate Nanocomposite Films for Active Food Packaging." *Nanomaterials* 8(11). doi:10.3390/nano8110930.

Cuellar, Tapiero, and Jose Libardo. 2017. "Evaluación de la Vida Útil de Quesos Semimaduros Con Recubrimientos Comestibles Utilizando Aceite Esencial de Jengibre (*Zingiber officinale*) Como Agente Antimicrobiano." *Revista Colombiana de Investigaciones Agroindustriales* . doi:10.23850/24220582.623.

Tavassoli-Kafrani, Elham, Hajar Shekarchizadeh, and Mahdieh Masoudpour-Behabadi. 2016. "Development of Edible Films and Coatings from Alginates and Carrageenans." *Carbohydrate Polymers* 137: 360–74. doi:10.1016/j.carbpol.2015.10.074.

Tawakkal, Intan S. M. A., Marlene J. Cran, and Stephen W. Bigger. 2017. "Effect of Poly(Lactic Acid)/Kenaf Composites Incorporated with Thymol on the Antimicrobial Activity of Processed Meat." *Journal of Food Processing and Preservation* 41(5). doi:10.1111/jfpp.13145.

Tecchio, Paolo, Catriona McAlister, Fabrice Mathieux, and Fulvio Ardente. 2017. "In Search of Standards to Support Circularity in Product Policies: A Systematic Approach." *Journal of Cleaner Production* 168: 1533–46. doi:10.1016/j.jclepro.2017.05.198.

Thakur, Rahul, Penta Pristijono, Christopher J. Scarlett, Michael Bowyer, S. P. Singh, and Quan V. Vuong. 2019. "Starch-Based Films: Major Factors Affecting Their Properties." *International Journal of Biological Macromolecules* 132: 1079–89. doi:10.1016/j.ijbiomac.2019.03.190.

Tomat, David, Marina Soazo, Roxana Verdini, Cecilia Casabonne, Virginia Aquili, Claudia Balagué, and Andrea Quiberoni. 2019. "Evaluation of an WPC Edible Film Added with a Cocktail of Six Lytic Phages against Foodborne Pathogens Such as Enteropathogenic and Shigatoxigenic *Escherichia coli*." *LWT* 113. doi:10.1016/j.lwt.2019.108316.

Trache, Djalal, M. Hazwan Hussin, M. K.Mohamad Haafiz, and Vijay Kumar Thakur. 2017. "Recent Progress in Cellulose Nanocrystals: Sources and Production." *Nanoscale* 9(5): 1763–86. doi:10.1039/c6nr09494e.

Tzeng, Jing Hua, Chih Huang Weng, Jenn Wen Huang, Ching Chang Shiesh, Yu. Hao Lin, and Lin Yao Tung. 2019. "Application of Palladium-Modified Zeolite for Prolonging Post-Harvest Shelf Life of Banana." *Journal of the Science of Food and Agriculture* 99(7): 3467–74. doi:10.1002/jsfa.9565.

Vera, Paula, Elena Canellas, and Cristina Nerín. 2018. "New Antioxidant Multilayer Packaging with Nanoselenium to Enhance the Shelf-Life of Market Food Products." *Nanomaterials* 8(10): 837. doi:10.3390/nano8100837.

Vimala Bharathi, S. K., M. Maria Leena, J. A. Moses, and C. Anandharamakrishnan. 2019. "Zein-Based Anti-Browning Cling Wraps for Fresh-Cut Apple Slices." *International Journal of Food Science and Technology*. doi:10.1111/ijfs.14401.

Vital, Ana Carolina Pelaes, Ana Guerrero, Jessica De Oliveira Monteschio, Maribel Velandia Valero, Camila Barbosa Carvalho, Benício Alves De Abreu Filho, Grasiele Scaramal Madrona, and Ivanor Nunes Do Prado. 2016. "Effect of Edible and Active Coating (with Rosemary and Oregano Essential Oils) on Beef Characteristics and Consumer Acceptability." *PLoS One* 11. doi:10.1371/journal.pone.0160535.

Wang, Hongxia, Yu Liao, Ailiang Wu, Bing Li, Jun Qian, and Fuyuan Ding. 2019. "Effect of Sodium Trimetaphosphate on Chitosan-Methylcellulose Composite Films: Physicochemical Properties and Food Packaging Application." *Polymers* 11(2). doi:10.3390/POLYM11020368.

Wang, Hongxia, Jun Qian, and Fuyuan Ding. 2018. "Emerging Chitosan-Based Films for Food Packaging Applications." *Journal of Agricultural and Food Chemistry* 66(2): 395–413. doi:10.1021/acs.jafc.7b04528.

Wang, Li, Xu Lu, and Liu Hu. 2019. "Preparation of Chitosan/Corn Starch/Cinnamaldehyde Films for Strawberry Preservation." *Foods* 8(9): 423. doi:10.3390/foods8090423.

Wang, Shuo, Peng Xia, Shaozhen Wang, Jin Liang, Yue Sun, Pengxiang Yue, and Xueling Gao. 2019. "Packaging Films Formulated with Gelatin and Anthocyanins Nanocomplexes: Physical Properties, Antioxidant Activity and Its Application for Olive Oil Protection." *Food Hydrocolloids* 96: 617–24. doi:10.1016/j.foodhyd.2019.06.004.

Wang, Xuejiao, Chaofan Guo, Wenhui Hao, Niamat Ullah, Lin Chen, Zhixi Li, and Xianchao Feng. 2018. "Development and Characterization of Agar-Based Edible Films Reinforced with Nano-Bacterial Cellulose." *International Journal of Biological Macromolecules* 118(A): 722–30. doi:10.1016/j.ijbiomac.2018.06.089.

Wang, Yifei, Yawen Xia, Pengyu Zhang, Lin Ye, Lianqiang Wu, and Shoukui He. 2017. "Physical Characterization and Pork Packaging Application of Chitosan Films Incorporated with Combined Essential Oils of Cinnamon and Ginger." *Food and Bioprocess Technology* 10(3): 503–11. doi:10.1007/s11947-016-1833-8.

Warsiki, E. 2018. "Application of Chitosan as Biomaterial for Active Packaging of Ethylene Absorber." In: *IOP Conference Series: Earth and Environmental Science*, Vol. 141. Institute of Physics Publishing. doi:10.1088/1755-1315/141/1/012036.

Watkins, Chris B. 2006. "The Use of 1-Methylcyclopropene (1-MCP) on Fruits and Vegetables." *Biotechnology Advances*. doi:10.1016/j.biotechadv.2006.01.005.

Woraprayote, Weerapong, Yutthana Kingcha, Pannawit Amonphanpokin, Jittiporn Kruenate, Takeshi Zendo, Kenji Sonomoto, Soottawat Benjakul, and Wonnop Visessanguan. 2013. "Anti-*Listeria* Activity of Poly(Lactic Acid)/Sawdust Particle Biocomposite Film Impregnated with Pediocin PA-1/AcH and Its Use in Raw Sliced Pork." *International Journal of Food Microbiology* 167(2): 229–35. doi:10.1016/j.ijfoodmicro.2013.09.009.

Wu, Chunhua, Yuan Li, Liping Wang, Yaqin Hu, Jianchu Chen, Donghong Liu, and Xingqian Ye. 2016. "Efficacy of Chitosan-Gallic Acid Coating on Shelf Life Extension of Refrigerated Pacific Mackerel Fillets." *Food and Bioprocess Technology* 9(4): 675–85. doi:10.1007/s11947-015-1659-9.

Wu, Hejun, Yanlin Lei, Rui Zhu, Maojie Zhao, Junyu Lu, Chun Jiao Di Xiao, Zhiqing Zhang, Guanghui Shen, Shanshan Li, and S. Li. 2019. "Preparation and Characterization of Bioactive Edible Packaging Films Based on Pomelo Peel Flours Incorporating Tea Polyphenol." *Food Hydrocolloids* 90: 41–9. doi:10.1016/j.foodhyd.2018.12.016.

Wu, Zhengguo, Xiujie Huang, Yi Chen Li, Hanzhen Xiao, and Xiaoying Wang. 2018. "Novel Chitosan Films with Laponite Immobilized Ag Nanoparticles for Active Food Packaging." *Carbohydrate Polymers* 199: 210–18. doi:10.1016/j.carbpol.2018.07.030.

Wu, Zhijun, Jingjing Wu, Tingting Peng, Yutong Li, Derong Lin, Baoshan Xing, Chunxiao Li, et al. 2017. "Preparation and Application of Starch/Polyvinyl Alcohol/Citric Acid Ternary Blend Antimicrobial Functional Food Packaging Films." *Polymers* 9(3). doi:10.3390/polym9030102.

Xanthos, Dirk, and Tony R. Walker. 2017. "International Policies to Reduce Plastic Marine Pollution from Single-Use Plastics (Plastic Bags and Microbeads): A Review." *Marine Pollution Bulletin* 118(1–2): 17–26. doi:10.1016/j.marpolbul.2017.02.048.

Xiao, Qian, Loong-tak Lim, Yujia Zhou, and Zhengtao Zhao. 2017. "Drying Process of Pullulan Edible Films Forming Solutions Studied by Low- Field NMR." *Food Chemistry* 230: 611–17. doi:10.1016/j.foodchem.2017.03.097.

Yadav, Mithilesh, Yu-kuo Liu, and Chiu Fang-chyou. 2019. "Fabrication of Cellulose Nanocrystal/Silver /Alginate Bionanocomposite Films with Enhanced Mechanical and Barrier Properties for Food Packaging Application." *Nanomaterials* 9(11): 1523. doi:doi:10.3390/nano9111523.

Yang, Dong, Xinwen Peng, Linxin Zhong, Xuefei Cao, Wei Chen, Xueming Zhang, Shijie Liu, and Runcang Sun. 2014. "Green Films from Renewable Resources: Properties of Epoxidized Soybean Oil Plasticized Ethyl Cellulose Films." *Carbohydrate Polymers* 103: 198–206. doi:10.1016/j.carbpol.2013.12.043.

Yang, Hyun Ju, Ji Hyeon Lee, Misun Won, and Kyung Bin Song. 2016. "Antioxidant Activities of Distiller Dried Grains with Solubles as Protein Films Containing Tea Extracts and Their Application in the Packaging of Pork Meat." *Food Chemistry* 196: 174–79. doi:10.1016/j.foodchem.2015.09.020.

Yang, So Young, Ka Yeon Lee, Song Ee Beak, Hyeri Kim, and Kyung Bin Song. 2017. "Antimicrobial Activity of Gelatin Films Based on Duck Feet Containing Cinnamon Leaf Oil and Their Applications in Packaging of Cherry Tomatoes." *Food Science and Biotechnology* 26(5): 1429–35. doi:10.1007/s10068-017-0175-2.

Yang, Weiqiao, Xihong Li, Jianan Jiang, Xuetong Fan, Meijun Du, Xianai Shi, and Ruizhi Cao. 2019. "Improvement in the Oxidative Stability of Flaxseed Oil Using an Edible Guar Gum-Tannic Acid Nanofibrous Mat." *European Journal of Lipid Science and Technology* 121(10): 1800438. doi:10.1002/ejlt.201800438.

Ye, Pengcheng, Zhaohua Fang, Baogen Su, Huabin Xing, Yiwen Yang, Yun Su, and Qilong Ren. 2010. "Adsorption of Propylene and Ethylene on 15 Activated Carbons." *Journal of Chemical and Engineering Data* 55(12): 5669–72. doi:10.1021/je100601n.

Yildirim, Selçuk, Bettina Röcker, Marit Kvalvåg Pettersen, Julie Nilsen-Nygaard, Zehra Ayhan, Ramune Rutkaite, Tanja Radusin, Patrycja Suminska, Begonya Marcos, and Véronique Coma. 2018. "Active Packaging Applications for Food." *Comprehensive Reviews in Food Science and Food Safety*. doi:10.1111/1541-4337.12322.

Yildız, Pınar Oğuzhan, and Filiz Yangılar. 2016. "Effects of Different Whey Protein Concentrate Coating on Selected Properties of Rainbow Trout (*Oncorhynchus mykiss*) During Cold Storage (4°C)." *International Journal of Food Properties* 19(9): 2007–15. doi:10.1080/10942912.2015.1092160.

Yin, Cheng, Chongxing Huang, Jun Wang, Ying Liu, Peng Lu, and Lijie Huang. 2019. "Effect of Chitosan- and Alginate-Based Coatings Enriched with Cinnamon Essential Oil Microcapsules to Improve the Postharvest Quality of Mangoes." *Materials* 12(13). doi:10.3390/ma12132039.

Younes, Islem, and Marguerite Rinaudo. 2015. "Chitin and Chitosan Preparation from Marine Sources. Structure, Properties and Applications." *Marine Drugs* 13(3): 1133–74. doi:10.3390/md13031133.

Zaman, Nurshahira Binti, N. K. Lin, and P. L. Phing. 2018. "Chitosan Film Incorporated with *Garcinia atroviridis* for the Packaging of Indian Mackerel (*Rastrelliger kanagurta*)." *Ciencia e Agrotecnologia* 42(6): 666–75. doi:10.1590/1413-70542018426019918.

Zhang, Le, Anjun Liu, Wenhang Wang, Ran Ye, Yaowei Liu, Jindong Xiao, and Kun Wang. 2017. "Characterisation of Microemulsion Nanofilms Based on Tilapia Fish Skin Gelatine and ZnO Nanoparticles Incorporated with Ginger Essential Oil: Meat Packaging Application." *International Journal of Food Science and Technology* 52(7): 1670–79. doi:10.1111/ijfs.13441.

Zhang, Qiang, Wenting Dai, Xinwen Jin, and Jixin Li. 2019. "Calcium Chloride and 1-Methylcyclopropene Treatments Delay Postharvest and Reduce Decay of New Queen Melon." *Scientific Reports* 9(1). doi:10.1038/s41598-019-49820-8.

Zhang, Yong, Lili Cui, Xiaoxia Che, Heng Zhang, Nianqiu Shi, Chunlei Li, Yan Chen, and Wei Kong. 2015. "Zein-Based Films and Their Usage for Controlled Delivery: Origin, Classes and Current Landscape." *Journal of Controlled Release* 206: 206–19. doi:10.1016/j.jconrel.2015.03.030.

Zhou, Xiaolong, Minchao Feng, Hui Niu, Yueqin Song, Chenglie Li, and Dongwen Zhong. 2016. "Adsorptive Recovery of Ethylene by $CuCl_2$ Loaded Activated Carbon via N-Complexation." doi:10.1177/0263617416658890.

Zhu, Kunkun, Shuo Shi, Yan Cao, Ang Lu, Jinlian Hu, and Lina Zhang. 2019. "Robust Chitin Films with Good Biocompatibility and Breathable Properties." *Carbohydrate Polymers* 212: 361–67. doi:10.1016/j.carbpol.2019.02.054.

Zhu, Yan, Bruno Guillemat, and Olivier Vitrac. 2019. "Rational Design of Packaging: Toward Safer and Ecodesigned Food Packaging Systems." *Frontiers in Chemistry* 7. doi:10.3389/fchem.2019.00349.

Zimet, Patricia, Álvaro W. Mombrú, Dominique Mombrú, Analía Castro, Juan Pablo Villanueva, Helena Pardo, and Caterina Rufo. 2019. "Physico-Chemical and Antilisterial Properties of Nisin-Incorporated Chitosan/Carboxymethyl Chitosan Films." *Carbohydrate Polymers* 219: 334–43. doi:10.1016/j.carbpol.2019.05.013.

Zink, Joël, Tom Wyrobnik, Tobias Prinz, and Markus Schmid. 2016. "Physical, Chemical and Biochemical Modifications of Protein-Based Films and Coatings: An Extensive Review. " *International Journal of Molecular Sciences* 17(9). doi:10.3390/ijms17091376.

2 Biodegradable Films for Food Packaging

Haniyeh Rostamzad

CONTENTS

2.1 INTRODUCTION

Maintaining nutritional value, quality, and freshness of food is of particular importance and packaging is one of the common ways to store food. Packaging is the art, science, and technology that prepare food for shipment and sale in the best conditions and at the right cost. Important aims of packaging can include physical protection, protection of goods within the package against extreme shaking, pressure, temperature, etc. and preventing the entrance of oxygen, water vapor, dust, and so on into the packaging. Information about the product and how to use it, shipping, recycling conditions, the best use time etc. is available on the package or its label. A good packaging can play the role of a silent salesman. Therefore, packaging is of particular importance in production and in marketing. With the advancement of various food, health, and cosmetics industries worldwide, sellers have significant competition for achieving a place in the market as well as customer satisfaction and the main factor for introduction of goods is their packaging. Various packaging show the different health and beauty characteristics of goods. A good packaging minimizes the communication between sellers and end consumers. However, the importance of storage and packaging of goods is no less important than their production, and packaging is the determining factor in maintaining goods until they reach the consumer and packaging is the best and largest promoter of a company's products. Nowadays, plastic, petroleum, glass, metal, and paper materials are used to pack the foods. The use of plastic containers and bottles is widespread due to the increasing tendency of people to use packaged food. The low weight, inexpensive raw material, and high mechanical strength of plastics are the reasons that these compounds were employed in the packaging industry (Qin et al., 2019). The use of plastic and metal materials in packaging has several disadvantages, including the transformation of compounds used in formulating food packaging, environmental contamination, and recycling problems. Plastic waste that is left over from food packaging has always been a major environmental problem. Therefore, new research has focused on finding suitable alternatives for conventional packaging. Synthetic plastics are lightweight and low cost, have high plasticity in automatic machines of packaging, and are resistant to a large number of acids and bases. Nevertheless, there is concern around them relating to their environmental problems, potential immigration of the ingredients and monomers in plastic (stabilizers and softeners) to the food and potentially causing disease in the consumer. Therefore, the low cost of recycling, the high cost of transporting the plastic waste, and the high cost of burning the plastics compounds led to the potential use of biodegradable and recyclable materials for packaging and storage of the food, attracting the attention of researchers. Biodegradable plastic is plastic that decomposes under the natural influence of microorganisms such as bacteria, fungus, and algae. The production and development of biodegradable polymer materials is needed by communities, and therefore research on biodegradable materials with controlled properties has been attracting the attention of material specialists worldwide.

2.2 TYPES OF BIODEGRADABLE POLYMERS

Biodegradable polymers are divided into two major groups:

1. Polymers that were made from natural renewable sources such as polylactic acid, polyhydroxybutyrate and their copolymers, starch, cellulose, gelatin, chitosan, and so on
2. Polymers that were made from petroleum sources and include biodegradable aliphatic polyesters, polybutylene succinate, polycaprolactone, and polyvinyl alcohol

Biodegradable polymers that were made from renewable natural resources have attracted much attention recently because of their economic and environmental appeal. These polymers are noteworthy for use due to the biological potential of these materials, reduction of the volume of created waste, fertilization in the natural cycle, reduction of carbon dioxide content, and ability of the agricultural resources to produce biodegradable polymers. Biopolymers are polymeric materials where at least one-step of their environmental degradation process is carried out by living organisms. In general, the benefits of biopolymers can be their natural degradation, the protection of environment their use entails, the creation of new markets for the sale of agricultural products, the possibility of controlling water vapor transmission, oxygen, carbon dioxide and fat in the food system and prevention of reduction of the taste of the food.

2.2.1 Properties that Affect Biodegradable Film Usability

Features that affect the usability of a biodegradable film as packaging different materials include:

- Mechanical features
- Inhibitory properties against gases and vapors
- Thermal characteristics
- Surface features
- Ability to convert to compost
- Organoleptic properties

The use of biopolymers in food packaging has several limitations. Fragility, rigidity, thermal instability, and high permeability to gases and water vapor are the major limiting factors of the use of biopolymers in the industry of packaging. Various methods are used to improve these properties, such as the use of nanomaterials, the use of different chemical materials, the biopolymer composition, the thermal processes, and so on, which are discussed below.

2.2.2 Biodegradable Films with Antimicrobial and Antioxidant Properties

One of the ways to improve the biodegradable films is to create antioxidant and antimicrobial properties in them using a variety of antimicrobials, antioxidants, extracted essential oils from plants or seasonings (Viuda-Martos et al., 2010). There is consumer demand for quality foods using environmentally friendly packaging systems and natural preservatives, such as edible/biodegradable films or bioactive coatings

There is a great tendency to use new types of natural antimicrobial compounds such as extracts of seasonings and herbs for food preservation. Today, some of the seasonings and herbal compounds used are valuable because of their antimicrobial activity and medicinal effects as well as the quality of their flavor and perfume. The composition of extracts of most plant species has been identified in recent years and efforts to identify their bioactive components for various pharmaceutical and food processing purposes have been accelerated.

Essential oils are volatile, so their effects on food shelf life are limited. Therefore, researchers are looking for a way to increase the effectiveness of these valuable materials. One of the newest ways in which these volatile substances can be controlled, is to encapsulate materials. In this way, by creating a wall around the essential oil, its release will be slow and will increase the duration of the effect of the substance. The use of the microencapsulation system in the production of edible films is a very new method and very little research has been done on it worldwide. The essential oils, through emulsions, are added to the film and edible covers and affect the physical and structural properties of the film (Perdones, 2014).

In another research study, Lessani et al. (2017) investigated the antimicrobial and antioxidant properties of biodegradable carrageenan film in combination with microcapsules of clove essential oil. They concluded that the addition of clove essential oil, as well as clove-maltodextrin microcapsules, to the carrageenan film led to antioxidant and antimicrobial activity against the bacterium.

Rostamzad and Zakipour (in press) investigated the functional and antimicrobial properties of chitosan film in combination with three concentrations of aqueous extract of licorice (*glycyrrhiza glabra*). The results indicated that the chitosan film containing 10% of licorice extract ($P<0.05$) had the best application properties, and the film showed resistance to bacteria. In addition, fillets of the fish coated with chitosan film containing the licorice extract showed the lowest amounts of peroxide, thiobarbituric acid, volatile nitrogenous bases, total microbial charge, and psychrophilic bacterial count compared to other treatments. According to the results of this study, the use of chitosan film containing licorice extract in packaging the fillets of fish can significantly increase their shelf life.

In addition, Rostamzad et al. (2019) and Hu et al. (2015) produced a film and coating of chitosan containing ginger extract. The results showed that the chitosan film containing ginger extract was resistant to some bacteria. In the study by Rostamzad et al. (2019), films were used to pack the fish fillets after making them. In another treatment, fillets were coated with chitosan-ginger solution to compare the effect of packaging with the film and coating with chitosan solution and extract. The coated fish specimens (chitosan-ginger film and chitosan-ginger solution) were kept in the refrigerator for 12 days (4°C) and the corruption factors (peroxide, TBA,* TVN,† and pH) were measured during the storage period every four days. Results indicated

* Tiobarbituric Acid
† Total Volatile Nitrogen

that the use of ginger extract in combination with chitosan had a significant effect on enhancing the quality of fish fillets ($P<0.05$) and the treatment of packaging with combined film of chitosan and ginger had the best result, preserving the quality of the fillets (Rostamzad et al., 2019).

2.2.3 Nanotechnology in the Production of Biodegradable Films

Among the available methods, the use of nanotechnology in the production of nanocomposites is one of the newest methods in the preparation of film and the improvement of its functional properties. The application of nanoparticles in the production of various materials results in the increased strength and thermal resistance of the materials. Nanoparticles are the most common elements of nanoscience and their remarkable properties have led to a variety of applications in the chemical, pharmaceutical, electronics, and agricultural industries (George et al., 2019; Indumathi et al., 2019; Roy et al., 2019; Saadat et al., 2019).

When nanocomposites are compared to pure polymers, nanocomposites exhibit better mechanical, thermal, optical, and physical-chemical properties. These properties include increased strength, thermal resistance, and reduced permeability to gases, and are achieved by adding small amounts of nanoparticles.

2.2.3.1 Types of Nanomaterials Used in Nanocomposites

- Carbon nanotubes
- Fullerenes
- Inorganic nano materials
- Nano clay (layer silicate)
- Cellulose N=nanocrystals (organic nanomaterials)

Extensive research has been done on the use of nanomaterials to improve the functional properties of biodegradable films (Dash et al., 2019; George et al., 2019; Indumathi et al., 2019; Saral Sarojini et al., 2019; Roy et al., 2019). Among nanomaterials, nanoclay is the most important and widely used material for the production of biopolymer nanocomposites.

This class of nanomaterials has two unique properties compared to other materials that have led to the expansion of their use in the production of nanocomposites:

1. The ability of nanoclay to diffuse as separate layers and the ability to modify the surface properties of these materials and to adapt to different types of polymers and biopolymers
2. Easy production, resulting in low cost, easy accessibility and compatibility with other materials

The most important and widely used layered silicate is montmorillonit (MMT), which is widely used in the production of nanocomposites due to its biocompatibility, easy accessibility, and low cost.

Rostamzad et al. (2016) investigated the use of different concentrations (1, 3, and 5%) of nanoclay in the production of biodegradable film of fish protein and they found that adding nanoclay improved the film quality up to 3% and the film strength significantly increased. It also significantly reduced the permeability of fish protein films to water vapor.

2.2.4 Use of Chemical Binders to Improve the Properties of Biodegradable Films

Another way to improve the quality of biodegradable films is to use chemical and enzymatic treatments. Researchers have used many types of enzymes and chemical materials to form suitable bonds in the polymer matrix and subsequently improve the functional properties of the films (Lei et al., 2007; Limpan et al., 2010). Polymers show different reactions to different chemical materials due to their nature (Rostamzad et al., 2016; Siriprom et al., 2018).

Rostamzad et al. (2016) investigated the effect of three concentrations of microbial transglutaminase on the functional properties of the film of fish protein. The results of scanning electron microscope (SEM) studies showed that the film containing 3% transglutaminase is clearer than the other concentrations, which could be due to cross-linking of transglutaminase enzyme with the polymeric agents group.

These results are consistent with the results of Marinillo et al. (2003) in investigation of the presence or absence of transglutaminase in the film of soy flour. In this film, the presence of transglutaminase also made the film structure smoother and more homogenous (Mariniello et al., 2003). In addition, in that study, the increase of the transglutaminase enzyme significantly reduced permeability of the films to the water vapor, which may be due to the cross-linking of amino acids in the presence of transglutaminase enzyme, which creates a better and more robust structure and greatly prevents the penetration of water vapor. DeCarvalho *et al.* (2004) also reported a decrease in water vapor permeability of gelatinous films in the presence of microbial transglutaminase (de Carvalho and Grosso, 2004).

The addition of microbial transglutaminase increases the cross-linking in the polymer matrix and the formation of these bonds increases the hardness and molecular weight and as a result of TS* increase (Rostamzad et al., 2016). Similar results were obtained in other investigations on clay-gelatin nanocomposites of fish (Marnillo et al., 2003). Also in a study on pectin polymer-soy flour, it was concluded that the use of microbial transglutaminase increased the film strength (Bae et al., 2009). Researchers have suggested that the use of chemical cross-linkers such as formaldehyde and glioxal also increase the TS of gelatinous films.

2.3 TYPES OF BIODEGRADABLE POLYMERS

Polymers used in packaging are divided into four categories based on their chemical structure:

1. Proteins such as gelatin, collagen, meat myofibrils, soy protein, whey protein, egg protein

* Tensile strength

2. Polysaccharides such as cellulose and its derivatives (methyl cellulose, carboxymethyl cellulose, hydroxypropyl cellulose), chitin and chitosan, alginate, carrageenan, xanthan, starch, guar, etc.
3. Polyesters such as polylactic acid, polyhydroxybutyrate, and polyhydroxyvalerate.
4. Lipids such as fats and vegetable oils, glycerol mono stearates and surfactants, waxes such as carnaubas and beeswaxes, etc.

2.3.1 Protein Films

Protein films are more effective in prevention of the gas penetration than lipid and polysaccharide films. The mechanical properties of edible coatings of the protein films are also better due to the tight intermolecular bonding of lipid- or polysaccharide-based films (Salgado et al., 2010). Protein films are made from various plant and animal sources. Materials used for the production of protein films and edible protein coatings include gelatin, wheat gluten, zein of corn, casein, whey protein, and meat proteins. Films of different types of proteins exhibit different properties.

2.3.1.1 Zein Protein

One of the proteins used in the production of biodegradable polymers is zein protein. Zein is one of the corn proteins. Commercial zein is yellow and contains less than 20% non-protein solids and small amounts of oil. Zein protein is insoluble in water but soluble in water in the presence of high concentrations of urea or alkali, anionic detergents such as SDS* and enzymatic modifications. In aqueous ethanol, polar solvents such as propylene glycol and organic acids such as acetic acid are soluble. Zein has an ability to form good film and has thermoplastic properties (it flows through heat and pressure). For the formation of the film by zein and other spherical proteins, the proteins must first be denatured by solvent, heat, acid, or base to open their structure, and then chain-to-chain interactions must occur. In recent years, significant research has been done in the field of producing zein films for packaging (Guiyun Chen et al., 2019; Mushtaq et al., 2018; L. Zhang et al., 2019).

2.3.1.2 Soy Protein

Soy protein-based films have some interesting properties, including being biodegradable, biocompatible, and inexpensive. However, their poor mechanical properties and their high sensitivity to moisture are major barriers to the use of this protein film for food packaging applications. As a product of the soy oil industry, soy protein isolates can be used for plastic production. Therefore, numerous studies have been carried out to improve the functional properties of soy-based films, including on the use of cellulose nano fibers (González et al., 2019) and on different acids such as acrylic acid (Zhao et al., 2016) and stearic acid (Ye et al., 2019) and types of oils such as colza (Galus, 2018) and galactomannan (González et al., 2019).

---

* Sodium Dodecyl Sulfate

2.3.1.3 Whey Protein

When the pH of milk reaches 4.6 at 20°C, the casein proteins, which make up 80% of the cow's milk proteins, precipitate, and soluble proteins remain. Whey proteins are composed of several constituents, such as β-ketoglobulin, alpha-lactoalbumin, immunoglobulin, proteosepitones, and serum albumin. Whey proteins, unlike casein proteins, are spherical and heat sensitive.

Because whey proteins are spherical, thermal denaturation must be done for production of their film. Thus, the aqueous solution of the protein must be heated (for example, 90°C for 30 minutes). Thermal denaturation opens the spherical structure, breaks the intermolecular disulfide bonds, and forms new intermolecular disulfide bonds and hydrophobic bonds. This causes polymerization and the film to form.

Using industrial dairy by-products to make biopolymer-based packaging is a sustainable strategy to reduce environmental problems and minimize food waste without competition in food resources (Cruz-Diaz et al., 2019). Therefore, numerous studies have been carried out on the production of whey-based polymers and in combination with various substances such as the water of Cornelian Cherry, essential oils such as rosemary and oregano (*Origanum vulgare* L.), garlic (Seydim and Sarikus, 2006), pomegranate seed oil (Sogut et al., 2019) and so on, and good results have been achieved.

2.3.1.4 Fish Protein

Among protein sources, fish protein can be a good choice for edible films. Edible films derived from fish protein can be sarcoplasmic, myofibril, and stroma proteins (Weng et al., 2007). Myofibrillar proteins have good ability for the formation of the film that results in the formation of polymers with interesting properties for the production of food packaging. Fish myofibril protein can be used to make films with good clarity and strength because myofibril proteins have the ability to form a continuous background during film drying (Limpan et al., 2010). Comparison of the synthetic films with films formed of myofibril protein shows that synthetic films have weaker mechanical properties than the myofibril protein films. Therefore, different methods such as chemical and enzymatic treatment (Rostamzad et al., 2016), thermal treatment (Lei et al., 2007), ultraviolet (UV) irradiation and the use of nanoparticles (Jo et al., 2005) are used to improve the functional properties of the films.

Rostamzad et al. studied the characteristics of a biodegradable protein-based film from silver carp and its application in silver carp fillets (Rostamzad et al., 2019). In another research study, they studied properties of fish protein film in combination with nanoclay and transglutaminase and their results showed that water vapor permeability decreased when nanoclay concentration was 1% and 3%, while nanoclay content greater than 5% increased the water vapor permeability, which indicates an inadequate dispersion of nanoclay particles at high concentrations (5%). In addition, the use of nanoclay up to 3% concentration and the transglutaminase enzyme up to 3% improved the mechanical properties of the fish protein films (Rostamzad et al., 2016).

Other methods that can improve the working properties of fish protein films are the combination of this polymer with other materials such as chitosan (Batista et al., 2019). In addition, the use of different pH (2 and 10) in the production of gels has led to good results in the production of coherent films, which reduced the permeability of fish protein films (Romani et al., 2018). In addition to the above, the new method of applying cold plasma has been effective in improving the working properties of fish protein film. According to studies of different researchers, the use of cold plasma reduces water vapor permeability and their solubility, which are important factors in food packaging. Consumers demand quality foods using environmentally friendly packaging systems and natural preservatives, such as edible/biodegradable films or bioactive coatings (Romani et al., 2019a; Romani et al., 2019b).

In this context, films made with by-proteins of fish products were combined with essential oils (garlic and clove) and they were characterized by their physical, mechanical, antioxidant and antibacterial properties. The control films were oil-free, homogeneous, transparent, slightly yellow, and mechanically resistant. The composition of the garlic and clove oil led, significantly, to reduction of the thickness, water solubility, fracture force, and elongation (Pires et al., 2013; Teixeira et al., 2014).

In addition, biodegradable polymers can be obtained from other sources such as wheat gluten (Ansorena et al., 2016; Nataraj et al., 2018; Rocca-Smith et al., 2016), milk casein (Picchio et al., 2018; Zhuang et al., 2018), albumen of egg, sunflower protein, and others, each of which has its own characteristics.

2.3.2 Polysaccharide Films

Films and coatings of polysaccharide such as chitosan, alginate, carrageenan, xanthan, starch, guar, etc. have many uses. The chemical structure and nature of the biopolymer plays a key role in the selection of the solvent type and film preparation conditions. Since polysaccharides are more uniform than proteins and because of the greater and stronger interactions between the polymer chains, the polysaccharide films usually have better mechanical strength than the protein films and the resulting films are stronger.

In addition, due to the polar nature of most polysaccharides, an inhibitory property of their films against gases is also desirable. Nevertheless, the hydrophilic nature of their constituent led to the water vapor permeability of polysaccharide films to be relatively higher than that of most protein films. Therefore, different methods are used in the production of polysaccharide films to improve their functional properties.

2.3.2.1 Chitin and Chitosan

Chitin and its acetylated compound, chitosan, are two well-known natural polymers whose history of extraction and use dates back 200 years. These two nitrogen polysaccharides are the most abundant polymers found in nature after cellulose. So far, more than 300 different sources of chitin have been studied including marine invertebrates, fungus, bacteria, diatoms, yeasts, and so on. Chitosan is a modified cationic carbohydrate that is obtained by chitin deacetylation.

Chitosan has attracted a great deal of attention in recent years and has proven its capability in the medical, agricultural, food, and chemical industries. Chitosan is a non-toxic, biocompatible, bioactive, biodegradable compound and several studies have reported its antimicrobial and antifungal properties. It is capable of forming gels and films due to its high molecular weight and solubility in acidic solutions.

The results of the comparison of chitosan with other biomolecule-based active films have shown chitosan to have more advantages due to its antibacterial activity and its ability to chelate bivalent minerals. Chitosan, due to its film-forming properties and unique viscosity-enhancing properties, can be used in food packaging especially as coating and edible film. In chitosan films, some useful properties including antioxidant, antimicrobial, and oxygen impermeability properties have been reported.

Rostamzad et al. (2019) studied the effect of film and coating of chitosan on shelf life of fish fillets. They reported that a fillet coated with chitosan solution as well as packed fillets with chitosan film had better quality (TVN* PV,† TC,‡ etc.) compared to the control treatment and were more durable. In addition to the above, the addition of ginger extract to the film and the chitosan solution showed better results in maintaining the fillet quality (Rostamzad et al., 2019).

Another study on the properties chitosan film enriched with cinnamon oil and soy oil revealed that by adding cinnamon oil and soy oil, water vapor permeability was significantly reduced. The film containing 2 or 3% cinnamon oil also inhibited the pathogens (Ma et al., 2016).

Investigation of the effect of chitosan nanoparticles with cinnamon oil on chilled pork showed that the antimicrobial and antioxidant properties of this film resulted in longer shelf life of the meat during refrigeration (Hu et al., 2015).

In another study on the shelf life of fish, the results showed that fillets packed with chitosan film compared to control treatment (regular packing) resulted in a fillet with higher quality and better preservation.

The results of studies of the antioxidant and antimicrobial properties of chitosan film with nanocapsules of liquid smoke with maltodextrin in maintaining the quality of tuna fish showed that nanocapsules of liquid smoke have been a preservative and effective agent for fresh fish. The nanocapsules were prepared in a mixture of chitosan (1.5 w/v) and maltodextrin (8.5 w/v) liquid smoke and with these nanocapsules, the fish stayed fresh for up to 48 hours at room temperature (Saloko et al., 2014).

Investigation of physical, antioxidant, and antimicrobial properties of chitosan film with cinnamon affected by oleic acid showed that the presence of oleic acid increased the surface area and dispersion of the particles in the film formation and helped to preserve the film during drying. Water vapor permeability was greatly reduced in chitosan films containing oleic acid compared to the film containing cinnamon alone. The results showed that the cinnamon-containing film showed good antioxidant and antifungal effects (Perdones et al., 2014). In addition, the physical and antibacterial properties of alginate film with cinnamon and soy oil were investigated and it was found that the presence of cinnamon oil improved uniformity and transparency of film, as well as increasing elasticity and reducing water vapor permeability. Soy oil also helped to improve the antibacterial effect of film (Zhang et al., 2015).

---

* Total Volatile Nitrogen
† Peroxide Value
‡ Total Coliform

In 2010, the biodegradable gelatin-chitosan film with essential oils as an antimicrobial agent in fish protection was investigated. That study found that clove essential oil has a great role in inhibition of the pathogenic microorganisms (Gómez-Estaca et al., 2010).

2.3.2.2 Carrageenan

Carrageenan is a polysaccharide that was extracted from the red algae known as Irish moss. This anionic polysaccharide has high molecular weight and its repeating units are galactose and sulfated or non-sulfated 3- and 4-anhydro-galactose. The three main types of carrageenan include kappa, iota, and lambda. Kappa carrageenan with molecular weights between 400 and 560 kDa is extracted from *Euchema cottonii*. Kappa has the ability to form solid gels and therefore is used in the food industry as a gelling and stabilizing agent.

In 2014, Shojaee et al. investigated the properties of carrageenan film that was enriched with thyme oil and pineapple. The result showed that the addition of thyme oil and pineapple oil reduced water vapor permeability as well as increasing elasticity of the films and films have ability to inhibit bacterial growth against *Staphylococcus aureus* (Shojaee-Aliabadi et al., 2014).

Lessani et al. investigated the properties of carrageenan film in combination with nanoclay particles and found that adding nanoclay up to 3% improved the mechanical properties as well as the inhibitory properties of the films. In addition, in this study, the addition of clove essential oil caused the antioxidant and antimicrobial properties against bacteria (Lessani et al., 2017).

Kamali et al. improved the properties of carrageenan film by adding the red cabbage extract to it. In addition, using the presence of anthocyanins in red cabbage extract, intelligent films have been produced where the color of packaging changes with spoilage of food (meat) and informed the consumer of the quality of the food in the package (Kamali et al., 2019).

Kanmani and Rhim investigated the potency of antimicrobial film of carrageenan that was enriched with grapefruit seed extract and their results showed that grapefruit extract was in good agreement with carrageenan. Tensile strength was also increased. Results indicate that carrageenan films have a high potential for use as antimicrobial or active packaging (Kanmani and Rhim, 2014).

In 2010, Alves et al. investigated the properties of kappa carrageenan-pectin film and the results showed that permeability of water vapor, CO_2, and O_2 gases decreased with its use. Permeability of gases of CO_2 and O_2 decreased by 73% and 27%, respectively (Alves, Costa, and Coelhoso, 2010).

In another research study, the properties of carrageenan-based nanocomposites enriched with minerals, nanoclay and silver nanoparticles were studied. Strong antibacterial properties against Gram-positive and Gram-negative bacteria were observed. Both nanomatters affected the film's properties, including smooth surface, color, and reduced water vapor permeability (Rhim and Wang, 2014).

2.3.2.3 Starch

Starch polymer consists of glucose molecules which consist of two types of polymer molecules: linear polymer called amylose and branched polymer called amylopectin.

Usually, about 70 to 80 percent of starch is amylopectin and the rest is amylose. Starch is used in various forms to produce bioplastics, including:

- As a filler
- As a blendable polymer
- As a grafted copolymerization
- As a single polymer

Starch produces the most resistant film among biopolymers and its mechanical properties are better than other polysaccharide films as well as protein films. Many researchers have done extensive research about the use of starch in combination with other polymers as well as the use of this biopolymer for food packaging (Hilmi et al., 2019; Liu et al., 2017; Lumdubwong and Namfone 2019; Noorbakhsh-Soltani et al., 2018; Peighambardoust et al., 2019).

2.3.2.4 Cellulose Derivatives

Cellulose is the most abundant organic matter in nature and is about one-third of the weight of all plant materials. Films formed from cellulose derivatives have the potential to reduce the exchange of moisture between food and the environment. These films have a normal mechanical resistance and they are resistant to the permeability of fats and oils. They are flexible and transparent, free of odor and taste, water-soluble, and have a good inhibitory property against moisture and gases. Many researchers have been investigating the use of cellulose and its derivatives in combination with other polymers or individually for food packaging (Cazón et al., 2018; Guo Chen et al., 2014; Deng et al., 2017).

2.3.3 Lipid Films

Films and coatings of lipid are another class of biodegradable biopolymers. These hydrophobic substances, due to their non-polar nature, are used as a moisture deterrent. Lipid films have low moisture permeability due to their hydrophobic nature, but these films are very fragile and to obtain lipid films with good mechanical properties, it is better to make a composite film. That is, they can be used in combination with other polymers such as protein or polysaccharide. Common lipid compositions in the preparation of films and coatings include a variety of waxes (carnauba wax, rice bran wax, beeswax, paraffin wax, etc.), triglycerides, monoglycerides, fatty acids, and fatty alcohols.

Investigation of adding different amounts of beeswax (B) or carnauba wax (C) to gelatin film showed that they can be used to increase thermal stability and inhibitory properties of film and protect food against deterioration. The results showed that beeswax used in combination with gelatin had a better effect than carnauba wax (Zhang et al., 2018)

In another research study, carnauba wax was used in combination with cassava starch and the results showed that water vapor permeability and water solubility were reduced by the addition of carnauba wax (Rodrigues et al., 2014). Therefore, its use in films with high permeability could have a good effect. Carnauba wax is widely used in food because of its physical and chemical properties. There has been a great deal of research recently on the use of this wax in preservation and processing of food, some of which have found it to be very effective in producing edible films and highly hydrophobic and degradable packaging (de Freitas et al., 2019; Rodrigues et al., 2014).

2.3.4 Polyesters

Polyesters are biodegradable polymers such as polyhydroxybutyrate, polyhydroxy valerate, polylactic acid, and polyglycolic acid. Most biodegradable polyesters have high potential to be used in packaging applications since they have desirable thermoplastic and biodegradable properties in addition to high strength, high modulus, and good processability. Biodegradable polyesters can be divided into aliphatic and aromatic groups, with members of each group being derived from renewable and non-renewable sources (Sin et al., 2013). Polylactic acid is an aliphatic polyester derived from renewable sources, mainly starch and sugar (Lim et al., 2008). PLA is the biodegradable polymer most present in the market. Its high stiffness and transparency make it a suitable material for manufacturing of plastic bottles, cups, and rigid trays. However, PLA is too stiff for most flexible packaging applications (Jiang et al., 2007; Marcos et al., 2016)

2.4 EDIBLE COATINGS

Films and coatings of biopolymer may be edible or biodegradable only, depending on the formulation, production method, and modifying treatments. Edible film refers to a thin layer of edible material that is placed on the surface of the food as a coating or in among a layer of the constituents of food to protect the food against damaging factors such as the presence of gases such as oxygen, carbon dioxide, moisture and to increase the shelf life of the food. Film and coatings of protein increase the nutritional value of food, but because some people are allergic to certain proteins, the type of protein used in the film or coating should be labeled on the packaging. This type of packaging has unique advantages compared to the use of non-edible polymers in food packaging and other conventional polymers. Film production requires at least one polymeric compound that can create a network structure with sufficient robustness and consistency.

Edible films are biodegradable and can be used as a component of food and reduce the consumption of basic petroleum polymers. Polysaccharides such as chitosan, starch and cellulose, proteins such as zein and collagen, and fats such as triglycerides and fatty acids can be used as edible films. These coatings and films reduce the interaction of nutrients with the environment and delay corruption by protecting the product against chemical, physical and microbial agents.

2.5 USE OF EDIBLE FILMS ON MEAT DISHES

With increasing exports and imports of meat, the importance of packaging under vacuum and modified atmosphere is increasing. However, this issue increases the use of plastic packaging and thereby increases the problem of plastic waste disposal. Edible films made from natural biopolymers such as proteins, polysaccharides, and lipids are more suitable than plastic polymers. They are biodegradable and can also prevent things like moisture loss, leakage, the growth of microbes, and lipid oxidation.

Edible coatings and films can prevent moisture loss, leakage, microbe's growth and swelling of lipid and meat products, and preserve their color and flavor and protect them from physical damage during transport.

2.6 USE OF EDIBLE COATINGS ON FRIED FOODS

Fried foods have a high level of dietary fat. Edible coatings of protein or other hydrocolloids can help reduce oil absorption during frying. Edible coatings can maintain and enhance the quality of food products due to their adhesion and ability to inhibit fat uptake.

2.7 ADVANTAGES OF EDIBLE COATING ON FRUITS AND VEGETABLES

The quality of fruits and vegetables is reduced by storage due to changes such as water loss, enzymatic browning, tissue spoilage, microbial growth, and so on. Edible coatings on fruits and vegetables have a long history. In the past, coatings were used to reduce water loss and improve the appearance of products such as apples and citrus fruits. Today, the main goal is to apply the coating on fruits that continue to ripen after harvest, to delay their ripening and increase their shelf life. Using the edible coatings on fruits and vegetables can reduce gas exchange, control the breathing rate, reduce nutrient loss, the evaporation of water and the growth of microorganisms in general, increasing shelf life, and maintaining the quality of fresh produce.

2.8 THE BENEFITS OF COATING EDIBLE NUTS

Nuts are prone to various kinds of corruption. Light, moisture, oxygen, high temperatures, microorganisms, and other factors can lead to a reduction in the quality of nuts and foods containing these products and decrease the safety of these foods for consumption.

Edible coatings for nuts can have the following benefits:

1. Inhibition of oxygen and preservation of the flavor and odor
2. Reduction of the use of packaging and its cost
3. Improvement of the appearance of the product
4. Inhibition of oil

REFERENCES

Alves, Vítor D., Costa, Nuno, and Coelhoso, Isabel M. (2010). Barrier properties of biodegradable composite films based on kappa-carrageenan/pectin blends and mica flakes. *Carbohydrate Polymers*, *79*(2), 269–276. doi: 10.1016/j.carbpol.2009.08.002.

Ansorena, María R., Zubeldía, Francisco, and Marcovich, Norma E. (2016). Active wheat gluten films obtained by thermoplastic processing. *LWT—Food Science and Technology*, *69*, 47–54. doi: 10.1016/j.lwt.2016.01.020.

Bae, Ho J., Darby, Duncan O., Kimmel, Robert M., Park, Hyun J., and Whiteside, William S. (2009). Effects of transglutaminase-induced cross-linking on properties of fish gelatin–nanoclay composite film. *Food Chemistry*, *114*(1), 180–189. doi: 10.1016/j.foodchem.2008.09.057.

Batista, J. T. S., Araújo, C. S., Peixoto Joele, M. R. S., Silva, J. O. C., and Lourenço, L. F. H. (2019). Study of the effect of the chitosan use on the properties of biodegradable films of myofibrillar proteins of fish residues using response surface methodology. *Food Packaging and Shelf Life*, *20*, 100306. doi: 10.1016/j.fpsl.2019.100306.

Cazón, Patricia, Vázquez, Manuel, and Velazquez, Gonzalo. (2018). Cellulose-glycerol-polyvinyl alcohol composite films for food packaging: Evaluation of water adsorption, mechanical properties, light-barrier properties and transparency. *Carbohydrate Polymers*, *195*, 432–443. doi: 10.1016/j.carbpol.2018.04.120.

Chen, Guiyun, Dong, Shuang, Zhao, Shuang, Li, Shuhong, and Chen, Ye. (2019). Improving functional properties of zein film via compositing with chitosan and cold plasma treatment. *Industrial Crops and Products*, *129*, 318–326. doi: 10.1016/j.indcrop.2018.11.072.

Chen, Guo, Zhang, Bin, Zhao, Jun, and Chen, Hongwen. (2014). Development and characterization of food packaging film from cellulose sulfate. *Food Hydrocolloids*, *35*, 476–483. doi: 10.1016/j.foodhyd.2013.07.003.

Cruz-Diaz, Karen, Cobos, Ángel, Fernández-Valle, María Encarnación, Díaz, Olga, & Cambero, María Isabel. (2019). Characterization of edible films from whey proteins treated with heat, ultrasounds and/or transglutaminase. Application in cheese slices packaging. *Food Packaging and Shelf Life*, *22*, 100397. doi: 10.1016/j.fpsl.2019.100397.

Dash, Kshirod K., Ali, N., Afzal, Das, Dipannita, and Mohanta, D. (2019). Thorough evaluation of sweet potato starch and lemon-waste pectin based-edible films with nano-titania inclusions for food packaging applications. *International Journal of Biological Macromolecules*, *139*, 449–458. doi: 10.1016/j.ijbiomac.2019.07.193.

de Carvalho, R. A., and Grosso, C. R. F. (2004). Characterization of gelatin based films modified with transglutaminase, glyoxal and formaldehyde. *Food Hydrocolloids*, *18*(5), 717–726. doi: 10.1016/j.foodhyd.2003.10.005.

de Freitas, Claisa Andréa Silva, de Sousa, Paulo Henrique Machado, Soares, Denise Josino, da Silva, José Ytalo Gomes, Benjamin, Stephen Rathinaraj, and Guedes, Maria Izabel Florindo. (2019). Carnauba wax uses in food—A review. *Food Chemistry*, *291*, 38–48. doi: 10.1016/j.foodchem.2019.03.133.

Deng, Zilong, Jung, Jooyeoun, and Zhao, Yanyun. (2017). Development, characterization, and validation of chitosan adsorbed cellulose nanofiber (CNF) films as water resistant and antibacterial food contact packaging. *LWT—Food Science and Technology*, *83*, 132–140. doi: 10.1016/j.lwt.2017.05.013.

Galus, Sabina. (2018). Functional properties of soy protein isolate edible films as affected by rapeseed oil concentration. *Food Hydrocolloids*, *85*, 233–241. doi: 10.1016/j.foodhyd.2018.07.026.

George, Archana, Shah, Priyanka A., and Shrivastav, Pranav S. (2019). Natural biodegradable polymers based nano-formulations for drug delivery: A review. *International Journal of Pharmaceutics*, *561*, 244–264. doi: 10.1016/j.ijpharm.2019.03.011.

Gómez-Estaca, J., López de Lacey, A., López-Caballero, M. E., Gómez-Guillén, M. C., and Montero, P. (2010). Biodegradable gelatin–chitosan films incorporated with essential oils as antimicrobial agents for fish preservation. *Food Microbiology*, *27*(7), 889–896. doi: 10.1016/j.fm.2010.05.012.

González, Agustín, Barrera, Gabriela N., Galimberti, Paola I., Ribotta, Pablo D., and Alvarez Igarzabal, Cecilia I. (2019). Development of edible films prepared by soy protein and the galactomannan fraction extracted from Gleditsia triacanthos (Fabaceae) seed. *Food Hydrocolloids*, *97*,. 105227. doi: 10.1016/j.foodhyd.2019.105227.

González, Agustín, Gastelú, Gabriela, Barrera, Gabriela N., Ribotta, Pablo D., and Álvarez Igarzabal, Cecilia I. (2019). Preparation and characterization of soy protein films reinforced with cellulose nanofibers obtained from soybean by-products. *Food Hydrocolloids*, *89*, 758–764. doi: 10.1016/j.foodhyd.2018.11.051.

Hilmi, F. F., Wahit, M. U., Shukri, N. A., Ghazali, Z., and Zanuri, A. Z. (2019). Physico-chemical properties of biodegradable films of polyvinyl alcohol/sago starch for food packaging. *Materials Today: Proceedings*, *16*, 1819–1824. doi: 10.1016/j.matpr.2019.06.056.

Hu, J., Wang, X., Xiao, Z., and Bi, W. (2015). Effect of chitosan nanoparticles loaded with cinnamon essential oil on the quality of chilled pork. *LWT—Food Science and Technology*, *63*(1), 519–526.

Indumathi, M. P., Saral Sarojini, K., and Rajarajeswari, G. R. (2019). Antimicrobial and biodegradable chitosan/cellulose acetate phthalate/ZnO nano composite films with optimal oxygen permeability and hydrophobicity for extending the shelf life of black grape fruits. *International Journal of Biological Macromolecules*, *132*, 1112–1120. doi: 10.1016/j.ijbiomac.2019.03.171.

Jiang, Long, Zhang, Jinwen, and Wolcott, Michael P. (2007). Comparison of polylactide/nano-sized calcium carbonate and polylactide/montmorillonite composites: Reinforcing effects and toughening mechanisms. *Polymer*, *48*(26), 7632–7644. doi: 10.1016/j.polymer.2007.11.001.

Jo, Cheorun, Kang, Hojin, Lee, Na Young, Kwon, Ho, Joong, Byun, and Woo, Myung. (2005). Pectin- and gelatin-based film: Effect of gamma irradiation on the mechanical properties and biodegradation. *Radiation Physics and Chemistry*, *72*(6), 745–750. doi: 10.1016/j.radphyschem.2004.05.045.

Saral Sarojini, K., Indumathi, M. P., and Rajarajeswari, G. R. (2019). Mahua oil-based polyurethane/chitosan/nano ZnO composite films for biodegradable food packaging applications. *International Journal of Biological Macromolecules*, *124*, 163–174. doi: 10.1016/j.ijbiomac.2018.11.195.

Kamali, Narjes, Rostamzad, Haniyeh, and Babakhani, Aria. (2019). Production and evaluation of smart biodegradable film based on carrageenan for fish fillet packaging. *Journal of Fisheries (Iranian Journal of Natural Resources)*, *72*(1), 85–95.

Kanmani, Paulraj, and Rhim, Jong-Whan. (2014). Development and characterization of carrageenan/grapefruit seed extract composite films for active packaging. *International Journal of Biological Macromolecules*, *68*, 258–266. doi: 10.1016/j.ijbiomac.2014.05.011.

Lei, Long, Zhi, Han, Xiujin, Zhang, Takasuke, Ishitani, and Zaigui, Li. (2007). Effects of different heating methods on the production of protein–lipid film. *Journal of Food Engineering*, *82*(3), 292–297. doi: 10.1016/j.jfoodeng.2007.02.030.

Lessani, Sajede. (2017). *Effect Clove Microcapsulate and Nano Clay of Applied Properties Carrageenan Film in Order to Packaging Fish.* (MSc). University of Guilan.

Lim, L. T., Auras, R., and Rubino, M. (2008). Processing technologies for poly(lactic acid). *Progress in Polymer Science*, *33*(8), 820–852. doi: 10.1016/j.progpolymsci.2008.05.004.

Limpan, Natthaporn, Prodpran, Thummanoon, Benjakul, Soottawat, and Prasarpran, Surasit. (2010). Properties of biodegradable blend films based on fish myofibrillar protein and polyvinyl alcohol as influenced by blend composition and pH level. *Journal of Food Engineering*, *100*(1), 85–92. doi: 10.1016/j.jfoodeng.2010.03.031.

Liu, Bin, Xu, Han, Zhao, Huiying, Liu, Wei, Zhao, Liyun, and Li, Yuan. (2017). Preparation and characterization of intelligent starch/PVA films for simultaneous colorimetric indication and antimicrobial activity for food packaging applications. *Carbohydrate Polymers*, *157*, 842–849. doi: 10.1016/j.carbpol.2016.10.067.

Lumdubwong, Namfone. (2019). Applications of starch-based films in food packaging. *Reference Module in Food Science*. doi: https://doi.org/10.1016/B978-0-08-100596-5.22481-5.

Ma, Qiumin, Davidson, P. Michael, Critzer, Faith, and Zhong, Qixin. (2016). Antimicrobial activities of lauric arginate and cinnamon oil combination against foodborne pathogens: Improvement by ethylenediaminetetraacetate and possible mechanisms. *LWT—Food Science and Technology*, *72*, 9–18. doi: 10.1016/j.lwt.2016.04.021.

Marcos, Begonya, Gou, Pere, Arnau, Jacint, and Comaposada, Josep. (2016). Influence of processing conditions on the properties of alginate solutions and wet edible calcium alginate coatings. *LWT*, *74*, 271–279. doi: 10.1016/j.lwt.2016.07.054.

Mariniello, Loredana, Di Pierro, Prospero, Esposito, Carla, Sorrentino, Angela, Masi, Paolo, and Porta, Raffaele. (2003). Preparation and mechanical properties of edible pectin–soy flour films obtained in the absence or presence of transglutaminase. *Journal of Biotechnology*, *102*(2), 191–198. doi: 10.1016/S0168-1656(03)00025-7.

Mushtaq, Mehvesh, Gani, Asir, Gani, Adil, Punoo, Hilal Ahmed, and Masoodi, F. A. (2018). Use of pomegranate peel extract incorporated zein film with improved properties for prolonged shelf life of fresh Himalayan cheese (Kalari/kradi). *Innovative Food Science and Emerging Technologies*, *48*, 25–32. doi: 10.1016/j.ifset.2018.04.020.

Nataraj, Divya, Sakkara, Seema, Hn, Meenakshi, and Reddy, Narendra. (2018). Properties and applications of citric acid crosslinked banana fibre-wheat gluten films. *Industrial Crops and Products*, *124*, 265–272. doi: 10.1016/j.indcrop.2018.07.076.

Noorbakhsh-Soltani, S. M., Zerafat, M. M., and Sabbaghi, S. (2018). A comparative study of gelatin and starch-based nano-composite films modified by nano-cellulose and chitosan for food packaging applications. *Carbohydrate Polymers*, *189*, 48–55. doi: 10.1016/j.carbpol.2018.02.012.

Peighambardoust, Seyed Jamaleddin, Peighambardoust, Seyed Hadi, Pournasir, Niloufar, and Mohammadzadeh Pakdel, Parisa. (2019). Properties of active starch-based films incorporating a combination of Ag, ZnO and CuO nanoparticles for potential use in food packaging applications. *Food Packaging and Shelf Life*, *22*, 100420. doi: 10.1016/j.fpsl.2019.100420.

Perdones, Ángela, Vargas, Maria, Atarés, Lorena, and Chiralt, Amparo. (2014). Physical, antioxidant and antimicrobial properties of chitosan–cinnamon leaf oil films as affected by oleic acid. *Food Hydrocolloids*, *36*, 256–264. doi: 10.1016/j.foodhyd.2013.10.003.

Picchio, Matías L., Paredes, Alejandro J., Palma, Santiago D., Passeggi, Mario C. G., Gugliotta, Luis M., Minari, Roque J., and Igarzabal, Cecilia I. Alvarez. (2018). pH-responsive casein-based films and their application as functional coatings in solid dosage formulations. *Colloids and Surfaces. Part A: Physicochemical and Engineering Aspects*, *541*, 1–9. doi: 10.1016/j.colsurfa.2018.01.012.

Pires, C., Ramos, C., Teixeira, B., Batista, I., Nunes, M. L., and Marques, A. (2013). Hake proteins edible films incorporated with essential oils: Physical, mechanical, antioxidant and antibacterial properties. *Food Hydrocolloids*, *30*(1), 224–231. doi: 10.1016/j.foodhyd.2012.05.019.

Qin, Yu, Wen, Peng, Guo, Hui, Xia, Dandan, Zheng, Yufeng, Jauer, Lucas, Poprawe, Reinhart, Voshag, Maximilian, Schleifenbaum, Johannes Henrich. (2019). Additive manufacturing of biodegradable metals: Current research status and future perspectives. *Acta Biomaterialia*. doi: 10.1016/j.actbio.2019.04.046.

Rhim, Jong-Whan, and Wang, Long-Feng. (2014). Preparation and characterization of carrageenan-based nanocomposite films reinforced with clay mineral and silver nanoparticles. *Applied Clay Science*, 97–98, 174–181. doi: 10.1016/j.clay.2014.05.025.

Rocca-Smith, Jeancarlo R., Marcuzzo, Eva, Karbowiak, Thomas, Centa, Jessica, Giacometti, Marco, Scapin, Francesco, Venir, E., Sensidoni, A., and Debeaufort, Frédéric. (2016). Effect of lipid incorporation on functional properties of wheat gluten based edible films. *Journal of Cereal Science*, *69*, 275–282. doi: 10.1016/j.jcs.2016.04.001.

Rodrigues, Delane C., Caceres, Carlos Alberto, Ribeiro, Hálisson L., de Abreu, Rosa F. A., Cunha, Arcelina P., and Azeredo, Henriette M. C. (2014). Influence of cassava starch and carnauba wax on physical properties of cashew tree gum-based films. *Food Hydrocolloids*, *38*, 147–151. doi: 10.1016/j.foodhyd.2013.12.010.

Romani, Viviane P., Machado, Ana Vera, Olsen, Bradley D., and Martins, Vilásia G. (2018). Effects of pH modification in proteins from fish (whitemouth croaker) and their application in food packaging films. *Food Hydrocolloids*, *74*, 307–314. doi: 10.1016/j.foodhyd.2017.08.021.

Romani, Viviane Patrícia, Olsen, Bradley, Pinto Collares, Magno, Meireles Oliveira, Juan Rodrigo, Prentice-Hernández, Carlos, and Guimarães Martins, Vilásia. (2019a). Improvement of fish protein films properties for food packaging through glow discharge plasma application. *Food Hydrocolloids*, *87*, 970–976. doi: 10.1016/j.foodhyd.2018.09.022.

Romani, Viviane Patrícia, Olsen, Bradley, Pinto Collares, Magno, Meireles Oliveira, Juan Rodrigo, Prentice, Carlos, and Guimarães Martins, Vilásia. (2019b). Plasma technology as a tool to decrease the sensitivity to water of fish protein films for food packaging. *Food Hydrocolloids*, *94*, 210–216. doi: 10.1016/j.foodhyd.2019.03.021.

Rostamzad, H., Abbasi Mesrdashti, R., Akbari Nargesi, E., and Fakouri, Z. (2019). Shelf life of refrigerated silver carp, Hypophthalmichthys molitrix, fillets treated with chitosan film and coating incorporated with ginger extract. *Caspian Journal of Environmental Sciences*, *17*(2), 143–153. doi: 10.22124/cjes.2019.3408.

Rostamzad, Haniyeh, Paighambari, Seyyed Yousef, Shabanpour, Bahareh, Ojagh, Seyyed Mahdi, and Mousavi, Seyyed Mahdi. (2016). Improvement of fish protein film with nanoclay and transglutaminase for food packaging. *Food Packaging and Shelf Life*, *7*, 1–7. doi: 10.1016/j.fpsl.2015.10.001.

Rostamzad, Haniyeh, and Rahimabadi, Eshagh Zakipour. (in press). Production and evaluation of chitosan film incorporated licorice extract for fish packaging. *Innovation in Food Science and Technology*.

Roy, Kripa, Thory, Rahul, Sinhmar, Archana, Pathera, Ashok Kumar, and Nain, Vikash. (2019). Development and characterization of nano starch-based composite films from mung bean (Vigna radiata). *International Journal of Biological Macromolecules*. doi: 10.1016/j.ijbiomac.2019.12.113.

Saadat, Saeida, Pandey, Gaurav, Tharmavaram, Maithri, Braganza, Vincent, and Rawtani, Deepak. (2019). Nano-interfacial decoration of halloysite nanotubes for the development of antimicrobial nanocomposites. *Advances in Colloid and Interface Science*, 102063. doi: 10.1016/j.cis.2019.102063.

Salgado, Pablo R., Molina Ortiz, Sara E., Petruccelli, Silvana, and Mauri, Adriana N. (2010). Biodegradable sunflower protein films naturally activated with antioxidant compounds. *Food Hydrocolloids*, *24*(5), 525–533. doi: 10.1016/j.foodhyd.2009.12.002.

Saloko, Satrijo, Darmadji, Purnama, Setiaji, Bambang, and Pranoto, Yudi. (2014). Antioxidative and antimicrobial activities of liquid smoke nanocapsules using chitosan and maltodextrin and its application on tuna fish preservation. *Food Bioscience*, *7*, 71–79. doi: 10.1016/j.fbio.2014.05.008.

Seydim, A. C., and Sarikus, G. (2006). Antimicrobial activity of whey protein based edible films incorporated with oregano, rosemary and garlic essential oils. *Food Research International*, *39*(5), 639–644. doi: 10.1016/j.foodres.2006.01.013.

Shojaee-Aliabadi, Saeedeh, Hosseini, Hedayat, Mohammadifar, Mohammad Amin, Mohammadi, Abdorreza, Ghasemlou, Mehran, Hosseini, Seyede Marzieh, and Khaksar, Ramin. (2014). Characterization of κ-carrageenan films incorporated plant essential oils with improved antimicrobial activity. *Carbohydrate Polymers*, *101*, 582–591. doi: 10.1016/j.carbpol.2013.09.070.

Sin, Lee Tin, Rahmat, Razak, Abdul, Rahman, and Abdul, Wan Aizan Wan. (2013). 5Mechanical properties of poly(lactic acid). In: L. T. Sin, A. R. Rahmat and W. A. W. A. Rahman (Eds.), *Polylactic Acid* (pp. 177–219): Oxford: William Andrew Publishing.

Siriprom, W., Sangwaranatee, N., Hidayat, R., Kongsriprapan, S., Teanchai, K., and Chamchoi, N. (2018). The physicochemical characteristic of biodegradable methylcellulose film reinforced with chicken eggshells. *Materials Today: Proceedings*, *5*(7, Part 1), 14836–14839. doi: 10.1016/j.matpr.2018.04.015.

Sogut, Ece, Ili Balqis, A. M., Nur Hanani, Z. A., and Seydim, Atif Can. (2019). The properties of κ-carrageenan and whey protein isolate blended films containing pomegranate seed oil. *Polymer Testing*, *77,* 105886. doi: 10.1016/j.polymertesting.2019.05.002.

Teixeira, Bárbara, Marques, António, Pires, Carla, Ramos, Cristina, Batista, Irineu, Saraiva, Jorge Alexandre, and Nunes, Maria Leonor. (2014). Characterization of fish protein films incorporated with essential oils of clove, garlic and origanum: Physical, antioxidant and antibacterial properties. *LWT—Food Science and Technology*, *59*(1), 533–539. doi: 10.1016/j.lwt.2014.04.024.

Viuda-Martos, M., Ruiz-Navajas, Y., Fernández-López, J., and Pérez-Álvarez, J. A. (2010). Effect of added citrus fibre and spice essential oils on quality characteristics and shelf-life of mortadella. *Meat Science*, *85*(3), 568–576. doi: 10.1016/j.meatsci.2010.03.007.

Weng, WuYin, Hamaguchi, Patricia Yuca, Osako, Kazufumi, and Tanaka, Munehiko. (2007). Effect of endogenous acid proteinases on the properties of edible films prepared from Alaska pollack surimi. *Food Chemistry*, *105*(3), 996–1002. doi: 10.1016/j.foodchem.2007.04.054.

Ye, Qianqian, Han, Yufei, Zhang, Jizhi, Zhang, Wei, Xia, Changlei, and Li, Jianzhang. (2019). Bio-based films with improved water resistance derived from soy protein isolate and stearic acid via bioconjugation. *Journal of Cleaner Production*, *214*, 125–131. doi: 10.1016/j.jclepro.2018.12.277.

Zhang, Liming, Liu, Zhanli, Wang, Xiangyou, Dong, Shuang, Sun, Yang, and Zhao, Zitong. (2019). The properties of chitosan/zein blend film and effect of film on quality of mushroom (Agaricus bisporus). *Postharvest Biology and Technology*, *155*, 47–56. doi: 10.1016/j.postharvbio.2019.05.013.

Zhang, Yi, Simpson, Benjamin K., and Dumont, Marie-Josée. (2018). Effect of beeswax and carnauba wax addition on properties of gelatin films: A comparative study. *Food Bioscience*, *26*, 88–95. doi: 10.1016/j.fbio.2018.09.011.

Zhang, Yue, Ma, Qiumin, Critzer, Faith, Davidson, P. Michael, and Zhong, Qixin. (2015). Physical and antibacterial properties of alginate films containing cinnamon bark oil and soybean oil. *LWT—Food Science and Technology*, *64*(1), 423–430. doi: 10.1016/j.lwt.2015.05.008.

Zhao, Yi, Xu, Helan, Mu, Bingnan, Xu, Lan, and Yang, Yiqi. (2016). Biodegradable soy protein films with controllable water solubility and enhanced mechanical properties via graft polymerization. *Polymer Degradation and Stability*, *133*, 75–84. doi: 10.1016/j.polymdegradstab.2016.08.003.

Zhuang, Fengchen, Li, Xiang, Hu, Jinhua, Liu, Xiaoming, Zhang, Shuang, Tang, Chunya, and Zhou, Peng. (2018). Effects of casein micellar structure on the stability of milk protein-based conjugated linoleic acid microcapsules. *Food Chemistry*, *269*, 327–334. doi: 10.1016/j.foodchem.2018.07.018.

3 Recent Advances in the Production of Multilayer Biodegradable Films

Daiane Nogueira, Gabriel da Silva Filipini and Vilásia Guimarães Martins

CONTENTS

3.1 INTRODUCTION

In the packaging industry, the use of multilayers is considered one of the greatest technological revolutions in this sector. This technique produces self-assembled and laminated materials in which its properties aggregate and improve, increasing the use in the most diverse industrial sectors (Fereydoon and Ebnesajjad, 2014).

The multilayer packages have been widely used for plastics from fossil origin, and the use of these large-scale polymers is associated with a wide disposal and a large degradation time, resulting in environmental and social problems (Avio, Gorbi, and Regoli, 2017). Based on this fact, the demand for new biodegradable packaging has been intensified in order to reduce damage to the environment.

The transfer of technology used in synthetic packaging has been extensively researched to try to replicate the same in the production of biodegradable packaging. However, there is as yet no method to extend the production of biodegradable multilayer packaging on an industrial scale due to the technological problems associated with the lack in mechanical and barrier properties (Flôres et al., 2017), as well as the greater time of production needed when compared to that of some synthetic polymers. However, in recent decades, several studies have been done on the use of different forms to produce multilayer films, as well as optimizing them to make possible the production on a larger scale to make it economically viable.

With regard to global pollution, a considerable increase has been observed in the pollution of the environment by the unlimited use, incorrect disposal of, and difficulties in the recycling of packages made of synthetic polymers from fossil origin such as petroleum (Rahmani et al., 2013).

In this context, the substitution of synthetic polymers with biodegradable and sustainable polymers is an alternative to aid the reduction of environmental pollution. However, the use of these polymers is not diffuse due to some impediments such as technological properties. Regarding the production of multilayer packaging, several parameters can influence the properties of the packages, such as materials used (proteins, polysaccharides, lipids) production methods (casting, extrusion and electrospinning), as well as different breeding strategies (plasma, UV and crosslinking).

Figure 3.1 shows an illustration of the overlapping polymer layers forming a multilayer.

In general, the production of multilayer polymers occurs due to the complementarities of the properties of the polymer layers forming a unique structure, in order to overcome the deficiencies of each polymer, trying to find a new and sustainable solution that can make bioplastics more profitable and competitive in terms of performance competing with the conventional plastics (Sanyang et al., 2016).

Studies about multilayer biodegradable films are still poorly investigated in the literature. In general, studies on biodegradable films are made for monolayer films, which limits their application as packaging due to their low technological properties. Currently, the use of multilayer films is still very directed to the elaboration of packaging from the use of synthetic materials. In this way, technology transfer is necessary so that biodegradable materials can be optimized in order to make them competitive, since it is already proven that the use of multilayers promotes improvements in their properties.

This chapter discusses recent information on biodegradable multilayer films for application as food packaging, as well as the main materials and techniques used in the development of these films in recent years. It also reviews the main properties evaluated in these films, strategies that can be used to improve their properties, advantages and disadvantages, as well as some applications of multilayers already evaluated as food packaging.

FIGURE 3.1 Multilayer film. (Produced by the co-author Gabriel da S. Filipini.)

3.2 MATERIALS USED IN THE DEVELOPMENT OF MULTILAYERS

Polymers have been playing an important role as materials in modern applications in almost all manufacturing sectors, including plastic for food packaging (Zhang et al., 2017). The advantages of polymers in comparison with other materials can be attributed to their adjustable properties, ease of processing, resistance to chemical and physical degradation and low cost (Colwell et al., 2017).

The industrial production of plastics is made from monomers supplied from petrochemical resources, which are not renewable and, in the long run, will not supply sufficient raw materials at economically viable costs (Zhang et al., 2017). Environmental impacts caused by non-degradable materials have raised concerns about air, water and soil pollution and pose significant challenges to sustainable development (Wang and Rhim, 2015). Because the demand for sustainability and environmentally safe materials is growing, several studies are being developed using biodegradable polymers applied in the development of packaging to produce materials that can rapidly degrade and mineralize completely (Bashir et al., 2018; Medina-Jaramillo et al., 2017).

According to their origin, biopolymers can be classified into three main classes: (1) biopolymers from agricultural resources, such as polysaccharides, lignocellulosic products, proteins, lipids and free fatty acids; (2) biopolymers obtained by microbial fermentation, such as pullulan and polyhydroxyalkanoates; and (3) chemically synthesized biopolymers employing monomers obtained from natural raw materials such as poly(lactic acid) (Garavand et al., 2017; Trinetta, 2016).

Numerous researches involving different biopolymers in the development of biodegradable plastics have been reported, including gelatin and chitosan (Rivero, García, and Pinotti, 2009), poly(lactic acid) (PLA) (Muller, González-Martínez, and Chiralt, 2017), collagen (Baba Ismael et al., 2017), soy protein (González and Igarzabal, 2013), zein (Zuo et al., 2017), agar (Vejdan, Mahdi, Adeli, and Abdollahi, 2016), corn starch and polyhydroxybutyrate (PHB) (Fabra, López-Rubio, Ambrosio-Martín, and Lagaron, 2016), sodium alginate (Zhuang et al., 2018), among others.

Vejdan et al. (2016) developed a biodegradable gelatin/agar bilayer for food packaging. Due to its low cost and film-forming ability, as well as high availability and biodegradability, gelatin is an approved protein by the Food and Drug Administration (FDA) and considered a good candidate in the preparation of food packaging (Deng et al., 2018). Gelatin is a water-soluble biopolymer obtained from the partial hydrolysis of collagen via an acid or alkaline process that divides it into type A and type B (Lin et al., 2017; Qiao et al., 2017). It can be obtained from the skins and bones of cattle, swine and fish waste. The use of fish gelatin has recently increased due to the increased risk of transmission of pathogens such as bovine spongiform encephalopathy (Vejdan et al., 2016). However, studies have found that fish gelatin films are fragile (Qiao et al., 2017) due to the polar amino acid groups present in their structure and the extensive intermolecular forces (hydrogen bonds and hydrophobic interactions), leading to high solubility in water and water vapor permeability, which limit their applications as packaging material (Lin et al., 2017).

Three-layer films containing bovine gelatin as the main component were produced by Martucci and Ruseckaite (2010). The results obtained indicate that the association of gelatin films modified with environmentally safe reagents such as dialdehyde starch (DAS), glycerol and sodium montmorillonite, can provide a new multilayer material with modulated properties and with the advantage of being biodegradable.

Polysaccharides have been extensively studied in the development of biodegradable monolayer and multilayer films. Elsabee and Abdou (2013) and Fajardo et al. (2010) reported the use of chitosan, Ferreira et al. (2017, 2014) used exopolysaccharide of FucoPol, Li et al. (2013) and López and García (2012) evaluated the use of starch, and Vejdan et al. (2016) used agar in the development of biodegradable films. This interest in part is due to its non-toxicity and high availability, but also because polysaccharides form films with excellent gas, aroma and lipid barriers, and have good mechanical properties; however, their water vapor barrier properties are low due to their hydrophilic nature (Ferreira et al., 2014).

Different strategies have already been suggested to overcome these weaknesses, including the development of polysaccharide bilayers. Thus, agar, a polysaccharide extracted from marine red algae of the Rhodophyceae class (Malagurski et al., 2017), which is biocompatible, has high mechanical resistance and good film-forming properties, making it a good alternative. Agar is composed of agarose and agaropectin molecules. Agarose is a neutral and linear molecule of 3,6-anhydro-L-galactose units linked to β-1,3-linked-D-galactose and α-1,4 and mediates the gelation of agar hydrogels, while agaropectin is a charged, sulfated, branched and non-gelling unit (Garrido et al., 2016; Malagurski et al., 2017). The relative insolubility and the ability of this polysaccharide to produce strong gels may help to reduce the water vapor permeability of the agar films (Vejdan et al., 2016).

Zuo et al. (2017) developed a bilayer film using a zein (Z) layer casted on the corn-wheat (CW) starch film. Zuo et al. reported that, based on the results obtained, CW/Z bilayer film is a potential packaging material for dry vegetables, oil sauces and spices during the production of instant food. Corn starch is a polysaccharide and consists of two major fractions, linear amylose and branched amylopectin; both compounds present in glucose residues, their physicochemical properties differ due to their structures (Masina et al., 2017). Films prepared with high amounts of amylose starch usually exhibit excellent mechanical performance because amylose has stronger gelation properties than amylopectin and linear amylose chains interact with hydrogen bonds to a greater degree than amylopectin branched chains (Wang et al., 2017).

Corn starch has been the most widely used raw material for the production of biodegradable plastics in recent years. This is possibly because corn is the main source of starch produced worldwide (approximately 65%), followed by sweet potatoes (13%) and cassava (11%). Corn is the most important starch source in the United States (99%), Europe (46%), Asia and Brazil (Luchese et al., 2017). The use of starch in the plastic sector has been motivated by low cost and wide availability, high purity, non-toxicity, biodegradability, and environmental compatibility, as well as being known for its excellent film-forming and oxygen barrier properties (Fabra et al., 2016a; Wang et al., 2017).

Although native starch is not thermoplastic, it can be processed with existing plastic processing technologies, while disrupting the granular structure of starch by mechanical and thermal energy in the presence of water or other plasticizers. Starch-based materials have limited applications due to their poor moisture barrier properties, brittleness and low tensile strength. A commonly used approach to overcoming such disadvantages and providing additional functional properties is to combine the starch with other natural biopolymers to form composite materials (Ren et al., 2017).

Obtaining polysaccharides from animal, algal or plant sources depends on climatic or seasonal impacts, so obtaining microbial source polysaccharides becomes a good alternative. These polysaccharides have become attractive because of their high variability of molecular structure and their availability and they have a wide range of applications, ranging from the chemical industry to food, medicines and cosmetics (Ferreira et al., 2014).

Many studies have been carried out to evaluate the properties of FucoPol polysaccharide and its use in the development of biodegradable films (Araújo et al., 2016; Ferreira et al., 2017, 2016, 2014; Torres et al., 2015). FucoPol is a newly published bacterial exopolysaccharide (EPS) produced by Enterobacter A47 (DSM23139) using glycerol as the only carbon source (Araújo et al., 2016). According to Freitas et al. (2011), it is a high-molecular-weight heteropolysaccharide composed of neutral sugars (fucose, galactose, glucose), an acidic sugar (glucuronic acid) and acyl groups (acetate, succinate and pyruvate). This biopolymer exhibits anionic character due to glucuronic acid, along with succinate and pyruvate. FucoPol has the ability to form films that are transparent with a brown tone, hydrophilic with high water vapor permeability and good gas barrier properties (CO_2 and O_2) (Ferreira et al., 2014).

Ferreira et al. (2016) developed and characterized FucoPol and chitosan bilayer films. The bilayers were dense and homogeneous and their properties have improved when compared to FucoPol monolayer films.

Among the biodegradable polymers used to produce multilayer films, poly(lactic acid) (PLA) has been one of the most promising, including for commercial use as a substitute for low density polyethylene (LDPE) and high density polyethylene (HDPE), polystyrene (PS) and polyethyleneterephthalate (PET) (Peelman et al., 2013). PLA is obtained from the controlled depolymerization of the lactic acid monomer obtained from the fermentation of carbohydrates, such as sugar and corn, which are renewable and biodegradable (Siracusa et al., 2008). It is a versatile, recyclable and compostable polymer with high transparency and molecular weight, good processing capacity and high resistance to water solubility (Sanyang et al., 2016).

All these characteristics make PLA one of the polymers most used in the elaboration of multilayers in order to improve the properties of the other polymers used. González and Igarzabal, (2013), Muller et al. (2017), Nguyen et al. (2016), Nilsuwan et al. (2017), Sanyang et al. (2016) and Scaffaro, Sutera, and Botta (2018) developed multilayer films using PLA as one of the layers adding improvements in the properties of the other polymers used in the multilayer production.

According to Li et al. (2013), chitosan is the second most abundant polysaccharide found in nature after cellulose, is a linear polysaccharide consisting of (1,4)-linked 2-amino-deoxy-b-d-glucan and is the deacetylated derivative from chitin. In addition, it is non-toxic, biodegradable, biofunctional, biocompatible polysaccharide with strong antimicrobial and antifungal properties. These functionalities, antimicrobial and general properties, are responsible for the wide application of chitosan in multilayer films, especially when the objective of the study is to develop active packaging to extend the shelf life. Valencia-Sullca et al. (2017) developed a starch and chitosan bilayer, Ferreira et al. (2018) studied the bilayer films composed of a bacterial exopolysaccharide (FucoPol) and chitosan, and Zhuang et al. (2017) used polyvinyl alcohol (PVA), sodium alginate (SA) and chitosan (CHI) in the development of multilayer films.

Alginate is a polysaccharide with an attractive ability to form films due to its non-toxicity, biodegradability, biocompatibility and low price (Vu and Won, 2013). Many studies have been carried out to evaluate its functional properties, thickness, stabilization, suspension, film formation, gel production and emulsion stabilization. Alginate is considered a good film-forming material because of its linear structure which can form strong films and suitable fibrous structures in the solid state (Tavassoli-Kafrani, Shekarchizadeh, and Masoudpour-Behabadi, 2016).

Chitosan films have advantages such as gas selective permeability (CO_2 and O_2) and good mechanical properties. However, they have the disadvantage of being highly permeable to water vapor, which limits their use, since effective moisture transfer control is a desirable property for most foods, especially in a humid environment (Elsabee and Abdou, 2013).

Among the polymers obtained through fermentation processes are the family of polyhydroxyalkanoates (PHA), which are biodegradable thermoplastic polymers produced by a wide range of microorganisms. Synthesis of the polymer occurs in the microbial cells through a fermentation process and then it is harvested using solvents such as chloroform, methylene chloride or propylene chloride. Polyhydroxybutyrate (PHB) is the most common and known PHA (Arrieta et al., 2014c). PHB is a highly crystalline and brittle material and can be degraded into different by-products under anaerobic (H_2O and CH_4) and aerobic (H_2O and CO_2) conditions.

PLA, PHB and their blends are biodegradable materials commonly investigated for food packaging applications (Arrieta et al., 2014b). Arrieta et al. (2015) used PHB in a blend with PLA to increase its crystallinity and improve the barrier properties of PLA films. Fabra et al. (2016) used PHB to reduce the hydrophilicity of multilayer films of thermoplastic starch. The study resulted in the improvement of the barrier properties of the multilayer, especially in reducing the water vapor permeability. Despite the fact that it presents these improvements in the properties of monolayer and multilayer films demonstrated by several studies, PHB has a high cost compared to traditional polymers, making it impossible to use on a large scale (Carofiglio et al., 2017).

According to Dehghani et al. (2018), polysaccharides and proteins are able to form cohesive films. In general, lipids are used as coatings or incorporated into biopolymers to form composite or multilayer films, providing a better water vapor barrier because of their low polarity.

3.3 PROPERTIES OF MULTILAYER

To be used as food packaging, multilayer films must have important characteristics, such as mechanical, optical, structural, thermal and barrier properties similar or equal to those presented by non-biodegradable films.

The integrity of films used as packaging materials must be maintained during the processing, transportation and handling of packaged foods (Hosseini et al., 2015). The most evaluated mechanical properties in multilayer films are tensile strength (MPa), that measures an applied force in a determined area unit breaking, and elongation (%), given as the percentage of elongation in a variation of film length (Wihodo and Moraru, 2012). In general, the properties of the biodegradable multilayer films depend on the nature of the material used and their structure, which is strongly related to the ability of the polymer to form strong or numerous bonds between the polymer chains.

Films based on biopolymers generally exhibit high water vapor permeability (WVP) (Garavand et al., 2017). In this way, the multilayer films approach allows the creation of new polymeric arrangements that can be established to prevent the migration of water into food packaging materials. Water solubility is considered an important feature of the dependence of films for food packaging applications, especially when the water activity is high or when the film is in contact with water and acts as a food protector (Vejdan et al., 2016).

The thickness of films generally varies from 80 μm to 200 μm (Garavand et al., 2017), however, in multilayer films the thickness is generally higher, and depends heavily on the method used to make the matrix, in addition to the concentration and arrangement structure of the polymer used. Thickness plays a key role in some barrier properties, including WVP and gas transportation. Thickness may influence film WVP due to differences between water vapor pressure and moisture accumulation at the film/air interface, since most materials used in multilayer manufacturing are from hydrophilic origin (Garavand et al., 2017). It is important to note that, when discussing and comparing the results of their studies, many authors do not take into account the thickness of the matrices, which may lead to possible errors in their evaluations.

Table 3.1 presents some properties of the multilayer films. Vejdan et al. (2016) evaluated gelatin/agar bilayer films with TiO_2 nanoparticles, observing an increase in tensile strength (TS) of the added bilayers of up to 0.5 g/100 g of TiO^2. However, the elongation (E) significantly reduced with the addition of nanoparticles. The observed improvement in TS of the bilayers may be related to the reinforcing effect of the high-strength metal nanoparticles dispersed homogeneously in the biopolymer.

Biodegradable three-layer gelatin film was obtained by heat compression of piled dialdehyde starch (DAS), cross-linked and plasticized-gelatin films (Ge-10DAS) outer layers and sodium montmorillonite (MMt) plasticized-gelatin film (Ge-5MMt) inner layer, prepared by Martucci and Ruseckaite (2010). They investigated the impact of lamination on mechanical and barrier properties compared to individual films. The TS values of the three-layer film increased significantly ($p < 0.05$) compared to single-layer films. This was attributed to the contribution of the inner layer of Ge-MMt. In addition, it was found that elongation (E) values did not vary significantly with lamination ($p > 0.05$), indicating that multilayer film retains extensibility and flexibility of individual components. The TS value of the multilayer prepared in this study was in the same range as the low density polyethylene (8.6–17 MPa). The lamination significantly improved the water vapor barrier properties, evidencing the decrease of the WVP value of the multilayer. According to Martucci and Ruseckaite, when the

TABLE 3.1
Mechanical and Barrier Properties of Some Multilayers

Multilayers	CPFS	Methods	T (μm)	TS (MPA)	E (%)	WVP	WS (%)	References
Cassava starch/ chitosan	100% Cassava starch/1.0% chitosan	Casting and compression molding	221	20	2.3	7.3 g mm kPa^{-1} h^{-1} m^{-2}	NR	Valencia-Sullca et al., 2017
Chitosan/gelatin	2% chitosan/ 2.5% gelatin	Casting	105	8.55	41	1.1×10^{13} kg/Pa s m	25.2	Pereda et al., 2011
Soy protein/ poly(lactic acid) PLA	0.75% soy protein/1.2% PLA	Casting	54	13.69	1.3	2.3×10^{-11} gm Pa^{-1} s^{-1} m^{-2}	40.1	González and Igarzabal, 2013
Corn-wheat starch/ zein	6% starch/6% zein	Casting	126.7	9.39	77.1	5.7×10^{-11} g/m s Pa	NR	Zuo et al., 2017b
Gelatin/agar	4% gelatin/1.5% agar	Casting	93.3	10.8	48.2	3.3×10^{-10} g/m^2 s Pa	29.8	Vejdan et al., 2016
Gelatin/cross-linked gelatin / reinforced gelatin	10% gelatin/10% gelatin + glicerol/10% gelatin + MMt	Heat compression	0.6	8	110	0.8×10^{13} kg m Pa^{-1} s^{-1} m^{-2}	31	Martucci and Ruseckaite, 2010
Corn starch/PHB	starch 1w/water 0.5 w/10% PHB	Heat compression	NR	6.9	5.8	2.6×10^{13} kg m Pa^{-1} s^{-1} m^{-2}	NR	Fabra et al., 2016b
Fish gelatin/ PLA	3.5% gelatin/5% PLA	Casting	101	27.23	11.6	2.7×10^{-11} g m^{-1} s^{-1} Pa^{-1}	NR	Nilsuwan; Benjakul and Prodpran, 2017
Thermoplastic corn starch/ PCL	5% starch/ starch + 5% PCL	Compression molding	325	8.5	32	0.9 g mm kPa^{-1} h^{-1}/m^{-2}	0.1	Ortega-Toro et al., 2015
PLA/MaterBi	7 g PLA/ 4.5 MaterBi	Co-extrusion	76	28	50	NR	NR	Scaffaro; Sutera and Botta, 2018
SPS/PLA	8% SPS/10% PLA	Casting	NR	13.65	15.5	$0.2\times10^{-10}\times$g $s^{-1}$$m^{-1}$ Pa^{-1}	20	Sanyang et al., 2016

(Continued)

TABLE 3.1 (CONTINUED)
Mechanical and Barrier Properties of Some Multilayers

Multilayers	CPFS	Methods	T (µm)	TS (MPA)	E (%)	WVP	WS (%)	References
Kafirin protein/ gelatin	5% kafirin/ 10% gelatin	Casting	143	8.72	43.5	$2{,}02 \times 10^{-10} \times g\ s^{-1}m^{-1}\ Pa^{-1}$	NR	Wang et al., 2018
PLA/whey protein	PLA Film/ 10% whey protein	Casting	68	8.6	222	NR	NR	Cinelli et al., 2014
BC/zein	(NR)BC/30% zein	Eletrospinning and compression molding	73	63.7	1.1	NR	NR	Wan et al., 2016
ALG/PVA+ chitosan	1% SA/ 2% chitosan+5% PLA	Casting	750	14.16	111.5	$188.1\ g.mm.m^{-2}\ d^{-1}\ kPa^{-1}$	43.5	Zhuang et al., 2018
FucoPol/chitosan	1.5% FucoPol/ 1% chitosan	Casting	NR	11.9	38.4	1.6×10^{-11} mol/ms Pa	33.6	Ferreira et al., 2016
ALG/BOPLA/PEI	0.2% ALG/BOPLA*/ 0.1% PEI	Dipped	3.9	137.38	91.6	$1.8 \times 10^{-6}\ g\ cm/cm^2\ s\ Pa$	NR	Gu et al., 2013
BI PI+Z	3% PI/20% Z	Casting	435	3.4	2.3	$24.1\ g.mm/day.m^2.kPa$	31.3	Nogueira and Martins, 2018
BI Z+WG	20% Z/5% WG	Casting	611	2.6	1.2	$13.7\ g.mm/day.m^2.kPa$	16.5	Nogueira and Martins, 2018
BI PI+WG	3% PI/5% WG	Casting	197	3.4	121.2	$24.1\ g.mm/day.m^2.kPa$	36.1	Nogueira and Martins, 2018

CPFS: concentration of the polymer in the filmogenic solution, T: Thickness, TS: tensile strength, E: elongation, WS: water solubility, WVP: water vapor permeability, PHB: polyhydroxybutyrate, PLA: polylactide, PCL: polycaprolactone, SPS: sugar palm starch, BC: bacterial celulose, PVA: polyvinyl alcohol, ALG: sodium alginate, BOPLA: biaxially oriented poly(lactic acid), PEI: polyethyleneimine, *The BOPLA films were provided by Shenzhen Esun Industrial Co., Ltd., China, BI PI+Z: bilayer of protein isolate from hake and zein, BI Z+WG: bilayer of zein and wheat gluten, BI PI+WG: bilayer of protein isolate of hake and wheat gluten, NR: no reported.

lamination is performed by heat compression of the stacked films, diffusion occurs between the layers, filling pores and irregularities and allowing the formation of a dense structure that decreases the permeability to the vapor. The Ge-MMt film disintegrated almost completely after immersion in water, while Ge-DAS exhibited the lowest solubility, not differing significantly from the water solubility (WS) of the multilayer.

Zuo et al. (2017) investigated the properties of bilayer films based on a blend of wheat starch and corn starch with the addition of a zein layer. It was found that TS values of the bilayers were decreased, while those of E were improved with the addition of the zein layer to the starch film. The addition of the zein layer significantly reduced the WVP of the starch films due to the hydrophobicity of the zein molecules.

The process of bilayer films production most frequently reported in the literature is the casting method, which in brief, involves dissolving the biopolymer and mixing it with plasticizers and/or additives to obtain a film-forming solution, which is cast onto plates and then dried by taking off the solvent. The extrusion is another method to produce films and for synthetic films is the process most used in the industry. This method is based on the thermoplastic behavior of proteins at low moisture levels. Films can be produced by extrusion followed by hot pressing at temperatures normally above 80°C. This may affect film properties, but their use would increase the commercial potential of the films, providing a number of advantages over casting solutions, for example, working in a continuous system with immediate control of process variables such as temperature, humidity and size/shape (Gómez-Guillén et al., 2009).

The comparison of gelatin films produced by hot extrusion/hot pressing or casting was investigated by Park, Whiteside and Cho (2008) who found that the extruded films had lower tensile strength and higher elongation and water vapor permeability values than casting films parameters.

The mechanical and barrier properties are the properties most studied in biodegradable films; however, many studies have carried out a more complete characterization of the obtained films, also evaluating their structural and thermal properties. Another important feature of these films is biodegradability. According to the "European Standard EN13432," for a packaging to be considered biodegradable, at least 90% of the material must decompose by biological action in a period of six months (Biodegradability, Organizations, and Standard, 1998). On the other hand, this characteristic is rarely evaluated. Many authors describe their films in the literature as biodegradable without evaluating them.

3.4 TECHNIQUES FOR PREPARATION OF MULTILAYER FILMS

Among the methods used to produce multilayer films in the last decades, we can mention the casting technique (Pereda et al., 2011), extrusion (Scaffaro, Sutero, and Botta, 2018), lamination (Cho et al., 2010) and electrospinning (Deng et al., 2018), as well as the use of technologies that can add functionality and improve the properties of the films through chemical modifications, such as addition of compounds and crosslinking, physics, with the use of cold plasma, UV lights and ultrasonic waves as well as the use of blends by the combination of different polymers.

3.4.1 Casting Method

Among the methods of multilayer films production in the laboratory, the method of casting stands out as the most used. This method consists of the solubilization of one or more polymers in a solvent, taking into account the addition of plasticizers as well as parameters of greater solubility of the polymers, such as pH, agitation and temperature (Nilsuwan, Benjakul, and Prodpran, 2017).

The production of multilayer films by the casting method consists in an adaptation of the original method, in which at the final moment of the drying of the first film a second film-forming solution is poured thereon and both are subjected to drying, resulting in a single composite material of different layers (Rivero et al., 2009), as shown in Figure 3.2.

Some studies that produced bilayers and multilayers films using the casting method are presented in Table 3.2.

The casting method was the first method used in film production, and it is still being used nowadays, due to its easy operation on a laboratory scale, its repeatability and reproducibility as well as the possibility of elaboration of functional packaging such as active (Pereda, Ponce, Marcovich, Ruseckaite, and Martucci, 2011) and intelligent packaging (Musso, Salgado, and Mauri, 2016).

However, the use of this method on a large scale is still a challenge to be overcome, given the longer production time required due to the drying process, which is around 24 hours, making it necessary to optimize the method so that it can be used industrially. One example of how to optimize this process is to use the casting tape that promotes drying in smaller periods of time (De Moraes, Scheibe, Sereno, and Laurindo, 2013). Taking into account the limitations of this method, other techniques can be used in the production of multilayer, such as extrusion (Scaffaro, Sutera, and Botta, 2018), lamination (Ghanbarzadeh and Oromiehi, 2009) and electrospinning (Deng, Kang, Liu, Feng, and Zhang, 2018).

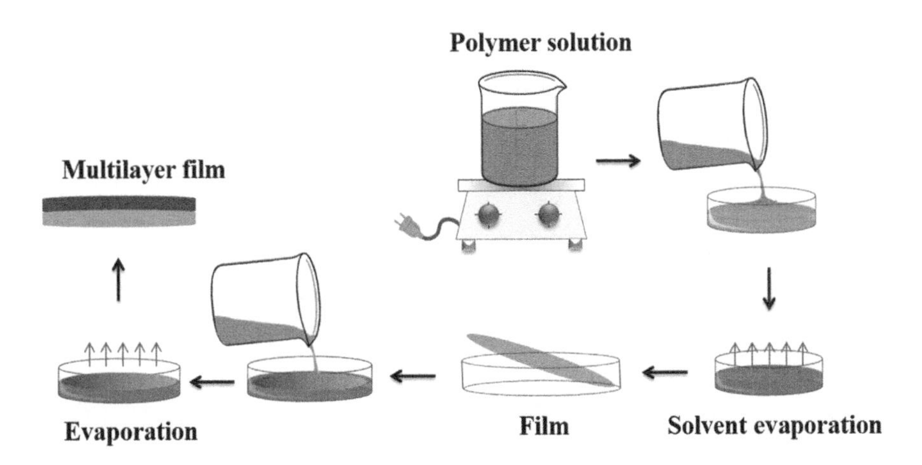

FIGURE 3.2 Production of multilayer films by casting. (Produced by the co-author Gabriel da S. Filipini.)

TABLE 3.2
Films Produced by the Casting Method

Polymers	References
PLA and microfibrillated cellulose (MFC)	Meriçer et al., 2016
Sugar palm starch and poly(lactic acid)	Sanyang et al., 2016
Cassava and starch-chitosan	Valencia-Sullca et al., 2017
Soy protein and poly(lactic acid)	González and Igarzabal, 2013
Chitosan and gelatin composites	Pereda et al., 2011
Fish gelatin and poly(lactic acid)	Nilsuwan et al., 2017

3.4.2 Extrusion Method

Thermoplastic extrusion is characterized as a technique widely used in the processing of plastic polymeric materials (Gilfillan, Moghaddam, Bartley, and Doherty, 2015), which consists of a thermomechanical process that transforms the materials physically and chemically due to the thermal and mechanical conditions generated by the heating. It is a processing technology based on multiple operations such as mixing, conveying, shearing, plastification, melting, baking, denaturing, fragmentation, texturing and modeling. (Gilfillan et al., 2015). This technique is a versatile and low-cost process that uses small volumes of water generating a small volume of effluents, and it is considered to be of low environmental impact. In contrast, the use of biodegradable materials in this process results in changes in the conformations of the polymers and often generate losses of the inherent properties (Gómez-Guillén et al., 2009).

The extruder is an item of equipment that operates through a thread inserted into a heated cylindrical cannon that carries the pelletized or powdered material, which is compacted and melted, forming a viscous polymer that is forced under high pressure through a draw and can produce different materials such as pellets, tubes and leaves (Briassoulis et al., 2017) as shown in Figure 3.3.

During the production of extrudates, a mix of polymers, plasticizers and water is generally used in the formulation. Operating parameters related to the flow ability of each material, such as temperature, time, torque and spin speed, are taken into account in the extrusion process (De Paoli, 2009). For the production of bilayers or

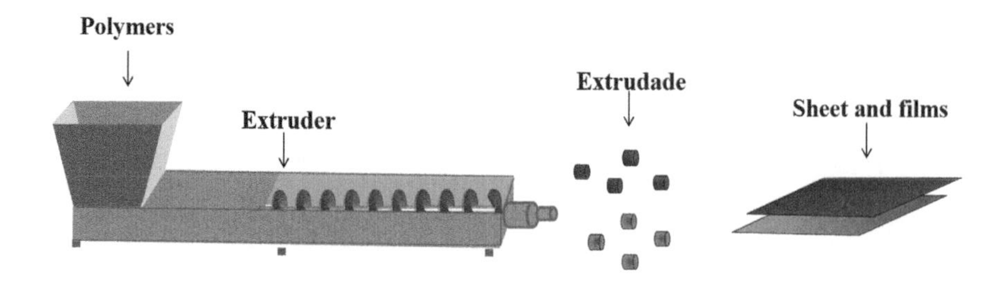

FIGURE 3.3 Process of extrusion. (Produced by the co-author Gabriel da S. Filipini.)

sheets from the extrusion process, a thermo pressing operation is used, in which the polymer pellets are melted and molded, taking into account operation parameters such as application of pressure, temperature and time in order to promote the melting between the materials (Valencia-Sullca et al., 2017).

In addition to the extrusion process, co-extrusion is a technique where there are two or more extruders linked in the output matrix, providing a multilayer final product. Figure 3.4 shows a multilayer production by co-extrusion.

In some cases, due to the incompatibility of the formed layers, it is necessary to use techniques to promote the joining of them; in this context the use of plasma on the surface of the films has shown good results in promoting the adhesion of the different layers and increasing mechanical properties (Honarvar et al., 2017). Another way of making the production of multilayers viable is by inserting an adhesive layer between the layers to promote adhesion (Sarantópoulos et al., 2002).

In search of innovative methods in the production of multilayer films, Scaffaro, Sutera, and Botta (2018) evaluated the produce of bilayer PLA/MaterBi using a co-extrusion blown film line with air cooling equipped with a Brabender in which it presented a good performance in the production of films, since the polymers were compatible with the method, showing good adhesion and a higher stiffness when compared to the monolayer films MaterBi, and a greater elongation in comparison with PLA.

In the production and characterization of bilayer films by thermo-compression using cassava starch and chitosan, Scaffaro, Sutera, and Botta observed that the films produced exhibited good interfacial adhesion and better mechanical resistance

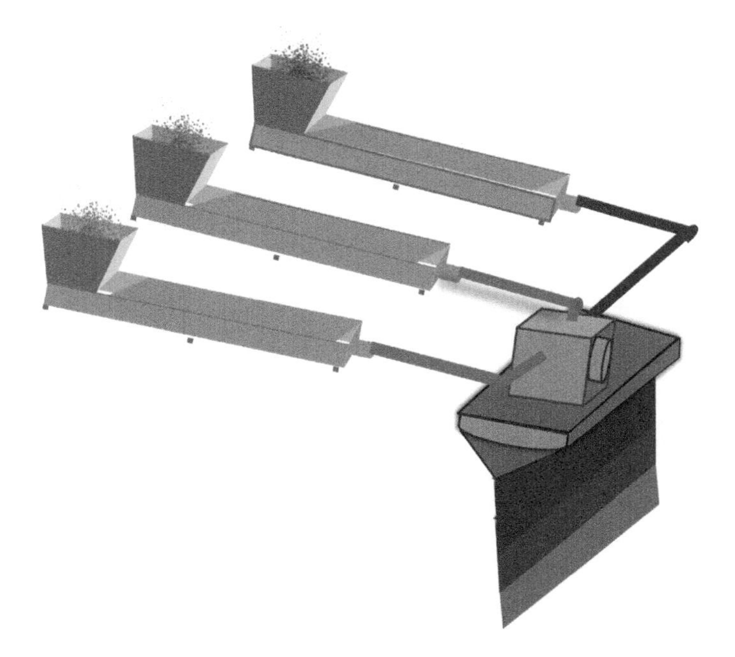

FIGURE 3.4 Multilayer production by co-extrusion. (Produced by the co-author Gabriel da S. Filipini.)

than the starch monolayers, although they presented lower elongation and less transparency (Chiralt et al., 2018).

3.4.3 Electrospinning Method

Among the innovations used in the production of multilayer bioplastics, the electrospinning technique is an alternative that can be used for the improvement of the barrier and the functional performance of the thermodynamically immiscible biodegradable polymers and the formation of coatings (Fabra, López-Rubio, and Lagaron, 2014).

The electrospinning system consists basically of a needle syringe, a collection screen and a high-voltage source as shown in Figure 3.5.

Among the features that comprise this equipment, a syringe is used for a polymer solution used, releasing the small droplets through the needle. In this system, there is an application of a certain level of separation power between a syringe and the collector, enabling a solution migration of polymers that are attracted to a needle collector and the more the solvent evaporates, the more the electrically polymeric fibers charged cells are formed without a collector (Li et al., 2004). In the production of multilayers by this method, some factors can influence the structure and morphology of the fibers, such as the applied voltage, needle-to-collector distance, polymer concentration, viscosity and solution flow (Dukali et al., 2014).

The use of the electrospinning technique for the production of bilayer films was investigated by Figueira et al. (2016) to obtain a dressing using hyaluronic acid and

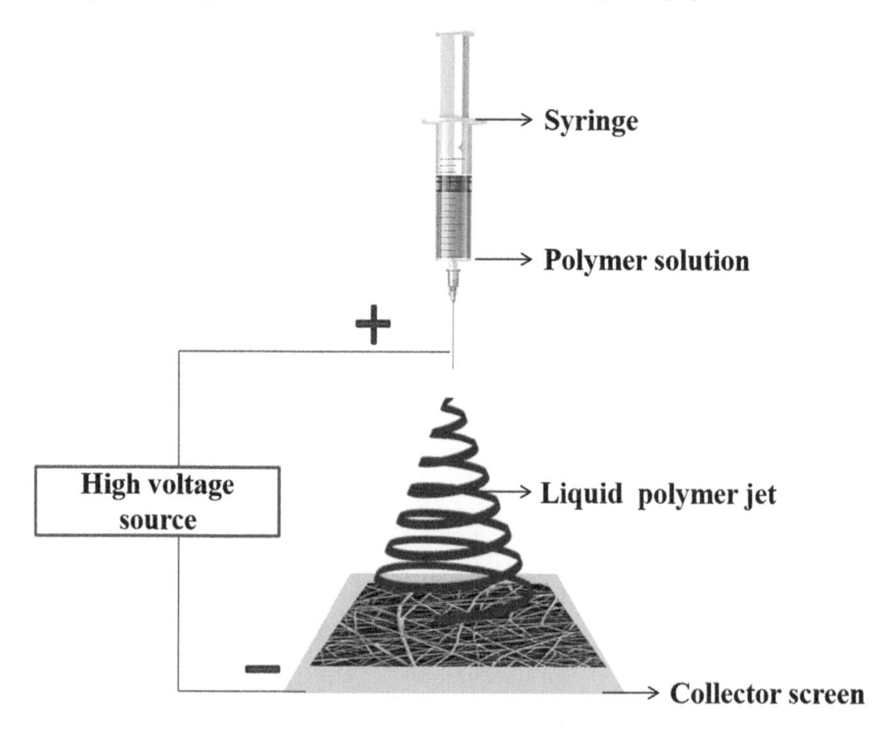

FIGURE 3.5 Electrospinning system. (Produced by the co-author Gabriel da S. Filipini.)

polycaprolactone in the top layer and chitosan, zein and salicylic acid in the lower layer. The study revealed appropriate mechanical properties, controlled water loss and a suitable salicylic acid release profile and the film showed anti-inflammatory and antimicrobial properties.

3.5 ADVANTAGES AND DISADVANTAGES OF MULTILAYER FILMS

The layer-by-layer technique has several advantages, as: (1) it is very easy to perform and an expensive or sophisticated instrument is not required; (2) different building blocks can be incorporated into the multilayer system to obtain the desired properties; and (3) it can be applied to any loaded substrate, of any shape or size (Gezgin et al., 2017), besides adding the different properties of the individual polymers.

According to Peelman et al. (2013), the use of biodegradable sources as food packaging materials is subject to different limitations that restrict their use. In addition to a higher price level compared to conventional plastics, concerns about availability as well as soil use to produce the raw materials lead to major constraints. The fragility, thermal instability, low resistance to heating, high permeability to water vapor and oxygen restricted the use of multilayers obtained from various agricultural and natural resources (Peelman et al., 2013).

3.6 STRATEGIES TO IMPROVE PROPERTIES

Similarly to individual films, multilayer films often need improvements in their properties in order to allow their application. The search for ways to improve polymers properties in recent decades has intensified and several research studies observed positive results. Many techniques can be used to improve the polymer properties, among them, we have chemical, physical and enzymatic methods, as well as the use of polymeric blends.

Rivero et al. (2009) evaluated the possibility of using blends in the production of bilayers from many materials, using gelatin and chitosan and they observed that the polymers showed compatibility and resulted in multilayer films with better mechanical properties and water vapor barrier. Ferreira et al. (2016) used FucoPol and chitosan in the production of multilayer films, and the results showed dense and homogeneous layers and presented improved properties when compared to films of FucoPol monolayer as higher tensile strength and higher elongation were attained compared to FucoPol monolayer films.

Among the physical methods, the use of atmospheric plasma treatment on the surface of films presented promising results, making it possible to join different polymers without the use of an adhesive. Studies performed by Meriçer et al. (2016) using PLA and microfibrillated cellulose showed that the plasma application on the surface promoted an increase in the tensile strength and in the oxygen barrier of bilayer films when compared to monolayer films. Crosslinking is the process of linking polymer chains by covalent or noncovalent bonds, forming tridimensional networks. Chemical crosslinking results in covalent bonding between polymer chains accomplished by irradiation, sulfur vulcanization, or chemical reactions, while physical crosslinking involves noncovalent bonds, such as ionic interactions, hydrogen bonds or hydrophobic interactions. Crosslinking may be intra- or intermolecular, and it is one of the main techniques for polymer modification. Also, it is possible to use grafting, blending and composite formation (Balaguer, Cerisuelo, Gavara, and Hernandez-Muñoz, 2013). Crosslinking is a promising technique for improving the performance and applicability of films produced from renewable sources, especially regarding that sensitivity to water, which hampers many of their possible applications as food packaging materials (Garavand, Rouhi, Razavi, Cacciotti, and Mohammadi, 2017).

Some aldehydes are very effective as crosslinking agents; however, their use is not allowed films that use direct contact with food due to the possible migration of aldehyde residues into food. Less toxic compounds such as phenolic acids, oxidized polysaccharides and enzymes have already been studied as crosslinking agents. Crosslinking techniques can make protein-based and polysaccharide-based materials more suitable for large-scale processing and applications in the future (Azeredo and Waldron, 2016).

Another strategy used to improve the properties of biopolymer films is non-ionizing radiation by ultraviolet (UV) light. UV radiation is absorbed by double bonds and aromatic rings, which causes the formation of free radicals in amino acids (such as tyrosine and phenylalanine), which can lead to the formation of intermolecular covalent bonds (Wihodo and Moraru, 2012). UV radiation at the dosage of 51.8 J/m^2 over 24 h exposure caused a significant increase in tensile strength of wheat gluten, corn zein and egg albumin films, but no effect on the caseinate films (Rhim, Gennadios, Fu, Weller, and Hanna, 1999).

3.7 APPLICATIONS OF THE MULTILAYERS FILMS

The use of synthetic plastics as food packaging is motivated by characteristics such as flexibility and mechanical resistance, which are associated with quality, safety and low final cost (Siracusa et al., 2008). In addition to reducing the risks of chemical and biological contamination, the packaging aims to contain food and protect against various injuries and ensure the recognition of brands, producers and relevant consumer information (Garavand et al., 2017). However, changes in habits and increased environmental awareness of consumers require practical and alternative packaging that, in addition to preserving food, is economical and environmentally viable (Ahmed et al., 2017).

In food preservation, biodegradable monolayer and multilayer films act as a barrier to external elements such as dirt, gases, moisture, aromatic compounds and controlling mass transfer (moisture, oxygen, carbon dioxide). In addition, these films also act as a vehicle for additives and thus can enrich, protect and increase the shelf life of food (Ribeiro-Santos et al., 2017). Biodegradable films can still be used as a complement to the synthetic packaging, as a coating of these packages and acting as a carrier for antimicrobial agents that will be released on the food surface, extending the shelf life and guaranteeing the quality of the final product. Besides this, it presents economic potential, with the possibility of using low-cost raw material (Fang et al., 2017; Ju et al., 2018).

Table 3.3 presents some studies carried out with multilayer films produced from different biodegradable resources, applied as packaging for food products. Ferreira

TABLE 3.3
Applications of the Multilayer Films in Food Products

Multilayer	Application	References
FucoPol/quitosan	Walnuts and walnut oil	Ferreira et al., 2018
Starch/chitosan	Sliced pork meat	Valencia-Sullca et al., 2017
Chitosan/gelatin	Rainbow trout fillets	Nowzari et al., 2013
PP/chitosan/pectin	Tomato	Elsabee et al., 2008

PP: polypropylene.

et al. (2018) developed bilayer films composed of a bacterial exopolysaccharide (FucoPol) and chitosan. They studied the barrier properties of the bilayer in the conservation of walnuts (*Juglans regia L.*) and walnut oil. The study was also performed using a non-biodegradable commercial film (polyamide cast flexible-polyethylene—PA/PE 90) for comparison purposes. Ferreira et al. observed that the behavior of the oil stored in the bilayer and commercial packaging presented similar behavior over the storage time (14 days). The sensory evaluation of the walnut, stored in the same conditions as the oil, showed no significant differences in rancidity. The results obtained in the study showed that the biodegradable bilayer FucoPol/chitosan preserved the products in a similar way to the non-biodegradable packaging, proving to be a promising alternative as packaging material for walnut and similar products.

Valencia-Sullca et al. (2017) produced a bilayer film of starch-chitosan thermocompressed containing or not containing essential oils, which were applied on sliced pork meat. The microbial growth was evaluated. According to Valencia-Sullca et al., the starch-chitosan bilayer was effective in the preservation of pork meat, however, the heat treatment applied to obtain the bilayers reduced its antimicrobial properties in comparison with the individual chitosan film. In addition, it was observed that the addition of essential oils to the films and bilayers did not increase the antimicrobial activity of the films and bilayers, confirming that the microbial growth of the meat was reduced by the presence of chitosan.

Nowzari et al. (2013) developed bilayer films, blends and monolayers from chitosan and gelatin and applied it on rainbow trout (*Oncorhynchus mykiss*) fillets stored at 4°C for 16 days. The application was made using the films and bilayers as packaging and with the use of the film-forming solutions as a cover of the fillets, in order to verify which form of application would be more efficient. According to Nowzari et al., all coatings, films and bilayers of chitosan-gelatin presented antioxidant activity and effect in reducing bacterial contamination of stored fillets, when compared to control samples, however, the protective effect of the compatible coatings against lipid oxidation was higher when compared to the effect presented by films and bilayers. According to Nowzari et al., the migration of chitosan active agents is more available when in solution, increasing its antioxidant effect. In terms of microbial control, as there was no significant difference between coatings and bilayers, Nowzari et al. suggest that these packaging systems can be used as active packaging to store fish under refrigerated storage.

Elsabee et al. (2008) developed multilayer polypropylene (PP) films coated with a multiple layer of chitosan/pectin, which were evaluated for their antifungal and antibacterial efficiency. This study was carried out with three treatments: (1) fresh tomatoes were stored in active multilayers; (2) the tomatoes were stored in a regular PP bag; and (3) the fruits were not packed. All fruits were stored refrigerated at 4°C for 13 days. At the end of the experiment, the fruits stored in the PP bag and the fruits without packaging were completely deteriorated, while the sample in the treated packaging remained almost intact without evidence of deterioration. According to Elsabee et al., the multilayer developed in this study resulted in antimicrobial films with good properties to make excellent packaging materials for post-harvest protection of different products.

Some studies concerning the production of multilayers present possibilities of application of these polymers in function of their inherent properties. In studies concerning the use of FucoPol and chitosan in the production of multilayer films a cohesive polymer was produced, with moderate resistance to humidity and with good properties of gas barrier, indicating a possibility of application in products with low moisture content. (Ferreira et al., 2016).

Films from PLA/microfibrillated cellulose (MFC) presented properties equivalent to synthetic polymers such as PET, indicating the possibility of use in biodegradable primary packaging since they had good oxygen barrier as well as water vapor barrier properties (Meriçer et al., 2016). Two-layer films produced from gelatin and zein showed low resistance to solvent due to the heterogeneous dispersion of gelatin and zein, and because of this it is suggested they have potentially promising applications in biological administration or controlled release of compounds in food, acting as active packaging (Rivero et al., 2009).

3.8 CONCLUSION

Based on what was previously described, it can be concluded that in general, individual biodegradable films are still the most studied films. However, they still have low technological properties that limit their use in the most diverse niches. In this context, the use of the multilayer technique presents itself as an alternative with great potential for improvements in the properties of these films.

Among the materials used in the preparation of biodegradable multilayer films, we have gelatin, agar, starch, chitosan, zein, PLA, PHB and wheat gluten, with emphasis on gelatin, agar and PLA, which are the most used polymers according to the survey accomplished.

The most studied properties of multilayer films are mechanical and barrier films, and some studies show improvements in these properties when compared to individual films of the same polymer.

Among the methods used to produce multilayer films in recent decades, we can mention the casting technique, extrusion, lamination and electrospinning with emphasis on the casting technique which is the most used method on laboratory scale.

The use of multilayers presents advantages such as the ease of execution and the possibility of production without the use of expensive or sophisticated instruments, including the possibility of using different polymers and combinations according to the properties required for application and the possibility of the addition of active compounds. Among the disadvantages is the cost of production, which is still higher when compared to non-renewable synthetic materials.

In the same way as in individual films, multilayer films often need improvements in their properties in order to allow their application. In this context, many techniques can be used to improve the polymer properties such as the use of blends, chemical crosslinking, atmospheric plasma treatment and ultraviolet (UV) light.

The production of multilayers allows the improvement of technological properties as well as increasing the application potential, some examples currently in the literature being biodegradable primary packaging for fruits and meat, active packaging with controlled release of compounds and combinations with synthetic materials promoting antimicrobial activity.

The diversity of materials and methods available for the production of biodegradable films, added to the use of innovative strategies such as the production of multilayer films, brings an interesting alternative for the production of new materials with improved properties that can be used as packaging materials and make them competitive to synthetic materials. However, there are still few studies in the literature related to the production of multilayer biodegradable films, making new research necessary for the optimization and insertion of these materials into the packaging market.

REFERENCES

Ahmed, I., Lin, H., Zou, L., et al. 2017. A comprehensive review on the application of active packaging technologies to muscle foods. *Food Control* 82:163–178. doi:10.1016/j.foodcont.2017.06.009.

Arrieta, M.P., Fortunati, E., Dominici, F., López, J., and Kenny, J.M. 2015. Bionanocomposite films based on plasticized PLA—PHB / cellulose nanocrystal blends. *Carbohydr. Polym.* 121:265–275. doi:10.1016/j.carbpol.2014.12.056.

Avio, C.G., Gorbi, S., and Regoli, F. 2017. Plastics and microplastics in the oceans: From emerging pollutants to emerged threat. *Mar. Environ. Res.* 128:2–11. doi:10.1016/j.marenvres.2016.05.012.

Azeredo, H.M.C., and Waldron, K.W. 2016. Crosslinking in polysaccharide and protein films and coatings for food contact: A review. *Trends Food Sci. Technol.* 52:109–122. doi:10.1016/j.tifs.2016.04.008.

Baba Ismael, Y.M., Ferreira, A.M., Bretcanu, O., Delgano, K., and El haj, A.J. 2017. Polyelectrolyte multi-layers assembly of SiCHA nanopowders and collagen type I on aminolysed PLA films to enhance cell-material interactions. *Colloids Surf. B Biointerfaces* 159:445–453.

Balaguer, M.P., Cerisuelo, J.P., Gavara, R., and Hernandez-Muñoz, P. 2013. Mass transport properties of gliadin films: Effect of cross-linking degree, relative humidity, and temperature. *J. Memb. Sci.* 428:380–392. doi:10.1016/j.memsci.2012.10.022.

Bashir, A., Jabeen, S., Gull, N., et al. 2018. Co-concentration effect of silane with natural extract on biodegradable polymeric films for food packaging. *Int. J. Biol. Macromol.* 106:351–359. doi:10.1016/j.ijbiomac.2017.08.025.

Biodegradability, D., Organizations, S., and Standard, T. 1998. Standard methods for testing the aerobic biodegradation of polymeric materials. *Rev Petsives* 135:263–281.

Briassoulis, D., Tserotas, P., and Hiskakis, M. 2017. Mechanical and degradation behaviour of multilayer barrier films. *Polym. Degrad. Stab.* 143:214–230. doi:10.1016/j.polymdegradstab.2017.07.019.

Carofiglio, V.E., Stufano, P., Cancelli, N., et al. 2017. Novel PHB/Olive mill wastewater residue composite based film: Thermal, mechanical and degradation properties. *J. Environ. Chem. Eng.* 5(6):6001–6007. doi:10.1016/j.jece.2017.11.013.

Chiralt, A., González-Martínez, C., Vargas, M., and Atarés, L. 2018. Edible films and coatings from proteins. In: *Proteins in Food Procesing.* Elsevier. 477–500 p.

Cho, S.Y., Lee, S.Y., and Rhee, C. 2010. Edible oxygen barrier bilayer film pouches from corn zein and soy protein isolate for olive oil packaging. *Food Sci. Technol.* 43(8):1234–1239. doi:10.1016/j.lwt.2010.03.014.

Cinelli, P., Schmid, M., Bugnicourt, E., et al. 2014. Whey protein layer applied on biodegradable packaging film to improve barrier properties while maintaining biodegradability. *Polym. Degrad. Stab.* 108:151–157. doi:10.1016/j.polymdegradstab.2014.07.007.

Colwell, J.M., Gauthier, E., Halley, P., et al. 2017. Lifetime prediction of biodegradable polymers. *Prog. Polym. Sci.* 71:144–189. doi:10.1016/j.progpolymsci.2017.02.004.

Dehghani, S., Hosseini, S.V., and Regenstein, J.M. 2018. Edible films and coatings in seafood preservation: a review. *Food Chem.* 240:505–513.

De Moraes, J.O., Scheibe, A.S., Sereno, A., and Laurindo, J.B. 2013. Scale-up of the production of cassava starch based films using tape-casting. *J. Food Eng.* 119(4):800–808. doi:10.1016/j.jfoodeng.2013.07.009

Deng, L., Kang, X., Liu, Y., Feng, F., and Zhang, H. 2018. Characterization of gelatin/zein films fabricated by electrospinning vs solvent casting. *Food Hydrocoll.* 74:324–332. doi:10.1016/j.foodhyd.2017.08.023.

De Paoli, M.A. 2009. *Degradação e estabilização de polímeros.* São Paulo: Artliber Editora. 286 p.

Dukali, R.M., Radovic, I.M., Stojanovic, D.B., et al. 2014. Electrospinning of the laser dye rhodamine b-doped poly (methyl methacrylate) nanofibers. *J. Serb. Chem. Soc.* 79(7):867–880.

Elsabee, M.Z., and Abdou, E.S. 2013. Chitosan based edible films and coatings: A review. *Mater. Sci. Eng. C* 33(4):1819–1841. doi:10.1016/j.msec.2013.01.010.

Elsabee, M.Z., Abdou, E.S., Nagy, K.S.A., and Eweis, M. 2008. Surface modification of polypropylene films by chitosan and chitosan/pectin multilayer. *Carbohydr. Polym.* 71(2):187–195. doi:10.1016/j.carbpol.2007.05.022.

Fabra, M.J., López-Rubio, A., Ambrosio-Martín, J., and Lagaron, J.M. 2016a. Improving the barrier properties of thermoplastic corn starch-based films containing bacterial cellulose nanowhiskers by means of PHA electrospun coatings of interest in food packaging. *Food Hydrocoll.* 61:261–268. doi:10.1016/j.foodhyd.2016.05.025.

Fabra, M.J., López-Rubio, A., Cabedo, L., and Lagaron, J.M. 2016b. Tailoring barrier properties of thermoplastic corn starch-based films (TPCS) by means of a multilayer design. *J. Colloid Interface Sci.* 483:8492.

Fabra, M.J., López-Rubio, A., and Lagaron, J.M. 2014. Use of electrospinning for developing nanostructured interlayers based on hydrocolloids to enhance barrier properties of food packaging multilayers films. *Food Hydrocoll.* 32:106–114.

Fajardo, P., Martins, J.T., Fuciños, C., Pastrana, L., Teixeira, J.A., and Vicente, A.A. 2010. Evaluation of a chitosan-based edible film as carrier of natamycin to improve the storability of Saloio cheese. *J. Food Eng.* 101(4):349–356. doi:10.1016/j.jfoodeng.2010.06.029.

Fang, Z., Zhao, Y., Warner, R.D., and Johnson, S.K. 2017. Active and intelligent packaging in meat industry. *Trends Food Sci. Technol.* 61:60–71. doi:10.1016/j.tifs.2017.01.002.

Fereydoon, M., and Ebnesajjad, S. 2014. Development of high-barrier film for food packaging. *J. Serb. Chem. Soc.* 79:867–880.

Ferreira, A.R.V., Bandarra, N.M., Moldão-Martins, M., Coelhoso, I.M., and Alves, V.D. 2018. FucoPol and chitosan bilayer films for walnut kernels and oil preservation. *Carbohydr. Polym.* doi:10.1016/j.lwt.2018.01.020.

Ferreira, A.R.V., Torres, C.A.V., Freitas, F. et al. 2017. Development and characterization of bilayer films of FucoPol and chitosan. *Carbohydr. Polym.* 147:8–15. doi:10.1016/j.carbpol.2016.03.089.

Ferreira, A.R.V., Torres, C.A.V., Freitas, F., Reis, M.A.M., Vítor, D., and Coelhoso, I.M. 2014. Biodegradable films produced from the bacterial polysaccharide FucoPol. *Int. J. Biol. Macromol.* 71:111–116.

Figueira, D.S.R., Miguel, S.A.P., Sá, K.D., and Correia, I.J.S. 2016. Production and characterization of polycaprolactone-hyaluronic acid/chitosan-zein electrospun bilayer nanofibrous membrane for tissue regeneration. *Int. J. Biol. Macromol.* 93:1100–1110.

Flôres, S.H., Rios, A.de O., Iahnke, A.O.S., et al. 2017. Films for food from ingredient waste. *Ref. Modul. Food Sci.* doi:10.1016/B978-0-08-100596-5.21366-8.

Garavand, F., Rouhi, M., Razavi, S.H., Cacciotti, I., and Mohammadi, R. 2017. Improving the integrity of natural biopolymer films used in food packaging by crosslinking approach: A review Running head: Modification of biopolymer films by crosslinking. *Int. J. Biol. Macromol.* 104(A):687–707. doi:10.1016/j.ijbiomac.2017.06.093.

Garrido, T., Etxabide, A., Guerrero, P., and De La Caba, K. 2016. Characterization of agar/soy protein biocomposite films: Effect of agar on the extruded pellets and compression moulded films. *Carbohydr. Polym.* 151:408–416. doi:10.1016/j.carbpol.2016.05.089.

Gezgin, Z., Lee, T.C., and Huang, Q. 2017. Nanoscale properties of biopolymer multilayers. *Food Hydrocoll.* 63:209–218. doi:10.1016/j.foodhyd.2016.08.040.

Ghanbarzadeh, B., and Oromiehi, A.R. 2009. Thermal and mechanical behavior of laminated protein films. *J. Food Eng.* 90(4):517–524. doi:10.1016/j.jfoodeng.2008.07.018.

Gilfillan, W.N., Moghaddam, L., Bartley, J., and Doherty, W.O.S. 2015. Thermal extrusion of starch film with alcohol. *J. Food Eng.* 170:92–99. doi:10.1016/j.jfoodeng.2015.09.023.

Gómez-Guillén, M.C., Pérez-Mateos, M., Gómez-Estaca, J., López-Caballero, E., Giménez, B., and Montero, P. 2009. Fish gelatin: A renewable material for developing active biodegradable films. *Trends Food Sci. Technol.* 20(1):3–16. doi:10.1016/j.tifs.2008.10.002.

González, A., and Igarzabal, A.C.I. 2013. Soy protein—Poly (lactic acid) bilayer films as biodegradable material for active food packaging. *Food Hydrocoll.* 33(2):289–296. doi:10.1016/j.foodhyd.2013.03.010.

Gu, C.H., Wang, J.J., Yu, Y., Sun, H., Shuai, N., and Wei, B. 2013. Biodegradable multilayer barrier films based on alginate/polyethyleneimine and biaxially oriented poly(lactic acid). *Carbohydr. Polym.* 92(2):1579–1585. doi:10.1016/j.carbpol.2012.11.004.

Honarvar, Z., Farhoodi, M., Khani, M.R., et al. 2017. Application of cold plasma to develop carboxymethyl cellulose-coated polypropylene films containing essential oil. *Carbohydr. Polym.* 176:1–10. doi:10.1016/j.carbpol.2017.08.054.

Ju, J., Xu, X., Xie, Y., et al. 2018. Inhibitory effects of cinnamon and clove essential oils on mold growth on baked foods. *Food Chem.* 240:850–855. doi:10.1016/j.foodchem.2017.07.120.

Li, B.D., Wang, Y., and Xia, Y. 2004. Electrospinning nanofibers as uniaxially aligned arrays and layer-by-layer stacked films. *Adv. Mater.* 4:361–366.

Li, T., Li, J., Hu, W., and Li, X. 2013. Quality enhancement in refrigerated red drum (*Sciaenops ocellatus*) fillets using chitosan coatings containing natural preservatives. *Food Chem.* 138(2–3):821–826. doi:10.1016/j.foodchem.2012.11.092.

Lin, J., Wang, Y., Pan, D., Sun, Y., Ou, C., and Cao, J. 2017. Physico-mechanical properties of gelatin films modified with lysine, arginine and histidine. *Int. J. Biol. Macromol.* 108:947–952.

Luchese, C.L., Spada, J.C., and Tessaro, I.C. 2017. Starch content affects physicochemical properties of corn and cassava starch-based films. *Ind. Crops Prod.* 109:619–626. doi:10.1016/j.indcrop.2017.09.020.

Malagurski, I., Levic, S., Nesic, A., Mitric, M., Pavlovic, V., and Dimitrijevic-Brankovic, S. 2017. Mineralized agar-based nanocomposite films: Potential food packaging materials with antimicrobial properties. *Carbohydr. Polym.* 175:55–62. doi:10.1016/j.carbpol.2017.07.064.

Martucci, J.F., and Ruseckaite, R.A. 2010. Biodegradable three-layer film derived from bovine gelatin. *J. Food Eng.* 99(3):377–383. doi:10.1016/j.jfoodeng.2010.02.023.

Masina, N., Choonara, Y.E., Kumar, P., et al. 2017. A review of the chemical modification techniques of starch. *Carbohydr. Polym.* 157:1226–1236. doi:10.1016/j.carbpol.2016.09.094.

Medina-Jaramillo, C., Ochoa-Yepes, O., Bernal, C., and Famá, L. 2017. Active and smart biodegradable packaging based on starch and natural extracts. *Carbohydr. Polym.* 176:187–194. doi:10.1016/j.carbpol.2017.08.079.

Meriçer, Ç., Minelli, M., Angelis, M.G.D., et al. 2016. Atmospheric plasma assisted PLA/microfibrillated cellulose (MFC) multilayer biocomposite for sustainable barrier application. *Ind. Crops Prod.* 93:235–243. doi:10.1016/j.indcrop.2016.03.020.

Muller, J., González-Martínez, C., and Chiralt, A. 2017. Poly(lactic) acid (PLA) and starch bilayer films, containing cinnamaldehyde, obtained by compression moulding. *Eur. Polym. J.* doi:10.1016/j.eurpolymj.2017.07.019.

Musso, Y.S., Salgado, P.R., and Mauri, A.N. 2016. Gelatin based films capable of modifying its color against environmental pH changes. *Food Hydrocoll.* 61:523–530.

Nguyen, D.M., do, T.V.V., Grillet, A.C., Ha Thuc, H., and Ha Thuc, C.N. 2016. Biodegradability of polymer film based on low density polyethylene and cassava starch. *Int. Biodeterior. Biodegrad.* 115:257–265. doi:10.1016/j.ibiod.2016.09.004.

Nilsuwan, K., Benjakul, S., and Prodpran, T. 2017. Physical/thermal properties and heat seal ability of bilayer films based on fish gelatin and poly(lactic acid). *Food Hydrocoll.*. doi:10.1016/j.foodhyd.2017.10.001.

Nogueira, D., and Martins, V.G. 2018. Biodegradable bilayer films prepared from individual films of different proteins. *J. Appl. Polym. Sci.*. doi:10.1002/app.46721.

Nowzari, F., Shábanpour, B., and Ojagh, S.M. 2013. Comparison of chitosan-gelatin composite and bilayer coating and film effect on the quality of refrigerated rainbow trout. *Food Chem.* 141(3):1667–1672. doi:10.1016/j.foodchem.2013.03.022.

Ortega-Toro, R., Morey, I., Talens, P., and Chiralt, A. 2015. Active bilayer films of thermoplastic starch and polycaprolactone obtained by compression molding. *Carbohydr. Polym.* 127:282–290. http://doi.org/10.1016/j.carbpol.2015.03.080.

Park, J.W., Scott Whiteside, W., and Cho, S.Y. 2008. Mechanical and water vapor barrier properties of extruded and heat-pressed gelatin films. *Food Sci. Technol.* 41(4):692–700. doi:10.1016/j.lwt.2007.04.015.

Peelman, N., Ragaert, P., De Meulenaer, B. et al. 2013. Application of bioplastics for food packaging. *Trends Food Sci. Technol.* 32(2):128–141. doi:10.1016/j.tifs.2013.06.003.

Pereda, M., Ponce, A.G., Marcovich, N.E., Ruseckaite, R.A., and Martucci, J.F. 2011. Chitosan-gelatin composites and bi-layer films with potential antimicrobial activity. *Food Hydrocoll.* 25(5):1372–1381. doi:10.1016/j.foodhyd.2011.01.001.

Qiao, C., Ma, X., Zhang, J., and Yao, J. 2017. Molecular interactions in gelatin /chitosan composite film. *Food Chem.* 235:45–50. doi:10.1016/j.foodchem.2017.05.045.

Rahmani, E., Dehestani, M., Beygi, M.H.A., Allahyari, H., and Nikbin, I.M. 2013. On the mechanical properties of concrete containing waste PET particles. *Constr. Build. Mater.* 47:1302–1308. doi:10.1016/j.conbuildmat.2013.06.041.

Ren, L., Yan, X., Zhou, J., Tong, J., and Su, X. 2017. Influence of chitosan concentration on mechanical and barrier properties of corn starch/chitosan films. *Int. J. Biol. Macromol.* 105(3):1636–1643. doi:10.1016/j.ijbiomac.2017.02.008.

Ribeiro-Santos, R., Andrade, M., Melo, N.R. de, and Sanches-Silva, A. 2017. Use of essential oils in active food packaging: Recent advances and future trends. *Trends Food Sci. Technol.* 61:132–140. doi:10.1016/j.tifs.2016.11.021.

Rhim, J.W., Gennadios, A., Fu, D., Weller, C.L., and Hanna, M.A. 1999. Properties of ultraviolet irradiated protein films. *Food Sci. Technol.* 32(3):129–133. doi:10.1006/fstl.1998.0516.

Rivero, S., García, M.A., and Pinotti, A. 2009. Composite and bi-layer films based on gelatin and chitosan. *J. Food Eng.* 90(4):531–539. doi:10.1016/j.jfoodeng.2008.07.021.

Sanyang, M.L., Sapuan, S.M., Jawaid, M., Ishak, M.R., and Sahari, J. 2016. Development and characterization of sugar palm starch and poly (lactic acid) bilayer films. *Carbohydr. Polym.* 146:36–45. doi:10.1016/j.carbpol.2016.03.051.

Sarantópoulos, C.I.G.L., Oliveira, L.M., Padula, M., Coltro, L., Alves, R.M.V., and Garcia, E.E.C. 2002. *Embalagens plásticas flexíveis: Principais polímeros e avaliação de propriedades.* Campinas: CETEA/ITAL.

Scaffaro, R., Sutera, F., and Botta, L. 2018. Biopolymeric bilayer films produced by co-extrusion film blowing. *Polym. Test.* 65:35–43. doi:10.1016/j.polymertesting.2017.11.010.

Siracusa, V., Rocculi, P., Romani, S., and Rosa, M.D. 2008. Biodegradable polymers for food packaging: A review. *Trends Food Sci. Technol.* 19(12):634–643. doi:10.1016/j.tifs.2008.07.003.

Tavassoli-Kafrani, E., Shekarchizadeh, H., and Masoudpour-Behabadi, M. 2016. Development of edible films and coatings from alginates and carrageenans. *Carbohydr. Polym.* 137:360–374. doi:10.1016/j.carbpol.2015.10.074.

Torres, C.A.V., Ferreira, A.R.V., Freitas, F. et al. 2015. Rheological studies of the fucose-rich exopolysaccharide FucoPol. *Int. J. Biol. Macromol.* 79:611–617. doi:10.1016/j.ijbiomac.2015.05.029.

Trinetta, V. 2016. Biodegradable packaging. *Ref. Modul. Food Sci.* 1–2. doi:10.1016/B978-0-08-100596-5.03351-5.

Valencia-Sullca, C., Chiralt, A., Vargas, M., and Atar, L. 2017. Thermoplastic cassava starch-chitosan bilayer films containing essential oils. *Food Hydrocoll.* 75:107–115. doi:10.1016/j.foodhyd.2017.09.008.

Vejdan, A., Mahdi, S., Adeli, A., and Abdollahi, M. 2016. Effect of TiO 2 nanoparticles on the physico-mechanical and ultraviolet light barrier properties of fish gelatin/agar bilayer film. *Food Sci. Technol.* 71:88–95. doi:10.1016/j.lwt.2016.03.011.

Vu, C.H.T., and Won, K. 2013. Novel water-resistant UV-activated oxygen indicator for intelligent food packaging. *Food Chem.* 140(1–2):52–56. doi:10.1016/j.foodchem.2013.02.056.

Wan, Z., Wang, L., Yang, X., Guo, J., and Yin, S. 2016. Enhanced water resistance properties of bacterial cellulose multilayer films by incorporating interlayers of electrospun zein fibers. *Food Hydrocoll.* 61:269–276.

Wang, K., Wang, W., Ye, R. et al. 2017. Mechanical properties and solubility in water of corn starch-collagen composite films: Effect of starch type and concentrations. *Food Chem.* 216:209–216. doi:10.1016/j.foodchem.2016.08.048.

Wang, L., and Rhim, J. 2015. Preparation and application of agar/alginate/collagen ternary blend functional food packaging films. *Int. J. Biol. Macromol.* 80:460–468. doi:10.1016/j.ijbiomac.2015.07.007.

Wang, W., Xiao, J., Chen, X., Luo, M., Liu, H., and Shao, P. 2018. Fabrication and characterization of multilayered kafirin/gelatin film with one-way water barrier property. *Food Hydrocoll.* 81:159–168. doi:10.1016/j.foodhyd.2018.02.044.

Wihodo, M., and Moraru, C.I. 2012. Physical and chemical methods used to enhance the structure and mechanical properties of protein films: A review. *J. Food Eng.* 114(3):292–302. doi:10.1016/j.jfoodeng.2012.08.021.

Zhang, C., Garrison, T.F., Madbouly, S.A., and Kessler, M.R. 2017. Recent advances in vegetable oil-based polymers and their composites. *Prog. Polym. Sci.* 71:91–143. doi:10.1016/j.progpolymsci.2016.12.009.

Zhuang, C., Jiang, Y., Zhong, Y. et al. 2018. Development and characterization of nano-bilayer films composed of polyvinyl alcohol, chitosan and alginate. *Food Control* 86. doi:10.1016/j.foodcont.2017.11.024.

Zuo, G., Song, X., Chen, F., and Shen, Z. 2017. Physical and structural characterization of edible bilayer films made with zein and corn-wheat starch. *J. Saudi Soc. Agric. Sci.* 18:324–331. doi:10.1016/j.jssas.2017.09.

4 Environmental Issues Related to Packaging Materials

Sandhya Alice Varghese, Sanjay M.R, Senthilkumar K., Sabarish Radoor, Suchart Siengchin, and Jyotishkumar Parameswaranpillai

CONTENTS

4.1 INTRODUCTION TO FOOD PACKAGING

In a study by the Food and Agricultural Organization of the United Nations, it was found that nearly one-third of the annual food produced worldwide is wasted (Ishangulyyev et al., 2019). In this context, the link between food wastage and food packaging can never be neglected. Packages serve various functions such as protection from the external environment, communicate with the customer various details such as date of packaging, expiry date, manufacturer details, content details, conditions of storage, etc. (Robertson, 2013). Proper food packaging is one of the main strategies suggested to reduce food wastage and loss. The packaging industry is ever-growing and has made tremendous progress to fit consumer needs. Smart packaging technologies such as intelligent and active packaging are the latest additions in the food packaging sector. Such developments can serve as a strong marketing tool for packaging material. The consumer demand for longer food shelf life has led to such innovations(Sonneveld, 2000). However, the production, use, and disposal of packaging have direct and indirect environmental impacts. The packaging waste poses a serious threat to the environment and the sustainability of the environment is affected.

4.1.1 Materials Used for Food Packaging

With the change in human lifestyle and growing demands, the growth of the packaging industry has reached the level we witness today. A wide variety of materials have been introduced as packaging material. Many innovative steps have been introduced in each sector to satisfy customer needs.

The materials used in food packaging can be broadly classified into two groups:

1. Traditional packaging materials
2. Industrial packaging materials

4.1.1.1 Traditional packaging materials

Primitive men had little need for food packaging as they mostly consumed food at the place it was found and storage/transportation of food was not considered. With socio-economic changes, there was a need for containers to carry food. The very early forms of packaging include covering in leaves, wood, bamboo, grass, shells, etc., and containers made out of these natural materials (Risch, 2009). In addition to these plant-based materials, materials of animal origin such as goatskin and cow's hide and stomach lining were used for storing drinking water (Jayathilakan, Sultana et al., 2011). A photograph of food wrapped in banana leaf is shown in Figure 4.1.

FIGURE 4.1 Photograph of traditional food packaging.

However, these traditional materials fail to satisfy the need to preserve food for a longer shelf life and are also unfit to the needs of commercial production processes. This paved the way for the development of industrial packaging materials.

4.1.1.2 Industrial Packaging Materials

In order to preserve the food for a longer period of time, new packaging materials were introduced such as glass, paper, metal, and plastic. Most of the packages of use today are made of any of these materials individually or in combinations. These materials are discussed in brief below:

1. *Glass*: The major ingredient of glass is silica, which is one of the most abundant natural raw materials available on earth. It is the only widely used packaging material that has been designated as fully safe by the U.S. Food and Drug Administration (Robertson, 2014). When used as packaging material, glass being inert serves to be an excellent barrier to gases. The fragility, heaviness, risk of breakage during freezing, and need for a large amount of energy during manufacture are the drawbacks of glass (Grayhurst and Girling, 2011, Kobayashi, 2016). Today, glass packaging mainly comes in two forms—as bottles and as jars, which are both easily reusable. Although glass is non-degradable, it can be 100% recycled without any loss of quality (Davis and Song, 2006).
2. *Wood, Cardboard, and Papers*: Wood, in various forms, is mainly used as packaging material for different food products such as fruit and vegetables, wine and spirits, oils, cheese and dairy, bread, etc. Products derived from wood such as paper and cardboard are widely used in packaging. (Raheem, 2013). In order to overcome the poor moisture resistance and low heat sealability of plain paper, it is often combined with additives like paraffin or used in combination with aluminum or plastic(Deshwal, Panjagari et al., 2019). Several layers of papers are superimposed over one another to make cardboard which is more moisture-resistant than paper. While paper and cardboard are considered to be eco-friendly, the production of these materials leads to deforestation and depletion of natural resources. Also, the degradation of the products in landfills may cause methane emission (Adhikari and Ozarska, 2018). Although these materials are recyclable, the chemicals used in recycling make them unfit to be used for food contact applications after the recycling process (Hernandez and Selke, 2001, Coles, 2013).
3. *Metals*: Steel, tin, and aluminum are the most widely used metals in food packaging applications. Metals are mostly used in the preservation of canned food and beverages. Metal packages offer advantages of compactness, toughness, malleability, and low cost. The packages can be produced in any desired shape and size and such packaging is resistant to physical shocks. The magnetic nature of the cans makes it easier to sort during waste collection and these cans maintain their original properties post recycling (Berk, 2018). In general, metal cans are given an internal organic coating

before being used for food contact applications. This coating functions to protect the food from the metal which may otherwise lead to food spoilage (eg., discoloration of dark-colored fruits like strawberries from metal contact) (Piergiovanni and Limbo, 2016). Examples of such coatings include white organosol coatings and coatings based on epoxy-ester, acrylic, and polyester resins (Beese and Ludwigsen, 1974).

4. *Plastics*: Plastics are widely used as food packaging material. The lightness and versatility of these materials along with their low cost and easy processability enable their use to replace most of the traditional food packaging materials. They also offer advantages of low energy consumption, easy handling, and flexibility and have an active surface for printing labels or brands. However, plastics have disadvantages including low melting point, absorption of flavors, and the improper use of plastic and irresponsible dumping of plastic packaging waste has led to serious environmental issues. Various types of plastics including polypropylene (PP), polystyrene (PS), polyethylene (PE), polyethylene terephthalate (PET), etc., are widely used in the packaging field (McKeen, 2013).

The current market share of packaging materials is shown in Figure 4.2.

4.2 LEVELS OF FOOD PACKAGING

Food packaging is often made in the form of a combination of several materials so as to obtain each component's functional/aesthetic advantages. The selection of

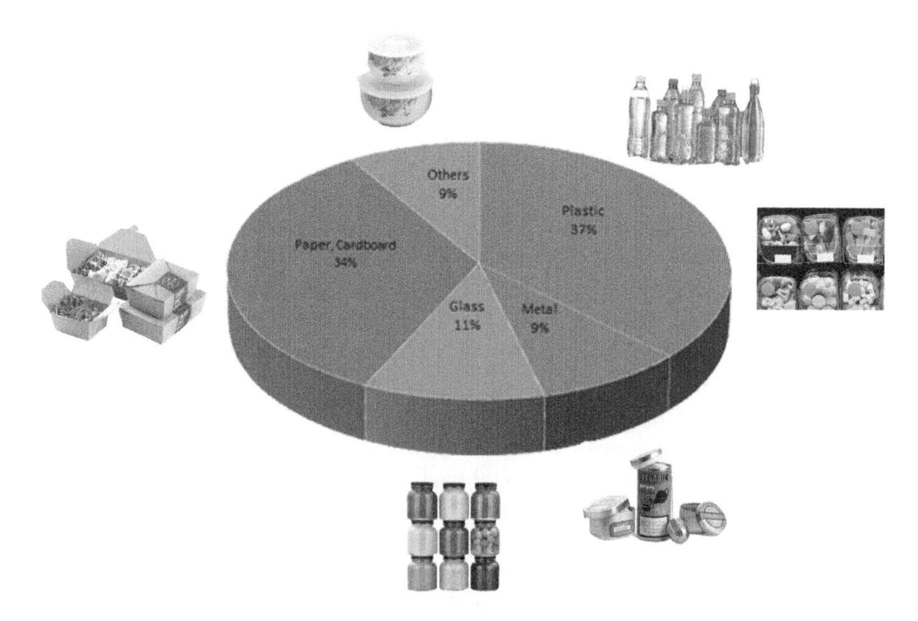

FIGURE 4.2 Market share of different food packaging materials. Statistics from https://www.foodpackagingforum.org/food-packaging-health/food-packaging-materials (2012).

these materials depends upon the nature of the food to be packed. Food packaging is classified into different levels as follows (Robertson, 1990; Roy, Saha et al., 2012):

1. *Primary packaging*, or consumer/retail packaging, is an essential packaging level to contain the food product and the package comes in direct contact with the food. Hence, the primary package material must be compatible with the food contained. Examples of primary packages include beverage can, jar containing jam, wrapper which covers a chocolate candy etc. (Berk, 2009).
2. *Secondary packaging*, or collection packaging, is usually used for bundling and storage purposes. It is a collection of two or more primary packages and serves to protect the primary packages during storage and distribution. Examples include cardboard boxes, wooden crates, or plastic containers used to carry bulk quantities of primary packaged products (Barlow and Morgan, 2013).
3. *Tertiary packaging*, or transport/ distribution packaging, is used during bulk transport of goods for the protection of primary and secondary packaging. These packages are often removed before it reaches the consumers. Examples include wooden pallets and plastic shrink-wrap etc. (Wikström, Williams et al., 2014). Figure 4.3 illustrates the various levels of food packaging.

Since secondary and tertiary packaging materials are generally made of a single packaging material, it is easy to collect and sort these packages. A large variety of materials are often combined in primary packaging and hence recycling and reuse issues are more prominent for primary packaging materials (Davis and Song, 2006).

The market of food packaging is estimated to cover one-third of the total packaging market. (Pauer, Wohner et al., 2019). Although appropriate food packaging saves resources and avoids food wastage, it must be balanced with energy and material cost along with the socio-environmental impacts. These environmental impacts of food packaging are discussed in the following section.

4.3 ENVIRONMENTAL IMPACTS OF FOOD PACKAGING MATERIALS

The used food package contributes to nearly two-thirds of the solid urban waste. This rising trend is a threat to environmental sustainability. Forest resources are

FIGURE 4.3 Various levels of food packaging.

highly depleted to produce many of the packaging products like paper and cardboard which may lead to ecological problems including soil erosion, desertification, and water shortage. Other than the consumption of resources, the waste production during the manufacture and disposal of packaging is a serious concern. Packaging occupies approximately 30% weight of municipal solid waste (MSW), but due to its bulkiness, it represents close to 65% of total waste volume. (Del Borghi, Gallo et al., 2014; Mahesh Kumar, Irshad et al., 2016; Simon, Amor et al., 2016; Molina-Besch et al., 2018). Hence, the environmental impact of the packaging material must be well examined. A tool developed to analyze the environmental impact of packaging material is the life cycle assessment (LCA).

4.3.1 Life Cycle Assessment of Packaging Material

LCA analyses the environmental impacts of the material from its production stage to disposal (Shapiro, 1993). The LCA of various packaging materials can be compared and improvement can be made on the products/processes (Silvenius, Katajajuuri et al., 2011). Although the life cycle of each packaging material is different, a general figure illustrating the life cycle of the packaging material is shown in Figure 4.4.

Taking the life cycle of packaging material into account, there are two main aspects of the environmental impact of packaging: (1) direct environmental impact and (2) indirect environmental impact (Pauer, Wohner et al., 2019).

4.3.1.1 Direct Environmental Impact

The direct environmental impact of food packaging is prominent in two stages: (1) during production and transportation and (2) after use.

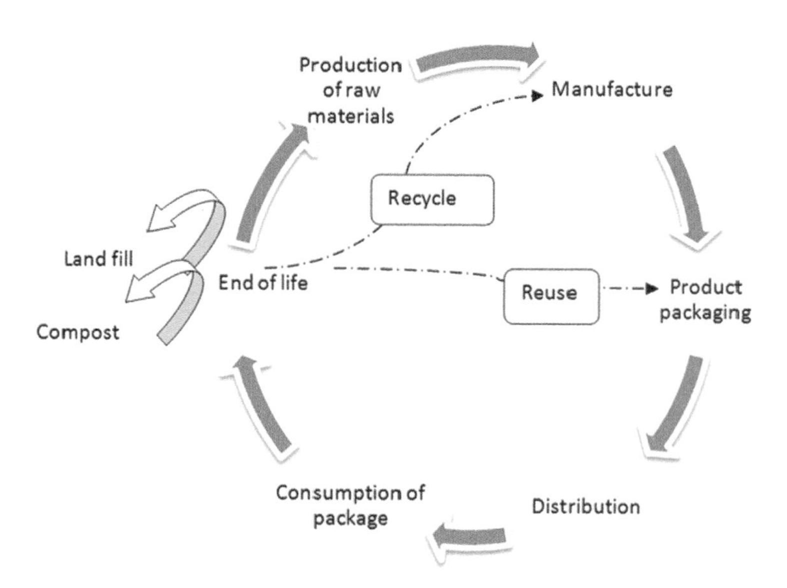

FIGURE 4.4 Life cycle assessment of food packaging.

1. *During Production and Transportation*: The most important resource for all packaging production is energy. The energy consumption leads to resource depletion as well as emission of greenhouse gases and toxic chemicals (North and Halden, 2013). For example, during glass and steel packaging production global greenhouse gases including carbon dioxide are emitted. The production of packaging materials also requires some non-renewable resources, which cannot be replaced once depleted, for example in the production of plastic packages like PS and PET, which have their origin in crude oil. Manufacturing of plastic packages also releases greenhouse gases that contribute to global warming (Jindal, 2010). Uncontrolled depletion of forest resources for wood and paper production causes an ecological imbalance (Zheng, 2012). Transportation-related emissions (e.g., CO_2, SO_2, hydrocarbons) must be considered when reuse or recovery is taken into account (Pongrácz, 2007).
2. *After Use*: Once the food within the package is consumed, the package no longer serves any function and is generally considered as waste. Environmental degradation that occurs due to the dumping of packaging waste is not a new topic. The direct disposal of the waste, especially of material such as plastic, adversely affects the ecosystem. These plastic wastes affect all the components of the environment including land, water, air, and living organisms. The waterways get polluted by the stamped or printed ink in packaging (Mahesh Kumar, Irshad et al., 2016). The presence of toxic heavy metal and volatile organic compounds (VOCs) in these inks are extremely harmful to the environment (Selke, 1994). Improper dumping of packaging waste into landfills also affects the soil and its fertility which in turn has a direct impact on agriculture and food security. Plastic waste accumulation is a threat to aquatic activities including fishing and tourism (Moore, 2008; Gregory, 2009). When plastic wastes are discarded into water bodies, they can remain on the sea surface for prolonged periods, and hence be a threat to the aquatic life and the waste might also affect the mating seasons of aquatic life (Thompson et al., 2009). When the plastic waste is irresponsibly dumped, it may be taken up by the wildlife which can even lead to their death. Many of the chemicals released during the manufacture as well as during uncontrolled burning of the packaging materials adversely affect human health (Talsness, Andrade et al., 2009).

4.3.1.2 Indirect Environmental Impact

Food production is associated with the utilization of a large amount of resources. The resources are wasted if the production does not reach its ultimate goal. Food wastage not only has an environmental impact, but the socio-ethical effects must also be considered. Packaging, although intended to minimize food loss, can sometimes be the cause of food loss and wastage in various levels of the food supply chain. Reasons for such inappropriate packaging include improper size of packages, use of inappropriate material for packaging, use of packaging which is not clean, and technical issues in the packaging process. Other issues include packaging that is too difficult to open, or packaging that is too loose so that its contents spill out (Wikström, Verghese et al., 2018; Wohner, Pauer et al., 2019). Hence, it is highly recommended that throughout the manufacturing and usage of food packages, environmentally responsible steps must be taken to reduce the impact of packaging on the ecosystem while at the same time, meet all the requirements of packaging.

4.4 MEASURES TO REDUCE THE ENVIRONMENTAL IMPACT OF PACKAGING

Proper waste management is extremely essential to reduce the adverse effect of packaging on the environment and human health. Several methods are being practiced to deal with packaging waste including discarding waste into landfills, incineration and conversion to energy, recovery, reuse, and recycling. Some of these practices are discussed below:

1. *Reduction and reuse*: Source reduction aims at reducing the use of packaging, thereby avoiding the production of waste in the first place. Source reduction can also cut down the material cost and minimize the use of labor resources associated with recycling, composting, or disposing of the waste (Mahesh Kumar, Irshad et al., 2016). Reusing of packaging is preferred to recycling as it uses a lesser amount of energy and resources. Although this method is the most powerful and effective way of waste management, it is considered only as a partial solution to environmental problems related to packaging materials (Beiras, 2018). This is mainly because only some materials such as glass and plastic carry bags can be repeatedly reused. In the case of food packaging, reuse is generally applicable only for refillable and cleanable containers (e.g., glass bottles) (Geueke, Groh et al., 2018).
2. *Combustion*: Combustion or incineration is the controlled burning of waste in an appropriate facility. Incineration can be done with or without energy recovery and is preferred for materials that cannot be recycled or composted efficiently. Incineration is considered to be the most hygienic method of solid waste management (Ferreira, Cabral et al., 2015). However, the high cost involved in the plant set up as well as operation along with the emission of hazardous chemicals involved in the burning of materials such as polyvinyl chloride (PVC) limits the widespread use of this method in waste management (Simoneit, Medeiros et al., 2005; Pongrácz, 2007).
3. *Composting*: Compostable materials are those which can be broken down into an organic state when placed in an appropriate environment with regulated level of heat, moisture, and oxygen. During composting, controlled biological degradation of organic materials by microorganisms takes place (Pagga, 1999). Examples of compostable materials include starch, cellulose, etc. (Ciriminna and Pagliaro, 2019). The main advantage of compostable materials is that they leave no toxic residue. Such materials in fact turn out to act as a fertilizer by adding nutrients to the soil. However, to be used as a food packaging material, such compostable packaging needs to meet the food packaging requirements (Avella, Bonadies et al., 2001).

4. *Recycling*: While composting is preferred for organic materials, recycling can be applied to all types of packaging materials. Recycling not only takes into account the environmental benefits, but also the socioeconomic factors (Dainelli, 2008). Waste reduction, maximum utilization of natural resources, and limitation of environmental impact are the primary advantages of recycling. However, the benefits of recycling should always be balanced with many factors such as environmental impact, cost effectiveness of the product, and consumer acceptance of the recycled materials in food packaging. All these elements determine the worthiness of the recycled material (da Cruz, Simões et al., 2014; Geueke, Groh et al., 2018).
5. *Land-filling*: Land-filling is considered as a way to dispose of remaining municipal waste along with the remainders of recycling and combustion operations. In this process, the waste is secured into the ground, avoiding contact with surrounding environment such as water. This method is preferred due to its comparatively low cost and most of the packaging waste materials are suitable for land-filling. However, landfills are considered a flawed technology because resources are wasted and groundwater pollution and greenhouse gas emissions are not eliminated (Marsh, 2007; Arvanitoyannis, 2008).
6. *Sustainable packaging*: The latest addition to methods to minimize the environmental impacts of packaging is the development of sustainable packaging. Sustainable packaging has several aspects. Firstly, it aims at minimizing food loss through the design of better and smarter packaging such as active and intelligent packaging to keep food materials fresher for an extended time period while accommodating consumer needs (Halloran et al., 2014). Secondly, sustainable packaging considers the long-term issue of plastic waste accumulation. One of the prominent developments in this area is the development of biodegradable polymers from agro-food waste residues. This will create a waste-based food packaging economy by minimizing the use of fossil resources. Such developments will reduce food waste and plastic waste accumulation (Russell, 2014).

4.5 REGULATIONS RELATED TO PACKAGING

Food packaging regulations refer to the safety requirements that are intended to preserve the physical and chemical integrity of the contents of the packaging. (Rijk and Veraart, 2010). Individual countries developed their own regulations for food packaging although most of these regulations based them on similar scientific principles. Hence, different countries have different regulations that may affect the trades between these countries. In some cases, the regulations of one country are also recognized in another country and accept the safety of a packaging material if it meets the needs of the regulations. Packaging industries take effort to demonstrate that their materials are developed in accordance with the regulatory compliance (Baughan, 2015). Two such important regulators are the European Union and the United States.

4.5.1 Regulations in the European Union

In the European Union, there are two legislations for food contact materials: (1) harmonized community legislation adopted by the EU and (2) non-harmonized national legislation adopted by the individual Member States (Rijk and Veraart, 2010).

1. *Harmonized Community Legislation Adopted by the EU*: Under the community legislation, Framework Regulation created safety standards and authorizes the European Commission (EC) to take measures for specific food contact materials or substances. The most important legislation is the "Plastics Regulation" which includes a union list of monomers, additives, and material aids that may be used in the production of food contact plastics. According to this list, all substances need to be clarified blank based on a toxicological evaluation published in regulations, and the use of substances out of this list is generally prohibited ("Regulation (EU) No. 10/2011," "Regulation (EC) No. 1935/2004"). Any changes to existing regulations are brought about by amendments.
2. *Non-Harmonized National Legislation Adopted by Individual Member States*: National legislation on packaging materials is adopted by individual EU countries in the absence of specific EU measures. National legislation may differ in each Member State.

   In addition to the above-mentioned legislations, the European Food Safety Authority (EFSA) was set up in 2002 to analyze the risks associated with the food chain (Geueke et al., 2014; Deepika Rani et al., 2018).

4.5.2 Regulations in the United States

In the United States, the Food and Drug Administration (FDA) regulates food safety and is in charge of both risk assessment and risk management. The FDA regulations aim to ensure the safety of food distributed throughout the United States and to generate consumer awareness about the food they are consuming. The FDA publishes databases related to food ingredients, food additives, color additives, etc. Another regulation passed by the FDA is related to irradiated food and packaging. The FDA also measures the environmental impact of these regulations (Marsh and Bugusu, 2007; Neltner, Kulkarni et al., 2011). Other than the above specified regulations, most of the individual countries have established their own standards and regulations for food contact materials and packaging.

4.6 CONCLUSION AND FUTURE OUTLOOK

Food packaging aims at the preservation and protection of food and minimizing food wastage. Today, most food packaging is intended for single use and is often discarded in a short period of time. Due to its high production volume and short usage time, this raises concern for the environment due to problems related to poor waste management and littering of packaging waste. Packaging is often considered as a source of waste and pollution. Efforts must be taken by industry, government, and consumers to promote the safety of the environment by introducing new materials and technology while maintaining the primary purpose of food packaging to ensure the safety, wholesomeness, and quality of food. The development of environmentally conscious packaging can satisfy human needs without affecting the environment. There is growing demand for sustainable packaging which provides optimal product protection and is safe for human health and cyclic, while having the smallest possible environmental footprint. The success of a good packaging material lies in its ability to satisfy consumer needs without compromising environmental and waste management issues and keeping the cost at a minimum. The indirect effects of food packaging including the food wastage and loss needs to be quantified effectively to take proper steps. New measures need to be taken to produce high-quality, affordable packaging products that are fully sustainable.

REFERENCES

Adhikari, S. and B. Ozarska (2018). "Minimizing environmental impacts of timber products through the production process 'From Sawmill to Final Products.'" *Environmental Systems Research* **7**(1): 6.

Arvanitoyannis, I. S. (2008). "Waste management in food packaging industries." In: *IS Arvanitoyannis, Waste Management for the Food Industries*: 941–1045. San Diego: Academic Press.

Avella, M., et al. (2001). "European current standardization for plastic packaging recoverable through composting and biodegradation." *Polymer Testing* **20**(5): 517–521.

Barlow, C. Y. and D. C. Morgan (2013). "Polymer film packaging for food: An environmental assessment." *Resources, Conservation and Recycling* **78**: 74–80.

Baughan, J. S. (2015). *Global Legislation for Food Contact Materials*. Boston, MA, Elsevier.

Beese, R. E. and R. J. Ludwigsen (1974). "Trends in the design of food containers." In: Swalm C. M (ed), *Chemistry of Food Packaging*: **135**: 1–14.

Beiras, R. (2018). "Plastics and other solid wastes." In: *Marine Pollution*: 69–88. Elsevier.

Berk, Z. (2009). "Food packaging." In: *Food Process Engineering and Technology*: 545–559. San Diego: Academic Press.

Berk, Z. (2018). "Food packaging." In *Food Process Engineering and Technology*: 625–641. San Diego: Academic Press.

Ciriminna, R. and M. Pagliaro (2019). "Biodegradable and compostable plastics: A critical perspective on the dawn of their global adoption." *Chemistry Open* **9**(1): 8–13.

Coles, R. (2013). "Paper and paperboard innovations and developments for the packaging of food, beverages and other fast-moving consumer goods." In: Farmer, N. (ed), *Trends in Packaging of Food, Beverages and Other Fast-Moving Consumer Goods (FMCG)*: 187–220. Cambridge: Woodhead Publishing.

da Cruz, N. F., et al. (2014). "Costs and benefits of packaging waste recycling systems." *Resources, Conservation and Recycling* **85**: 1–4.

Dainelli, D. (2008). "Recycling of food packaging materials: An overview." In: Chiellini, E. (ed), *Environmentally Compatible Food Packaging*: 294–325. Cambridge: Woodhead Publishing.

Davis, G. and J. H. Song (2006). "Biodegradable packaging based on raw materials from crops and their impact on waste management." *Industrial Crops and Products* **23**(2): 147–161.

Deepika Rani, V. S., et al. (2018). "A review on regulatory aspects of food contact materials (FCM's)." *The Indian Journal of Nutrition and Dietetics* **55**(4): 500.

Del Borghi, A., et al. (2014). "An evaluation of environmental sustainability in the food industry through life cycle assessment: The case study of tomato products supply chain." *Journal of Cleaner Production* **78**: 121–130.

Deshwal, G. K., et al. (2019). "An overview of paper and paper based food packaging materials: Health safety and environmental concerns." *Journal of Food Science and Technology* **56**(10): 4391–4403.

Ferreira, S., et al. (2015). "Life cycle assessment and valuation of the packaging waste recycling system in Belgium." *Journal of Material Cycles and Waste Management* **19**(1): 144–154.

Geueke, B., et al. (2014). "Food contact substances and chemicals of concern: A comparison of inventories." *Food Additives and Contaminants: Part A* **31**(8): 1438–1450.

Geueke, B., et al. (2018). "Food packaging in the circular economy: Overview of chemical safety aspects for commonly used materials." *Journal of Cleaner Production* **193**: 491–505.

Grayhurst, P. and P. J. Girling (2011). "Packaging of food in glass containers." In: Coles, R., Kirwan, M. (eds), *Food and Beverage Packaging Technology*: 137–156. London, UK: Blackwell Publishing, CRC Press.

Gregory, M. R., (2009). "Environmental implications of plastic debris in marine settings—entanglement, ingestion, smothering, hangers-on, hitch-hiking and alien invasions." Philosophical Transactions of the Royal Society B: *Biological Sciences* **364**(1526): 2013–2025.

Halloran, A., et al. (2014). "Addressing food waste reduction in Denmark." *Food Policy* **49**: 294–301.

Hernandez, R. J. and S. E. Selke (2001). "Packaging: Corrugated paperboard." *Encyclopedia of Materials: Science and Technology*: 6637–6642. https://doi.org/10.1016/B0-08-043152-6/01173-6.

Ishangulyyev, R., et al. (2019). "Understanding food loss and waste—why are we losing and wasting food?" *Foods* **8**(8): 297.

Jayathilakan, K., et al. (2011). "Utilization of byproducts and waste materials from meat, poultry and fish processing industries: A review." *Journal of Food Science and Technology* **49**(3): 278–293.

Jindal, M. K., (2010). "Unpacking the packaging: Environmental impact of packaging wastes." *Journal of Environmental Research And Development* **4**(4):1084–1092.

Kobayashi, M. L. (2016). "Glass packaging properties and attributes." In: *Reference Module in Food Science*. Elsevier. doi: 10.1016/B978-0-08-100596-5.03191-7.

Mahesh Kumar, G., et al. (2016). "Waste management in food packaging industry." In: Prashanthi M., Sundaram R. (eds), *Integrated Waste Management in India*: 265–277. Environmental Science and Engineering. Springer, Cham.

Marsh, K. and B. Bugusu (2007). "Food packaging? Roles, materials, and environmental issues." *Journal of Food Science* **72**(3): R39–R55.

McKeen, L. W. (2013). "Introduction to use of plastics in food packaging." In: Ebnesajjad, S. (ed), *Plastic Films in Food Packaging*: 1–15. doi: 10.1016/B978-1-4557-3112-1.00001-6.

Molina-Besch, K., et al. (2018). "The environmental impact of packaging in food supply chains—Does life cycle assessment of food provide the full picture?" *The International Journal of Life Cycle Assessment* **24**(1): 37–50.

Moore, C. J. (2008). "Synthetic polymers in the marine environment: A rapidly increasing, long-term threat." *Environmental Research* **108**(2): 131–139.

Neltner, T. G., et al. (2011). "Navigating the U.S. Food additive regulatory program." *Comprehensive Reviews in Food Science and Food Safety* **10**(6): 342–368.

North, E. J. and R. U. Halden (2013). "Plastics and environmental health: The road ahead." *Reviews on Environmental Health* **28**(1): 1–8.

Pagga, U. (1999). "Compostable packaging materials—Test methods and limit values for biodegradation." *Applied Microbiology and Biotechnology* **51**(2): 125–133.

Pauer, E., et al. (2019). "Assessing the environmental sustainability of food packaging: An extended life cycle assessment including packaging-related food losses and waste and circularity assessment." *Sustainability* **11**(3): 925.

Piergiovanni, L. and S. Limbo (2016). "Metal packaging materials." In: *Food Packaging Materials*: 13–22. Springer Briefs in Molecular Science. Springer, Cham.

Pongrácz, E. (2007). "The environmental impacts of packaging." In: Kutz, M. (ed), *Environmentally Conscious Materials and Chemicals Processing*: 237–278. John Wiley & Sons.

Raheem, D. (2017). "Application of plastics and paper as food packaging materials? An overview." *Emirates Journal of Food and Agriculture* **25**(3): 177–188.

Rijk, R. and R. Veraart (2010). *Global Legislation for Food Packaging Materials*. John Wiley & Sons.

Risch, S. J. (2009). "Food packaging history and innovations." *Journal of Agricultural and Food Chemistry* **57**(18): 8089–8092.

Robertson, G. L. (1990). "Good and bad packaging: Who decides?" *International Journal of Physical Distribution and Logistics Management* **20**(8): 37–40.

Robertson, G. L. (2013). *Food Packaging: Principles and Practice*. Boca Raton, FL, CRC Press.

Robertson, G. L. (2014). "Food packaging." *Encyclopedia of Agriculture and Food Systems*: 232–249.

Roy, N., et al. (2012). "Biodegradation of PVP–CMC hydrogel film: A useful food packaging material." *Carbohydrate Polymers* **89**(2): 346–353.

Russell, D. A. M. (2014). "Sustainable (food) packaging—An overview." *Food Additives and Contaminants: Part A* **31**(3): 396–401.

Selke, S. E. M. (1994). *Packaging and the Environment: Alternatives, Trends, and Solutions*. Lancaster, PA: Technomic Pub. Co.

Shapiro, K. (1993). "Life-cycle evaluation of packaging materials." *Proceedings of the 1993 IEEE International Symposium on Electronics and the Environment*: 106–111. doi: 10.1109/ISEE.1993.302827.

Silvenius, F., et al. (2011). "Role of packaging in LCA of food products." In: Finkbeiner, M. (ed), *Towards Life Cycle Sustainability Management*: 359–370. Springer, Dordrecht.

Simon, B., et al. (2016). "Life cycle impact assessment of beverage packaging systems: Focus on the collection of post-consumer bottles." *Journal of Cleaner Production* **112**: 238–248.

Simoneit, B. R. T., et al. (2005). "Combustion products of plastics as indicators for refuse burning in the atmosphere." *Environmental Science and Technology* **39**(18): 6961–6970.

Sonneveld, K. (2000). "What drives (food) packaging innovation?" *Packaging Technology and Science* **13**(1): 29–35.

Talsness, C. E., et al. (2009). "Components of plastic: Experimental studies in animals and relevance for human health." *Philosophical Transactions of the Royal Society Series B: Biological Sciences* **364**(1526): 2079–2096.

Thompson, R. C., et al. (2009). "Plastics, the environment and human health: Current consensus and future trends." *Philosophical Transactions of the Royal Society Series B: Biological Sciences* **364**(1526): 2153–2166.

Wikström, F., et al. (2014). "The influence of packaging attributes on consumer behaviour in food-packaging life cycle assessment studies—A neglected topic." *Journal of Cleaner Production* **73**: 100–108.

Wikström, F., et al. (2018). "Packaging strategies that save food: A research agenda for 2030." *Journal of Industrial Ecology* **23**(3): 532–540.

Wohner, B., et al. (2019). "Packaging-related food losses and waste: An overview of drivers and issues." *Sustainability* **11**(1): 264.

Zheng, B. J. (2012). "Green packaging materials and modern packaging design." *Applied Mechanics and Materials* **271–272**: 77–80.

5 Antimicrobial Studies on Food Packaging Materials

Mohd Nor Faiz Norrahim, S.M. Sapuan, Tengku Arisyah Tengku Yasim-Anuar, Farah Nadia Mohammad Padzil, Nur Sharmila Sharip, Lawrence Yee Foong Ng, Liana Noor Megashah, Siti Shazra Shazleen, Noor Farisha Abd. Rahim, R. Syafiq, and R.A. Ilyas

CONTENTS

5.1 INTRODUCTION

Recently, there has been great interest in the production of packaging plastics with various applications such as foods, pharmaceuticals, chemicals, detergents, cosmetics, compost bags, grocery bags, shipping bags, cutlery, plates and toys (Abral, Basri et al., 2019a; Jumaidin et al., 2019a, 2019b, 2019c; Norrahim et al., 2013; Sanyang et al., 2018). Packaging is a means of ensuring the safe delivery of a product to the consumer at the right time in every condition at optimum cost. Currently, many scientists and polymer engineers focus on sustainable eco-friendly packaging development based on renewable natural biopolymers due to health and current environmental issues caused by non-biodegradable food packaging (Atikah et al., 2019; Ilyas et al., 2018a, 2018b, 2018c, 2018d, 2018e, 2019a, 2020a, 2020b, 2020c). Besides this, to overcome their disadvantages such as low mechanical and water barrier properties, various physical and chemical treatments have been used to strengthen the biobased polymer for food packaging application including blending with other synthetic polymers, chemical modification, graft copolymerization, and incorporating fillers (e.g., lignin, clay, multi-walled carbon nanotubes, cellulose, and nanocellulose) (Abral et al., 2020a; Halimatul et al., 2019a; Ilyas et al., 2017, 2018f, 2019b, 2019c, 2019d; Syafri et al., 2019). The design of packaging material must consider three environments, which are physical, ambient and human. Failure to consider all three environments during packaging development will result in poorly designed packages, increased costs, consumer complaints, and even avoidance or rejection of the product by the consumer.

The areas of food packaging materials as well as active and modified atmosphere packaging have seen considerable advancement and application in the last two decades (Siracusa, 2016). The main concern of food packaging material is to protect and preserve a product from several factors such as oxidizing agents and microorganisms. Protection enables preservation of the food product from chemical, biological and physical deterioration factors (Abral et al., 2019b, 2020b; Atikah et al., 2019; Azammi et al., 2020; Halimatul et al., 2019b; Syafri et al., 2019).

The presence of microbial pathogens in foods is one of the important issues that researchers must deal with since it can cause a serious health threat to humans. The food industries are constantly developing and implementing new procedures in order to minimize the possibility than even a single pathogenic microbial cell survives in the food (Villa and Rama, 2016). Antimicrobial packaging presents an excellent innovative solution in food technology to reduce the microbial growth in food systems and extend food shelf life.

Antimicrobial packaging is classified as active packaging. Active packaging interacts with the product or the headspace between the package and the food system, to obtain a desired outcome. An example of active packaging is the combination of food-packaging materials with antimicrobial substances to control microbial surface contamination of foods. Several antimicrobial compounds such as enzymes, bacteriocins, natural extracts, organic acids, chelators and metal ions, have been applied in food packaging systems (Taylor et al., 2010). The incorporation of the antimicrobial agents into the packaging material can be done using several strategies. Several recent strategies developed for antimicrobial packaging materials include:

1. Uses of antimicrobial sachet pads
2. Incorporation of volatile and non-volatile antimicrobial agents
3. Antimicrobials surface coating
4. Immobilization of antimicrobials
5. Use of antimicrobial polymers as packaging material

5.2 STRATEGY 1: USES OF ANTIMICROBIAL SACHET PADS

Sachet pads are the first design of active packaging by which active compounds are either loosely enclosed or attached inside antimicrobial packages. Developed in Japan in the 1970s, they are considered to be the most successful antimicrobial packaging due to their high activity rate with simple packaging procedure by which no complicated machineries or modification are needed (Appendini and Hotchkiss, 2002; López-Rubio et al., 2004). Even though the recent development of active packaging has shifted toward uses of biodegradable/edible film, uses of sachet pads are still remarkably relevant especially for commercial application (Otoni, Espitia, Avena-Bustillos, and McHugh, 2016).

In general, sachet pads comprised of two major components of carrier and active ingredients. For example, ethanol as antimicrobial agent is absorbed into silicone dioxide powder and encapsulated in sachets that will act as ethanol emitting sachets such as the one established by Freund Industrial Co. Ltd in their product Ethicap™ package. The material used can be polymer and gels, calcium alginate beads, diatomaceous earth, finely divided silica, micro-celllulose foam, starch zeolite and porous high-density polyethylene (Otoni et al., 2016). In turns, the active ingredient can be oxygen scavengers, carbon dioxide scavengers/emitters, ethylene scavengers, ethanol emitters and moisture absorbers (Fang et al., 2017; López-Rubio et al., 2004; Mangalassary, 2019). These two main components (carrier and active ingredients) are enclosed in sachet pads of packaging barrier including metalized polypropylene, paper bag, polyethylene/nylon laminate, etc. Some commercially manufactured sachet pads for antimicrobial packaging are listed in Table 5.1 (adapted from Day (2008)).

Among the successful form of sachets, include oxygen and moisture absorbers as well as ethanol vapor generators. Sachet pads that act as moisture and oxygen absorbers are often used for pastries and bread, pasta and meat by which the oxidation and condensation of water can be restrained (Jideani and Vogt, 2016). It is important to note that sachet pads for antimicrobial packages must meet five specific requirements for ensuring preservation of food products which are: (1) able to take in its structure and the exuded liquid rapidly whether in its horizontal or inclined position; (2) able to hold the absorbed exuded liquid, thus visually and physically isolating the liquid from the product; (3) able to preserve visual presentation and sensorial properties of products without substantial cost to consumers; (4) possess certain mechanical resistance allowing uses at high-speed automation, where it is appropriate and when it is possible; and lastly (5) affords to extend shelf life and inhibit microbial growth of the products (Otoni et al., 2016).

It is well known that moisture is the main root of food spoilage where reduced moisture can greatly extend the shelf life of foods by microorganisms' growth inhibition and prevention of moisture-related degradation. Moisture absorber sachet pads usually contain desiccants such as calcium oxide (CaO), silica gel and natural clay. Its application, however, is not restricted to food but also can be used for pharmaceutical, electronic and electric goods (Day, 2008; Kerry and Butler, 2008). In addition, oxygen scavenger sachets contain various iron powders with catalysts that react with moisture within the package hence producing reactive hydrated metallic reducing agent capable of scavenging oxygen and turning it into stable oxide (Akelah, 2013; Jideani and Vogt, 2016; López-Rubio et al., 2004). Other mechanisms can be used for oxygen scavenger sachets including ascorbate, sulfites, catechol, photosensitive dyes and enzymes (Kour et al., 2013; Otoni et al., 2016). This kind of sachet is reported to be useful mainly for dried meat products to withstand oxidation (Fang et al., 2017).

In other systems, carbon dioxide emitters/scavenging sachets containing a mixture of calcium oxide and activated charcoal have been used for scavenging carbon dioxide in packages. This is because some products such as roasted coffee beans

TABLE 5.1
Summary of Some Commercially Manufactured Sachet Pads for Antimicrobial Packaging

Active system	Mechanism	Manufacturer (trade name)
Oxygen scavenger	Iron based	Mitsubishi Gas Chemical Co. Ltd. (Ageless®); Toppan Printing Co. Ltd. (Freshilizer™); Toagosei Chem. Industry Co. Ltd. (Vitalon™); Nippon Soda Co. Ltd. (Seagul™); Finetec Co. Ltd. (Sanso-Cut™); Ueno Seiyaku Co. Ltd. (Oxyeater™) Standa Industrie (ATCO®).
	Enzyme based (glucose oxidase)	Bioka Ltd (Bioka™).
Carbon dioxide scavengers/emitters	Iron based (usually used together for oxygen scavengers as dual-action system)	Mitsubishi Gas Chemical Co. Ltd. (Ageless® type E, Ageless® type G, Fresh Lock™); Toppan Printing Co. Ltd. (Freshilizer™ type CV); EMCO Packaging Systems Ltd.; SARL Codimer (Verifrais™); Multisorbs Technologies Inc. (FreshPax® type M).
Ethylene scavengers	Potassium permanganate	Air Repair Products Inc.; Ethylene Control Inc.; Extenda Life Systems.
	Activated carbon	Sekisui Jushi Ltd. (Hatofresh™); Mitsubishi Gas Chemicals Co. Ltd. (Sendo-Matte™).
Ethanol emitters	Ethanol	Freund Industrial Co. Ltd. (Ethicap™, Antimold 102™, Negamold™); Nippon Kayaku Co. Ltd. (Oitech™); Ueno Seiyaku Co. Ltd. (ET Pack™); Ohe Chemicals Co. Ltd. (Oytech L); Mitsubishi Gas Chemical Co. Ltd. (Ageless® type SE).
Moisture absorbers	Dessicants (silica gel, calcium oxide, activated clay, minerals)	Dupont Chemicals (Tyvek™); Multisorb Technologies Inc.; United Dessicants; Baltimore Chemicals; Mark & Spencer Plc.

emit carbon dioxide by which gas accumulation can lead to packages bursting (Day, 2008), aside from causing pH, color and flavor changes in products (López-Rubio et al., 2004). Besides calcium oxide and activated carbon, uses of calcium oxide and iron-based systems have been reported for carbon dioxide scavenging sachets (Kour et al., 2013). Applying a similar concept to oxygen scavengers, an iron-based system in fact allows harnessing the reaction of both oxygen scavengers and carbon dioxide emitters and is commercially used as dual-action sachets (Biji, Ravishankar, Mohan, and Srinivasa Gopal, 2015; Mangalassary, 2019). Carbon dioxide scavenger used with oxygen scavenger can also help in preventing package collapse or vacuum-generated condition from too much oxygen scavenger reaction. In a way, it stabilizes the act of oxygen scavenging sachets that absorb oxygen and generates an equal volume of carbon dioxide, producing balance that can prolong the shelf life of products (Akelah, 2013).

Ethylene scavengers help prolong shelf life of fruits and vegetables. The presence of ethylene can lead to accelerated maturing and ripening, softening yellowing and loss of chlorophyll in vegetables and development of bitter flavors (Biji et al., 2015; López-Rubio et al., 2004). In sachets containing ethylene scavengers, potassium permanganate has commonly been used, however, it is not suitable for food-contact packages due to the toxicity of KMnO4 and its purple color. Similar to dual-action oxygen/carbon dioxide scavengers, ethylene scavenger sachets are usually applied in packages together with moisture absorbers that consist of activated carbon, metal catalyst and silica gel (Akelah, 2013).

The function of sachet pads as active packaging has also been extended for antimicrobial-embedded volatile compounds. These include ethanol, chlorine dioxide and several kinds of plant essential oils (Eos). Figure 5.1 shows the generally used active compounds in emitting sachets and absorbent pads for food packaging. Briefly, non-volatile antimicrobial agents are directly incorporated into carrier or packaging materials for applications through direct contact. On the other hand, volatile antimicrobial agents can be easily released in a gaseous form to the headspace of packages that contains the sachets (Otoni et al., 2016).

The uses of essential oil-containing sachets have been reported which include uses of garlic (Ayala-Zavala and González-Aguilar, 2010), oregano (Oral et al., 2009), lemongrass (Espitia et al., 2011; Medeiros, Soares, Polito, de Sousa, and Silva, 2011), cinnamon (Jo et al., 2013), rosemary and thymes (Han et al., 2014) that are able to inhibit growth of a wide range of microbes including aerobic mesophilics, coliforms, yeasts, molds and bacteria through the action of phenolic compounds. Moreover, uses of sachets containing allyl isothiocyanate (AITC) were found to be effective in growth retardation of yeast, mold and bacteria. As an aldehyde compound, AITC induces a chemical reaction that is responsible for damaging cell membranes of microbes and fungi (Gonçalves, Pires, Soares, and Araújo, 2009; Otoni et al., 2014;

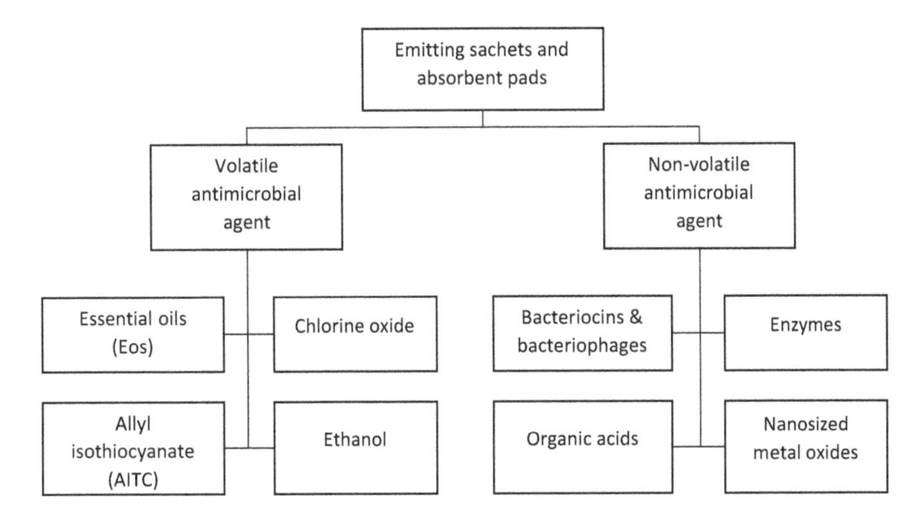

FIGURE 5.1 Classification of sachet pads for antimicrobial packaging. (Redrawn from Otoni et al., 2016.)

Pires et al., 2009; Seo et al., 2012). Similarly, chlorine dioxide and ethanol react with plasma membranes of microbes. For instance, ethanol vapor generator or ethanol emitter sachets, and releases of ethanol (absorbed or embedded within packages) through packages' selective barrier could inhibit growth of microorganisms including molds and bacteria. This however depends on moisture contents within the package, reflected through water activity (a_w) by which only at less than 0.92 a_w could mold inhibition reaction be allowed by a relatively small amount of ethanol emitted (Appendini and Hotchkiss, 2002). In its application for food packaging, the ethanol odor is often masked with a hint of vanilla or other flavors (Akelah, 2013).

Despite being successfully commercialized and used for antimicrobial packaging, the use of sachet pads can gain consumers' rejection primarily due to the uses of toxic products that are incorporated into sachet pads. The inclusion of sachet pads in the packages possesses the risk of their leaking out and thus contaminating the products (Akelah, 2013). This also explains why it is more suitable to be used for solid packaging rather than liquid or semiliquid products. As such, the uses of antimicrobial plastics where active ingredients are incorporated into the wall of package materials are intensively studied in search of an alternative to replacing sachet pads (López-Rubio et al., 2004). In fact, concerns over accidental ingestion of toxic active ingredients from sachets pads hindered thr commercial success of this application especially in North America and Europe (Kerry and Butler, 2008). However, continuous improvement of technology using emitting sachets and absorbent pads might cater to this disadvantage. In fact, the use of sachets is now being expanded not only for antimicrobial purposes, but also in flavoring, antioxidant, nutrient releasing and anti-insect activity (Otoni et al., 2016).

5.3 STRATEGY 2: INCORPORATION OF VOLATILE AND NON-VOLATILE ANTIMICROBIAL AGENTS

In this strategy, antimicrobial agents were used as an antimicrobial film in a polymer matrix for food packaging. In general, antimicrobial agents could also be classified into two categories: volatile and non-volatile compounds. Depending on their volatility, these types of active substances that are incorporated in polymers are either in direct contact with the food surface or in the package headspace. An antimicrobial film with a volatile substance is suitable for dry foods, as it is vaporized in the headspace packages and condensed into the food whereas, the non-volatile antimicrobial film needs to be in close contact with food as it can only pass into the food by having a film-food interface (Smith, 1993).

5.3.1 Volatile Antimicrobial Agents in Polymer Matrix

Volatile antimicrobial compounds in active packaging can migrate to the surface of the food without the polymer film being in direct contact with the food product. As shown in Figure 5.2, the polymer film acts as a carrier for antimicrobial agents to be released from the empty space left above the contents in a sealed container (Han, 2000). In general, the use of volatile antimicrobial compounds can be found in dried foods that are not packaged in a vacuum where internal interaction between food

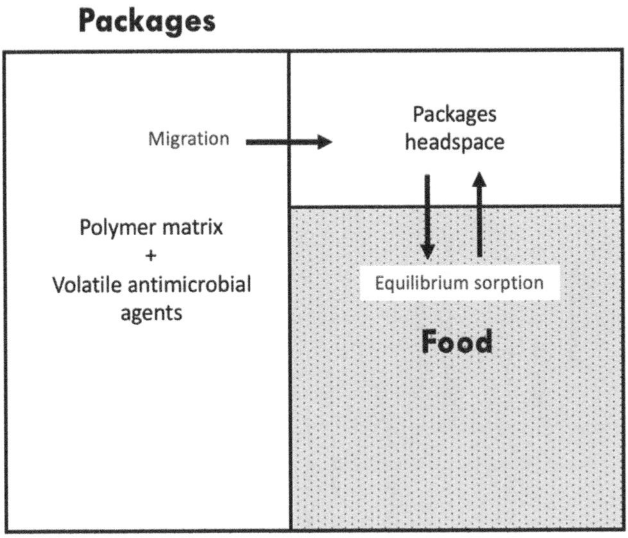

Packages

Migration

Packages headspace

Polymer matrix
+
Volatile antimicrobial agents

Equilibrium sorption

Food

FIGURE 5.2 Volatile antimicrobial packaging via packages headspace. (Redrawn from Han, 2002.)

and packaging can be avoided. Table 5.2 presents several antimicrobial packaging studies involving the incorporation of volatile antimicrobial compounds into polymer matrix. Examples of volatile compounds that have been subject to antimicrobial activity in food packaging include chlorine dioxide, lysozymes, essential oils (e.g., cinnamon, carvacrol, tea tree oil), ethylene vinyl alcohol (e.g., oregano, citral), lactic acid, nisin etc.

Chlorine dioxide is a gas-formed antimicrobial agent that directly incorporates into polymer. For the reaction of diffusion of chlorine dioxide to occur, the embedded chlorine dioxide within polymer matrix is dependent on moisture and temperature (Appendini and Hotchkiss, 2002). The natural antimicrobial lysozymes from the egg white of chicken eggs were found to have a beneficial effect when applied in several food packages. Previously, the lysozyme was successfully incorporated into a polymer matrix by several techniques including immobilization, absorption and entrapment process (Silvetti, Morandi, Hintersteiner, and Brasca, 2017). Other natural antimicrobial agents are essential oils which can be extracted from plants, herbs and spices. The most common method for extracting this aromatic oil without losing its properties is steam distillation. It is reported that essential oil has strong antimicrobial properties and can replace the synthetic preservatives in food packaging (Saeed, Afzaal, Tufail, and Ahmad, 2019). Carvacrol has been classified as generally recognized as safe (GRAS) by the Food and Drug Administration (FDA). Meanwhile, lactic acid bacteria can be obtained from seafood products or fermented foods. This type of antimicrobial agent is highly temperature dependent. Usually, lactic acid bacteria are stable at low temperatures for food spoilage preservation. In a review by Irkin and Esmer (Irkin and Esmer, 2015), nisin was found to have a wide range of antimicrobial activity in several food products. It has

TABLE 5.2
Volatile Antimicrobial Compound Incorporated in Polymer Matrix.

Volatile antimicrobial compounds	Types of polymer matrix	Function	Application	Sources
Chlorine dioxide	Polylactic acid (PLA)	• Preserve the quality of the food • Inhibit the growth of *Salmonella* and *Escherichia coli*. • Extend shelf life of fresh food product	Fruits like grape and tomato/vegetables	Ray, Jin, Fan, Liu, and Yam (2013)
Lysozymes	Chitosan, polyvinyl alcohol	• Extend shelf life of eggs • Maintain the freshness of internal quality • Chitosan coatings decrease eggshell breakage during handling and storage • Prevent the growth of *Clostridium tyrobutyricum* in cheese • Control the growth of lactic acid bacteria in beer and wine	Chicken eggs, cheese, beer and wine	Silvetti et al. (2017); Yuceer and Caner (2014)
Essential oils (e.g., cinnamon)	PLA and chitosan	• Has antifungal, antioxidant, antiviral, antimycotic, antioxygenic, antiparasitic, antibiotic, and antiseptic properties • Enable extension of shelf life in food packaging • Improve performance of water barrier properties • Provide environmental benefits	Fruits, meats	Liu et al. (2017)
Essential oils (e.g. carvacrol, tea tree oil)	Microcrystalline cellulose and PLA	• Effective against a variety of pathogenic and spoilage microorganisms including gram-negative and gram-positive bacteria (e.g., *Escherichia coli*, *Staphylococcus aureus*), yeast and molds • Inhibit the fungal growth • Extending shelf life of food	Fruits and vegetables	Rozenblit et al. (2018)

(*Continued*)

TABLE 5.2 (CONTINUED)
Volatile Antimicrobial Compound Incorporated in Polymer Matrix

Volatile antimicrobial compounds	Types of polymer matrix	Function	Application	Sources
Essential oil (e.g., oregano, and citral)	Ethylene vinyl alcohol copolymer/polypropylene (PP)	• Reduce the spoilage and inhibit the growth of pathogenic microorganisms such as *Escherichia coli*, *Salmonella enterica*, *Listeria monocytogenes* and natural microflora • Provide high barrier characteristics • Increase the storage period of food product	Packaged salad	Muriel-Galet et al. (2012)
Lactic acid	PLA	• Inhibit the bacteria growth by changing the pH to acidic • Extending shelf life to several days	Chilled poultry, fish	Ibrahim and El-Khawas (2019)
Nisin	Polyethylene (PE) and PLA	• Act as depolarizing agent on cytoplasmic membrane • Improve permeability of outer membrane against gram-negative bacteria • Inactivate the growth of microorganism like *Salmonella spp., Listeria monocytogenes, Escherichia coli and Lactobacillus plantarum.* • Extend shelf life of egg products • Prevent the organoleptic degradation	Dairy eggs, cereal-based products, vegetables	Jin and Gurtler (2011); Yuceer and Caner (2014)

also generally recognized as safe by the Food and Agriculture Organization of the United Nations (FAO), World Health Organization (WHO) and the US Food and Drug Administration. The incorporation of nisin with polymer film was effective in functional food packages, extending shelf life and providing good inhibition against pathogenic bacteria.

Table 5.3 shows starch-based polymer film incorporated with antimicrobial plant essential oil for food packaging applications. The result of the inhibition zone shows that if the content of essential oil is increased, a larger area can be inhibited by the essential oil. Increasing the content of essential oil shows that the inhibition activity can cover an extended area. Besides that, a better inhibition was observed with higher content of cinnamon oil. Even at minimum concentration applied into the film formulation, EO showed inhibition against microorganisms, which was considered an important result since higher concentrations could imply a sensorial impact, altering the natural taste of the food packaged by exceeding the acceptable flavor threshold.

5.3.2 Non-Volatile Antimicrobial Agents in Polymer Matrix

In contrast to the volatile antimicrobial compounds, that do not require close contact with food, the application of non-volatile antimicrobial compounds in packaging requires intimate contact with the food, to allow them to diffuse to the surface and onto the food (Kerry and Butler, 2008). Hence, to ensure successful diffusion as shown in Figure 5.3, surface concentration and diffusion kinetics are crucial, as these two factors will influence the diffusion of antimicrobial toward foods. The surface concentration will influence the antimicrobial release from the polymer; hence, it needs to be maintained above the minimal inhibitory concentration (MIC), to ensure the antimicrobial can be diffused from the packaging onto the food (Kerry and Butler, 2008; Limbo and Khaneghah, 2015). A previous study by Lü et al. (2009) revealed that the use of multilayer films was able to control the antimicrobial released to the food. The theory is, the inner layer may control the diffusion rate of the antimicrobial, while both matrix and barrier layers may prevent migration of the microbial toward the outside of a package.

This method is usually applied to food packaging containing semi-solid or liquid foods, where the foods can intimately get in contact with the package, and indirectly the antimicrobial can effectively be diffused into the food (Han, 2000). Organic acids, enzymes, bacteriocins and nanosized metal extracts are among the common antimicrobials used for this method. Table 5.4 lists some examples of non-volatile antimicrobial compounds incorporated into various polymer matrices for various applications. Nevertheless, the objectives were to protect foods from microbial attack, which indirectly restrict the growth of microorganisms, sustain the food's quality, slow the deterioration process and prolong the food's shelf life (Carbone et al., 2016).

5.4 STRATEGY 3: ANTIMICROBIALS SURFACE COATING

Antimicrobial agents have been widely used especially in many industries due to its ability to kill pathogenic microorganisms on material surfaces. However, due to its many advantages, such as short-term antimicrobial ability, most of the research

TABLE 5.3
Starch-Based Polymer Film Incorporated with Antimicrobial Plant Essential Oil for Food Packaging Application

Starch	Additive	Essential oil	Inhibition zone (%)		References
			S. aureus Gram **(+)**	***E. coli Gram*** **(−)**	
Corn/wheat	Tween80	Lemon	47.72 ± 2.87	45.56 ± 3.15	Song, Zuo, and Chen (2018)
Corn/wheat	Span 80	Lemon	45.89 ± 3.36	45.34 ± 3.48	Song et al. (2018)
			L. monocytogenes		
Modified starch	—	Peppermint oil	—	—	Liang et al. (2012)
			P. commune	***E. amstelodami***	
Cassava	—	Cinnamon	25.94 ± 5.72	91.06 ± 15.48	Souza, Goto, Mainardi, Coelho, and Tadini 2013)
			Escherichia coli	***Staphylococcus aureus***	
Cassava	—	Cinnamon	12.17 ± 0.29	11.17 ± 0.29	Iamareerat, Singh, Sadiq, and Anal (2018)
			E.coli ATCC 25922	***S.aureus ATCC 25923***	
Potato	—	Lavender	—	—	Jamróz, Juszczak, and Kucharek (2018)
			Aspergillus niger	***Penicillium spp.***	
Amaranth/chitosan/starch	—	Mexican oregano	—	—	Avila-Sosa et al. (2010)
			A. niger	***P. expansum***	
Pea	—	Oregano	—	—	Cano et al. (2015)
Pea	—	Neem	4.35 ± 0.10	4.34 ± 0.06	Cano et al. (2015)
			Pseudomonas fluorescens	***Aspergillus niger***	
Cassava	—	Lemon grass (0.5%)	25.89 ± 0.12	11.97 ± 0.176	Resianingrum (2016)

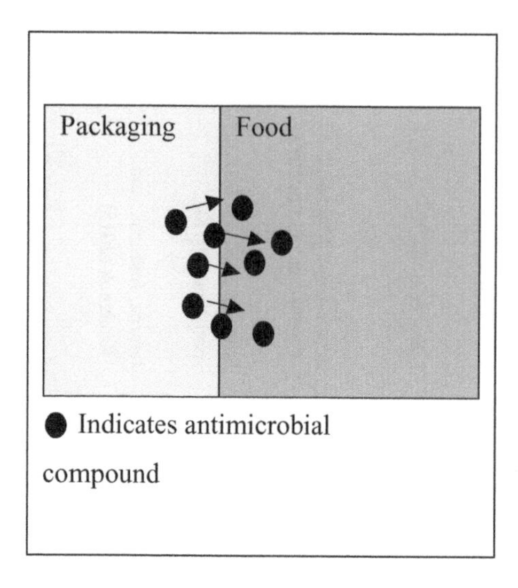

FIGURE 5.3 Mode of releasing non-volatile antimicrobials onto foods. (Redrawn from Limbo and Khaneghah, 2015.)

direction has been diverted to investigate the potential of polymeric antimicrobial agents. Among the major requirements of being a good antimicrobial polymer should be included: (1) long-term activity, (2) stability, (3) cost effective, (4) biocidal to a broad spectrum of pathogenic microorganisms, (5) can be re-generated upon loss of activity, and (6) not soluble in water. The food packaging is considered as the carrier for antimicrobial and antioxidant compounds which act as preservatives that avoid moisture loss, reduce the amount of spoilage on the surface of the food and restrict the volatile flavor loss. Therefore, it is important to make sure that appropriate coating methods can be delivered to ensure compatibility of the antimicrobial agent with the packaging material. Synthesis of antimicrobial polymer usually involves first synthesizing the polymers, followed by modification with an active species.

Antimicrobial polymers are commonly obtained either by synthesizing monomeric biocide moieties and then polymerizing subsequently or copolymerizing with another monomer. These reactions would result in antimicrobial material with low surface energy chemistry that minimizes the microbial attachment while antimicrobial additives kill bacteria or inhibit their growth. With current advanced deposition techniques, antimicrobial agent can be incorporated into surface coating by vapor deposition, ion implantation, and sputtering, electrochemical deposition from solution and can even be further designed to release the active agents over a prolonged time period. Preparation of polyethylene-co-methacrylic acid films coated with benzoic or sorbic acids has been reported to exhibit an antimicrobial property. It was also shown that sodium hydroxide-treated films have dominant antimicrobial properties for fungal growth compared to hydrochloric acid-treated films due to the higher amount of preservatives released from the films.

An effective antimicrobial suture coating has been successfully demonstrated by Yan Li et al. (2012), where an amphiphilic polymer poly [(aminoethyl

TABLE 5.4
Non-Volatile Antimicrobial Compound Incorporated into Polymer Matrix

Non-volatile antimicrobial compounds	Types of polymer matrix	Function	Applications	Sources
Silver nanoparticles (AgNPs) and zinc oxide nanoparticles (ZnONPs)	LDPE	• AgNPs and ZnONPs can provide close contact with the microorganism due to their extremely large surface area • AgNPs and ZnONPs are stable at high temperature (55–65°C) • AgNPs and ZnONPs were able to restrict the growth of yeast and mold in orange juice • The antimicrobial activity of AgNPs was able to reduce the pasteurization temperature of orange juice by 10°C	Orange juice	Emamifar, Kadivar, Shahedi, and Soleimanian-zad (2010)
Silver nanoparticles (AgNPs) and titanium oxide nanoparticles (TiO_2NPs)	PE	• AgNPs and TiO_2NPs were able to reduce the rate of deterioration and extend the shelf life of foods	Fresh orange juice	Metak (2015)
Propionic acid, acetic acid and lactic acid	Chitosan	• Inhibit microbial growth • Commonly been used for food preservation because of their antimicrobial properties	Water	Adapted from Matche and Jagadish (2011)
Nisin	SPI film, corn zein films	• Protect foods from foodborne pathogens and spoilage microorganisms by inhibiting the cell wall biosynthesis and spore outgrowth • Nisin is harmless, heat stable and does not lose its antimicrobial activity after pasteurization, sterilization and other processing methods	Dairy products, baked goods, mayonnaise and milk shakes	
Lysozyme	SPI film, corn zein films	• Useful to preserve foods from Gram-positive bacteria and thermophilic spores	Culture media	
Lacticin	—	• Inhibit growth of *L. monocytogenes Scott A* and *B. cereus* • Effective in eliminating pathogenic microorganisms, for example, 99% death rate of *L. monocytogenes* cells was recorded after the addition on lacticin 3147 to the milk	Yogurt, soup, milk	Santos et al. (2018)

methacrylate)-co-(butyl methacrylate)] (PAMBM) was bactericidal against *Staphylococcus aureus*. Coating of this antimicrobial peptide, PAMBM onto a variety of polymer blends showed excellent antimicrobial activity at low concentration. Similar antimicrobial coating formula of PAMBM applied to the commercial vicryl plus sutures has effectively killed bacteria (98% reduction at 0.75%). Further investigation proved that higher concentration of PAMBM increased the zone of inhibition.

Jaiswal et al. (2012) have successfully described the preparation of antibacterial silver, copper and zinc doped sol–gel surfaces where metal nitrate (Ag, Cu, Zn) doped methyltriethoxysilane (MTEOS) coating was synthesized. This liquid-sol–gel was tested for its antimicrobial properties by coating the material on the microtitre plate wells and challenged with biofilm-forming and antibiotic-resistant strains. The broad-spectrum antibacterial activity of the silver doped sol–gel showed its potential for use as a coating for biomaterials according to the following order: silver > zinc > copper. This order is affected by the presence of silver nanoparticles at the sol–gel coating surface, which lead to higher elution rates.

Han et al. (2007) reported that the antimicrobial activity of sorbic acid/carva crol/trans-cinnamaldehyde/thymol/rosemary oleoresin coated LDPE/polyamide films can be improved with the influence of electron beam irradiation. The results showed successful inhibition zones in agar diffusion test against *Listeria innocua* ATCC 33090 and *Escherichia coli* ATCC 884. Interestingly, neither the film's tensile strength, oxygen permeability nor toughness were affected with the presence of active compound nor dose. The film even became more ductile and had improved moisture barrier functionality.

The use of antimicrobial agents in food packaging has been widely emerging and contributing a significant impact on shelf life extension and food safety. The food packaging is considered as the carrier for antimicrobial and antioxidant compounds which act as preservatives that avoid moisture loss, reduce the amount of spoilage on the surface of the food and restrict the volatile flavor loss. Therefore, it is important to make sure that appropriate coating methods can be delivered to ensure compatibility of the antimicrobial agent with the packaging material. Preparation of polyethylene-co-methacrylic acid films coated with benzoic or sorbic acids has been reported to exhibit an antimicrobial property. It was also shown that sodium hydroxide-treated films have dominant antimicrobial properties for fungal growth compared to hydrochloric acid-treated films due to a higher amount of preservatives being released from the films.

Few studies of antibacterial packaging bags developed using plasma treatment and coating an antibacterial agent such as lysozyme or sodium benzoate have been performed. Wong et al. (2012) have reported that synthesis of gallic acid (3, 4, 5-trihydroxybenzoic acid, GA) agent-coated polymer can be improved by physically modifying the surface of the polyethylene (PE) film by plasma treatment. The effect of antimicrobial properties has shown its inhibition activities on *E. coli* and S. aureus, especially with the increase in concentration of GA. This result is consistent with the result of Karam et al. (2016) which showed antibacterial activity on plasma-treated nisin coated film only. It was also emphasized that plasma treatment has strengthened the antibacterial activity of the film as the concentration of the GA increase.

Gogliettino et al. (2020) investigated the effects of a food contact surface of polyethylene terephthalate (PET) coated with antimicrobial peptide mitochondrial-targeted peptide 1 (MTP1), in reducing the microbial population related to spoilage and in providing the shelf-life stability of different types of fresh foods such as ricotta cheese and buffalo meat. They have successfully shown that MTP1-PETs provided a strong antimicrobial effect for spoilage microorganisms with no cytotoxicity on a human colon cancer cell line, and the materials have also shown high storage stability and good reusability.

5.5 STRATEGY 4: IMMOBILIZATION OF ANTIMICROBIALS

The immobilization of antimicrobial agents (AMAs) to polymers typically involves the formation of intramolecular bonds (covalent or ionic) between the AMA and the polymer chains. There are many materials with known antimicrobial properties. However, their use in food-related applications is limited due to the cytotoxicity of some antimicrobial agents in addition to their potential to migrate into food, increasing the risk of harmful effects on the consumer (Zheng et al., 2016). Silver nanoparticles (AgNPs) for example, are among the most well-characterized nanoparticles and their antimicrobial properties have been well studied (Rai, Yadav et al., 2009). However, despite the obvious advantages of AgNPs in terms of their antimicrobial capabilities, studies have shown that AgNPs are toxic not just to mammalian cells but also to the environment (Lekamge et al., 2018; Sambale et al., 2015). Therefore, physical incorporation of antimicrobials such as AgNPs may not be stable enough to prevent the migration of said particles into food content (Addo Ntim, Thomas, Begley, and Noonan, 2015). To alleviate this limitation, researchers turned to immobilization of AMAs onto polymers with the help of covalent or ionic bonds.

In order to attach AMAs covalently to polymers, oftentimes, the polymer would have to undergo surface modification in order to facilitate the covalent bonding of the AMAs. For instance, Zheng et al. 2016, reported the covalent attachment of AgNPs to electrospun cotton fibers. In their study, electrospun cotton fibers were first modified by replacing the surface hydroxyl groups of the fibers with trityl-3-mercaptopropionic in a process of esterification before the complex was further modified by removing the trityl protecting group to allow the AgNPs to bind covalently to the newly exposed sulfhydryl groups (Zheng et al., 2016). Sometimes, the target polymer may not possess the required polar groups that would allow metallic particles such as AgNPs to bind to them. In such cases, other polymers that contain these polar groups may be covalently attached as a thin layer onto the surface of the target polymer. A good example of such a case is polyethylene terephthalate, a well-known and incredibly versatile polymer material that is used in a variety of applications. However, PET does not contain strong polar groups that would allow the binding of AgNPs onto its surface. To overcome this problem, Fragal et al. 2016, covalently attached ultrathin layers of polyvinyl alcohol (PVA) and polyacrylic acid (PAA) onto the surface of PET. The new composite would then have the required polar groups to form ionic bonds with the AgNPs.

Another type of AMA that is also suitable for covalent immobilization onto polymer surfaces are the antimicrobial peptides (AMPs). In addition to the benefit of reducing the toxicity of the AMPs, covalent immobilization of AMPs onto polymer surfaces increases their duration of effectiveness due to higher stability as compared to other methods of surface coating (Silva et al., 2016). Many studies on the immobilization of AMPs involve the use of self-assembled monolayers (SAMs). Monteiro et al. (2019), did a study on the covalent immobilization of Chain201D, a type of AMP that is effective against bacteria like *E. coli* and *S. aureus*, onto the surface of an EG4-thiol monolayer. It was found that the EG4-thiol monolayer containing Chain201D had a killing percentage of approximately 80% for *E. coli* and approximately 99% for *S. aureus* proving that the antimicrobial activity of the Chain201D AMP was not hindered by their immobilization onto a surface (Monteiro et al., 2019). In another study, researchers covalently immobilized MSI-78, a synthetic analog of the AMP magainin 2, onto a maleimide-terminated SAM (Xiao et al., 2018). They found that, when attached to a surface, the mechanism of bacterial killing of the AMPs was altered. Typically, most AMPs kill bacteria by disrupting the membrane of the bacteria, forming pores (Zhang and Gallo, 2016). However, due to the lack of freedom of the bound AMPs, the MSI-78 used by Xiao et al., 2018, was theorized to kill bacteria through charge interactions instead. In another study by the same authors, they compared the long-term effectiveness of cecropin-melittin hybrid peptide covalently attached to a SAM and a dibromomaleimide (DBM) which was polymerized via chemical vapor deposition (CVD) (Xiao, Jasensky, Gerszberg et al., 2018). They found that the CVD polymerized substrate was a superior candidate when it comes to long-term antibacterial activity due to its significant antibacterial activity as compared to the SAM substrate which lost its antibacterial function after five days (Xiao, Jasensky, Gerszberg et al., 2018). Covalent immobilization of AMAs could also be taken a step further by immobilizing them on highly porous materials, further increasing the scope of potential applications of antimicrobial materials. KR-12, an AMP with broad-spectrum antimicrobial activity, was covalently immobilized onto modified-starch sponges and was found to be extremely effective against a wide range of bacteria including methicillin-resistant *S. aureus* (MRSA) (Yang et al., 2019). This was achieved by a straightforward click chemistry reaction between the thiol groups in KR-12 and the ethylene group in the modified starch (Yang et al., 2019).

Either through covalent or ionic linkages, the immobilization of antimicrobials to polymers needs the interaction of functional groups on both the antimicrobial as well as the polymer. The antimicrobials on polymers will slowly be released into the food via ionic bonding and in the cationic polyelectrolytes, it easily interacts with the bacterial membrane. For instance, quaternization of ammonium/phosphonium and N-alkyl chitosan/methyl iodide was able to produce water-soluble cationic polyelectrolytes. According to Kenawy et al. (2007), the increase in the chain length of the alkyl substituent contributed to the increase of the antimicrobial activities of the chitosan quaternary ammonium salts. These increments are due to the increased lipophilic properties of the derivatives. The cationic polyelectrolytes are described where polycations interact with the cytoplasmic membrane of bacteria. Ikeda and co-workers are the pioneers in this study by utilizing two types of bacterial strains (*E.coli* and *S.aureus*) to explore biocidal activity of polymethacrylates functionalized with biguanide groups and polyvinylbenzyl ammonium chloride. For the past few years, several groups have carried out studies on incorporation of quaternary ammonium and phosphonium groups within the polymer chain to observe their antimicrobial activity.

Sanches et al. (2015) work on the synthesis of nanoparticles over several concentration ranges of poly (diallyldimethylammonium) chloride (PDDA)/ methylmethacrylate(MMA) via emulsion polymerization. The PMMA/PDDA dispersions are characterized by zeta-potential, dynamic light scattering, scanning electron microscopy, and others in order to determine the minimal microbial concentration (MMC) against *C. albicans*, *S. aureus* and *E. coli*, including hemolysis evaluation towards mammalian erythrocytes. There is a high colloidal stability for the cationic PMMA/PDDA nanoparticles over a range of PDDA. Nanoparticles' diverse antimicrobial activity against the aforementioned microorganisms reduces cell viability and totally kills *E. coli* over a range of PDDA that are innocuous to the erythrocytes. Hemolysis is measured higher than 15% for immobilized PDDA at the minimal microbicidal concentrations for both *C. albicans* and *S. aureus*. Therefore, the access to the inner layers of the cell wall and the cell membrane which are the major sites of PDDA antimicrobial action is determined by the mobility of the cationic antimicrobial polymer PDDA (Sanches et al., 2015).

In another study, coating or films embedded with chitosan and poly-L-lysine (cationic polymers) where it is already known as antimicrobial were used. Both cationic polymers will react with negative charges on the cell membrane, resulting in leakage of their intracellular components. Some of the synthetic antimicrobial macromolecules are effectively used as antimicrobial polymer because of their unique biochemical properties where they also include antimicrobial peptides, polymers and peptide polymer hybrids. These kinds of antimicrobial macromolecules are a promising approach to reducing the tendency of pathogen-resistant development.

Mostly two functional groups are involved in these immobilizations of antimicrobial to polymers and which one it is depends on the interactions between the antimicrobial and the polymer. The functional groups that have been utilized usually consist of enzymes, peptides, polyzmines, or organic acids for the antimicrobials. While for the polymers, using functional monomers or modifying nonfunctional polymer has been employed as the other alternative. In addition to this, a particular aspect must be taken into consideration in this strategy in which the activity of the immobilized compound needs to be reduced. Proteins and enzymes may lower the antimicrobial activity due to their denaturation processes including alteration of secondary structure. Therefore, to cater to this issue, spacers or small molecules are required to be placed between the bioactive agent and the polymer surface. They improve the freedom of motion and favor the contact between the microorganism and antimicrobial agent.

5.6 STRATEGY 5: USE OF ANTIMICROBIAL POLYMERS AS PACKAGING MATERIAL

Some polymers that are inherently antimicrobial have been used as antimicrobial packaging in films and coatings for various food applications today. Direct addition of antimicrobials is well recognized to involve a certain loss of activity due to leaching into the food matrix and reaction with other food components such as lipids or proteins (Wang et al., 2015). Hence, the use of antimicrobial polymer-based packaging film has been proven to be more effective by maintaining high concentrations on the food surface with a low level of migration of the active substance into the food and maintaining antimicrobial activity in transportation and storage over time (Coma et al., 2002). The applications of inherently antimicrobial polymers as antimicrobial packaging in food industries are summarized in Table 5.5.

Cationic polymers such as chitosan and poly-L-lysine have shown antimicrobial properties by promotion of cell adhesion whereby these polymers interact with negative charges on the cell membrane and cause the leakage of their intracellular components. Other than chitosan and poly-L-lysine, calcium alginate, bactericidal acrylic polymers made by co-polymerizing acrylic protonated amine co-monomer and nylon films treated with UV irradiation also have been proposed as antimicrobial packaging materials for meat, fruits and vegetables (Appendini and Hotchkiss, 2002).

Among them, chitosan has been favored to be used in antimicrobial packaging because it can be formed into edible films and coatings or blended with other polymers or incorporated with micro- or nanoparticles to form bio-based composites or nanocomposites in order to improve their functional properties such as mechanical and barrier properties. Apart from being a polymer substrate, chitosan also has been widely used as an antimicrobial agent owing to its efficacy in inhibiting the growth of gram-negative and gram-positive bacteria alongside the yeast and molds (Dıblan and Kaya, 2018). According to a study conducted by Lin et al. (2018), it was reported that the number of *Salmonella typhimurium* and *Salmonella enteritidi* in Ɛ-polylysine/chitosan nanofibers were decreased as compared in control from 8.21 Log CFU/g and 8.37 Log CFU/g to 5.03 Log CFU/g and 5.25 Log CFU/g, respectively. This finding demonstrated that the Ɛ-polylysine/chitosan nanofibers were successful in inhibiting *Salmonella* on chicken. Also, Ali et al. (2014) had produced edible antimicrobial coating from chitosan combined with lemongrass essential oil. They have observed that edible coating from chitosan and lemongrass essential oil at concentrations of 1.0% and 0.5% was significantly better at maintaining the bell pepper quality.

Nevertheless, inherently antimicrobial polymers generally have the advantage of film-forming properties and biodegradability but the downside of such a system is the direct contact between packaging and food (Gonçalves, 2012).

5.7 CONCLUSIONS

Food quality and freshness products may cause by happenstance numerous unwanted physical and chemical properties changes before, during or after processing stages. Packaging is one of the preventions for food damage. In order to retain food quality and freshness, selecting materials and packaging technologies correctly is required. The biodegradable film keeps food quality and is an environmentally friendly packaging. Good packaging is necessary to prevent food spoilage before and after processing. The introduction of antimicrobial agent into food packaging films form strong prevention of bacterial and fungi properties, positively affecting the storage of

TABLE 5.5
Inherently Antimicrobial Polymers Used in Food Packaging

Antimicrobial polymers	Packaging materials	Food applications	Targeted microorganisms	References
Chitosan	PLA/chitosan	Turkey meat	*Listeria innocue*	Guo et al. (2014)
	Chitosan	Bologna, cooked ham, beef pastrami	*Lactobacillus sakei*, *Serratia liquefaciens*	Ouattara et al. (2000)
	Chitosan	Culture media	*Listeria monocytogenes*	Coma et al. (2002)
	Ɛ-polylysine/chitosan nanofibers	Chicken	*Salmonella*	Lin et al. (2018)
	Chitosan/poly(L-lactic acid)	—	*Staphylococcus aureus*	Wang et al. (2015)
	Gelatin-chitosan-oregano essential oil	Fish	*Escherichia coli*, *Staphylococcus aureus*, *Bacillus subtilis*, *Salmonella enteriditis*, *Shigabacillus*	Wu et al. (2014)
	Chitosan-lemongrass essential oil	Bell pepper	*Colletotrichum capsici*	Ali et al. (2014)
	Chitosan-essential oil	Processed meat (bologna slices)	*Listeria monocytogenes*, *Escherichia coli*	Zivanovic et al. (2005)
	Chitosan acetate	—	*Escherichia coli*, *Staphylococcus aureus*	Liu et al. (2004)
	Chitosan-olive oil	Apple, strawberry	*Penicillium expansum*, *Rhizopus stolonifer*, *Bacillus cereus*, *Escherichia coli*, *Listeria monocytogenes*, *Staphylococcus aureus*, *Yersinia Spp.*, *Salmonella typhimurium*	Khalifa et al. (2016)
	Chitosan-oregano essential oil	Cheese slices	*Listeria monocytogenes*, *Staphylococcus aureus*	Torlak and Nizamlioglu (2011)
Poly-L-lysine	Starch/ Ɛ-poly-L-lysine	—	*Escherichia coli*, *Bacillus subtilis*, *Aspergillus niger*	Zhang et al. (2015)
	Ɛ -polylysine	Bread	*Aspergillus parasiticus*, *Penicillium expansum*	Luz et al. (2018)
	Ɛ-poly-L-lysine/nisin A	—	*Listeria monocytogenes*, *Bacillus cereus*	Najjar et al. (2007)
	Ɛ-poly-L-lysine	—	*Saccharomyces cerevisiae*	Bo et al. (2016)
	Ɛ-poly-L-lysine	Ready-to-eat (RTE) meats	*Listeria monocytogenes*	Brandt et al. (2010)

(Continued)

TABLE 5.5 (CONTINUED)
Inherently Antimicrobial Polymers Used in Food Packaging

Antimicrobial polymers	Packaging materials	Food applications	Targeted microorganisms	References
	Carboxymethyl cellulose/ montmorillonite/ε-poly-L-lysine	—	*Staphylococus aureus*, *Escherichia coli*, *Botrytis cinerea*, *Rhizopus oligosporus*	He et al. (2019)
Calcium alginate	Calcium alginate gels	Raw beef	*Listeria monocytogenes*	Siragusa and Dickson (1992)
	Calcium alginate-garlic oil	Lean beef	*Staphylococcus aureus*, *Bacillus cereus*	Pranoto et al. (2005)
	Calcium alginate	Refrigerated lamb meat	—	Koushki et al. (2015)
Bacterial acrylic polymers	Bacterial acrylic polymers	—	*Aspergillus niger*, *Saccharomyces cerevisiae*, *Cladosporium Sp.*, *Klebsiella pneumonia*, *Escherichia coli*, *Enterobacter agglomerans*, *Staphylococcus aureus*, *Staphylococcus epidermidis*, *Bacillus subtilis*, *Pseudomonas aeruginosa*	Pardini, (1987)
Polyamide	UV-irradiated nylon film	—	*Staphylococcus aureus*, *Pseudomonas fluorescens*, *Enterococcus faecalis*	Paik et al. (1998)

food products. The reinforcement of the antimicrobial agents within the packaging material can be achieved in several ways such as (1) use of antimicrobial sachet pads; (2) incorporation of volatile and non-volatile antimicrobial agents; (3) antimicrobials surface coating; (4) immobilization of antimicrobials; and (5) use of antimicrobial polymers as packaging material.

REFERENCES

Abral, H., Ariksa, J., Mahardika, M., Handayani, D., Aminah, I., Sandrawati, N., Sapuan, S. M., and Ilyas, R. A. (2019a). Highly transparent and antimicrobial PVA based bionanocomposites reinforced by ginger nanofiber. *Polymer Testing*, 106186. doi:10.1016/j.polymertesting.2019.106186.

Abral, H., Ariksa, J., Mahardika, M., Handayani, D., Aminah, I., Sandrawati, N., Pratama, A. B., Fajri, N., Sapuan, S. M., and Ilyas, R. A. (2020a). Transparent and antimicrobial cellulose film from ginger nanofiber. *Food Hydrocolloids*, *98*, 105266. doi:10.1016/j.foodhyd.2019.105266.

Abral, H., Atmajaya, A., Mahardika, M., Hafizulhaq, F., Kadriadi, Handayani, D., Sapuan, S. M., and Ilyas, R. A. (2020b). Effect of ultrasonication duration of polyvinyl alcohol (PVA) gel on characterizations of PVA film. *Journal of Materials Research and Technology*, 1–10. doi:10.1016/j.jmrt.2019.12.078.

Abral, H., Basri, A., Muhammad, F., Fernando, Y., Hafizulhaq, F., Mahardika, M., Sugiarti, E., Sapuan, S. M., Ilyas, R. A., and Stephane, I. (2019b). A simple method for improving the properties of the sago starch films prepared by using ultrasonication treatment. *Food Hydrocolloids*, *93*, 276–283. doi:10.1016/j.foodhyd.2019.02.012.

Addo Ntim, S., Thomas, T. A., Begley, T. H., and Noonan, G. O. (2015). Characterisation and potential migration of silver nanoparticles from commercially available polymeric food contact materials. *Food Additives and Contaminants—Part A: Chemistry, Analysis, Control, Exposure and Risk Assessment*, *32*(6), 1003–1011. doi:10.1080/19440049.2015.1029994.

Akelah, A. (2013). Polymers in food packaging and protection. In: *Functionalized Polymeric Materials in Agriculture and the Food Industry* (pp. 293–347). Springer US. doi:10.1007/978-1-4614-7061-8_6.

Ali, A., Noh, N. M., and Mustafa, M. A. (2014). ScienceDirect antimicrobial activity of chitosan enriched with lemongrass oil against anthracnose of bell pepper. *Food Packaging and Shelf Life*, *3*, 56–61. doi:10.1016/j.fpsl.2014.10.003.

Appendini, P., and Hotchkiss, J. H. (2002). Review of antimicrobial food packaging. *Innovative Food Science and Emerging Technologies*, *3*(2), 113–126. doi:10.1016/S1466-8564(02)00012-7.

Atikah, M. S. N., Ilyas, R. A., Sapuan, S. M., Ishak, M. R., Zainudin, E. S., Ibrahim, R., Atiqah, A., Ansari, M. N. M., and Jumaidin, R. (2019). Degradation and physical properties of sugar palm starch/sugar palm nanofibrillated cellulose bionanocomposite. *Polimery*, *64*(10), 27–36. doi:10.14314/polimery.2019.10.5.

Avila-Sosa, R., Hernández-Zamoran, E., López-Mendoza, I., Palou, E., Munguía, M. T. J., Nevárez-Moorillón, G. V., and López-Malo, A. (2010). Fungal inactivation by Mexican oregano (Lippia berlandieri Schauer) essential oil added to amaranth, chitosan, or starch edible films. *Journal of Food Science*, *75*(3). doi:10.1111/j.1750-3841.2010.01524.x.

Ayala-Zavala, J. F., and González-Aguilar, G. A. (2010). Optimizing the use of garlic oil as antimicrobial agent on fresh-cut tomato through a controlled release system. *Journal of Food Science*, *75*(7), M398–M405. doi:10.1111/j.1750-3841.2010.01723.x.

Azammi, A. M. N., Ilyas, R. A., Sapuan, S. M., Ibrahim, R., Atikah, M. S. N., Asrofi, M., and Atiqah, A. (2020). Characterization studies of biopolymeric matrix and cellulose fibres based composites related to functionalized fibre-matrix interface. In: *Interfaces in Particle and Fibre Reinforced Composites* (1st ed., pp. 29–93). London: Elsevier. doi:10.1016/B978-0-08-102665-6.00003-0.

Biji, K. B., Ravishankar, C. N., Mohan, C. O., and Srinivasa Gopal, T. K. (2015, October). Smart packaging systems for food applications: A review. *Journal of Food Science and Technology*. Springer India. doi:10.1007/s13197-015-1766-7.

Bo, T., Han, P., Su, Q., Fu, P., Guo, F., Zheng, Z.-X., Tan, Z., Zhong, C., and Jia, S.-R. (2016). Antimicrobial ε-poly-L-lysine induced changes in cell membrane compositions and properties of Saccharomyces cerevisiae. *Food Control*, *61*, 123–134. doi:10.1016/j.foodcont.2015.09.018.

Brandt, A. L., Castillo, A., Harris, K. B., Keeton, J. T., Hardin, M. D., and Taylor, T. M. (2010). Inhibition of Listeria monocytogenes by food antimicrobials applied singly and in combination. *Journal of Food Science*, *75*(9), 557–563. doi:10.1111/j.1750-3841.2010.01843.x.

Cano, A., Cháfer, M., Chiralt, A., and González-Martínez, C. (2015). Physical and antimicrobial properties of starch-PVA blend films as affected by the incorporation of natural antimicrobial agents. *Foods*, *5*(1), 3. doi:10.3390/foods5010003.

Carbone, M., Tommasa, D., Sabbatella, G., and Antiochia, R. (2016). Silver nanoparticles in polymeric matrices for fresh food packaging. *Journal of King Saud University—Science*, *28*(4), 273–279. doi:10.1016/j.jksus.2016.05.004.

Coma, V., Martial-Gros, A., Garreau, S., Copinet, A., Salin, F., and Deschamps, A. (2002). Edible antimicrobial films based on chitosan matrix. *Journal of Food Science*, *67*(3), 1162–1169. doi:10.1111/j.1365-2621.2002.tb09470.x.

Day, B. P. F. (2008). Active packaging of food. In: *Smart Packaging Technologies for Fast Moving Consumer Goods* (pp. 1–18). John Wiley & Sons, Ltd. doi:10.1002/9780470753699.ch1.

Dıblan, S., and Kaya, S. (2018). Antimicrobials used in active packaging films. *Food and Health*, *4*(1), 63–79. doi:10.3153/jfhs18007.

Emamifar, A., Kadivar, M., Shahedi, M., and Soleimanian-zad, S. (2010). Evaluation of nanocomposite packaging containing Ag and ZnO on shelf life of fresh orange juice. *Innovative Food Science and Emerging Technologies*, *11*(4), 742–748. doi:10.1016/j.ifset.2010.06.003.

Espitia, P. J. P., De Soares, N. F. F., Botti, L. C. M., Da Silva, W. A., De Melo, N. R., and Pereira, O. L. (2011). Active sachet: Development and evaluation for the conservation of Hawaiian papaya quality. *Italian Journal of Food Science*, *23*(SUPPL.), 107–110.

Fang, Z., Zhao, Y., Warner, R. D., and Johnson, S. K. (2017, March). Active and intelligent packaging in meat industry. *Trends in Food Science and Technology*. Elsevier Ltd. doi:10.1016/j.tifs.2017.01.002.

Fragal, V. H., Cellet, T. S. P., Pereira, G. M., Fragal, E. H., Costa, M. A., Nakamura, C. V., Asefa, T., Rubira, A. F., and Silva, R. (2016). Covalently-layers of PVA and PAA and in situ formed Ag nanoparticles as versatile antimicrobial surfaces. *International Journal of Biological Macromolecules*, *91*, 329–337. doi:10.1016/j.ijbiomac.2016.05.056.

Gogliettino, M., Balestrieri, M., Ambrosio, R. L., Anastasio, A., Smaldone, G., Proroga, Y. T. R., Moretta, R., Rea, I., De Stefano, L., Agrillo, B. and Palmieri, G. (2020, January). Extending the shelf-life of meat and dairy products via PET-modified packaging activated with the antimicrobial peptide MTP1. *Frontiers in Microbiology*, *10*, 1–11. doi:10.3389/fmicb.2019.02963.

Gonçalves, A. A. (2012). Packaging for chilled and frozen seafood. In: L. Nollet (Ed.), *Handbook of Meat, Poultry and Seafood Quality* (2nd ed., pp. 510–545). John Wiley & Sons, Inc. doi:10.1002/9781118352434.ch32.

Gonçalves, M. P. J. C., Pires, A. C. D. S., Soares, N. de F. F., and Araújo, E. A. (2009). Use of allyl isothiocyanate sachet to preserve cottage cheese. *Journal of Foodservice*, *20*(6), 275–279. doi:10.1111/j.1748-0159.2009.00150.x.

Guo, M., Jin, T. Z., Wang, L., Scullen, O. J., and Sommers, C. H. (2014). Antimicrobial films and coatings for inactivation of Listeria innocua on ready-to-eat deli turkey meat. *Food Control*, *40*(1), 64–70. doi:10.1016/j.foodcont.2013.11.018.

Halimatul, M. J., Sapuan, S. M., Jawaid, M., Ishak, M. R., and Ilyas, R. A. (2019a). Effect of sago starch and plasticizer content on the properties of thermoplastic films: Mechanical testing and cyclic soaking-drying. *Polimery*, *64*(6), 32–41. doi:10.14314/polimery.2019.6.5.

Halimatul, M. J., Sapuan, S. M., Jawaid, M., Ishak, M. R., and Ilyas, R. A. (2019b). Water absorption and water solubility properties of sago starch biopolymer composite films filled with sugar palm particles. *Polimery*, *64*(9), 27–35. doi:10.14314/polimery.2019.9.4.

Han, J., and Moreira, R. G. (2007). The influence of electron beam irradiation of antimicrobial-coated LDPE / polyamide films on antimicrobial activity and film properties. *LWT—Food Science and Technology* *40*(9), 1545–1554. doi:10.1016/j.lwt.2006.11.012.

Han, J. H. (2000). Antimicrobial food packaging. *Food Technology*, *54*, 56–65. doi:10.1533/9781855737020.1.50.

Han, J. H. (2002). *Antimicrobial Food Packaging: Novel Food Packaging Techniques: An Introduction*. Woodhead Publishing Limited. doi:10.1533/9781855737020.1.50.

Han, Jung H., Patel, D., Kim, J. E., and Min, S. C. (2014). Retardation of *Listeria monocytogenes* growth in mozzarella cheese using antimicrobial sachets containing rosemary oil and thyme oil. *Journal of Food Science*, *79*(11), E2272–E2278. doi:10.1111/1750-3841.12659.

He, Y., Fei, X., and Li, H. (2019). Carboxymethyl cellulose-based nanocomposites reinforced with montmorillonite and ε-poly-L-lysine for antimicrobial active food packaging. *Journal of Applied Polymer Science*, *48782*, 1–12. doi:10.1002/app.48782.

Iamareerat, B., Singh, M., Sadiq, M. B., and Anal, A. K. (2018). Reinforced cassava starch based edible film incorporated with essential oil and sodium bentonite nanoclay as food packaging material. *Journal of Food Science and Technology*. doi:10.1007/s13197-018-3100-7.

Ibrahim, S., and El-Khawas, K. M. (2019). Development of eco-environmental nano-emulsified active coated packaging material. *Journal of King Saud University—Science*, *31*(4), 1485–1490. doi:doi:10.1016/j.jksus.2019.09.010.

Ilyas, R. A., Sapuan, S. M., Atiqah, A., Ibrahim, R., Abral, H., Ishak, M. R., Zainudin, E. S., Nurazzi, N. M., Atikah, M. S. N., Ansari, M. N. M., Asyraf, M. R. M., Supian, A. B. M., and Ya, H. (2020a). Sugar palm (Arenga pinnata [Wurmb.] Merr) starch films containing sugar palm nanofibrillated cellulose as reinforcement: Water barrier properties. *Polymer Composites*, *41*(2), 459–467. doi:10.1002/pc.25379.

Ilyas, R. A., Sapuan, S. M., Ishak, M. R., and Zainudin, E. S. (2018a). Water transport properties of bio-nanocomposites reinforced by sugar palm (Arenga pinnata) nanofibrillated cellulose. *Journal of Advanced Research in Fluid Mechanics and Thermal Sciences Journal*, *51*(2), 234–246.

Ilyas, R. A., and Sapuan, S. M. (2020b). The preparation methods and processing of natural fibre bio-polymer composites. *Current Organic Synthesis*, *16*(8), 1068–1070. doi:10.2174/157017941608200120105616.

Ilyas, Rushdan Ahmad, Sapuan, S. M., Ibrahim, R., Abral, H., Ishak, M. R., Zainudin, E. S., Asrofi, M., Atikah, M. S. N., Huzaifah, M. R. M., Radzi, A. M., Azammi, A. M. N., Shaharuzaman, M. A., Nurazzi, N. M., Syafri, E., Sari, N. H., Norrrahim, M. N. F., and Jumaidin, R. (2019d). Sugar palm (Arenga pinnata (Wurmb.) Merr) cellulosic fibre hierarchy: A comprehensive approach from macro to nano scale. *Journal of Materials Research and Technology*, *8*(3), 2753–2766. doi:10.1016/j.jmrt.2019.04.011.

Ilyas, R. A., Sapuan, S. M., Ibrahim, R., Abral, H., Ishak, M. R., Zainudin, E. S., Atiqah, A., Atikah, M. S. N., Syafri, E., Asrofi, M., and Jumaidin, R. (2020c). Thermal, biodegradability and water barrier properties of bio-nanocomposites based on plasticised sugar palm starch and nanofibrillated celluloses from sugar palm fibres. *Journal of Biobased Materials and Bioenergy*, *14*(2), 234–248. doi:10.1166/jbmb.2020.1951.

Ilyas, R. A., Sapuan, S. M., Ibrahim, R., Abral, H., Ishak, M. R., Zainudin, E. S., Atikah, M. S. N., Mohd Nurazzi, N., Atiqah, A., Ansari, M. N. M., Syafri, E., Asrofi, M., Sari, N. H., and Jumaidin, R. (2019a). Effect of sugar palm nanofibrillated cellulose concentrations on morphological, mechanical and physical properties of biodegradable films based on agro-waste sugar palm (Arenga pinnata (Wurmb.) Merr) starch. *Journal of Materials Research and Technology*, *8*(5), 4819–4830. doi:10.1016/j.jmrt.2019.08.028.

Ilyas, R. A., Sapuan, S. M., Ibrahim, R., Atikah, M. S. N., Atiqah, A., Ansari, M. N. M., and Norrrahim, M. N. F. (2019b). Production, processes and modification of nanocrystalline cellulose from agro-waste: A review. In: *Nanocrystalline Materials* (pp. 3–32). Intech Open. doi:10.5772/intechopen.87001.

Ilyas, R. A., Sapuan, S. M., and Ishak, M. R. (2018b). Isolation and characterization of nanocrystalline cellulose from sugar palm fibres (Arenga pinnata). *Carbohydrate Polymers*, *181*, 1038–1051. doi:10.1016/j.carbpol.2017.11.045.

Ilyas, R. A., Sapuan, S. M., Ishak, M. R., and Zainudin, E. S. (2017). Effect of delignification on the physical, thermal, chemical, and structural properties of sugar palm fibre. *BioResources*, *12*(4), 8734–8754. doi:10.15376/biores.12.4.8734-8754.

Ilyas, R. A., Sapuan, S. M., Ishak, M. R., and Zainudin, E. S. (2018c). Development and characterization of sugar palm nanocrystalline cellulose reinforced sugar palm starch bionanocomposites. *Carbohydrate Polymers*, *202*, 186–202. doi:10.1016/j.carbpol.2018.09.002.

Ilyas, R. A., Sapuan, S. M., Ishak, M. R., and Zainudin, E. S. (2018d). Sugar palm nanocrystalline cellulose reinforced sugar palm starch composite: Degradation and water-barrier properties. In: *IOP Conference Series: Materials Science and Engineering* (Vol. 368). doi:10.1088/1757-899X/368/1/012006.

Ilyas, R. A., Sapuan, S. M., Ishak, M. R., and Zainudin, E. S. (2019c). Sugar palm nanofibrillated cellulose (Arenga pinnata (Wurmb.) Merr): Effect of cycles on their yield, physic-chemical, morphological and thermal behavior. *International Journal of Biological Macromolecules*, *123*, 379–388. doi:10.1016/j.ijbiomac.2018.11.124.

Ilyas, R. A., Sapuan, S. M., Ishak, M. R., Zainudin, E. S., and Atikah, M. S. N. (2018e). Characterization of sugar palm nanocellulose and its potential for reinforcement with a starch-based composite. In: *Sugar Palm Biofibers, Biopolymers, and Biocomposites* (1st ed., pp. 189–220). Boca Raton, FL : CRC Press/Taylor and Francis Group. CRC Press. doi:10.1201/9780429443923-10.

Ilyas, R. A., Sapuan, S. M., Sanyang, M. L., Ishak, M. R., and Zainudin, E. S. (2018f). Nanocrystalline cellulose as reinforcement for polymeric matrix nanocomposites and its potential applications: A review. *Current Analytical Chemistry*, *14*(3), 203–225. doi:10.2174/1573411013666171003155624.

Irkin, R., and Esmer, O. K. (2015). Novel food packaging systems with natural antimicrobial agents. *Journal of Food Science and Technology*, *52*(10), 6095–6111. doi:10.1007/s13197-015-1780-9.

Jaiswal, S., Mchale, P., and Duffy, B. (2012). Colloids and Surfaces B : Biointerfaces Preparation and rapid analysis of antibacterial silver , copper and zinc doped sol–gel surfaces. *Colloids and Surfaces, Part B: Biointerfaces*, *94*, 170–176. doi:10.1016/j.colsurfb.2012.01.035.

Jamróz, E., Juszczak, L., and Kucharek, M. (2018). Investigation of the physical properties, antioxidant and antimicrobial activity of ternary potato starch-furcellaran-gelatin films incorporated with lavender essential oil. *International Journal of Biological Macromolecules*. doi:10.1016/j.ijbiomac.2018.04.014.

Jideani, V. A., and Vogt, K. (2016). Antimicrobial packaging for extending the shelf life of bread—A review. *Critical Reviews in Food Science and Nutrition*, *56*(8), 1313–1324. doi:10.1080/10408398.2013.768198.

Jin, T., and Gurtler, J. (2011). Inactivation of Salmonella in liquid egg albumen by antimicrobial bottle coatings infused with allyl isothiocyanate, nisin and zinc oxide nanoparticles. *Journal of Applied Microbiology*, *110*(3), 704–712. doi:10.1111/j.1365-2672.2011.04938.x.

Jo, H.-J., Park, K.-M., Min, S. C., Na, J. H., Park, K. H., and Han, J. (2013). Development of an anti-insect sachet using a polyvinyl alcohol–cinnamon oil polymer strip Against *Plodia interpunctella*. *Journal of Food Science*, *78*(11), E1713–E1720. doi:10.1111/1750-3841.12268.

Jumaidin, R., Ilyas, R. A., Saiful, M., Hussin, F., and Mastura, M. T. (2019a). Water transport and physical properties of sugarcane bagasse fibre reinforced thermoplastic potato starch biocomposite. *Journal of Advanced Research in Fluid Mechanics and Thermal Sciences*, *61*(2), 273–281.

Jumaidin, R., Khiruddin, M. A. A., Asyul Sutan Saidi, Z., Salit, M. S., and Ilyas, R. A. (2019b). Effect of cogon grass fibre on the thermal, mechanical and biodegradation properties of thermoplastic cassava starch biocomposite. *International Journal of Biological Macromolecules*. doi:10.1016/j.ijbiomac.2019.11.011.

Jumaidin, R., Saidi, Z. A. S., Ilyas, R. A., Ahmad, M. N., Wahid, M. K., Yaakob, M. Y., Maidin, N. A., Ab Rahman, M. H. and Osman, M. H. (2019c). Characteristics of cogon grass fibre reinforced thermoplastic cassava starch biocomposite: Water absorption and physical properties. *Journal of Advanced Research in Fluid Mechanics and Thermal Sciences*, *62*(1), 43–52.

Karam, L., Casetta, M., Chihib, N. E., Bentiss, F., Maschke, U., and Jama, C. (2016). Optimization of cold nitrogen plasma surface modification process for setting up antimicrobial low density polyethylene films. *Journal of the Taiwan Institute of Chemical Engineers*, 1–7. doi:10.1016/j.jtice.2016.04.018.

Kenawy, E. R., Worley, S. D., and Broughton, R. (2007). The chemistry and applications of antimicrobial polymers: A state-of-the-art review. *Biomacromolecules*, *8*(5), 1359–1384. doi:10.1021/bm061150q.

Kerry, J., and Butler, P. (2008). *Smart Packaging Technologies for Fast Moving Consumer Goods: Smart Packaging Technologies for Fast Moving Consumer Goods*. John Wiley & Sons. doi:10.1002/9780470753699.

Khalifa, I., Barakat, H., El-mansy, H. A., and Soliman, S. A. (2016). Improving the shelf-life stability of apple and strawberry fruits applying chitosan-incorporated olive oil processing residues coating. *Food Packaging and Shelf Life*, *9*, 10–19. doi:10.1016/j.fpsl.2016.05.006.

Kour, H., Towseef Wani, N. A., Malik, A., Kaul, R., Chauhan, H., Gupta, P., Bhat, A., and Singh, J. (2013). Advances in food packaging—A review. *Stewart Postharvest Review*, *9*(4). doi:10.2212/spr.2013.4.7.

Koushki, M. R., Azizi, M. H., Koohy-Kamaly, P., and Azizkhani, M. (2015). Effect of calcium alginate edible coatings on microbial and chemical properties of lamb meat during refrigerated storage. *Journal of Food Quality and Hazards Control*, *2*, 6–10.

Lekamge, S., Miranda, A. F., Abraham, A., Li, V., Shukla, R., Bansal, V., and Nugegoda, D. (2018). The toxicity of silver nanoparticles (AgNPs) to three freshwater invertebrates with different life strategies: Hydra vulgaris, Daphnia carinata, and Paratya australiensis. *Frontiers in Environmental Science*, *6*(DEC), 1–13. doi:10.3389/fenvs.2018.00152.

Li, Y., Kumar, K. N., Dabkowski, M., Corrigan, M., Scott, R. W., Nu, K., and Tew, G. N. (2012). New bactericidal surgical suture coating. *Langmuir, 28*(33), 12134–12139. doi:10.1021/la302732w.

Liang, R., Xu, S., Shoemaker, C. F., Li, Y., Zhong, F., and Huang, Q. (2012). Physical and antimicrobial properties of peppermint oil nanoemulsions. *Journal of Agricultural and Food Chemistry*. doi:10.1021/jf301129k.

Limbo, S., and Khaneghah, A. M. (2015). *Active Packaging of Foods and Its Combination with Electron Beam Processing: Electron Beam Pasteurization and Complementary Food Processing Technologies*. Woodhead Publishing Limited. doi:10.1533/97817824 21085.2.195.

Lin, L., Liao, X., Surendhiran, D., and Cui, H. (2018). Preparation of ε -polylysine/chitosan nanofibers for food packaging against Salmonella on chicken. *Food Packaging and Shelf Life, 17*, 134–141. doi:10.1016/j.fpsl.2018.06.013.

Liu, H., Du, Y., Wang, X., and Sun, L. (2004). Chitosan kills bacteria through cell membrane damage. *International Journal of Food Microbiology, 95*(2), 147–155. doi:10.1016/j.ijfoodmicro.2004.01.022.

Liu, Y., Wang, S., Zhang, R., Lan, W., and Qin, W. (2017). Development of poly(lactic acid)/chitosan fibers loaded with essential oil for antimicrobial applications. *Nanomaterials, 7*(7). doi:10.3390/nano7070194.

López-Rubio, A., Almenar, E., Hernandez-Muñoz, P., Lagarón, J. M., Catalá, R., and Gavara, R. (2004). Overview of active polymer-based packaging technologies for food applications. *Food Reviews International*. Taylor and Francis Inc. doi:10.1081/FRI-200033462.

Lü, F., Ye, X., and Liu, D. (2009). Review of antimicrobial food packaging. *Nongye Jixie Xuebao/Transactions of the Chinese Society of Agricultural Machinery, 40*(6), 138–142.

Luz, C., Calpe, J., Saladino, F., Luciano, F. B., Fernandez-Franzon, M., Manes, J., and Meca, G. (2018). Antimicrobial packaging based on ε-polylysine bioactive film for the control of mycotoxigenic fungi in vitro and in bread. *Journal of Food Processing and Preservation*, 1–6. doi:10.1111/jfpp.13370.

Mangalassary, S. (2019). Advances in packaging of poultry meat products. In: *Food Safety in Poultry Meat Production* (pp. 139–159). Springer International Publishing. doi:10.1007/978-3-030-05011-5_7.

Matche, B. R. R. S., and Jagadish, R. S. (2011). Incorporation of chemical antimicrobial agents into polymeric films for food packaging. In: *Multifunctional and Nanoreinforced Polymers for Food Packaging* (pp. 368–420). doi:10.1533/9780857092786.3.368.

Medeiros, E. A. A., Soares, N. de F. F., Polito, T. de O. S., de Sousa, M. M., and Silva, D. F. P. (2011). Sachês antimicrobianos em pós-colheita de manga. *Revista Brasileira de Fruticultura, 33*(SPEC. ISSUE 1), 363–370. doi:10.1590/s0100-29452011000500046.

Metak, A. M. (2015). Effects of nanocomposite based nano-silver and nano-titanium Dioxideon food packaging materials. *International Journal of Applied Science and Technology, 5*(2), 26–40.

Monteiro, C., Costa, F., Pirttilä, A. M., Tejesvi, M. V., and Martins, M. C. L. (2019). Prevention of urinary catheter-associated infections by coating antimicrobial peptides from crowberry endophytes. *Scientific Reports, 9*(1), 1–14. doi:10.1038/s41598-019-47108-5.

Muriel-Galet, V., Cerisuelo, J. P., Lopez-Carballo, G., Lara, M., Gavara, R., and Hernandez-Munoz, P. (2012). Development of antimicrobial films for microbiological control of packaged salad. *International Journal of Food Microbiology, 157*(2), 195–201. doi:10.1016/j.ijfoodmicro.2012.05.002.

Najjar, M. B., Kashtanov, D., and Chikindas, M. L. (2007). ε-Poly-L-lysine and nisin A act synergistically against Gram-positive food-borne pathogens Bacillus cereus and Listeria monocytogenes. *Letters in Applied Microbiology, 45*(1), 13–18. doi:10.1111/j.1472-765X.2007.02157.x.

Norrrahim, M. N. F., Ariffin, H., Hassan, M. A., Ibrahim, N. A., and Nishida, H. (2013). Performance evaluation and chemical recyclability of a polyethylene/poly(3- hydroxybutyrate-co-3-hydroxyvalerate) blend for sustainable packaging. *RSC Advances*, *3*(46), 24378–24388. doi:10.1039/c3ra43632b.

Oral, N., Vatansever, L., Sezer, Ç., Aydin, B., Güven, A., Gülmez, M., Başer, K. H., and Kürkçüoğlu, M. (2009). Effect of absorbent pads containing oregano essential oil on the shelf life extension of overwrap packed chicken drumsticks stored at four degrees Celsius. *Poultry Science*, *88*(7), 1459–1465. doi:10.3382/ps.2008-00375.

Otoni, Caio G., Espitia, P. J. P., Avena-Bustillos, R. J., and McHugh, T. H. (2016, May). Trends in antimicrobial food packaging systems: Emitting sachets and absorbent pads. *Food Research International*. Elsevier Ltd. doi:10.1016/j.foodres.2016.02.018.

Otoni, Caio G., Moura, M. R. de, Aouada, F. A., Camilloto, G. P., Cruz, R. S., Lorevice, M. V., Soares, NdF. F., and Mattoso, L. H. C. (2014). Antimicrobial and physical-mechanical properties of pectin/papaya puree/cinnamaldehyde nanoemulsion edible composite films. *Food Hydrocolloids*, *41*, 188–194. doi:10.1016/j.foodhyd.2014.04.013.

Otoni, Caio Gomide, Soares, N. de F. F., da Silva, W. A., Medeiros, E. A. A., and Baffa Junior, J. C. (2014). Use of Allyl Isothiocyanate-containing sachets to reduce *Aspergillus flavus* Sporulation in Peanuts. *Packaging Technology and Science*, *27*(7), 549–558. doi:10.1002/pts.2063.

Ouattara, B., Simard, R. E., Piette, G., Begin, A., and Holley, R. A. (2000). Inhibition of surface spoilage bacteria in processed meats by application of antimicrobial films prepared with chitosan. *International Journal of Food Microbiology*, *62*(1–2), 139–148. doi:10.1016/S0168-1605(00)00407-4.

Paik, J. S., Dhanasekharan, M., and Kelly, M. J. (1998). Antimicrobial activity of UV-irradiated nylon film for packaging applications. *Packaging Technology and Science*, *11*(4), 179–187. doi:10.1002/(SICI)1099-1522(199807/08)11:4<179::AID-PTS429>3.0.CO;2-J.

Pardini, S. P. (1987). Method for imparting antimicrobial activity from acrylics. U.S. Patent 4,708,870.

Pires, A. C., Soares, dos S., de Andrade, N. F. F., da Silva, N. J. da M, Camilloto, G. P., and Bernardes, P. C. (2009). Increased preservation of sliced mozzarella cheese by antimicrobial sachet incorporated with allyl isothiocyanate. *Brazilian Journal of Microbiology*, *40*(4), 1002–1008. doi:10.1590/S1517-83822009000400036.

Pranoto, Y., Salokhe, V. M., and Rakshit, S. K. (2005). Physical and antibacterial properties of alginate-based edible film incorporated with garlic oil. *Food Research International*, *38*(3), 267–272. doi:10.1016/j.foodres.2004.04.009.

Rai, M., Yadav, A., and Gade, A. (2009). Silver nanoparticles as a new generation of antimicrobials. *Biotechnology Advances*, *27*(1), 76–83. doi:10.1016/j.biotechadv.2008.09.002.

Ray, S., Jin, T., Fan, X., Liu, L., and Yam, K. L. (2013). Development of chlorine dioxide releasing film and its application in decontaminating fresh produce. *Journal of Food Science*, *78*(2), M276–M284. doi:10.1111/1750-3841.12010.

Resianingrum, R. (2016). Characterization of cassava starch-based edible film enriched with lemongrass oil (Cymbopogon citratus). *Nusantara Bioscience*. doi:10.13057/nusbiosci/n080223.

Rozenblit, B., Tenenbaum, G., Shagan, A., Salkmon, E. C., Shabtay-Orbach, A., and Mizrahi, B. (2018). A new volatile antimicrobial agent-releasing patch for preserving fresh foods. *Food Packaging and Shelf Life*, *18*, 184–190. doi:10.1016/j.fpsl.2018.11.003.

Saeed, F., Afzaal, M., Tufail, T., and Ahmad, A. (2019). Use of natural antimicrobial agents: A safe preservation approach. *Active Antimicrobial Food Packaging*, 1–18. doi:10.5772/intechopen.80869.

Sambale, F., Wagner, S., Stahl, F., Khaydarov, R. R., Scheper, T., and Bahnemann, D. (2015). Investigations of the toxic effect of silver nanoparticles on mammalian cell lines. *Journal of Nanomaterials*, *2015*. doi:10.1155/2015/136765.

Sanches, L. M., Petri, D. F. S., de Melo Carrasco, L. D., and Carmona-Ribeiro, A. M. (2015). The antimicrobial activity of free and immobilized poly (diallyldimethylammonium) chloride in nanoparticles of poly (methylmethacrylate). *Journal of nanobiotechnology, 13*(58), 1–13. doi:10.1186/s12951-015-0123-3.

Santos, J. C. P., Sousa, R. C. S., Otoni, C. G., Moraes, A. R. F., Souza, V. G. L., Medeiros, E. A. A., Espitia, P. J. P., Pires, A. C. S., Coimbra, J. S. R., and Soares, N. F. F. (2018). Nisin and other antimicrobial peptides: Production, mechanisms of action, and application in active food packaging. *Innovative Food Science and Emerging Technologies, 48*, 179–194. doi:10.1016/j.ifset.2018.06.008.

Sanyang, M. L., Ilyas, R. A., Sapuan, S. M., and Jumaidin, R. (2018). Sugar palm starch-based composites for packaging applications. In: *Bionanocomposites for Packaging Applications* (pp. 125–147). Cham: Springer International Publishing. doi:10.1007/978-3-319-67319-6_7.

Seo, H. S., Bang, J., Kim, H., Beuchat, L. R., Cho, S. Y., and Ryu, J. H. (2012). Development of an antimicrobial sachet containing encapsulated allyl isothiocyanate to inactivate Escherichia coli O157:H7 on spinach leaves. *International Journal of Food Microbiology, 159*(2), 136–143. doi:10.1016/j.ijfoodmicro.2012.08.009.

Silva, R. R., Avelino, K. Y. P. S., Ribeiro, K. L., Franco, O. L., Oliveira, M. D. L., and Andrade, C. A. S. (2016). Chemical immobilization of antimicrobial peptides on biomaterial surfaces. *Frontiers in Bioscience—Scholar, 8*(1), 129–142. doi:10.2741/S453.

Silvetti, T., Morandi, S., Hintersteiner, M., and Brasca, M. (2017). *Use of Hen Egg White Lysozyme in the Food Industry* (pp. 233–242). doi:10.1016/B978-0-12-800879-9.00022-6.

Siracusa, V. (2016). *Packaging Material in the Food Industry: Antimicrobial Food Packaging*. Elsevier Inc. doi:10.1016/B978-0-12-800723-5.00007-3.

Siragusa, G. R., and Dickson, J. S. (1992). Inhibition of Listeria monocytogenes on beef tissue by application of organic acids immobilized in a calcium alginate gel. *Journal of Food Science, 57*(2), 293–296. doi:10.1111/j.1365-2621.1992.tb05479.x.

Smith, J. (1993). Antimicrobial preservative-reduced foods. In: Jim Smith (Ed.), *Technology of Reduced-Additive Foods* (pp. 123–138). Boston, MA: Springer US. doi:10.1007/978-1-4615-2115-0_6.

Song, X., Zuo, G., and Chen, F. (2018). Effect of essential oil and surfactant on the physical and antimicrobial properties of corn and wheat starch films. *International Journal of Biological Macromolecules, 107*(PartA), 1302–1309. doi:10.1016/j.ijbiomac.2017.09.114.

Souza, A. C., Goto, G. E. O., Mainardi, J. A., Coelho, A. C. V., and Tadini, C. C. (2013). Cassava starch composite films incorporated with cinnamon essential oil: Antimicrobial activity, microstructure, mechanical and barrier properties. *LWT—Food Science and Technology*. doi:10.1016/j.lwt.2013.06.017.

Syafri, E., Sudirman, Mashadi, Yulianti, E., Deswita, Asrofi, M., Abral, H., Sapuan, S. M., Ilyas, R. A., and Fudholi, A. (2019). Effect of sonication time on the thermal stability, moisture absorption, and biodegradation of water hyacinth (Eichhornia crassipes) nanocellulose-filled bengkuang (Pachyrhizus erosus) starch biocomposites. *Journal of Materials Research and Technology, 8*(6), 6223–6231. doi:10.1016/j.jmrt.2019.10.016.

Taylor, P., Vermeiren, L., Devlieghere, F., Debevere, J., Vermeiren, L., Devlieghere, F., and Debevere, J. (2010). Effectiveness of some recent antimicrobial packaging concepts. *Food Additives* and *Contaminants*, 37–41. doi:10.1080/02652030110010485.

Torlak, E., and Nizamlıoglu, M. (2011). Antimicrobial effectiveness of chitosan-essential oil coated plastic films against foodborne pathogens. *Journal of Plastic, Film and Sheeting, 27*(3), 235–248. doi:10.1177/8756087911407391.

Villa, T. G., and Rama, J. L. R. (2016). *Resistant and Emergent Pathogens in Food Products: Antimicrobial Food Packaging*. Elsevier Inc. doi:10.1016/B978-0-12-800723-5.00002-4.

Wang, H., Liu, H., Chu, C., She, Y., Jiang, S., Zhai, L., Jiang, S., and Li, X. (2015). Diffusion and antibacterial properties of nisin-loaded chitosan/poly (L-lactic acid) Towards development of active food packaging film. *Food and Bioprocess Technology*, *8*(8), 1657–1667. doi:10.1007/s11947-015-1522-z.

Wong, S., Kasapis, S., and Huang, D. (2012). Food Hydrocolloids molecular weight and crystallinity alteration of cellulose via prolonged ultrasound fragmentation. *Food Hydrocolloids*, *26*(2), 365–369. doi:10.1016/j.foodhyd.2011.02.028.

Wu, J., Ge, S., Liu, H., Wang, S., Chen, S., Wang, J., Li, J., and Zhang, Q. (2014). Properties and antimicrobial activity of silver carp (Hypophthalmichthys molitrix) skin gelatin-chitosan films incorporated with oregano essential oil for fish preservation. *Food Packaging and Shelf Life*, *2*(1), 7–16. doi:10.1016/j.fpsl.2014.04.004.

Xiao, M., Jasensky, J., Foster, L., Kuroda, K., and Chen, Z. (2018). Monitoring antimicrobial mechanisms of surface-immobilized peptides in situ. *Langmuir: The ACS Journal of Surfaces and Colloids*, *34*(5), 2057–2062. doi:10.1021/acs.langmuir.7b03668.

Xiao, M., Jasensky, J., Gerszberg, J., Chen, J., Tian, J., Lin, T., Lu, T., Lahann, J., and Chen, Z. (2018). Chemically immobilized antimicrobial peptide on polymer and self-assembled monolayer substrates. *Langmuir: The ACS Journal of Surfaces and Colloids*, *34*(43), 12889–12896. doi:10.1021/acs.langmuir.8b02377.

Yang, X., Liu, W., Xi, G., Wang, M., Liang, B., Shi, Y., Feng, Y., Ren, X., and Shi, C. (2019). Fabricating antimicrobial peptide-immobilized starch sponges for hemorrhage control and antibacterial treatment. *Carbohydrate Polymers*, *222*(June), 115012. doi:10.1016/j.carbpol.2019.115012.

Yuceer, M., and Caner, C. (2014). Antimicrobial lysozyme-chitosan coatings affect functional properties and shelf life of chicken eggs during storage. *Journal of the Science of Food and Agriculture*, *94*(1), 153–162. doi:10.1002/jsfa.6322.

Zhang, L. J., and Gallo, R. L. (2016). Antimicrobial peptides. *Current Biology: CB*, *26*(1), R14–R19. doi:10.1016/j.cub.2015.11.017.

Zhang, L., Li, R., Dong, F., Tian, A., Li, Z., and Dai, Y. (2015). Physical, mechanical and antimicrobial properties of starch films incorporated with e -poly- L -Lysine. *Food Chemistry*, *166*, 107–114. doi:10.1016/j.foodchem.2014.06.008.

Zheng, Y., Cai, C., Zhang, F., Monty, J., Linhardt, R. J., and Simmons, T. J. (2016). Can natural fibers be a silver bullet? Antibacterial cellulose fibers through the covalent bonding of silver nanoparticles to electrospun fibers. *Nanotechnology*, *27*(5), 55102. doi:10.1088/0957-4484/27/5/055102.

Zivanovic, S., Chi, S., and Draughon, A. (2005). Antimicrobial activity of chitosan films enriched with essential oils. *Journal of Food Science*, *70*(1), 45–51. doi:10.1111/j.1365-2621.2005.tb09045.x.

6 Biodegradable Eco-Friendly Packaging and Coatings Incorporated of Natural Active Compounds

Josemar Gonçalves de Oliveira Filho, Ailton Cesar Lemes, Anna Rafaela Cavalcante Braga, and Mariana Buranelo Egea

CONTENTS

6.1 OVERVIEW OF BIODEGRADABLE PACKAGING AND COATINGS TECHNOLOGIES

The packaging and coating market for the food industries is based on the use of several materials, such as paper, glass, metal, wood, and mainly plastic materials such as polyethylene (PE), polypropylene (PP), and polystyrene (PS) that are based on fossil sources (Song et al., 2009). Plastics, used alone or in combination, are formed from the construction of a network of monomers combined to compose macromolecules with varied characteristics and properties to meet market needs (Roohi et al., 2017). The conventional plastic industry is distant from sustainable development due to the resistance of plastic to degradation, with consequent accumulation in the environment, which is compromising ecosystems (Vroman and Tighzert, 2009; Chen and Yan, 2019).

This concern with pollution, mainly due to the use of plastic, together with the adoption of public policies that encourage consumers to purchase products with a low environmental impact, are factors responsible for the growth of the eco-friendly packaging and coatings market. In this sense, products developed from biodegradable or biobased materials are preferred by consumers and by industries interested in attracting customers concerned with this cause (Nazareth et al., 2019). The improvement of technologies and processes will permit lower production costs to be achieved, allowing the development of products with more appropriate characteristics and the growth of bioplastics processing industries.

The growing demand for biodegradable plastics has been driven, mainly, by the growth of the market in emerging countries like India, China, and Brazil. It is estimated that, worldwide, the biodegradable plastics market will reach sales of around $12.4 billion in 2027, with a compound annual growth rate (CAGR) of 15.1% (Singh, 2019).

6.1.1 Bioplastics

Bioplastic materials, defined as a biodegradable plastic and/or biobased content, appear as an alternative that could be used to replace conventional plastics because their use could minimize the environmental problems associated with the use of plastics derived from petrochemicals and can be part of the solution in the fight against climate change (Philp et al., 2013; Jiménez-Rosado et al., 2019).

Bioplastics applied in containers and packaging can be classified according to their structure, their resources, and also by their chemical and technological properties. During their development, it is necessary to consider issues related to their reuse, recyclability, and efficiency, which can determine the feasibility of applying the packaging.

Regardless of the material used for their production, bioplastics must provide ideal compatibility and specific characteristics to the packaging to fit the products to which they are intended, such as mechanical properties, barrier, permeability, and flexibility, among others. In addition to its technological function, packaging must be a strategic tool for business competitiveness related to efficiency, preservation, distribution, and sale of products. It is crucial to ensure that the product packaged in this material has an expiration date so as not to pose risks to consumers. Thus, it is essential that the packaging produced with bioplastics (1) complies with the same safety requirements and regulations as conventional plastics regarding the migration of compounds from packaging materials to food; (2) protects food from external contamination; and (3) preserves the nutritional value and the physical and sensory quality of food (Ubeda et al., 2020).

The classification of bioplastics can occur with the source of material used in their manufacture, in biodegradable and biobased materials, which are related to the sustainability of plastic materials and which also have different characteristics and applications (see Figure 6.1).

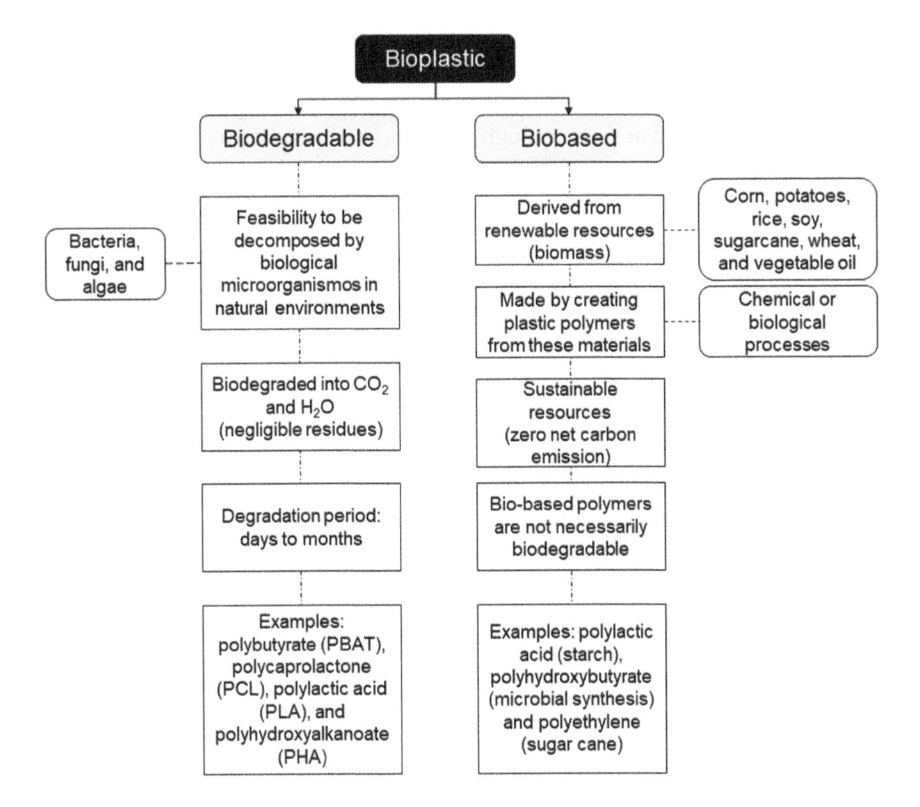

FIGURE 6.1 Main characteristics of biodegradable and biobased material used for creating eco-friendly plastics.

Bioplastics, both biodegradable and biobased, still have a higher cost compared to conventional sources, but can become more advantageous through the improvement and optimization of specific properties of the material and, also, by reducing costs in the final stage of use, considering its elimination. On the other hand, the price of these materials is much more stable when compared to the price of fossil-based plastics, which vary according to the fluctuation of oil prices (Van-den-Oever et al., 2017).

Starch blends and polylactic acid (PLA) are the dominant types in the current market of biobased and biodegradable renewable plastics. Thus, other materials such as cellulose, protein, polymers produced by microbial synthesis, e.g., polyhydroxybutyrate (PHB) and polyhydroxyalkanoates (PHA), and synthetic biopolymers, e.g., polycaprolactone (PCL) and others, are utilized (Sudesh and Iwata, 2008; Imre and Pukánszky, 2013).

6.1.1.1 Biobased and Biodegradable Plastics

Biobased plastics are produced wholly or partially from renewable resources, and this term refers to the origin of the material, not how it will be disposed of at the end of the use chain. The monomers used for the polymerization processes are obtained from plant or wood biomass (starch, cellulose, hemicellulose, and lignin) or plant oil, produced by the photosynthesis of atmospheric carbon dioxide (Siracusa and Lotti, 2018). Biobased plastics are produced and synthesized by creating plastic polymers from these materials, through bacteria, or chemical processes (Gross and Kalra, 2002).

Among the advantages related to the use of biobased plastic is the fact that it reduces the dependence on non-renewable raw materials and contributes to the reduction of greenhouse gases (Muthusamy and Pramasivam, 2019). In addition, materials produced with agricultural waste provide valorization of agricultural by-products, since it directs products that would be improperly discarded or destined for animal feed to uses with high-value products (Lemes et al., 2016).

The resources used to produce biobased plastic can be sustainable, and therefore the decomposition induces zero net carbon emission. When this material is discarded (incineration, for example), the CO_2 produced is converted into biomass by the process of photosynthesis (Siracusa and Lotti, 2018; Chen and Yan, 2019). Thus, the biobased material does not necessarily present biodegradability, but it generates practically no environmental impact. The biodegradability of the biobased plastics generally depends on the chemical nature, not on the source (Reddy et al., 2013). All of the biobased plastics are synthesized either by production within bacteria or by chemical polymerization (Gross and Kalra, 2002).

Biodegradable plastics are polymeric materials that have the ability to undergo the decomposition process and can be thoroughly degraded, regardless of whether oil or renewable resources are used as raw material (Siracusa and Lotti, 2018). In addition, they can be produced from low-cost renewable resources, such as plant starch or derived from bacterial fermentation of vegetables (Liliani et al., 2020).

The biodegradable plastics decomposition process occurs by converting the material into CO_2, methane, water, and other inorganic compounds or biomass, through exposure to microorganisms (bacteria, fungi, and algae) that occur naturally and the action of their respective enzymes. Meanwhile, conventional petroleum-based plastics are not entirely decomposed and assimilated by microorganisms in a biodegradation process.

The biodegradability of these materials, as well as their classification, are tested in a standardized measurement, in a controlled period and environment, reflecting the condition of available disposal (Song et al., 2009; Urbanek et al., 2018; Kjeldsen et al., 2019). The rate of biodegradation of different materials can vary exponentially, with biodegradables designed to deteriorate and cease to exist in specific environments.

Among the advantages of using biodegradable plastics is the fact that they contribute to the reduction of plastic waste in marine environments (Kumar et al., 2020). They are ideal for application in single-use and short-term packaging, such as utensils for food services, as they can be later discarded in composting systems without the need for recycling, reducing the material destined for landfills and, still, having the capacity to increase productivity in horticultural applications (APM, 2017).

The biodegradation process takes place in three stages, as described in Figure 6.2. Plastics undergo the process of biodegradation (Stage 1) due to abiotic

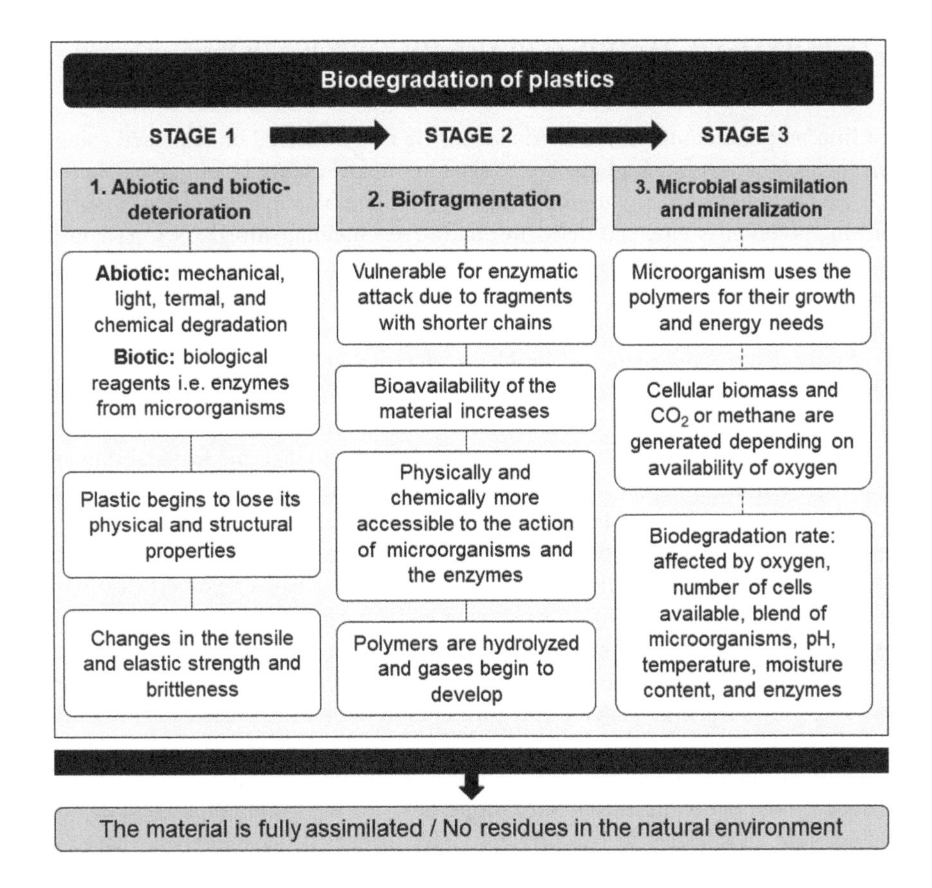

FIGURE 6.2 Generic mechanism of the biodegradation process of plastics.

(mechanical, light, thermal, and chemical degradation) and biotic (enzymes and microorganisms) factors, where the material begins the process of losing its physical and structural properties, including changes in tensile strength, elasticity, and fragility. In Stage 2, fragmented plastic, in a size that allows hydrolysis, becomes more vulnerable to microbial and enzymatic activity. The biofragmentation stage is effectively influenced by the shape and linearity of the material, the concentration of enzymes, and the diversity of microorganisms present in the place where biodegradation occurs. In the last stage (Stage 3), the microorganisms use the polymers to meet their growth and energy needs, generating cell biomass and carbon dioxide or methane, depending on the availability of oxygen, pH, temperature, moisture content, and other environmental factors. Finally, the material is wholly assimilated without leaving residues (Van-den-Oever et al., 2017; Kjeldsen et al., 2019).

Several natural polymeric matrices such as lipids, proteins, and carbohydrates can be used during the development of biodegradable packaging (films and coatings) and will be addressed in the following section.

6.2 POLYMERIC MATRIX FOR THE INCORPORATION OF ACTIVE AGENT IN FOOD PACKAGING

For film preparation, the raw materials must be first dissolved or dispersed using a solvent such as water or/and alcohol. Adjusting the pH and/or heating the solutions may be necessary to facilitate the solubility of some biopolymers. Then, the film-forming solution is cast and dried to obtain free-standing films. As a packaging material, film-forming solutions could be applied to food as a coating by several methods, including dipping, spraying, brushing, and panning, followed by drying (Cazón et al., 2017).

Mainly due to environmental problems, materials that are partially or entirely soluble in water, and therefore considered biodegradable have been extensively studied for packaging applications. Biopolymers such as polysaccharides, lipids, and proteins could be used to solve environmental hazards due to their biodegradability and non-toxicity. Table 6.1 shows the characteristics and sources of isolated biopolymers.

The biodegradable films based on polysaccharides are known to be an effective barrier to gas transference like O_2 and CO_2. However, these materials are generally very hydrophilic, resulting in reduced water vapor barrier properties (Cazón et al., 2017). To improve the results obtained and reported using the cited polymers, research has studied a combination of polymeric compounds for film or food coating.

Films formed only by a purely polymeric ingredient tend to be brittle, and therefore plasticizers are added to the formulation in order to decrease its fragility. Plasticizers are defined as substances with a high melting point, capable of altering the physical, chemical, and mechanical properties of the material to which they will be added. They act on the mechanical properties, making the material more flexible and resistant. Plasticizers such as glycerol, polyethylene glycol, and sorbitol decrease the brittleness and improve the flexibility and elasticity of the material (Khan et al., 2017). These thermoplastic polymers present excellent compatibility with natural antimicrobial compounds such as essential oils, nisin, and lysozyme, which can positively impact the morphology and crystallinity of the polymers and indirectly affect the release rate of the active compounds (Sanyang et al., 2016; Khaneghah et al., 2018).

In addition to controlling the rate of diffusion of preservative substances from the surface into the food, the coatings can act as carriers of antimicrobial and antioxidant agents. Mainly, bioactive compounds and natural agents with unique properties can be introduced in order to improve the shelf life of the food products. The choice of material for forming the film or coating for the food will depend on the nature and amount of compound to be incorporated (Huang et al., 2019).

6.3 NATURAL AGENTS AND THEIR APPLICATION IN FOOD PACKAGING

The types of the polymeric matrix and the wide range of additives that could be incorporated during the preparation of biodegradable films could allow enhancing the shelf life of ready-to-eat foods (Cazón et al., 2017). The addition of natural active agents in polymers of biological origin is a valuable strategy to increase the shelf

TABLE 6.1
Characteristics and Source of Isolated Biopolymers for Natural Agent in Food Packaging

Biopolymers	Characteristics	Sources	References
Alginate	The alginate has hydrophilic carboxyl groups and ability to react with polyvalent metal cations, specifically calcium ions, which will form a rapid cold setting and heat stable gel.	Brown algae (*Laminaria digitata* and *Ascophyllum nodosum*) cell walls or microorganisms	(Tavassoli-Kafrani et al., 2016; Cazón et al., 2017)
Carboxymethyl cellulose	Hydrophobic polysaccharide backbone and many hydrophilic carboxyl groups; biocompatibility with substances such as water-soluble polysaccharides, proteins, surfactants, and plasticizers	Agro-industrial waste, rice, sugarcane, corn, and cassava, among others	(Ballesteros et al., 2018; Tavares et al., 2019)
Carrageenan	Carrageenans are highly hydrophilic and present limited barrier to moisture with good barrier to fats and oils, as well as oxygen barriers and can provide protection against lipid oxidation	Cell walls of various red seaweeds such as *Chondrus crispus* species, *Kappaphycus alvarezii*, and *Eucheuma denticulatum*	(Tavassoli-Kafrani et al., 2016; Cazón et al., 2017)
Cellophane	The film produced with this biopolymer has good mechanical properties and elasticity, great purity, and can be used with a variety of layers of coatings to keep foods from moisture, light, bacteria, and additional hazards	Hydroxyethyl cellulose and cellulose acetate	(Youssef and El-Sayed, 2018)
Cellulose	This biopolymer is abundantly available because it is present in plant material and has good potential to achieve controlled release in antimicrobial packaging	Agro-industrial waste, rice, sugarcane, corn, cassava, and microorganisms, among others	(Gallegos et al., 2016; Gutiérrez and Alvarez, 2016; Khan et al., 2017)
Chitosan	It has antibacterial activity that varies according to the degree of deacetylation, molecular weight, the target organism, and the conditions of the medium in which it is applied	Chitin	(Aider, 2010; Hu et al., 2016; Mujtaba et al., 2019)
Fiber	This compound has low weight due to low density, wide availability, low cost, restricted processing temperature, lower strength properties (impact strength) and durability than in 100% polymers	Agro-industrial waste	(Sydow and Bieńczak, 2019)

(*Continued*)

TABLE 6.1 (CONTINUED)
Characteristics and Source of Isolated Biopolymers for Natural Agent in Food Packaging

Biopolymers	Characteristics	Sources	References
Starch	Limited range of applications because of the unfavorable properties such as brittleness, uncontrolled/high viscosity, retrogradation, low mechanical properties, and insolubility in cold water. The large amylose content provides the better mechanical properties when there is no plasticizer. Native starches can be modified both chemically and physically.	Cassava, corn, and rice, among others	(Khan et al., 2017; Piñeros-Hernandez et al., 2017; Aung et al., 2018)
Pectin	Excellent barrier to oxygen, barrier to oil, aroma preservation, and good mechanical properties with disadvantage of ineffectiveness against moisture transfer through the films by their hydrophilic nature	Agro-industrial waste from fruit industry	(Adeyeye et al., 2019)
Polylactic acid (PLA)	The film has good moisture barrier properties and has a low gas transmittance; derived from renewable resources however, the price is still very high	Corn starch and lactic acid	(Khan et al., 2017)
Protein films	Protein films are generally better than polysaccharide film in mechanical and barrier properties and also provide higher nutritional value. Has limitations in packaging with protein (film or coating) like low water vapor barrier due to being hydrophilic in nature and poor mechanical properties compared with synthetic polymers. Protein-based packaging films have high potential for proper linkages as they can develop the bonds at several different locations	Wheat gluten and corn zein or fibrous form like casein and gelatin, soy protein, whey from milk protein	(Aung et al., 2018)

life of food products. For this application, the active compounds must be released at a controlled and gradual rate. In addition, the concentration of active compound released should not be too high or too low to avoid adverse effects on the toxicological and sensory properties of the food (Mastromatteo et al., 2010).

6.3.1 Essential Oils (EOs)

Essential oils (EOs) receive special attention as active ingredients applicable in films and food coatings due to their potent antimicrobial and antioxidant activities (Atarés and Chiralt, 2016). EOs are volatile aromatic substances of low molecular weight (e.g., phenolic compounds, such as monoterpenes, flavonoids, and phenolic acids) produced by plants (e.g., cinnamon, thyme, lavender, ginger, palmarosa, lemongrass, mint, citrus, and fennel) or their isolated components (e.g., eugenol, geraniol, menthol, limonene, carvacrol, and linalool) that can decrease lipid oxidation and reduce microbial growth in food and have therefore been used as additives in active packaging for cheese, fish, meat, fruits, and vegetables (Maisanaba et al., 2017).

Many EOs such as those obtained from cloves, oregano, thyme, nutmeg, mint, garlic, lemongrass, basil, oregano, mustard, and cinnamon, among others, are classified as generally recognized as safe (GRAS) by the Food and Drug Administration of the United States (Ruiz-Navajas et al., 2013) and are gaining popularity due to their bioactive properties (Sivakumar and Bautista-Baños, 2014). However, some of its characteristics, such as intense aroma, toxicity problems (when used in high concentrations), low water solubility, and possible changes in the sensory properties of food, have limited its use in food preservation. One strategy to solve this problem is to incorporate essential oils in edible films and coatings. This application can minimize the necessary doses for encapsulating inside of the polymeric matrix, which limits their volatilization, controls their release (thus reducing the negative impact of these ingredients). It can also prolong the time of action of EOs through the effect of gradual release, which effectively promotes the application of EO in food, minimizing its impact on sensory properties (Ju and Song, 2019). Table 6.2 shows applications of essential oils in films and coatings prepared from biopolymers from different sources.

The development of active films and coatings using polymeric matrices and EO has attracted considerable attention in recent years as they are conducive to reducing oxidative degradation and inhibiting microbial growth, which extends the shelf life of food (Ju and Song, 2019).

Hydroxypropylmethylcellulose (HPMC) coatings incorporated with essential oils of oregano and bergamot are very effective in reducing respiration rate, ethylene production, weight loss, and microbial load, in addition to improving the sensory aspects of plums. Plums coated with HPMC and 2% oregano EO were firmer and showed less change in surface color than the control fruit at the end of storage. The results indicated that the use of HPMC coating and 2% oregano EO showed the potential to extend the storage quality of plums (Choi et al., 2016).

The application of carboxymethylcellulose (CMC) combined with garlic EO showed a beneficial impact on mass loss, percentage of decay, soluble solids content, titratable acidity, and ascorbic acid, as well as positive effects in maintaining higher

TABLE 6.2
Application of Essential Oils in Films and Coatings Produced with Different Biopolymers

Matrix/Coating	EOs	Foods	Function/results	References
Hydroxypropyl methylcellulose (HPMC)	Oregano and Bergamot	"Formosa" plum	Better appearance and quality	(Choi et al., 2016)
Carboxymethylcellulose (CMC)	Garlic	Strawberry	Maintenance of nutritional content and reduction of senescence	(Dong and Wang, 2017)
Basil Seed gum	Oregano	Fresh cut apricots	Reduced water vapor permeability of films. Reduction in the total count of bacteria, yeast, and fungi. Phenolic content and antioxidant activity were increased in refrigerated storage.	(Hashemi et al., 2017)
Sodium alginate and mandarin fiber	Oregano	Low-fat cut cheese	Decreased *Staphylococcus aureus* and inhibition of growth of mold and yeast. Preserved cheese quality and extended shelf life.	(Artiga-Artigas et al., 2017)
Sodium caseinate	Ginger	Chicken breast (fillet)	Increased shelf life and reduced total content of aerobic psychrophilic bacteria, mold, and yeast	(Noori et al., 2018)
Arabica gum/sodium caseinate	Cinnamon and lemongrass	Guava	Reduction of enzymatic activity, increased antioxidant activity, and extended shelf life (up to 40 days)	(Murmu and Mishra, 2018)
Chitosan	Clove	Tambaqui (*Colossoma macropomum*)	Delayed lipid oxidation, inhibition of growth of psychrotrophic bacteria, reduction of pH, and humidity of tambaqui fillets for 120 days	(Vieira et al., 2019)
			Films	
Chitosan fish gelatin	Oregano	—	Growth inhibition of *Staphylococcus aureus*, *Listeria monocytogenes*, *Salmonella enteritidis*, and *Escherichia coli*. Influence on mechanical properties, water vapor permeability, transparency, and microstructure	(Hosseini et al., 2015)

(*Continued*)

TABLE 6.2 (CONTINUED)
Application of Essential Oils in Films and Coatings Produced with Different Biopolymers

Matrix/Coating	EOs	Foods	Function/results	References
Zein	*Zataria multiflora* Boiss.	Meat	Increased antioxidant and antimicrobial activity. Inhibition of bacteria with film application on meat.	(Moradi et al., 2016)
Chitosan	Apricot seed (*Prunus armeniaca*)	Bread slices	Modified films with better water resistance and moisture barrier, mechanical, antioxidant, and antimicrobial properties, as well as increased inhibition of fungal growth and increased shelf life	(Priyadarshi et al., 2018)
Chitosan/gelatin	Cinnamon, citronella, pink carnation, nutmeg, and thyme	—	Growth inhibition of *Campylobacter jejuni, Escherichia coli, Listeria monocytogenes*, and *Salmonella typhimurium* (most effective EO: thyme). Films with barriers against UV light. EO incorporation increased thickness, moisture content, and water vapor permeability, as well as decreased the transparency.	(Haghighi et al., 2019)

concentrations of total phenols and anthocyanins in strawberries. The composite coating of CMC with 2% garlic EO showed the best results that improved the quality of strawberries (Dong and Wang, 2017).

Hashemi et al. (2017) related that the addition of oregano EO decreased the water vapor permeability (WVP) of basil seed gum films significantly. The coatings reduced the total count of bacteria, yeasts, and molds of the apricot. Among the treatments evaluated, the combination of 6% EO was the most effective in reducing the microbial populations of apricot cuts. In comparison with the control, the apricot coated with film with EO added showed a significant increase for phenolic content and the total antioxidant activity at the end of the cold storage. Edible coatings of sodium alginate and mandarin fiber with at least 2% oregano EO improved the microbial stability of pieces of cheese, being efficient in inhibiting the growth of external pathogens, such as *S. aureus*. Consequently, the incorporation of nanoemulsions containing EO on edible coatings for low-fat cheeses has prolonged the shelf life of this product. The results showed the advantages of using oregano EO as a natural antimicrobial, acting as preservatives, and improving the safety, quality, and nutritional properties of high-fat cheese (Artiga-Artigas et al., 2017).

The edible coating based on nanoemulsion of essential oil of ginger (6%) and sodium caseinate increased the shelf life of chicken breast, with a significant decrease in the total of aerobic psychrophilic bacteria, molds, and yeasts. Samples coated with ginger essential oil nanoemulsion achieved the highest total acceptance during storage. The results confirmed the potential of the coatings incorporated by ginger oil nanoemulsion to improve the shelf life of raw poultry meat (Noori et al., 2018).

Coatings of gum arabic + sodium caseinate incorporated with cinnamon and lemongrass EOs applied in guavas resulted in decreased activity of the enzymes polyphenol oxidase and peroxidase, increased significant activity of elimination of DPPH radicals, and retention of ascorbic acid, phenol content, flavonoids, and total sugar in guava pulp. The formulations extended the guava shelf life by 40 days compared with seven days for uncoated guava (Murmu and Mishra, 2018).

Tambaqui fillets with 2% chitosan and chitosan with clove EO (0.16% and 0.08%, respectively) delayed lipid oxidation, inhibited the growth of psychrotrophic bacteria, and reduced the pH and humidity of fillets at 120 days. The combination (chitosan with EO) proved to be promising as a natural preservative, as it was able to inhibit the chemical and microbial deterioration of frozen tambaqui fillets and can be applied to preserve fish fillets of different species (Vieira et al., 2019).

The addition of essential oils to the films can increase their antioxidant and antimicrobial properties. Also, the incorporation affects the continuity of the polymeric matrix, leading to physical changes depending on the specific interactions of the polymer-oil components. Generally, the structure of the film is weakened by the addition of oil, while the water barrier properties are improved, and transparency is reduced. The composition of the oil and the specific interactions with the polymer determine its effectiveness as an active ingredient (Atarés and Chiralt, 2016).

The addition of oregano EO directly influenced the mechanical properties, the permeability to water vapor, and the transparency of fish gelatin/chitosan films. The microstructure of the film revealed that the aggregation and creation of oil droplets occurred in several extensions (depending on the proportion of EO added) during the drying of the film. Films were generally more effective against Gram-positive bacteria (*S. aureus* and *L. monocytogenes*) than Gram-negative bacteria (*S. enteritidis* and *E. coli*). Therefore, the incorporation of EO of oregano into the film composed of fish gelatin/chitosan proved to be adequate for the development of antibacterial films (Hosseini et al., 2015). The antimicrobial and antioxidant properties of zein films have been improved with the addition of EO from *Z. multiflora*. When applied as an active meat packaging, the films significantly inhibited the growth of *L. monocytogenes* (Moradi et al., 2016).

Active films of chitosan and gelatin incorporated with cinnamon, citronella, pink cloves, nutmeg, and thyme EOs inhibited the growth of four leading food bacterial pathogens, including *Campylobacter jejuni*, *E. coli*, *L. monocytogenes*, and *S. typhimurium*, and higher effectiveness was attributed to thyme EO. The films presented practical barriers against UV light, and the incorporation of EOs increased the thickness, the moisture content, the water vapor permeability, the values of b* and ΔE* (yellow color and total color difference, respectively) while decreasing the value of L* (luminosity), light transparency, and opacity. Overall, the characterization of the functional properties revealed that the chitosan-gelatin films incorporated into the EO could be used as active packaging with antimicrobial properties and the potential to prolong the shelf life of food products (Haghighi et al., 2019).

The addition of apricot seed EO (*Prunus armeniaca*) in chitosan films improved water resistance and moisture barrier properties of the films. The mechanical, antioxidant, and antimicrobial properties of the films have also improved, resulting in an increase of bread shelf life by inhibiting the fungi growth (Priyadarshi et al., 2018).

6.3.2 Plant Extracts

Plant extracts (PEs) are composed of various compounds characteristic of secondary plant metabolisms such as phenols, phenolic acids, quinones, flavones, flavonoids, flavonols, tannins, and coumarins. The compounds with phenolic structure have high antioxidant and antimicrobial efficiency (Šernaitė, 2017; Shabana et al., 2017).

The compounds obtained from vegetables are a potential source of natural bioactive ingredients, especially antioxidants and antimicrobials, which can be exploited in various areas of the food industry. PEs represent new ingredients for the development of edible films and coatings, mainly due to their natural origin and phytochemical properties, which enable obtaining active materials capable of prolonging the useful life and value of products (Luchese et al., 2018; Mir et al., 2018).

Table 6.3 shows the information on incorporating PEs in films and coatings prepared from various sources of biopolymers. Multiple PEs from different sources, including leaves (Tesfay et al., 2017; Zhang et al., 2019c), peels (Licciardello et al., 2018; Zhang et al., 2019b), and fruits (Ju and Song, 2019; Staroszczyk et al., 2020), among others, have been used as active agents in films based on biodegradable polymers to achieve properties of interest.

The application of PEs of different species as active agents in different materials for food coatings, mainly fruits, and meats, is an effective alternative to guarantee the quality and safety of these foods. Several studies have shown the promising nature of coatings with bioactive PEs to preserve food products (Das et al., 2013;

TABLE 6.3
Incorporation of Plant Extracts in Films and Coatings Prepared from Various Sources of Biopolymers

Matrix coating	Extract	Foods	Function/results of films and coatings	References
Rice starch	Coconut oil, green tea extract	Tomatoes	Ripening control; reduced weight loss, pH changes, titratable acidity, ascorbic acid, soluble solids content, and reducing sugars. Antioxidant coating and good microbial barrier property.	(Das et al., 2013)
Carboxymethylcellulose (CMC)	Moringa Leaf extract	Avocado	Disease suppression; increasing shelf life and maintaining quality characteristics during postharvest	(Tesfay et al., 2017)
Carboxymethylcellulose (CMC)	Rosemary extracts	Smoked eel fillets	Antioxidant and antimicrobial activity on smoked eels; delay in the formation of primary and secondary oxidation products; decrease in total viable count rate of *Pseudomonas* spp. and lactic acid bacteria growth	(Choulitoudi et al., 2017)
Chitosan and carob gum	Pomegranate bark extract	White shrimps (*Parapenaeus longirostris*)	Reduction of microbial spoilage during storage	(Licciardello et al., 2018)
Chitosan	Banana peel extract	Apple	Improved water barrier properties of films; excellent antioxidant activity in different food simulators; improvement of postharvest quality of apples	(Zhang et al., 2019b)
κ–, ι– and λ- carrageenan	Green tea extract	Raspberries and Blueberries	Extended shelf life of fruits under refrigeration; preservation of firmness and better appearance; increased carrageenan antiviral activity in fruits	(Falcó et al., 2019)

(Continued)

TABLE 6.3 (CONTINUED)
Incorporation of Plant Extracts in Films and Coatings Prepared from Various Sources of Biopolymers

Matrix coating	Extract	Foods	Function/results of films and coatings	References
Alginate	Rhubarb extract	Peaches	Extension of shelf life; preservation of sensory quality; decay inhibition caused by *Penicillium expansum*	(Li et al., 2019)
Furcellaran-gelatin	Green and pu-erh tea extracts	Salmon sushi	No increased shelf life of stored product	(Kulawik et al., 2019)
Agar/alginate	*Larrea nitida* extract	Blueberries	Color change and insignificant effects on mechanical and barrier properties. Antiviral activity against murine norovirus	(Moreno et al., 2020)
		Films		
Teff starch	Camu-camu extracts	—	Excellent antioxidant activity, decreased tensile strength, and increased elongation at break; complete blockage of ultraviolet light; surface of the film became rough; increased contact angle of films	(Ju and Song, 2019)
Chitosan	Purple rice extract and black rice extract	Pork	Increased thickness; increased light barrier and antioxidant activity; Increased water vapor barrier and tensile strength (may vary with progressive addition of extract content)	(Yong et al., 2019)
Chitosan	Mango leaf extract	—	Increased thickness; lower moisture content and hydrophilicity; better tensile strength, stretching, and antioxidant activity	(Rambabu et al., 2019)

(*Continued*)

TABLE 6.3 (CONTINUED)
Incorporation of Plant Extracts in Films and Coatings Prepared from Various Sources of Biopolymers

Matrix coating	Extract	Foods	Function/results of films and coatings	References
Fish gelatin	Mango leaf extract	—	Best barrier properties and antioxidant activity; less flexible and more rigid films	(Adilah et al., 2018)
Fish gelatin	Extracts from rowanberry, blue-berried honeysuckle, and chokeberry pomace	—	Increased antioxidant and antimicrobial activity; unaffected water barrier properties	(Staroszczyk et al., 2020)
Chitosan	Jujube, pine nut, and peanut shell extract	—	Better antioxidant capacity and gas permeability; higher porosity of films by alteration of microstructure	(Zhang et al., 2019c)
Soy protein	*Cortex Phellodendri* extracts	Beef tallow	Enhanced antioxidant activity; effective in lipid protection due to antioxidant activity, barrier properties and opacity (retard lipid oxidation)	(Ma et al., 2019)

Choulitoudi et al., 2017; Tesfay et al., 2017; Licciardello et al., 2018; Li et al., 2019; Zhang et al., 2019c; Moreno et al., 2020).

The CMC coating associated with the rosemary extract provided an antioxidant and antimicrobial activity to smoked eels, delaying the formation of primary and secondary oxidation products and also to decrease the total viable count rate of *Pseudomonas* spp. and the growth of lactic acid bacteria (Choulitoudi et al., 2017).

Moringa extract associated with the carboxymethylcellulose coating improved the postharvest quality of avocado fruit, reducing the ripening rate and inhibiting the growth of postharvest pathogens. CMC mixed with moringa extracts has the potential to be used as an edible coating in the avocado industry (Tesfay et al., 2017).

An edible coating based on chitosan and locust bean gum incorporated with pomegranate peel extract was effective in reducing microbial deterioration in white shrimp (*Parapenaeus longirostris*) during storage. Although the extract has shown a positive and synergistic effect from the perspective of inhibiting the microbial and chemical deterioration of shrimp, its use as an additive in the coating can affect the visual quality due to the presence of pigmented compounds in the crude extract, which does not necessarily represent a disadvantage, because extract purification steps can overcome it. Thus, pomegranate peel extract can be successfully exploited to maintain the quality of seafood through an appropriate formulation with film-forming compounds, such as chitosan, which can act synergistically to control deterioration (Licciardello et al., 2018).

The incorporation of banana peel extract in chitosan film improved the mechanical and water barrier properties of the films, increasing their hydrophobicity. Also, the composite film exhibited excellent antioxidant activity in different food simulators. The best formulation was identified and applied to coat apple fruits, and the results showed that the chitosan coating with banana peel extract was more effective in improving the postharvest quality of fruits than the chitosan coating. These findings highlight the promising nature of the film and coating with PEs as a desirable option for active packaging and are believed to be conducive to the enhancement of banana peel by-products (Zhang et al., 2019b).

The alginate coatings with rhubarb extract showed efficiency to prolong the useful life of peaches, preserving their sensory quality. Peach fruits treated with a coating containing the extract maintained good physiological and sensory quality, compared to the control. Besides, the extract coating had a very beneficial effect on the inhibition of decay caused by *P. expansum*. The alginate coating containing rhubarb extract showed the potential to improve postharvest quality and prolong the life of peach (Li et al., 2019).

Larrea nitida extract, when applied to agar, alginate or alginate/agar films made the films darker, with a more saturated orange-brown color, but did not affect the mechanical and barrier properties of the films. Besides, when applied to the coating of blueberry fruits, they showed antiviral activity against murine norovirus, a cultivable substitute for norovirus. These edible polysaccharide coatings that contain *Larrea nitida* extract are an option to reduce or eliminate foodborne viruses and protect food against the oxidative process (Moreno et al., 2020).

The incorporation of extracts in active films, besides the fact of increasing their antioxidant and antimicrobial activities, can also affect their properties positively or negatively. The changes in the technological and functional properties of films containing PEs depend on the physical or chemical interactions between biopolymers and extracts that affect their structure and, subsequently, their functionality. The nature of the interaction between biopolymers and extracts depends on the nature, chemical characteristics, and concentration of polymer and extract, as well as the structural properties of the active components, including stereochemistry and conformational flexibility (Luchese et al., 2018; Mir et al., 2018). Interactions occur between the hydrogen bonds of the phenolic constituents of PEs and the hydrocolloid component (polysaccharide or protein molecules), as a result of increasing intermolecular interactions within polymers (Adilah et al., 2018).

The addition of PEs can influence the thickness, color, water vapor permeability, tensile strength, solubility, antioxidant, and antimicrobial properties of the films (see Table 6.3). Data on improvement (López-de-Dicastillo et al., 2016; Rambabu et al., 2019; Yong et al., 2019), worsening, or lack of changes (Ju and Song, 2019; Moreno et al., 2020) in the tensile strength of films containing extracts of plants or herbs were reported. Different directions of changes in WVP of these films were also shown (Rambabu et al., 2019; Yong et al., 2019; Staroszczyk et al., 2020). Thus, when formulating active films, their functional properties must be considered.

The addition of ginseng extract to the alginate film decreased the tensile strength and the modulus of elasticity but increased the percentage elongation at break (Norajit et al., 2010). The addition of myrtle extract improved the mechanical properties of methylcellulose films. The incorporation of myrtle fruit extracts induced films to have a considerable increase in elongation at break values. This effect may be due to the increased mobility of polymer chains caused by the presence of some components of myrtle extract between polymer chains that increase flexibility (López-de-Dicastillo et al., 2016).

The mango peel extract, when incorporated into fish gelatin films, improved the water barrier properties and made the films more rigid and less flexible. Also, the color of the films was altered, and the transparency reduced due to the hydrogen bonds between the fish gelatin molecules and the phenolic content in the film matrix. Higher free radical scavenging activities have also been observed for films with higher extract concentrations, which highlights the benefits of mango by-products incorporated into gelatin-based films as a potential material for active packaging (Adilah et al., 2018).

Films formulated with camu-camu extract and teff starch showed excellent radical scavenging activities. With the increase in the extracted content, the tensile strength of the films decreased, while the elongation at break increased. The films completely blocked ultraviolet light in the 200 to 360 nm range. In addition, the contact angle of the films increased from 36° to 69°, and the surface of the films became rough as the incorporated amount of extract increased. The results indicated that the films of teff starch containing camu-camu extract are suitable as antioxidant packaging material (Ju and Song, 2019).

Active and smart packaging films have been successfully developed, incorporating purple rice extract and black rice extract in the chitosan matrix. Due to the hydrogen bonds formed between chitosan and the extract, the incorporation of a low content (1% for weight) significantly increased the water barrier property and the tensile strength of the films. Besides, the addition of the extracts significantly increased the light barrier, and the antioxidant capacity of the chitosan film due to the functionality of the polyphenols in the extract. Due to the abundant amounts of anthocyanins in the extracts, the films with the extracts were sensitive to pH in different buffer solutions. When used to monitor the deterioration of pork, purple rice extract films showed more significant color changes than films with black rice extract. The results of the study suggested that films with purple rice extract could be used as antioxidants and smart packaging films in the food industry (Yong et al., 2019).

Aqueous extracts of rowanberry, honeysuckle of blueberries, and chokeberry bagasse showed antioxidant and antimicrobial activity against *E. coli*, *Pseudomonas fluorescens*, *S. aureus*, and *L. innocua*, as well as improving the mechanical and water barrier properties of the fish gelatin films. A positive effect of the extracts on mechanical strength was only observed at a concentration of 20%. None of the extracts affected the water barrier properties of the films (Staroszczyk et al., 2020).

6.3.3 Bioactive Peptides and Protein Hydrolysates from Food Proteins

Bioactive peptides derived from food proteins have great potential for application as active agents in the development of biodegradable and active food packaging. These biomolecules are defined as short sequences of proteins, 2 to 20 amino acid residues, which have bioactive properties such as antioxidant, antimicrobial, anti-cancer, anti-hypertensive, and immunomodulatory activity, among others (Lemes et al., 2016; Chalamaiah et al., 2019).

Bioactive peptides are inactive while encrypted inside of the protein molecule native sequence, but can be released using a variety of methods, such as enzymatic hydrolysis in vitro, fermentation by proteolytic microorganisms and gastrointestinal digestion (Lemes et al., 2016). However, the most used mechanism for obtaining bioactive peptides in vitro is through enzymatic hydrolysis of protein molecules (Opheim et al., 2015). Enzymatic hydrolysis (in vitro) of food proteins is commonly performed using proteolytic enzymes such as pepsin, trypsin, papain, alcalase, pancreatin, chymotrypsin, thermolysin, Flavourzyme, Neutrase, and Protamex; at the same time, fermentation is generally carried out using different microorganisms generally recognized as safe, for example, lactic acid bacteria (Bhat et al., 2015; Opheim et al., 2015; Rizzello et al., 2016; Chalamaiah et al., 2019).

Several peptides with different bioactivities have already been identified and isolated from various food sources, including milk, whey, meat, egg, fish, rice, soybeans, marine species, peanuts, chickpeas, corn, and algae, among others (see Table 6.4).

In the past few decades, several studies have reported that protein hydrolysates from different food sources, in addition to their nutritional properties, have exhibited various biological functions, including being antioxidant and antimicrobial (Vaštag et al., 2011; Zarei et al., 2012; Saidi et al., 2014; Vieira and Ferreira, 2017). The antioxidant and antimicrobial properties of these biomolecules have been explored for application as active agents in the development of active food packaging since foods are susceptible to deterioration processes mediated by free radicals and microorganisms.

TABLE 6.4
Bioactive Peptides from Different Food Sources

Protein source	Peptide sequences	Biological activity	References
Soy	MITLAIPVALPGA	Immunostimulator	(Tsuruki et al., 2003)
Pacific oyster (*Crassostrea gigas*)	CgPep33	Antifungal	(Liu et al., 2007)
Egg yolk	LAPSLPGLPLPA	Antidiabetic	(Zambrowicz et al., 2015)
Milk	LVNPHDHQNLK	Antioxidant	(Capriotti et al., 2015)
Red scorpion fish (*Scorpaena notata*)	FPIGMGHGSRPA	Antibacterial	(Tang et al., 2015)
Bovine casein	TGLPPGTLGT	Antihypertensive	(Xue et al., 2018)

The antioxidant property of bioactive peptides is dependent on the source of the proteins, the pre-treatment of the protein substrate, the type of proteases used, and the hydrolysis conditions applied. Purified and crude enzymes can be used to produce antioxidant peptides. However, to reduce the cost of production, mixtures of crude enzymes are preferred (Zarei et al., 2012). According to Liu et al. (2007), most antioxidant peptides have between 4 and 16 amino acids, with a molecular mass of about 400–2000 Da. Molecular weight affects the routes used to reach target sites and the ability to undergo additional digestion by gastrointestinal enzymes that could increase antioxidant capacity in vivo (Rubas and Grass, 1991).

The type of amino acid plays an essential role in determining the antioxidant activity of the peptides. In this context, aromatic amino acids such as tyrosine, histidine, tryptophan, and phenylalanine can donate protons contributing to the radical scavenging properties. On the other hand, hydrophobic amino acids have been described as relevant for increasing the presence of peptides at the water-lipid interface and then accessing the free radicals to eliminate the lipid phase. Finally, charged amino acids use carbonyl and side-chain amino groups as metal ion chelators (Toldrá et al., 2018).

Bioactive peptides that exhibit antimicrobial activity are generally defined as peptides with molecular weight less than 100 kDa, with a primary character due to the presence of positively charged amino acids, such as lysine and arginine, together with a substantial portion of the hydrophobic residues. The structural and physicochemical properties of these peptides play an essential role in determining their specificity for target cells (Pushpanathan et al., 2013).

The mechanism of action of antimicrobial peptides usually involves changes in biological membranes. Initially, there is an electrostatic attraction between the peptide molecules, usually positively charged, and the anionic lipids found on the surface of the microbial plasma membrane. Then, due to the amphipathic structure of these peptides, the interaction between the peptides and the membrane surface occurs, with subsequent structural degradation of the plasma membrane through the formation of ion channels or by the production of transmembrane pores. This process causes an imbalance of cellular contents, thereby deregulating the process of replication, transcription, and translation of the DNA sequence by binding to specific intracellular targets, preventing the multiplication and growth of microbial cells (Naghmouchi et al., 2007; Zhao et al., 2012).

Currently, efforts have been made to make bioactive peptides available as active agents in active food packaging (see Table 6.5). The addition of these compounds can attribute several bioactivities, such as antioxidant and antimicrobial activity to the films and have a positive or negative effect on the various properties of the system, such as optics and traction, among others (Giménez et al., 2009; Salgado et al., 2011; Rostami et al., 2017; Oliveira-Filho et al., 2019; Zhang et al., 2019b).

Giménez et al. (2009) demonstrated that the incorporation of increasing percentages of gelatin hydrolysates in squid skin gelatin films increased the antioxidant activity of the films (measured by the iron reduction capacity—FRAP) and the ability to eliminate the ABTS radical. On the other hand, drilling deformation and water vapor permeability increased, and the shape of drilling decreased.

TABLE 6.5
Bioactive Peptides and Protein Hydrolysates in Biodegradable Films

Matrix	Peptide source	Effect of bioactive peptides	References
Skin of squid gelatin	Fish gelatin	Increased antioxidant activity (ABTS and FRAP methods); reduced punching force; increased drilling deformation and water vapor permeability	(Giménez et al., 2009)
Soy and sunflower protein	Bovine plasma	Antioxidant activity (ABTS method); reduction of tensile strength, modulus of elasticity, and glass transition temperature of the films; increased elongation at break and water vapor permeability	(Salgado et al., 2011)
Fish gelatin	Silver carp	Antioxidant activity (DPPH, FRAP, and ABTS methods); increased elongation at break, color difference, water vapor permeability, and opacity; reduction of tensile strength, elastic modulus, and contact angle	(Rostami et al., 2017)
Alginate	Cotton by-product	Antioxidant activity (DPPH, FRAP, and ABTS methods) and antimicrobial activity against *S. aureus*, *Colletotrichum gloeosporioides* and *Rhizopus oligosporus*; increased thickness, color difference; reduction of UV/Vis light transmission rate; change the microstructure of the films	(Oliveira-Filho et al., 2019)
Chitosan	Rapeseed by-product	Increased antimicrobial activity against *E. coli, Bacillus subtilis, and S. aureus*; improvement of mechanical properties; Water vapor permeability reduction	(Zhang et al., 2019a)

The addition of bovine plasma hydrolysates improved antioxidant properties, increased elongation at break and water vapor permeability, and decreased tensile strength, elasticity modulus, and glass transition temperature of films based on soy and sunflower proteins (Salgado et al., 2011). Rostami et al. (2017) observed that the addition of silver carp protein hydrolysate in fish gelatin films increased the antioxidant activity measured by the DPPH and ABTS methods, and the iron-reducing antioxidant potency (FRAP), elongation at break, color difference, permeability water vapor, and opacity, while the tensile strength, modulus of elasticity, and contact angle decreased. Films prepared with 20% hydrolysate had a lower glass transition temperature and a more homogeneous structure, compared to the control film. In addition, Fourier transform infrared spectroscopy (FTIR) confirmed the increase in free hydrolysate groups and less interaction between film chains incorporated in the higher hydrolysate content. The results indicated that the hydrolysate increased the antioxidant activity and affected some characteristics due to less interaction with gelatin

In another study, Oliveira-Filho et al. (2019) demonstrated that the addition of the protein hydrolysate of cotton by-product in alginate films increased the barrier properties against visible light, and the color of the film became darker, reddish, and yellowish. The total phenolic content and antioxidant activity (tested by the DPPH, FRAP, and ABTS methods) also increased. Hydrolyzed films showed an inhibitory effect against *S. aureus*, *C. gloeosporioides*, and *R. oligosporus* but not against *E. coli*. In aqueous migration tests, the active films released more than 60% of their peptides in 30 min. However, there was a controlled and gradual diffusion of the compounds incorporated in the fatty food simulator. The results showed that the alginate films with protein hydrolysate of cotton by-products have the potential for application as an active packaging to preserve fatty foods. Chitosan composite films and rapeseed protein hydrolysate with 12% hydrolysis grade showed better antibacterial activity against *E. coli*, *S. aureus*, and *B. subtilis* compared to the chitosan film. The mechanical properties of the films were improved, and the water vapor permeability reduced. This study created the possibility for the practical application of a film composed of chitosan and rapeseed protein hydrolysate in antimicrobial food packaging (Zhang et al., 2019a).

6.3.4 Bacteriocins

Bacteriocins are small peptides or proteins, usually 30 to 60 amino acids, with antimicrobial capacity, synthesized by different bacteria at the ribosomal level and released extracellularly (Klaenhammer, 1988; Ahmad et al., 2017). Bacteriocins produced by lactic acid bacteria (BAL) are the most studied because most bacteriocin-producing BALs are isolated from food and generally recognized as safe for consumption—BAL bacteriocins are also expected to be safe (Woraprayote et al., 2016).

The bacteriocins produced by BAL have adequate characteristics to be used as food preservatives, as they are: (1) GRAS, (2) inactive and non-toxic in eukaryotic cells and for consumers, (3) inactivated by digestive protease, with little influence on the consumer's intestinal microbiota, (4) generally active over a wide range of pH and temperature, (5) have an antimicrobial spectrum against many pathogenic and food-damaging bacteria, and (6) do not affect sensory characteristics of food (Woraprayote et al., 2016; Baptista et al., 2020).

These molecules exhibit shallow antimicrobial activity (picomolar to nanomolar) and, although there are several bacteriocins in nature, for example, nisin, pediocin, reuterin, natamycin, plantaricin, and helveticin, among others, nisin is the only GRAS registered by the FDA and widely used in the food industry (Fu et al., 2016).

Table 6.6 presents the main bacteriocins studied as biopreservatives in foods and their antimicrobial characteristics. The antimicrobial activity of bacteriocins presents particularities depending on each class. However, in general, the mechanism of action of bacteriocins is mainly due to the interruption of cell wall integrity or inhibition of protein or nucleic acid synthesis. Bacteriocins bind to cell wall components, including lipid or surface molecular binding sites, specific or nonspecific binding to the receptor, which facilitates the formation of pores or directs cell lysis, resulting in cell death via force dissipation proton driving the bacterial system (Ariyapitipun et al., 2000; Ahmad et al., 2017).

The incorporation of bacteriocins in films and coatings is an alternative technology since the direct application of these antimicrobials in food products can cause undesirable interactions with food components and an inefficient distribution in the food matrix, reducing its effectiveness. The application of bacteriocins in films and coatings can minimize the amount needed for antimicrobial effect and also prolong the action time by the controlled release effect (Yildirim, 2011).

The positive effect of bacteriocins on films and coatings as antimicrobials for better food quality and safety has been confirmed in several studies (Jiang et al., 2013; Narsaiah et al., 2015; Aguayo et al., 2016; Song et al., 2017; Diblan and Kaya, 2018; González-Forte et al., 2019; Guitián et al., 2019; Sun et al., 2019).

Table 6.6 shows the effects of films and coatings incorporated with bacteriocins on food preservation and the effect of these compounds on the properties of these materials. The shitake mushroom coating with arabic gum and natamycin showed better efficiency in maintaining the firmness of the tissues, sensory quality, and reduced microbial count of yeasts and molds when compared to the application of only arabic gum, or natamycin, individually. Also, the bacteriocin coating delayed changes in the concentration of soluble solids, total sugar, and ascorbic acid during the storage period. The combination of arabic gum with natamycin has proved to be a useful technique for maintaining the quality of shitake mushrooms and extending their postharvest life (Jiang et al., 2013).

For minimally processed papaya, the alginate coating with pediocin acted as a barrier to the transmission of water vapor and gas exchange, making it difficult to change the values of soluble solids, firmness, and weight loss. The microbial counts, at the end of the storage period, were 10^3 CFU/g for the sample coated with alginate and incorporating bacteriocin compared to 10^7 CFU/g in the case of the control. Alginate (2%) with bacteriocin film showed good results for use in the storage of minimally processed papaya for three weeks, without compromising the physical-chemical quality or microbial safety (Narsaiah et al., 2015). A study conducted by Aguayo et al. (2016) showed that enterocins-based coatings AS-48 and natural polymeric substances were effective in inhibiting the growth of *L. monocytogenes* on minimally processed apple surfaces.

TABLE 6.6
Antimicrobial Characteristics of Bacteriocins Studied as Biopreservatives in Food

Bacteriocin	Microorganism	Antimicrobial potential	References
Enterocin	*Enterococcus faecalis*	Antibacterial activity against Gram-positive microorganisms (*S. aureus*, *L. monocytogenes*) and spore-forming bacteria (*B. cereus* and *Clostridium*)	(Egan et al., 2016; Yildirim, 2016)
Nisin	*Lactococcus lactis* subsp. *Lactis*	Antibacterial activity against Gram-positive microorganisms (*S. aureus*, *Micrococcus luteus*) and spore-forming bacteria (*B. cereus* and *Clostridium*). Generally, shows no activity against Gram-negative bacteria	(Hugo and Hugo, 2015; Modugno et al., 2018)
Pediocin	*Pediococcus* like *P. acidilactici* (Ach, PA-1, JD), *P. pentosaceus* (A, N5p, ST18 e PD1)	Antibacterial activity against *L. monocytogenes*, *E. faecalis*, *S. aureus* and *C. perfringens*	(Pisoschi et al., 2018)
Natamycin	*Streptomyces natalensis*, *Streptomyces chmanovgensis*, and *Streptomyces gilvosporeu*	Antifungal activity against yeasts and molds. No effects on bacteria, viruses, and protozoa	(Carocho et al., 2015; Duchateau and van Scheppingen, 2018)
ε-Polylysine (ε-PL)	*Streptomyces albulus*	Strong antimicrobial activity against a wide range of Gram-positive and Gram-negative bacteria, yeasts, fungi, and bacteriophages	(Shukla et al., 2012; Lopez-Pena and McClements, 2015)
Reuterin	*Lactobacillus reuteri*	Strong antimicrobial activity against a wide range of Gram-positive and Gram-negative bacteria, yeasts, and molds	(Garde et al., 2016; Montiel et al., 2016)

The effects of combining nisin and ε-polylysine with chitosan coating on maintaining quality and inhibiting white blush were also investigated in freshly cut carrots. The coatings controlled the white blush on the carrots and significantly reduced the respiration rate, the decline in ascorbic acid, and the growth of microorganisms (yeasts and fungi, total viable cell count, total coliform count, *S. aureus*, and *Pseudomonas spp.*) and increased the total phenolic content compared to the control after nine days of storage. The results revealed that the use of the combination of nisin and ε-polylysine with chitosan coating was efficient as an alternative preservation method for freshly cut carrots (Song et al., 2017).

The agar coating with enterocins produced by *Enterococcus avium* DSMZ17511 inhibited the growth of *L. monocytogenes* in samples of commercial Tybo cheese (semi-hard and semi-fat cheese), and artisanal goat cheese (soft and semi-hard cheese, semi-fat, and fat) that were artificially inoculated. In addition, antimicrobial peptides exhibited mild cytotoxicity against the L929 and Caco-2 cell lines only at very high concentrations. The application of agar coatings with enterocins is an effective, low-cost, natural, and safe option for the control of *L. monocytogenes* in cheeses (Guitián et al., 2019). Corn starch-based coatings with natamycin were also effective in controlling environmental fungi (mainly *Penicillium* spp.) in semi-hard cheese during ripening (González-Forte et al., 2019).

The addition of bacteriocins can influence the properties of the films (positively or negatively) such as thickness, color, water vapor permeability, tensile strength, solubility, thermal performance, microstructure, and antimicrobial properties (see Table 6.7).

Films produced with cellulose acetate and pediocin inhibited the growth of *L. innocua* and *Salmonella* sp. in sliced hams during the storage period, demonstrating potential as barrier technology in the storage period (Santiago-Silva et al., 2009).

In another study, the addition of natamycin to alginate/pectin films affected the properties of the films, increasing water solubility, water vapor permeability, and film opacity. The mechanical properties of the films were affected by the addition of the antimicrobial agent, resulting in decreased tensile strength (Bierhalz et al., 2012).

Starch films incorporated with nisin or pediocin also showed the ability to inhibit the growth of *L. monocytogenes* and *C. perfringens*. The addition of bacteriocins affected the crystallinity and morphology of the films, causing irregularities on the surface. Thermal performance and water barrier properties have also improved with the addition of bacteriocins. Additionally, physical interactions between additives and starch matrix via hydrogen bonding were evidenced by FTIR, indicating the potential of films as food packaging material (Meira et al., 2017).

The incorporation of nisin and ε-polylysine in corn starch/nanocrystalline cellulose phosphate-based films via the casting method improved the antibacterial properties of the films against *S. aureus* and *E. coli*. Scanning electron microscopy (SEM) showed that the microstructure of the films with the bacteriocins was more continuous and compact, and FTIR revealed the intermolecular interactions between the hydroxyl groups and the amino groups that occurred after the incorporation of nisin and ε-polylysine in the film. Meanwhile, the addition of nisin and ε-polylysine affected the color parameters of the films by increasing the L^*, b^*, and ΔE^* values (luminosity, yellow color, and total color difference, respectively) and decreased significantly the values of a^*. The water barrier and mechanical properties of the films were also improved with the addition of bacteriocins (Sun et al., 2019).

6.4 FINAL CONSIDERATIONS AND OUTLOOK

In recent years, several polymeric materials have been extensively investigated, aiming at replacing conventional plastics produced from non-renewable sources, with plastics produced from renewable sources that cause less environmental impacts, such as the accumulation of plastics and microplastics in ecosystems, and the reduction of greenhouse gases.

TABLE 6.7
Incorporation of Bacteriocins in Films and Coatings Prepared from Various Sources of Biopolymers

Matrix	Bacteriocin	Foods	Function/results	References
Gum arabic	Natamycin	Shitake mushroom	Better overall mushroom quality; better sensory properties; extended shelf life up to 16 days	(Jiang et al., 2013)
Alginate	Pediocin	Minimally processed papaya	Minimizing weight loss; Maintenance of physical and chemical characteristics; inhibition of microbial growth for up to 21 days	(Narsaiah et al., 2015)
Chitosan, caseinate, alginate, k-carrageenan, xanthan gum, pectin, starch, carboxymethyl cellulose, and methyl cellulose (alone and in combination)	Enterocin	Minimally processed apple	Inhibition of the growth of *L. monocytogenes* on the surface of apples (jointly inoculated)	(Aguayo et al., 2016)
Chitosan	Nisina and ε- polylysine	Minimally processed carrot	Inhibition of white blush; positive effect on the physicochemical and microbiological properties of freshly cut carrots	(Song et al., 2017)
Agar	Enterocin	Cheese	Growth inhibition of *L. monocytogenes*	(Guitián et al., 2019)
Corn starch	Natamycin	Semi-hard cheese	Control of the development of environmental fungi on the surface of cheeses	(González-Forte et al., 2019)
		Films		
Cellulose acetate	Pediocin	Sliced ham	Growth inhibition of *L. innocua* and *Salmonella sp.*	(Santiago-Silva et al., 2009)
Alginate/pectin	Natamycin	—	Increased solubility, water vapor permeability, and opacity; reduced tensile strength	(Bierhalz et al., 2012)
Corn starch	Nisina, pediocin	—	Antibacterial activity against *L. monocytogenes* and *C. perfringens*. Better thermal performance and water barrier properties	(Meira et al., 2017)
Corn starch phosphate/nanocrystalline cellulose	Nisin and ε-polylysine	Port Salut cheese	Antibacterial activity against *S. aureus* and *E. coli*. Microstructure of the films more continuous and compact; better mechanical and water barrier properties	(Sun et al., 2019)

Despite the numerous advantages, biodegradable and biobased plastics still have a high cost when compared to conventional sources. However, the production of these materials can become more advantageous through the improvement and optimization of specific properties of the materials, as well as the development of new tools and production technologies, which allows improving the competitiveness of these materials against the materials produced from non-renewable resources.

Another factor that can encourage the use of bioplastics, especially biodegradable ones, is the fact that the disposal costs are reduced at the end of the consumption chain, as well as the possibility of incorporating active food components in order to enhance the mechanical and chemical properties of packaging and coatings, making them similar or better than conventional plastics. The incorporation of active compounds consists of the addition of components with specific activities to food packaging and coatings, to promote or perform particular actions, influencing the quality of food or its safety. Among the active components added are essential oils, peptides, plant extracts, and bacteriocins, which act to inhibit oxidative processes and microbial attacks on food. Currently, efforts are focused on the search for new active molecules capable of performing efficiently in packaging and coatings and, still finding the correct combination between the mixed materials and the applied food products so that the developed packages have similar or superior technological properties, making them excellent alternatives to synthetic plastics and so to replace them.

REFERENCES

Adeyeye, O.A., Sadiku, E.R., Reddy, A.B., Ndamase, A.S., Makgatho, G., Sellamuthu, P.S., et al., 2019. *The Use of Biopolymers in Food Packaging: Green Biopolymers and Their Nanocomposites*, Springer, Singapore, pp. 137–158.

Adilah, A.N., Jamilah, B., Noranizan, M.A., Hanani, Z.A.N., 2018. Utilization of mango peel extracts on the biodegradable films for active packaging. *Food Packag. Shelf Life* 16, 1–7. doi:10.1016/j.fpsl.2018.01.006.

Aguayo, M.C.L., Burgos, M.J.G., Pulido, R.P., Gálvez, A., López, R.L., 2016. Effect of different activated coatings containing enterocin AS-48 against Listeria monocytogenes on apple cubes. *Innov. Food Sci. Emerg. Technol.* 35, 177–183. doi:10.1016/j.ifset.2016.05.006.

Ahmad, V., Khan, M.S., Jamal, Q.M.S., Alzohairy, M.A., Al Karaawi, M.A., Siddiqui, M.U., 2017. Antimicrobial potential of bacteriocins: In therapy, agriculture and food preservation. *Int. J. Antimicrob. Agents* 49(1), 1–11. doi:10.1016/j.ijantimicag.2016.08.016.

Aider, M., 2010. Chitosan application for active bio-based films production and potential in the food industry: Review. *LWT Food Sci. Technol.* 43(6), 837–842. doi:10.1016/j.lwt.2010.01.021.

APM, 2017. *Biodegradable Plasticis*, Assocaiation of Plastics Manufactures, Brussels, Belgium.

Ariyapitipun, T., Mustapha, A., Clarke, A.D., 2000. Survival of Listeria monocytogenes Scott A on vacuum-packaged raw beef treated with polylactic acid, lactic acid, and nisin. *J. Food Prot.* 63(1), 131–136. doi:10.4315/0362-028x-63.1.131.

Artiga-Artigas, M., Acevedo-Fani, A., Martín-Belloso, O., 2017. Improving the shelf life of low-fat cut cheese using nanoemulsion-based edible coatings containing oregano essential oil and mandarin fiber. *Food Control* 76, 1–12. doi:10.1016/j.foodcont.2017.01.001.

Atarés, L., Chiralt, A., 2016. Essential oils as additives in biodegradable films and coatings for active food packaging. *Trends Food Sci. Technol.* 48, 51–62. doi:10.1016/j.tifs.2015.12.001.

Aung, S.P.S., Shein, H., Aye, K.N., Nwe, N., 2018. *Environment-Friendly Biopolymers for Food Packaging: Starch, Protein, and Poly-Lactic Acid (PLA): Green and Sustainable Advanced Packaging Materials*, Springer, Singapore, pp. 173–195.

Ballesteros, L.F., Michelin, M., Vicente, A.A., Teixeira, J.A., Cerqueira, M.Â., 2018. *Food Applications of Lignocellulosic-Based Packaging Materials. Lignocellulosic Materials and Their Use in Bio-Based Packaging*, Springer, Chan, pp. 87–94.

Baptista, R.C., Horita, C.N., Sant'Ana, A.S., 2020. Natural products with preservative properties for enhancing the microbiological safety and extending the shelf-life of seafood: A review. *Food Res. Int.* 127, 108762. doi:10.1016/j.foodres.2019.108762.

Bhat, Z.F., Kumar, S., Bhat, H.F., 2015. Bioactive peptides of animal origin: A review. *J. Food Sci. Technol.* 52(9), 5377–5392. doi:10.1007/s13197-015-1731-5.

Bierhalz, A.C.K., da Silva, M.A., Kieckbusch, T.G., 2012. Natamycin release from alginate/pectin films for food packaging applications. *J. Food Eng.* 110(1), 18–25. doi:10.1016/j.jfoodeng.2011.12.016.

Capriotti, A.L., Caruso, G., Cavaliere, C., Samperi, R., Ventura, S., Zenezini Chiozzi, R., Laganà, A., 2015. Identification of potential bioactive peptides generated by simulated gastrointestinal digestion of soybean seeds and soy milk proteins. *Food Compos. Anal.* 44, 205–213. doi:10.1016/j.jfca.2015.08.007.

Carocho, M., Morales, P., Ferreira, I.C.F.R., 2015. Natural food additives: Quo vadis? *Trends Food Sci. Technol.* 45(2), 284–295. doi:10.1016/j.tifs.2015.06.007.

Cazón, P., Velazquez, G., Ramírez, J.A., Vázquez, M., 2017. Polysaccharide-based films and coatings for food packaging: A review. *Food Hydrocoll.* 68, 136–148. doi:10.1016/j.foodhyd.2016.09.009.

Chalamaiah, M., Keskin Ulug, S., Hong, H., Wu, J., 2019. Regulatory requirements of bioactive peptides (protein hydrolysates) from food proteins. *J. Funct. Foods* 58, 123–129. doi:10.1016/j.jff.2019.04.050.

Chen, X., Yan, N., 2019. A brief overview of renewable plastics. *Mater. Today* 7–8, 1–7. doi:10.1016/j.mtsust.2019.100031.

Choi, W.S., Singh, S., Lee, Y.S., 2016. Characterization of edible film containing essential oils in hydroxypropyl methylcellulose and its effect on quality attributes of "Formosa" plum (Prunus salicina L.). *LWT Food Sci. Technol.* 70, 213–222. doi:10.1016/j.lwt.2016.02.036.

Choulitoudi, E., Ganiari, S., Tsironi, T., Ntzimani, A., Tsimogiannis, D., Taoukis, P., Oreopoulou, V., 2017. Edible coating enriched with rosemary extracts to enhance oxidative and microbial stability of smoked eel fillets. *Food Packag. Shelf Life* 12, 107–113. doi:10.1016/j.fpsl.2017.04.009.

Das, D.K., Dutta, H., Mahanta, C.L., 2013. Development of a rice starch-based coating with antioxidant and microbe-barrier properties and study of its effect on tomatoes stored at room temperature. *LWT Food Sci. Technol.* 50(1), 272–278. doi:10.1016/j.lwt.2012.05.018.

Diblan, S., Kaya, S., 2018. Antimicrobials used in active packaging films. *Food Health* 4, 63–79. doi:10.3153/JFHS18007.

Dong, F., Wang, X., 2017. Effects of carboxymethyl cellulose incorporated with garlic essential oil composite coatings for improving quality of strawberries. *Int. J. Biol. Macromol.* 104(A), 821–826. doi:10.1016/j.ijbiomac.2017.06.091.

Duchateau, A.L.L., van Scheppingen, W.B., 2018. Stability study of a nisin/natamycin blend by LC-MS. *Food Chem.* 266, 240–244. doi:10.1016/j.foodchem.2018.05.121.

Egan, K., Field, D., Rea, M.C., Ross, R.P., Hill, C., Cotter, P.D., 2016. Bacteriocins: Novel solutions to age old spore-related problems? *Front. Microbiol.* 7, 461. doi:10.3389/fmicb.2016.00461.

Falcó, I., Randazzo, W., Sánchez, G., López-Rubio, A., Fabra, M.J., 2019. On the use of carrageenan matrices for the development of antiviral edible coatings of interest in berries. *Food Hydrocoll.* 92, 74–85. doi:10.1016/j.foodhyd.2019.01.039.

Fu, Y., Sarkar, P., Bhunia, A.K., Yao, Y., 2016. Delivery systems of antimicrobial compounds to food. *Trends Food Sci. Technol.* 57, 165–177. doi:10.1016/j.tifs.2016.09.013.

Gallegos, A.M.A., Herrera Carrera, S., Parra, R., Keshavarz, T., Iqbal, H.M.N., 2016. Bacterial cellulose: A sustainable source to develop value-added products—A review. *Biores* 11(2), 5641–5655. doi:10.1007/978-981-13-1909-9_8.

Garde, S., Gomez-Torres, N., Delgado, D., Gaya, P., Avila, M., 2016. Influence of reuterin-producing Lactobacillus reuteri coupled with glycerol on biochemical, physical and sensory properties of semi-hard ewe milk cheese. *Food. Res. Int.* 90, 177–185. doi:10.1016/j.foodres.2016.10.046.

Giménez, B., Alemán, A., Montero, P., Gómez-Guillén, M.C., 2009. Antioxidant and functional properties of gelatin hydrolysates obtained from skin of sole and squid. *Food Chem.* 114(3), 976–983. doi:10.1016/j.foodchem.2008.10.050.

González-Forte, L.S., Amalvy, J.I., Bertola, N., 2019. Corn starch-based coating enriched with natamycin as an active compound to control mold contamination on semi-hard cheese during ripening. *Heliyon* 5(6), 1–8. doi:10.1016/j.heliyon.2019.e01957.

Gross, R.A., Kalra, B., 2002. Biodegradable polymers for the environment. *Science* 297(5582), 803–807. doi:10.1126/science.297.5582.803.

Guitián, M.V., Ibarguren, C., Soria, M.C., Hovanyecz, P., Banchio, C., Audisio, M.C., 2019. Anti-Listeria monocytogenes effect of bacteriocin-incorporated agar edible coatings applied on cheese. *Int. Dairy J.* 97, 92–98. doi:10.1016/j.idairyj.2019.05.016.

Gutiérrez, T., Alvarez, V., 2016. Cellulosic materials as natural fillers in starch-containing matrix-based films: A review. *Polym. Bull.* 74(6), 2401–2430. doi:10.1007/s00289-016-1814-0.

Haghighi, H., Biard, S., Bigi, F., De Leo, R., Bedin, E., Pfeifer, F., et al., 2019. Comprehensive characterization of active chitosan-gelatin blend films enriched with different essential oils. *Food Hydrocoll.* 95, 33–42. doi:10.1016/j.foodhyd.2019.04.019.

Hashemi, S.M.B., Mousavi Khaneghah, A., Ghaderi Ghahfarrokhi, M., Eş, I., 2017. Basil-seed gum containing Origanum vulgare subsp. viride essential oil as edible coating for fresh cut apricots. *Postharvest Biol. Technol.* 125, 26–34. doi:10.1016/j.postharvbio.2016.11.003.

Hosseini, S.F., Rezaei, M., Zandi, M., Farahmandghavi, F., 2015. Bio-based composite edible films containing Origanum vulgare L. essential oil. *Ind. Crop. Prod.* 67, 403–413. doi:10.1016/j.indcrop.2015.01.062.

Hu, D., Wang, H., Wang, L., 2016. Physical properties and antibacterial activity of quaternized chitosan/carboxymethyl cellulose blend films. *LWT Food Sci. Technol.* 65, 398–405. doi:10.1016/j.lwt.2015.08.033.

Huang, T., Qian, Y., Wei, J., Zhou, C., 2019. Polymeric antimicrobial food packaging and its applications. *Polymers (Basel)* 11(3), 560. doi:10.3390/polym11030560.

Hugo, C.J., Hugo, A., 2015. Current trends in natural preservatives for fresh sausage products. *Trends Food Sci. Technol.* 45(1), 12–23. doi:10.1016/j.tifs.2015.05.003.

Imre, B., Pukánszky, B., 2013. Compatibilization in bio-based and biodegradable polymer blends. *Eur. Polym. J.* 49(6), 1215–1233. doi:10.1016/j.eurpolymj.2013.01.019.

Jiang, T., Feng, L., Zheng, X., Li, J., 2013. Physicochemical responses and microbial characteristics of shiitake mushroom (Lentinus edodes) to gum arabic coating enriched with natamycin during storage. *Food Chem.* 138(2–3), 1992–1997. doi:10.1016/j.foodchem.2012.11.043.

Jiménez-Rosado, M., Zarate-Ramírez, L.S., Romero, A., Bengoechea, C., Partal, P., Guerrero, A., 2019. Bioplastics based on wheat gluten processed by extrusion. *J. Clean. Prod.* 239, 1–8. doi:10.1016/j.jclepro.2019.117994.

Ju, A., Song, K.B., 2019. Development of teff starch films containing camu-camu (Myrciaria dubia Mc. Vaugh) extract as an antioxidant packaging material. *Ind. Crop. Prod.* 141, 111737. doi:10.1016/j.indcrop.2019.111737.

Khan, B., Niazi, Bilal Khan, M., Samin, G., Jahan, Z., 2017. Thermoplastic starch: A possible biodegradable food packaging material—A review. *J. Food Process Eng.* 40(3), e12447. doi:10.1111/jfpe.12447.

Khaneghah, A.M., Hashemi, S.M.B., Limbo, S., 2018. Antimicrobial agents and packaging systems in antimicrobial active food packaging: An overview of approaches and interactions. *Food Bioprod. Process.* 111, 1–19. doi:10.1016/j.fbp.2018.05.001.

Kjeldsen, A., Price, M., Lilley, C., Guzniczak, E., 2019. *A Review of Standards for Biodegradable Plastics*, Retrieved from https://assets.publishing.service.gov.uk/government/uploads/system/uploads/attachment_data/file/817684/review-standards-for-biodegradable-plastics-IBioIC.pdf, accessed 23 Jan 2020.

Klaenhammer, T.R., 1988. Bacteriocins of lactic acid bacteria. *Biochimie* 70(3), 337–349. doi:10.1016/0300-9084(88)90206-4.

Kulawik, P., Jamróz, E., Zając, M., Guzik, P., Tkaczewska, J., 2019. The effect of furcellaran-gelatin edible coatings with green and pu-erh tea extracts on the microbiological, physicochemical and sensory changes of salmon sushi stored at 4 °C. *Food Control* 100, 83–91. doi:10.1016/j.foodcont.2019.01.004.

Kumar, A.G., Anjana, K., Hinduja, M., Sujitha, K., Dharani, G., 2020. Review on plastic wastes in marine environment—Biodegradation and biotechnological solutions. *Mar. Pollut. Bull.* 150, 110733. doi:10.1016/j.marpolbul.2019.110733.

Lemes, A.C., Sala, L., Ores Jda, C., Braga, A.R., Egea, M.B., Fernandes, K.F., 2016. A review of the latest advances in encrypted bioactive peptides from protein-rich waste. *Int. J. Mol. Sci.* 17(6), 1–24. doi:10.3390/ijms17060950.

Li, X.-y., Du, X.-l., Liu, Y., Tong, L.-j., Wang, Q., Li, J.-l., 2019. Rhubarb extract incorporated into an alginate-based edible coating for peach preservation. *Sci. Hortic.* 257, 1–7. doi:10.1016/j.scienta.2019.108685.

Licciardello, F., Kharchoufi, S., Muratore, G., Restuccia, C., 2018. Effect of edible coating combined with pomegranate peel extract on the quality maintenance of white shrimps (Parapenaeus longirostris) during refrigerated storage. *Food Packag. Shelf Life* 17, 114–119. doi:10.1016/j.fpsl.2018.06.009.

Liliani, Tjahjono, B., Cao, D., 2020. Advancing bioplastic packaging products through co-innovation: A conceptual framework for supplier-customer collaboration. *J. Clean. Prod.* 252, 119861. doi:10.1016/j.jclepro.2019.119861.

Liu, Z., Zeng, M., Dong, S., Xu, J., Song, H., Zhao, Y., 2007. Effect of an antifungal peptide from oyster enzymatic hydrolysates for control of gray mold (Botrytis cinerea) on harvested strawberries. *Postharvest Biol. Technol.* 46(1), 95–98. doi:10.1016/j.postharvbio.2007.03.013.

López-de-Dicastillo, C., Bustos, F., Guarda, A., Galotto, M.J., 2016. Cross-linked methyl cellulose films with murta fruit extract for antioxidant and antimicrobial active food packaging. *Food Hydrocoll.* 60, 335–344. doi:10.1016/j.foodhyd.2016.03.020.

Lopez-Pena, C.L., McClements, D.J., 2015. Impact of a food-grade cationic biopolymer (ε-polylysine) on the digestion of emulsified lipids: In vitro study. *Food Res. Int.* 75, 34–40. doi:10.1016/j.foodres.2015.05.025.

Luchese, C.L., Garrido, T., Spada, J.C., Tessaro, I.C., de la Caba, K., 2018. Development and characterization of cassava starch films incorporated with blueberry pomace. *Int. J. Biol. Macromol.* 106, 834–839. doi:10.1016/j.ijbiomac.2017.08.083.

Ma, Q., Liang, S., Xu, S., Li, J., Wang, L., 2019. Characterization of antioxidant properties of soy bean protein-based films with Cortex Phellodendri extract in extending the shelf life of lipid. *Food Packag. Shelf Life* 22. doi:10.1016/j.fpsl.2019.100413. http://www.ncbi.nlm.nih.gov/pubmed/100413.

Maisanaba, S., Llana-Ruiz-Cabello, M., Gutiérrez-Praena, D., Pichardo, S., Puerto, M., Prieto, A.I., et al., 2017. New advances in active packaging incorporated with essential oils or their main components for food preservation. *Food Rev. Int.* 33(5), 447–515. do i:10.1080/87559129.2016.1175010.

Mastromatteo, M., Mastromatteo, M., Conte, A., Del Nobile, M.A., 2010. Advances in controlled release devices for food packaging applications. *Trends Food Sci. Technol.* 21(12), 591–598. doi:10.1016/j.tifs.2010.07.010.

Meira, S.M.M., Zehetmeyer, G., Werner, J.O., Brandelli, A., 2017. A novel active packaging material based on starch-halloysite nanocomposites incorporating antimicrobial peptides. *Food Hydrocoll.* 63, 561–570. doi:10.1016/j.foodhyd.2016.10.013.

Mir, S.A., Dar, B.N., Wani, A.A., Shah, M.A., 2018. Effect of plant extracts on the techno-functional properties of biodegradable packaging films. *Trends Food Sci. Technol.* 80, 141–154. doi:10.1016/j.tifs.2018.08.004.

Modugno, C., Loupiac, C., Bernard, A., Jossier, A., Neiers, F., Perrier-Cornet, J.-M., Simonin, H., 2018. Effect of high pressure on the antimicrobial activity and secondary structure of the bacteriocin nisin. *Innov. Food Sci. Emerg. Technol.* 47. doi:10.1016/j.ifset.2018.01.006.

Montiel, R., Martín-Cabrejas, I., Peirotén, Á., Medina, M., 2016. Reuterin, lactoperoxidase, lactoferrin and high hydrostatic pressure treatments on the characteristics of cooked ham. *Innov. Food Sci. Emerg. Technol.* 35, 111–118. doi:10.1016/j.ifset.2016.04.013.

Moradi, M., Tajik, H., Razavi Rohani, S.M., Mahmoudian, A., 2016. Antioxidant and antimicrobial effects of zein edible film impregnated with Zataria multiflora Boiss. essential oil and monolaurin. *LWT Food Sci. Technol.* 72, 37–43. doi:10.1016/j.lwt.2016.04.026.

Moreno, M.A., Bojorges, H., Falcó, I., Sánchez, G., López-Carballo, G., López-Rubio, A., et al., 2020. Active properties of edible marine polysaccharide-based coatings containing Larrea nitida polyphenols enriched extract. *Food Hydrocoll.* 102, 1–11. doi:10.1016/j.foodhyd.2019.105595.

Mujtaba, M., Morsi, R.E., Kerch, G., Elsabee, M.Z., Kaya, M., Labidi, J., Khawar, K.M., 2019. Current advancements in chitosan-based film production for food technology; A review. *Int. J. Biol. Macromol.* 121, 889–904. doi:10.1016/j.ijbiomac.2018.10.109.

Murmu, S.B., Mishra, H.N., 2018. The effect of edible coating based on Arabic gum, sodium caseinate and essential oil of cinnamon and lemon grass on guava. *Food Chem.* 245, 820–828. doi:10.1016/j.foodchem.2017.11.104.

Muthusamy, M.S., Pramasivam, S., 2019. Bioplastics—An eco-friendly alternative to petrochemical plastics. *Curr. World Environ.* 61, 49–59. doi:10.12944/cwe.14.1.07.

Naghmouchi, K., Drider, D., Fliss, I., 2007. Action of divergicin M35, a class IIa bacteriocin, on liposomes and Listeria. *J. Appl. Microbiol.* 102(6), 1508–1517. doi:10.1111/j.1365-2672.2006.03206.x.

Narsaiah, K., Wilson, R.A., Gokul, K., Mandge, H.M., Jha, S.N., Bhadwal, S., et al., 2015. Effect of bacteriocin-incorporated alginate coating on shelf-life of minimally processed papaya (Carica papaya L.). *Postharvest Biol. Technol.* 100, 212–218. doi:10.1016/j.postharvbio.2014.10.003.

Nazareth, M., Marques, M.R.C., Leite, M.C.A., Castro, Í.B., 2019. Commercial plastics claiming biodegradable status: Is this also accurate for marine environments? *J. Hazard. Mater.* 366, 714–722. doi:10.1016/j.jhazmat.2018.12.052.

Noori, S., Zeynali, F., Almasi, H., 2018. Antimicrobial and antioxidant efficiency of nanoemulsion-based edible coating containing ginger (Zingiber officinale) essential oil and its effect on safety and quality attributes of chicken breast fillets. *Food Control* 84, 312–320 .doi:10.1016/j.foodcont.2017.08.015.

Norajit, K., Kim, K.M., Ryu, G.H., 2010. Comparative studies on the characterization and antioxidant properties of biodegradable alginate films containing ginseng extract. *J. Food Eng.* 98(3), 377–384. doi:10.1016/j.jfoodeng.2010.01.015.

Oliveira-Filho, J.G., Rodrigues, J.M., Valadares, A.C.F., Almeida, A.B., Lima, T.M., Takeuchi, K.P., et al., 2019. Active food packaging: Alginate films with cottonseed protein hydrolysates. *Food Hydrocoll.* 92, 267–275. doi:10.1016/j.foodhyd.2019.01.052.

Opheim, M., Šližytė, R., Sterten, H., Provan, F., Larssen, E., Kjos, N.P., 2015. Hydrolysis of Atlantic salmon (Salmo salar) rest raw materials—Effect of raw material and processing on composition, nutritional value, and potential bioactive peptides in the hydrolysates. *Process Biochem.* 50(8), 1247–1257. doi:10.1016/j.procbio.2015.04.017.

Philp, J.C., Ritchie, R.J., Guy, K., 2013. Biobased plastics in a bioeconomy. *Trends Biotechnol.* 31(2), 65–67. doi:10.1016/j.tibtech.2012.11.009.

Piñeros-Hernandez, D., Medina-Jaramillo, C., López-Córdoba, A., Goyanes, S., 2017. Edible cassava starch films carrying rosemary antioxidant extracts for potential use as active food packaging. *Food Hydrocoll.* 63, 488–495. doi:10.1016/j.foodhyd.2016.09.034.

Pisoschi, A.M., Pop, A., Georgescu, C., Turcuş, V., Olah, N.K., Mathe, E., 2018. An overview of natural antimicrobials role in food. *Eur. J. Med. Chem.* 143, 922–935. doi:10.1016/j.ejmech.2017.11.095.

Priyadarshi, R., Sauraj, Kumar, B., Deeba, F., Kulshreshtha, A., Negi, Y.S., 2018. Chitosan films incorporated with apricot (Prunus armeniaca) kernel essential oil as active food packaging material. *Food Hydrocoll.* 85, 158–166. doi:10.1016/j.foodhyd.2018.07.003.

Pushpanathan, M., Gunasekaran, P., Rajendhran, J., 2013. Antimicrobial peptides: Versatile biological properties. *Int. J. Pept.* 2013, 675391–675391. doi:10.1155/2013/675391.

Rambabu, K., Bharath, G., Fawzi, B., Show, P.L., Cocoletzi, H.H., 2019. Mango leaf extract incorporated chitosan antioxidant film for active food packaging. *Int. J. Biol. Macromol.* 126, 1234–1243. doi:10.1016/j.ijbiomac.2018.12.196.

Reddy, M.M., Vivekanandhan, S., Misra, M., Bhatia, S.K., Mohanty, A.K., 2013. Biobased plastics and bionanocomposites: Current status and future opportunities. *Prog. Polym. Sci.* 38(10–11), 1653–1689. doi:10.1016/j.progpolymsci.2013.05.006.

Rizzello, C.G., Tagliazucchi, D., Babini, E., Sefora Rutella, G., Taneyo Saa, D.L., Gianotti, A., 2016. Bioactive peptides from vegetable food matrices: Research trends and novel biotechnologies for synthesis and recovery. *J. Funct. Foods* 27, 549–569. doi:10.1016/j.jff.2016.09.023.

Roohi, Bano, K., Kuddus, M., Zaheer, M.R., Zia, Q., Khan, M.F., et al., 2017. Microbial enzymatic degradation of biodegradable plastics. *Curr. Pharm. Biotechnol.* 18(5), 429–440. doi:10.2174/1389201018666170523165742.

Rostami, A.H., Motamedzadegan, A., Hosseini, S.E., Rezaei, M., Kamali, A., 2017. Evaluation of plasticizing and antioxidant properties of silver carp protein hydrolysates in fish gelatin film. *J. Aquat. Food Prod. Technol.* 26(4), 457–467. doi:10.1080/10498850.2016.1213345.

Rubas, W., Grass, G.M., 1991. Gastrointestinal lymphatic absorption of peptides and proteins. *Adv. Drug Del. Rev.* 7(1), 15–69. doi:10.1016/0169-409X(91)90047-G.

Ruiz-Navajas, Y., Viuda-Martos, M., Sendra, E., Perez-Alvarez, J.A., Fernández-López, J., 2013. In vitro antibacterial and antioxidant properties of chitosan edible films incorporated with Thymus moroderi or Thymus piperella essential oils. *Food Control* 30(2), 386–392. doi:10.1016/j.foodcont.2012.07.052.

Saidi, S., Deratani, A., Belleville, M., Ben Amar, R., 2014. Antioxidant properties of peptide fractions from tuna dark muscle protein by-product hydrolysate produced by membrane fractionation process. *Food Res. Int.* 65, 329–336. doi:10.1016/j.foodres.2014.09.023.

Salgado, P.R., Fernández, G.B., Drago, S.R., Mauri, A.N., 2011. Addition of bovine plasma hydrolysates improves the antioxidant properties of soybean and sunflower protein-based films. *Food Hydrocoll.* 25(6), 1433–1440. doi:10.1016/j.foodhyd.2011.02.003.

Santiago-Silva, P., Soares, N.F.F., Nóbrega, J.E., Júnior, M.A.W., Barbosa, K.B.F., Volp, A.C.P., et al., 2009. Antimicrobial efficiency of film incorporated with pediocin (ALTA® 2351) on preservation of sliced ham. *Food Control* 20(1), 85–89. doi:10.1016/j.foodcont.2008.02.006.

Sanyang, M.L., Sapuan, S.M., Jawaid, M., Ishak, M.R., Sahari, J., 2016. Effect of plasticizer type and concentration on physical properties of biodegradable films based on sugar palm (Arenga pinnata) starch for food packaging. *J. Food Sci. Technol.* 53(1), 326–336. doi:10.1007/s13197-015-2009-7.

Šernaitė, L., 2017. Plant extracts: Antimicrobial and antifungal activity and appliance in plant protection (Review). *Sodininky Daržinink* 36, 58–68. http://www.lsdi.lt/straipsniai/36-3ir.

Shabana, Y.M., Abdalla, M.E., Shahin, A.A., El-Sawy, M.M., Draz, I.S., Youssif, A.W., 2017. Efficacy of plant extracts in controlling wheat leaf rust disease caused by Puccinia triticina. *Egypt. J. Basic Appl. Sci.* 4(1), 67–73. doi:10.1016/j.ejbas.2016.09.002.

Shukla, S., Singh, A., Pandey, A., Mishra, A., 2012. Review on production and medical applications of ε-Polylysine. *Biochem. Eng. J.* 65, 70–81. doi:10.1016/j.bej.2012.04.001.

Singh, S., 2019. *Biodegradable Plastics Market Worth $12.4 Billion by 2027. Biodegradable Plastics Market*, Merkets and Markets, Northbrook.

Siracusa, V., Lotti, N., 2018. *Biobased Plastics for Food Packaging. Reference Module in Food Science*, Elsevier, doi.org/10.1016/B978-0-08-100596-5.22413-X.

Sivakumar, D., Bautista-Baños, S., 2014. A review on the use of essential oils for postharvest decay control and maintenance of fruit quality during storage. *Crop Protect.* 64, 27–37. doi:10.1016/j.cropro.2014.05.012.

Song, J.H., Murphy, R.J., Narayan, R., Davies, G.B.H., 2009. Biodegradable and compostable alternatives to conventional plastics. *Philos. Trans. R. Soc. Lond. B* 364(1526), 2127–2139. doi:10.1098/rstb.2008.0289.

Song, Z., Li, F., Guan, H., Xu, Y., Fu, Q., Li, D., 2017. Combination of nisin and ε-polylysine with chitosan coating inhibits the white blush of fresh-cut carrots. *Food Control* 74, 34–44. doi:10.1016/j.foodcont.2016.11.026.

Staroszczyk, H., Kusznierewicz, B., Malinowska-Pańczyk, E., Sinkiewicz, I., Gottfried, K., Kołodziejska, I., 2020. Fish gelatin films containing aqueous extracts from phenolic-rich fruit pomace. *LWT Food Sci. Technol.* 117, 1–9. doi:10.1016/j.lwt.2019.108613.

Sudesh, K., Iwata, T., 2008. Sustainability of biobased and biodegradable plastics. *Clean* 36(5–6), 433–442. doi:10.1002/clen.200700183.

Sun, H., Shao, X., Zhang, M., Wang, Z., Dong, J., Yu, D., 2019. Mechanical, barrier and antimicrobial properties of corn distarch phosphate/nanocrystalline cellulose films incorporated with Nisin and epsilon-polylysine. *Int. J. Biol. Macromol.* 136, 839–846. doi:10.1016/j.ijbiomac.2019.06.134.

Sydow, Z., Bieńczak, K., 2019. The overview on the use of natural fibers reinforced composites for food packaging. *J. Nat. Fibers* 16(8), 1189–1200. doi:10.1080/15440478.2018.1455621.

Tang, W., Zhang, H., Wang, L., Qian, H., Qi, X., 2015. Targeted separation of antibacterial peptide from protein hydrolysate of anchovy cooking wastewater by equilibrium dialysis. *Food Chem.* 168, 115–123. doi:10.1016/j.foodchem.2014.07.027.

Tavares, K.M., Campos, A.d., Mitsuyuki, M.C., Luchesi, B.R., Marconcini, J.M., 2019. Corn and cassava starch with carboxymethyl cellulose films and its mechanical and hydrophobic properties. *Carbohydr. Polym.* 223, 1–11. doi:10.1016/j.carbpol.2019.115055.

Tavassoli-Kafrani, E., Shekarchizadeh, H., Masoudpour-Behabadi, M., 2016. Development of edible films and coatings from alginates and carrageenans. *Carbohydr. Polym.* 137, 360–374. doi:10.1016/j.carbpol.2015.10.074.

Tesfay, S.Z., Magwaza, L.S., Mbili, N., Mditshwa, A., 2017. Carboxyl methylcellulose (CMC) containing Moringa plant extracts as new postharvest organic edible coating for Avocado (Persea americana Mill.) fruit. *Sci. Hortic.* 226, 201–207. doi:10.1016/j.scienta.2017.08.047.

Toldrá, F., Reig, M., Aristoy, M.C., Mora, L., 2018. Generation of bioactive peptides during food processing. *Food Chem.* 267, 395–404. doi:10.1016/j.foodchem.2017.06.119.

Tsuruki, T., Kishi, K., Takahashi, M., Tanaka, M., Matsukawa, T., Yoshikawa, M., 2003. Soymetide, an immunostimulating peptide derived from soybean beta-conglycinin, is an fMLP agonist. *FEBS Lett.* 540(1–3), 206–210. doi:10.1016/s0014-5793(03)00265-5.

Ubeda, S., Aznar, M., Rosenmai, A.K., Vinggaard, A.M., Nerín, C., 2020. Migration studies and toxicity evaluation of cyclic polyesters oligomers from food packaging adhesives. *Food Chem.* 311, 1–10. doi:10.1016/j.foodchem.2019.125918.

Urbanek, A.K., Rymowicz, W., Mirończuk, A.M., 2018. Degradation of plastics and plastic-degrading bacteria in cold marine habitats. *Appl. Microbiol. Biotechnol.* 102(18), 7669–7678. doi:10.1007/s00253-018-9195-y.

Van-den-Oever, M., Molenveld, K., Zee, M., Bos, H., 2017. *Bio-Based and Biodegradable Plastics–Facts and Figures. Focus on Food Packaging in the Netherlands,* Wageningen Food & Biobased Research, *Wageningen.*

Vaštag, Ž., Popović, L., Popović, S., Krimer, V., Peričin, D., 2011. Production of enzymatic hydrolysates with antioxidant and angiotensin-I converting enzyme inhibitory activity from pumpkin oil cake protein isolate. *Food Chem.* 124(4), 1316–1321. doi:10.1016/j.foodchem.2010.07.062.

Vieira, B.B., Mafra, J.F., Bispo, A.S.D.R., Ferreira, M.A., Silva, F.d.L., Rodrigues, A.V.N., Evangelista-Barreto, N.S., 2019. Combination of chitosan coating and clove essential oil reduces lipid oxidation and microbial growth in frozen stored tambaqui (Colossoma macropomum) fillets. *LWT Food Sci. Technol.* 116, 1–7. doi:10.1016/j.lwt.2019.108546.

Vieira, E.F., Ferreira, I.M.P.L.V.O., 2017. Antioxidant and antihypertensive hydrolysates obtained from by-products of cannery sardine and brewing industries. *Int. J. Food Prop.* 20(3), 662–673. doi:10.1080/10942912.2016.1176036.

Vroman, I., Tighzert, L., 2009. Biodegradable polymers. *Materials (Basel)* 2(2), 307–344. doi:10.3390/ma2020307.

Woraprayote, W., Malila, Y., Sorapukdee, S., Swetwiwathana, A., Benjakul, S., Visessanguan, W., 2016. Bacteriocins from lactic acid bacteria and their applications in meat and meat products. *Meat Sci.* 120, 118–132. doi:10.1016/j.meatsci.2016.04.004.

Xue, L., Wang, X., Hu, Z., Wu, Z., Wang, L., Wang, H., Yang, M., 2018. Identification and characterization of an angiotensin-converting enzyme inhibitory peptide derived from bovine casein. *Peptides* 99, 161–168. doi:10.1016/j.peptides.2017.09.021.

Yildirim, S., 2011. Active packaging for food biopreservation. In: Lacroix, C. (Ed.), *Protective Cultures, Antimicrobial Metabolites and Bacteriophages for Food and Beverage Biopreservation*, Woodhead Publishing, Oxford, pp. 460–489.

Yildirim, Z., Öncül, N., Yildirim, M., Karabiyikli, Ş, 2016. Application of lactococcin BZ and enterocin KP against Listeria monocytogenes in milk as biopreservation agents. *Acta Aliment.* 45(4), 486–492. doi:10.1556/066.2016.45.4.4.

Yong, H., Liu, J., Qin, Y., Bai, R., Zhang, X., Liu, J., 2019. Antioxidant and pH-sensitive films developed by incorporating purple and black rice extracts into chitosan matrix. *Int. J. Biol. Macromol.* 137, 307–316. doi:10.1016/j.ijbiomac.2019.07.009.

Youssef, A.M., El-Sayed, S.M., 2018. Bionanocomposites materials for food packaging applications: Concepts and future outlook. *Carbohydr. Polym.* 193, 19–27. doi:10.1016/j.carbpol.2018.03.088.

Zambrowicz, A., Eckert, E., Pokora, M., Bobak, Ł., Dąbrowska, A., Szołtysik, M., et al., 2015. Antioxidant and antidiabetic activities of peptides isolated from a hydrolysate of an egg-yolk protein by-product prepared with a proteinase from Asian pumpkin (Cucurbita ficifolia). *RSC Adv.* 5(14), 10460–10467. doi:10.1039/C4RA12943A.

Zarei, M., Ebrahimpour, A., Abdul-Hamid, A., Anwar, F., Saari, N., 2012. Production of defatted palm kernel cake protein hydrolysate as a valuable source of natural antioxidants. *Int. J. Mol. Sci.* 13(7), 8097–8111. doi:10.3390/ijms13078097.

Zhang, C., Wang, Z., Li, Y., Yang, Y., Ju, X., He, R., 2019a. The preparation and physiochemical characterization of rapeseed protein hydrolysate-chitosan composite films. *Food Chem.* 272, 694–701. doi:10.1016/j.foodchem.2018.08.097.

Zhang, W., Li, X., Jiang, W., 2019b. Development of antioxidant chitosan film with banana peels extract and its application as coating in maintaining the storage quality of apple. *Int. J. Biol. Macromol.*, 1–10. doi:10.1016/j.ijbiomac.2019.10.275.

Zhang, X., Lian, H., Shi, J., Meng, W., Peng, Y., 2019c. Plant extracts such as pine nut shell, peanut shell and jujube leaf improved the antioxidant ability and gas permeability of chitosan films. *Int. J. Biol. Macromol.* 148, 1242–1250. doi:10.1016/j.ijbiomac.2019.11.108.

Zhao, J., Guo, L., Zeng, H., Yang, X., Yuan, J., Shi, H., et al., 2012. Purification and characterization of a novel antimicrobial peptide from Brevibacillus laterosporus strain A60. *Peptides* 33(2), 206–211. doi:10.1016/j.peptides.2012.01.001.

7 Biodegradable Polymer Composites with Reinforced Natural Fibers for Packaging Materials

B. Ashok and A. Varada Rajulu

CONTENTS

7.1 INTRODUCTION

Due to environmental awareness and the legislations to control pollution, the demand for environmentally benign materials is ever increasing. Further, society is looking for green methods in the development of materials. Polymers and their composites are finding many uses in our daily life due to their outstanding properties; unfortunately, however, because of their non degradation, they are causing environmental problems. Researchers are currently putting their efforts into developing alternative biodegradable polymers and their composites to widen their applications in the fields of transportation, building, packaging etc. [1, 2]. The components of the biodegradable composites (also referred to as biocomposites)—matrix and reinforcement are derived from nature and so cause only very few environmental problems. Further, these biocomposites have balanced properties [3–6]. In conventional polymers, the reinforcement components are either synthetic or natural. Synthetic fibers are expensive, cause skin irritation, are nonbiodegradable and nonrenewable. On the other hand, the natural fibers are lighter, non-toxic, have cellular structure, and are cheaper and renewable. Also, the natural fibers are either cellulose- or protein-based. As cellulose is the most abundant source in nature, the usage of natural fibers as reinforcing components in the biocomposites is ever increasing [7–8]. Matrix materials, the major components of biocomposites include cellulose, polylactic acid (PLA), polypropylene carbonate (PPC), polyhydroxybutarate-co-valerate (PHBV) etc. Many natural fibers such as bamboo, sisal, borassus fruit fibers, tamarind fruit fibers etc. are widely used as discrete components of biocomposites. Another new approach in developing biocomposites is development of self-reinforced all-cellulose composite films for packaging applications [9] in which both matrix and reinforcement are cellulose-based. Cellulose-based biocomposites have the advantages of light weight, high transparency, higher mechanical properties, oxygen barrier properties, biodegradable nature etc. [10]. However, due to its lower thermal degradation temperature over its melting point, cellulose cannot be melt processed and hence the cellulose-based films can be processed by regeneration method only [11, 12]. The regenerated cellulose composites, due to their outstanding properties, find applications in biomedicine, water treatment, food packaging etc. fields [13–15].

The other inherently biodegradable polymer matrices such as PLA, PPC, PHBV, etc. find limited applications as they lack sufficient strength and thermal stability. Further, they are expensive when compared to the synthetic polymer matrices. In order to improve these properties and reduce the cost, they are often reinforced with natural fibers and agricultural waste materials. The salient features of some biodegradable matrices like cellulose and PLA are presented below.

7.2 CELLULOSE

Cellulose is an abundantly available transparent biopolymer with excellent user temperatures. The structure of cellulose pulp (preferably with the degree of polymerization Dp<1200) used for making films is presented in Figure 7.1.

For preparing the films by regeneration method, initially cellulose has to be dissolved in an appropriate solvent. Due to the presence of abundant OH groups, the intermolecular forces are very strong in cellulose pulp and hence difficult to dissolve it in normal solvents. Further, the conventional solvents used for the dissolution of cellulose are toxic and pose many environmental problems. Hence, an environmental friendly solvent is used for the dissolution of cellulose in preparing the cellulose-based composite systems in this chapter. For this, an ionic liquid with a high boiling point is used besides the alkali-based solvent. In the case of environmentally safe solvents, an aqueous solution containing 7% sodium hydroxide and 12% urea cooled to −12.5°C is employed as the solvent [11]. To this cooled alkali solution, 4 g of cotton linter pulp is added and stirred well by a mechanical stirrer at a speed of 1000 rpm at room temperature for 2 minutes. The cellulose solution thus obtained is centrifuged at 5°C at a speed of 7200 rpm for 15 minutes to remove the undissolved pulp and impurities if any. The clear cellulose solution obtained is stored in a refrigerator at 5°C until used [12]. Cellulose, due to its abundance, low cost, and excellent properties, is attracting the attention of many scientists [16].

7.3 POLYLACTIC ACID

PLA is a biodegradable thermoplastic derived from natural sources like kiwi fruits and sugarcane bagasse. It has the following chemical formula:

$$\left(C_3H_4O_2\right)_n$$

Polylactic acid is a predominantly amorphous polymer with a melting temperature range of 150°–170°C and is soluble in many organic solvents. The preferable range of the molecular weight for PLA for its applications is 128–152 kDa. Though it is biocompatible and biodegradable, PLA has certain drawbacks such as its high cost, rapid aging, brittleness, low impact strength etc., which limit its industrial applications. In order to overcome these problems, it is often advisable to make the biocomposites of PLA with biodegradable natural low-cost fillers [17–23].

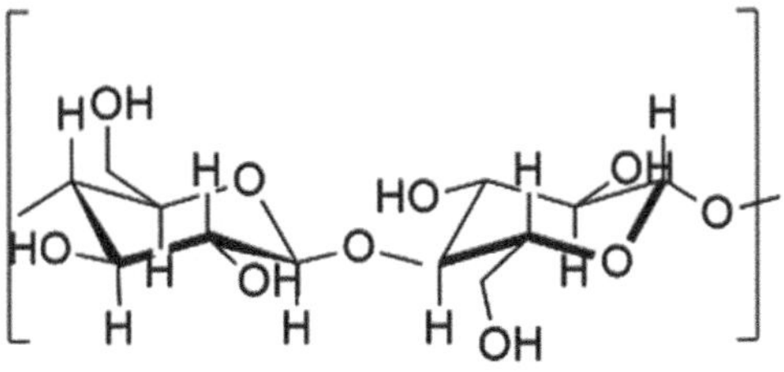

FIGURE 7.1 Structure of cellulose.

In the biodegradable polymer composite systems reported in this chapter, the natural fibers like *Sterculia urens*, *Agave sp.*, and *Thespesia lampas* have been used as the reinforcements along with biodegradable matrices like cellulose (Figure 7.1) and PLA. The extraction and properties of these natural fibers are briefly discussed in the following sections.

7.4 *STERCULIA URENS* NATURAL FIBERS

The natural fibers derived from the uniaxial *Sterculia urens* fabrics collected from the Nallamala forest area of Kurnool District of Andhra Pradesh, India are utilized. The *Sterculia urens* is a natural uniaxial fabric belonging to the Malvaceae family. The uniaxial fabric is extracted from the branches of the tree by retting process as follows [24]. The branches are slit opened and put in water for several days with regular changing of water. After two weeks, the stacked multilayered uniaxial fabrics are separated and are washed thoroughly with water and air dried. The length and breadth of the fabrics are in the range of 300–700 mm and 70–100 mm, respectively with an average thickness of 0.18 mm. In order to remove the oily impurities and to decrease the amorphous hemicelluose content, these fabrics have been treated with 5 wt.% aqueous NaOH solution for 30 minutes. These alkali-treated fabrics are washed thoroughly with water and air dried. The untreated fabrics are composed of 62.9% α-cellulose, 24.3% hemicelluose, and 12.0% lignin whereas the alkali-treated fabrics are 81.5% α-cellulose, 7.5% hemicellulose, and 10.% lignin. On the other hand, the tensile strength, Young's modulus, and % elongation at break of the untreated fabrics are found to be 10.0 MPa, 640.7 Mpa, and 2.0, respectively, while these values for the alkali-treated fabrics are 18.9 MPa, 2018.7 Mpa, and 2.5, respectively [24]. These observations indicate that the hemicelluloses and lignin contents of the fabric under study are decreased on alkali treatment resulting in higher α- cellulose content. Further, the tensile properties of these fabrics have also been improved on alkali treatment of the fabrics [24]. Using the short fibers (both untreated and treated) of *Sterculia urens* as reinforcement, completely biodegradable biocomposites have been prepared, employing cellulose and PLA as matrices. Let us discuss the preparation and properties of these composites.

7.5 PREPARATION AND PROPERTIES OF CELLULOSE/*STERCULIA URENS* SHORT FIBER COMPOSITES

7.5.1 Preparation

The uniaxial *Sterculia* fabrics (both untreated and alkali-treated) are cut into short fibers with mean length of 2 mm. Cellulose stock solution, made as described in Section 7.2, is taken in different flasks and to each, *Sterculia* short fibers in different loadings (5 to 20 wt.%) are added and stirred well by a mechanical stirrer for uniform distribution of fibers in the matrix. These cellulose/*Sterculia* short fibers solutions are then degassed. For film formation, the regeneration method is employed. In this process, the prepared individual solutions are poured on clean glass plates and spread using a glass tube with wire spacers at the end. The glass plates with spread up solutions are immersed in a 5% aq.H_2SO_4 acid regeneration bath. The regenerated composite gel films are separated from the glass plates and rinsed in distilled water repeatedly and dried. Using the same procedure, the cellulose matrix films are also prepared. The dried cellulose and cellulose/*Sterculia* short fiber composite films are stored in desiccators until tested.

7.5.2 Testing

Tensile testing of the matrix and the cellulose/*Sterculia* short fiber composite films has been carried out employing an Instron 3369 Universal testing machine fitted with 10 kN load cell at a stretching rate of 5 mm/minute. Film strips with dimensions of 100 mm length and 10 mm width (thickness of 20 μm) have been used in the test. The test is conducted at 30°C and 70% relative humidity (RH). In each case, 5 specimens were tested and the average values of tensile strength and tensile modulus are reported.

7.5.3 Tensile Properties

The determined tensile strength and modulus of the cellulose matrix are found to be 85.1 MPa and 6223 MPa, respectively. In the case of the cellulose-based biocomposites with untreated short *Sterculia* fibers as reinforcement with a loading of 5%, 10%, 15%, and 20% the tensile strength values are 16.4 MPa, 17.2 MPa, 16.3 Mpa, and 15.3 MPa, respectively, while these values with alkali-treated fibers are found to be 26.9 MPa, 18.3 MPa, 17.0 Mpa, and 17.7 MPa, respectively. On the other hand, the tensile modulus values of the composites with untreated short *Sterculia* fibers as reinforcement with a loading of 5%, 10%, 15%, and 20% are 3124 MPa, 2196 MPa, 1146 Mpa, and 1002 MPa, respectively, while these values for the composites with alkali-treated short fibers are 4820 MPa, 3701 MPa, 2980 Mpa, and 2093 MPa, respectively. These observations indicate that the tensile strength and modulus of the cellulose/*Sterculia* short fiber composites are lower than those of the matrix. This may be due to the short lengths of the reinforcing fibers and their random orientation in the composites. However, these values for the composite films with alkali-treated fibers are higher when compared to those with untreated fibers. Even though the tensile parameters of the biocomposites are lower than those of the matrix, they are still sufficient for packaging applications. Further, the biocomposites prepared are found to be stable up to 340°C [25]. Hence, these completely biodegradable composite films of cellulose/*Sterculia* short fibers can be considered for packaging applications.

7.6 PREPARATION AND PROPERTIES OF PLA/*STERCULIA URENS* SHORT FIBER COMPOSITES

7.6.1 Preparation

As described in Section 7.4, some portion of the short *Sterculia* fibers are treated with 5% alkali. One more portion of the fibers are treated with a silane coupling agent (1% in acetone) for 6 hours and later dried for 1 hour at 60°C. The alkali and coupling agent treatments are carried out to improve the bonding between the matrix and the reinforcement. The *Stercurlia* fabric is cut into short fibers of ~4 mm length. These short fibers (20 wt.%) are mixed with PLA (20 wt.%) using a two-roll mixer and then molded into laminates in a compression molder at a temperature of 180°C and a pressure of 20 MPa for 4 minutes and then allowed to cool at a reduced pressure of 5 MPa [26].

7.6.2 Characterization

The prepared PLA/*Sterculia* short fiber composites are tested for their tensile properties like maximum stress, Young's modulus, and % elongation at break using an Instron 3369 Universal testing machine at an extension rate of 5 mm/minute.

7.6.3 Tensile Properties

As reported in Section 7.4, the tensile strength, Young's modulus, and % elongation at break of the untreated fabrics are found to be 10.0 MPa, 640.7 Mpa, and 2.0, respectively, while these values for the alkali-treated fabrics are 18.9 MPa, 2018.7 MPa and 2.5, respectively [26]. In order to study the effect of the alkali treatment and the coupling agents of the *Sterculia* fibers on the properties of the PLA/*Sterculia* short fibers composites, the tensile properties are determined. The maximum stress (MPa), Young's modulus (MPa) and % elongation at break of the PLA composites with 20 wt.% untreated fibers are found to be 58.2, 3180.9, and 3.7, respectively. For the composites with alkali-treated fibers, these values (maximum stress (MPa), Young's modulus (MPa), and %elongation at break) are 69.1, 4207.5, and 5.0, respectively. These observations indicate that the tensile properties of the composites are improved when alkali-treated fibers are used as the reinforcement. Similarly, the tensile parameters of maximum stress (MPa), Young's modulus (MPa) and % elongation at break of the PLA/*Sterculia* short fiber composites with untreated but coupling agent-treated fibers are found to be 62.1, 4120.6, and 4.2, respectively, and at the same time, these values for the composites with alkali- and coupling agent-treated fibers are 78.4, 5904.7 and 5.4, respectively. Thus, it is evident that the tensile properties of the composites are improved when the alkali- and coupling agent-treated fibers are utilized as the reinforcement. The improved interfacial bonding between the fibers and the matrix may be responsible for these improved properties. The biodegradable PLA/*Sterculia* short fiber composites with improved tensile properties can be considered for packaging applications.

7.7 *AGAVE* (CENTURY) NATURAL FIBERS

These fibers are extracted from the leaves of the plant *Agave Americana* which is a native of Mexico and is also widely grown in India. In India, these fibers are used to make twines as well as ropes in agricultural areas.

7.7.1 Extraction of the Fibers

The century fibers intended for making of biocomposites are obtained from the Mahaboobabad town of Warangal district, Telangana, India. The fibers from the leaves of *Agave* plant are extracted by the conventional retting process. In this process, the mature leaves are cut from the plants which are then kept in water for three weeks with periodic change of water. After this period, the leaves are subjected to light beating. In this process, the bound fibers in the leaves are loosened and are separated. Then these fibers are shaken well in water to remove the fleshy gel, dust, dirt etc. The obtained separated fibers are air dried in the sun for about a week and then subsequently in hot air oven maintained at 105°C for one day. The dried fibers are stored in desiccators until tested. In order to examine the effect of alkali treatment on the properties of the century fibers, some of the dried fibers are treated with 5 wt.% sodium hydroxide in water for 1 hour with a fiber to liquor ratio of 25:1 and then again washed with distilled water and air dried.

7.7.2 Characterization

The chemical composition of the century fibers (both untreated and alkali-treated) is studied by the conventional Technical Association of the Pulp and Paper Industries (TAPPI) method [27]. The tensile properties of the untreated and the alkali-treated century fibers are determined by employing an Instron 3369 Universal testing machine maintaining a gauge length (distance between the grip) of 50 mm at an extension rate of 5 mm/minute. In each case, at least ten fibers are tested and the average values are reported.

7.7.3 Properties

The average length (mm), diameter (mm), aspect ratio (length to diameter), and moisture (%) of the untreated century fibers are found to be 300, 0.20, 1428, and 5.3, respectively while these values for the alkali-treated fibers are 295, 0.18, 1638, and 4.9, respectively. Regarding the chemical composition, the percent values of the α–cellulose, hemicelluloses, and lignin of the untreated century fibers are found to be 71.7, 22.2, and 6.1, respectively, whereas these values for the alkali-treated fibers are 81.7, 7.0, and 11.3, respectively. From these observations, it is clear that on alkali treatment, the century fibers became fine with reduced diameter, thus increasing the aspect ratio. Further, the moisture content of these fibers is also decreased by alkali treatment.

The tensile parameters such as the maximum stress (MPa), Young's modulus (GPa), and elongation at break (%) of the untreated century fibers are 275, 8.6, and 3.4, respectively, while for the alkali-treated fibers these parameters are found to be 338, 7.7, and 7.5, respectively [27]. From the data of the tensile properties of the century fibers, it can be observed that the tensile stress and the % elongation increased on alkali treatment. This may be due to the removal of a part of the amorphous hemicellulose from the fibers on alkali treatment. Though the modulus decreased slightly on alkali treatment, still it is higher when compared to other conventional natural fibers.

7.8 SELF REINFORCED CELLULOSE FILMS BY PARTIAL DISSOLUTION OF *AGAVE* MICRO FIBRILS

A novel method of preparing self-reinforced cellulose films is followed in this system. Both the matrix and reinforcement are derived from the single source of *Agave* microfibrils. In the initial step, the microfibrils are prepared from the *Agave* natural fibers. Then the microfibrils are partially dissolved in an environmentally friendly ionic liquid solvent. The dissolved cellulose part of the microfibrils acts as the matrix while the undissolved part acts as the reinforcement. In this process, greater compatibility is expected between the matrix and the reinforcement of the composite films prepared as both are derived from the same source.

7.8.1 PREPARATION OF THE *AGAVE* MICRO FIBRILS

The extracted and alkali-treated *Agave* natural fibers are chopped into fine fibers that are able to pass through a 250-mesh sieve and are subsequently dried for 24 hours at 105°C. In the second step, these chopped fibers are dewaxed by refluxing in a Soxhlet apparatus using toluene-ethonol mixture (taken in 2:1 (v/v) ratio for 6 hours). The resulting dewaxed fibers are subjected to delignification at 100°C using 0.7% sodium chlorite for 2 hours in acidic medium (at a pH between 4.0 and 4.2 adjusted with 10% acetic acid). A liquor ratio of 1:50 is maintained during this process. After the delignification process, the samples are filtered and are periodically washed using 2% sodium bisuphite followed by distilled water. Subsequently, the obtained delignified fibers are dried in an oven at 105°C until constant weight is achieved. In this way, the lignocelluloses fibers are converted into holocellulose (mixture of α-cellulose and hemicelluloses) fibers. In order to remove the amorphous hemicelluloses, the holocellulose fibers are reacted with NaOH of concentration 17.5% (w/v) at a temperature of 20°C for a period of 45 minutes and are subsequently filtered and then neutralized using 10% acetic acid and then washed with distilled water. The fibers with the removed hemicellulose component are then dried in an oven at a temperature of 105°C until constant weight is achieved. In this way the crude α-cellulose fibers are obtained. In order to purify, the crude cellulose fibers are reacted for 20 minutes at 120°C with 80% acetic acid and 70% HNO_3 mixture taken in a ratio of 10:1. These fibers are then cooled and washed with 95% ethanol followed by distilled water to remove the unreacted acid mixture. In the last step, the pure microfibrils of cellulose are dried in an oven at 105°C and stored in desiccators until used.

7.8.2 PREPARATION OF SELF-REINFORCED CELLULOSE COMPOSITE FILMS EMPLOYING *AGAVE* MICROFIBRILS

Initially, the *Agave* microfibrils are partially dissolved in an ionic liquid (AmimCl) [28] as described in what follows. In the first step, in a round bottom flask, a certain quantity of AmimCl is taken that is maintained at 80°C on an oil bath. To this ionic liquid, the *Agave* microfibrils are added by 4 wt.% of the ionic liquid taken in the flask. The nitrogen gas is passed through the flask at constant stirring during the partial dissolution process (for a duration of 6 hours). In this process, a major portion of the *Agave* microfibrils is dissolved and a minor high-molecular part remains undissolved in the form of fine *Agave* microfibrils. This solution mixture with dissolved part (which can be used as a matrix) and the undissolved microfibrils (which can act as reinforcement) is poured on glass plates and evenly spread using a spreader. The spread solutions with liquid film form are regenerated in a distilled water bath and the obtained composite gel films are separated, washed thoroughly with distilled water until they are free from the unreacted ionic liquid, and then dried to get transparent self-reinforced cellulose composite films. These dry composite films are stored in desiccators until they are tested.

7.8.3 Tensile Testing

The tensile test of the prepared self-reinforced cellulose films is carried out using an Instron 3369 Universal testing machine equipped with 10 kN load cell at 30°C temperature and 70% RH. In this test, the composite films are cut into rectangular strips of 100 mm length and 10 mm width and are used as specimens. A gauge length of 50 mm and an extension rate of 5 mm/minute are maintained during the test. In each case, ten specimens are tested and the average values of tensile strength, Young's modulus, and the % elongation at break are reported. The tensile strength (MPa), Young's modulus (MPa), and % elongation at break of the regenerated self-reinforced cellulose composite films using *Agave* microfibrils are found to be 135, 8150, and 3.2, respectively. On the other hand, the tensile strength (MPa), Young's modulus (MPa), and % elongation at break of other biodegradable polymer films reported are (the values are reported in the parenthesis): polylactic acid (16.8, 1152, and 402) [29], polyvinyl alcohol (83, 1900, and 45) [30], polypropyene carbonate (12.8, 823, and 821) [31], polyhydroxybutyrate (28, 3570, and 10.2) [32], banana starch (5.01, 46.86, and 14.3) [33], cassava starch (2.56, 30.27, and 43.45) [33], sweet potato starch (3.72, 106.01, and 33.83) [34], and chitosan (8.4, 134.6, and 19.6) [34]. These observations indicate that the tensile parameters of the regenerated self-reinforced cellulose composite films using *Agave* microfibrils are far more than the rest of the above. Hence, the films prepared in this system can be utilized as completely biodegradable high-strength packaging materials.

7.9 *THESPESIA LAMPAS* NATURAL FIBERS

Thespesia lampas, which is also known as Adavi benda locally and by the synonym *Hibiscus lampas Cav.*, corresponds to the family of Malvaceae. It is abundantly grown in India and tropical areas of East Africa [35]. The plant grows to 2.3m in height. Its stems are composed of soft and long fibers.

7.9.1 Extraction and Alkali Treatment of the *Thespesia* Fibers

The stems of the *Thespesia* plant are kept in water for about two weeks with occasional change of water. The stems of the plant are slit opened and the separated fibers are removed and subsequently washed several times to remove the materials adhering to the individual fibers. The clean fibers are then dried in the sun for a week and then at 105°C in an oven. Part of the fibers is subjected to alkali treatment as described in what follows.

Some of the dried fibers are immersed in 2% sodium hydroxide in water (w/v) for a period of 24 hours at ambient conditions. For this purpose, a liquor ratio of 25:1 is maintained. In this process, a part of the hemicelluloses content of the fibers and impurities are removed. The alkali-treated fibers are subsequently neutralized with acetic acid (1% v/v) in water and then again washed with distilled water to remove the traces of acetic acid if any. Finally, the alkali-treated fibers are dried in an oven for 24 hours at a temperature of 105°C until constant weight is achieved.

7.9.2 Chemical Composition of the Fibers

The chemical composition of both untreated and alkali-treated fibers is determined separately by gradual removal of the components and gravimetrically determining their content. In the first step, the lignin component of the fibers (both untreated and alkali-treated) is removed. Some quantity (3 grams) of the fibers is dewaxed using 0.7% sodium chlorite solution that is adjusted to have a pH between 4.0 and 4.2 (adjusted utilizing a buffer) for 2 hours at a temperature in the range of 70 to 80°C. The fibers are separated by filtering and washed extensively with dilute sodium bisulphite solution (2% w/v) in distilled water. Then these fibers are washed thoroughly with distilled water and subsequently dried in an oven at 105°C until constant weight is achieved. The difference between the weight of the fibers before and after the lignin removal gives the weight of lignin present in the fibers. In this process, the lignin component of the fibers is removed resulting in holocellulose (α-cellulose + hemicelluloses). In order to determine the α-cellulose content in the fibers, the hemicelluose component is removed by treating the pre-weighed holocellulose sample with 17.5% NaOH (in water) solution and after a period of 35 minutes, the required quantity of distilled water is added and stirred well. In this process the alkali-soluble hemicelluloses component gets dissolved. The remaining α-cellulose content is filtered and then neutralized with acetic acid (10%) to remove unreacted NaOH if any and filtered. Then the resulting pure α-cellulose sample is washed thoroughly with distilled water, dried and re-weighed. The difference in the initial weight of holocellulose and the α-cellulose gives the hemicelluose content.

7.9.3 Chemical Composition and Tensile Properties of *Thespesia* Fibers

The wt.% of α-cellulose, hemicelluloses, and lignin of the untreated fibers is found to be 60.7, 26.6, and 12.7, respectively, while these values for alkali-treated fibers are 64.2, 20.8, and 15.0, respectively. These results indicate that the % hemicellulose content in the fibers is decreased on alkali treatment as this component is sensitive for the action of an alkali [36]. The α-cellulose, hemicelluose and lignin values (%) of other important natural fibers are (the values are presented in the parenthesis): abaca (56–63, 20–25, and 7–9) [37]; bamboo (26–25, 30, and 21–31) [37]; banana (63–64, 19, and 5) [37]; coir (32–43, 0.15–0.25, and 40–45) [37]; flax (71, 18.6–20.6, and 2.2) [37]; hemp (68, 15, and 10) [37]; jute (61–71, 14–20, and 12–13) [37]; kenaf (72, 20.3, and 9) [37]; napier (45.6, 33.67, and 20.6) [38]; ramie (68.6, 13–16, and 0.6–0.7) [37]; sisal (66–78,10–14, and 10–14) [37]; and *Thespesia lampas* (60.6, 26.6, and 12.7) [36]. These values indicate that the main α-cellulose content which is responsible for most of the properties of the *Thespesia* fibers is found to be higher than those of abaca, bamboo, *Borassus*, coir, and napier grass and are comparable with other fibers.

The tensile parameters maximum stress (MPa), Young's modulus (GPa), and elongation at break (%) of the untreated *Thespesia lampas* fibers are 573, 61.2, and 0.79, respectively, and at the same time these values for alkali-treated fibers are found to be 780, 94.5, and 1.05, respectively. These values for other important natural fibers are (the values are presented in the parenthesis): abaca (400, 12, and 3–10) [37]; bamboo (140–230, 11–17) [37]; banana (500, 12, 5.9) [37]; borassus (70.8, 10.8, 34.8) [39]; coir (175, 4–6, 30) [37]; flax (345–1500, 27.6, 2.7–3.2) [37]; hemp (690, 70, 1.6) [37]; jute (393–773, 26.5, 1.5–1.8) [37]; kenaf (930, 53, 1.6) [37]; napier (75, 6.8, 2.8) [38]; oil palm (248, 3.2, 25) [37]; pineapple (1.44, 400–627, 14.5) [37]; ramie (560, 24.5, 2.5) [37]; and sisal (468–700, 9.4–22, 3–7) [37]. These values indicate that the tensile properties of *Thespesia lampas* fibers are higher than those of abaca, bamboo, banana, borassus, coir, flax, hemp, jute, napier, oil palm, pineapple, ramie, and sisal while comparable with other fibers. Hence, based on the favorable tensile properties, *Thespesia lampas* fibers can be considered as reinforcement in making polymer composites.

7.10 CELLULOSE/*THESPESIA LAMPAS* SHORT FIBER COMPOSITE FILMS

Akali-treated *Thespesia lampas* fibers discussed in Section 7.8.1 are cut into short fibers and used as reinforcement in cellulose matrix. The tensile properties of the cellulose/*Thespesia lampas* short fiber composite films are studied. The effect of fiber loadings on the tensile properties of these composite films is studied [40].

7.10.1 Preparation of Cellulose/*Thespesia* Composite Films

Initially, the alkali-treated *Thespesia lampas* fibers are cut to a length of ~1 mm and dried in the oven and stored separately in desiccators. The cellulose matrix is prepared by dissolving the cotton linter pulp with a degree of polymerization of 620 in a pre-cooled (–12.5°C) aqueous solution of 7 wt.% NaOH and 12 wt.% urea under stirring for 2 minutes at a speed of 1000 rpm. The cellulose solution thus obtained is centrifuged at 5°C and a speed of 7200 rpm for 20 minutes to remove the undissolved cellulose and other impurities if any. The clear cellulose solution is decanted and stored at 5°C until used. Now the clear cellulose solution is taken in different 100-ml glass beakers and in each beaker the short *Thespesia* fibers are added in the quantity of 1 to 5% by the weight of cellulose taken in the beakers. These solutions with short fibers are stirred well and degassed to remove any air bubbles. The clear solutions are then poured on glass plates separately and spread uniformly using glass spreaders with spacers to obtain uniform thickness. Then the wet films are regenerated in an aqueous 5 wt.% H_2SO_4 bath. The separated gel composite films are then washed thoroughly in distilled water and then air dried to obtain the dry cellulose/*Thespesia* short fiber composite films. Finally, the cellulose/*Thespesia* composite films with different loadings of the alkali-treated short fibers obtained with a thickness of ~30 μm are stored in desiccators until they are tested.

7.10.2 Testing

The tensile and migration tests are conducted on the cellulose/*Thespesia* composite films. For tensile test, the film strips in rectangular shape with length of 100 mm and width of 10 mm are employed. The test is conducted employing an Instron 3369 Universal testing machine. Maintaining a gauge length of 50 mm, the tensile test is conducted at an extension rate of 5 mm/minute. In each case, ten specimens are used and the average values of the tensile parameters—tensile modulus, maximum stress, and % elongation at break are reported. For food packaging applications, the composite films should be non-toxic. In order to assess the toxicity of the cellulose/*Thespesia* composite films, the migration test is conducted. For this, the procedure given in the standard migration test [41] is followed. Briefly, in this method, the film strips of 100 mm × 100 mm in each case are subjected to the migration test. n-heptane is employed as the stimulant. In different beakers, certain quantity of n-heptane (for the sample to immerse completely) is taken and the weight of each beaker with n-heptane W1 is found. Now in each beaker, one specimen is immersed for a period of 30 minutes maintaining a temperature of 38°C. After this period, the samples are removed and in each case, the weight of the beaker with n-heptane and extract is re-weighed (W_2). The difference of these weights i.e., $M=(W_2-W_1)$ in grams, represents the weight of the extracted residue. Using M and the area of the specimen A in cm^2, the amount of extract (E_x) is calculated by Equation 7.1 as:

$$E_x = (M/A)\times 100 \text{ g/dm}^2 \tag{7.1}$$

In each case, three specimens are used and the average values reported.

7.10.3 Tensile and Toxicity Properties

The tensile modulus (MPa), maximum stress (MPa), and elongation at break (%) of the cellulose matrix are found to be 2541, 95.8, and 5.6, respectively. The tensile parameters—tensile modulus (MPa), maximum stress (MPa), and elongation at break (%) of the cellulose/*Thespesia* composite films for each fiber loadings are (the values are presented in parenthesis): 1 wt.% (2036, 93.7, and 4.8); 2 wt.% (1972, 89.2, and 4.5); 3 wt.% (1895, 74.6, and 4.3); 4 wt.% (1443, 72.5, and 3.), and 5 wt.% (963, 63.5, and 3.6) while the values reported for the conventional synthetic polymer films polyethylene are 117, 15.9, and 790, respectively, whereas for polypropylene they are 320, 28.6, and 948 [42]. These results indicate that the tensile parameters of the cellulose/*Thespesia* composite films are lower than those of the cellulose matrix and decreasing with increasing fiber loadings. The lower aspect ratio (due to the usage of short fibers) of the fibers and their random orientation are responsible for the decreased tensile properties of the composite films under study. However, these values are still higher than those of the conventional non-biodegradable polymer films such as polyethylene and polypropylene. Hence, the cellulose/*Thespesia* fiber composite films can be used as biodegradable packaging and wrapping materials.

In order to probe the toxicity of the films under study, the migration test is conducted. The amount of extract for the cellulose matrix is found to be 0.40 g/dm^2. The values for the cellulose/*Thespesia* composite films with fiber loadings of 1 wt.%, 2 wt.%, 3 wt.%, 4 wt.%, and 5 wt.% are found to be 0.48 g/dm^2, 0.48 g/dm^2, 0.56 g/dm^2, 0.56 g/dm^2, and 0.64 g/dm^2, respectively. However, these values are far below the allowed amount of extract of 10 g/dm^2 [41]. These observations indicate that the cellulose/*Thespesia lampas* short fiber composite films are non-toxic and hence, can be safely used as packaging materials.

REFERENCES

1. Ashori, A., 2006. "Nonwood fibers a potential source of raw material in paper making." *Polymer-Plastic and Technology and Engineering* 45(10), 1133–1136. doi:10.1080/03602550600728976.
2. Reddy, N., Yang, Y., 2005. "Biofibers from agricultural byproducts for industrial applications." *Trends in Biotechnology* 23(1), 22–27. doi:10.1016/j.tibtech.2004.11.002.
3. Cheung, Hoi-yan, Mei-po, Ho, Lau, Kin-tak, Cardona, Francisco, Hui, David, 2009. "Natural fibre- reinforced composites for bioengineering and environmental engineering applications." *Composites Part B: Engineering* 40, 655–663. doi:10.1016/j.compositesb.2009.04.014.
4. Mohanty, A.K., Misra, M., Drza, L.T., 2002. "Sustainable biocomposites from renewable resources: Opportunities and challenges in green materials world." *Journal of Polymers and the Environment* 10(1/2), 19–26. doi:1566-2543/02/0400-0019/0.
5. Omer Faruka, D., Andrzej, K., Bledzkia, C., Fink, Hans Peter, Mohini, B.S., 2012. "Biocomposites reinforced with natural fibers: 2000–2010." *Progress in Polymer Science* 37(11), 1552–1596. doi:10.1016/j.progpolymsci.2012.04.003.
6. Satyanarayana, K.G., Arizaga, G.G.C., Wypych, F., 2009. "Biodegradable composites based on lignocellulosic fibers an overview." *Progress in Polymer Science* 34(9), 982–1021. doi: 10.1016/j.progpolymsci.2008.12.002.
7. Eichhorn, S.J., 2011. "Cellulose nanowhiskers: Promising materials for advanced applications." *Soft Matter* 7(2), 303–315. doi: 10.1039/C0SM00142B.
8. Klemm, D., Kramer, F., Moritz, S., Lindstrom, T., Ankerfors, M., Gray, D., Dorris, A., 2011. "Nanocelluloses: A new family of nature-based materials." *Angewandte Chemie International Edition* 50(24), 5438–5466. doi:10.1002/anie.201001273.
9. Obireddy, K., Zhang, Jinming, Zhang, Jun, Varada Rajulu, A., 2014. "Preparation and properties of self-reinforced cellulose composite films from Agave microfibrils using an ionic liquid." *Carbohydrate Polymers* 114, 537–545. doi:10.1016/j.carbpol.2014.08.054.
10. Takagi, H., Asano, A., 2008. "Effects of processing conditions on flexural properties of cellulose nanofiber reinforced 'green' composites." *Composites—Part A: Applied Science and Manufacturing* 39(4), 685–689. doi: 10.1016/j.compositesa.2007.08.019.
11. Nadhan, A., Varada Rajulu, A., Li, R., Cai, J., Zhang, L., 2012. "Properties of waste silk short fiber/ cellulose green composite films." *Journal of Composite Materials* 46(1), 123–127. doi:10.1177/0021998311410507.
12. Nadhan, A., Varada Rajulu, A., Li, R., Cai, J., Zhang, L., 2012. "Properties of regenerated cellulose short fibers/ cellulose green composite films." *Journal of Polymers and the Environment* 20(2), 454–458. doi:10.1007/s10924-011-0398-x.

13. Han, D., Yan, L., Chen, W., Li, W., Bangal, P.R., 2011. "Cellulose/graphite oxide composite films with improved mechanical properties over a wide range of temperature." *Carbohydrate Polymers* 83(2), 966–972. doi:10.1016/j.carbpol.2010.09.006.
14. Delhom, C.D., White-Ghoorahoo, L.A., Pang, S.S., 2010. "Development and characterization of cellulose/clay nanocomposites." *Composites, Part B: Engineering* 41(6), 475–481. doi:10.1016/j.compositesb.2009.10.007.
15. Obireddy, K., Guduri, B.R., Varada Rajulu, A., 2009. "Structural characterization and tensile properties of Borassus fruit fibers." *Journal of Applied Polymer Science* 114(1), 603–611. doi:10.1002/app.30584.
16. Klemm, D., Heublein, B., Fink, H.P., Bohn, A., 2005. "Cellulose: Fascinating biopolymer and sustainable raw material." *Angewandte Chemie International Edition* 44(22), 3358–3393. doi:10.1002/anie.200460587.
17. Ganster, J., Ljungberg, H.P.N., Wesslen, B., 2003. "Tributyl citrate oligomers as plasticizers for poly (lactic acid): Thermo-mechanical film properties and aging." *Polymer* 44(25), 7679–7688. doi:10.1016/j.polymer.2003.09.055.
18. Afrifah, K.A., Matuana, L.M., 2010. "Impact modification of polylactide with a biodegradable ethylene/acrylate copolymer." *Macromolecular Materials and Engineering* 295(9), 802–811. doi:10.1002/mame.201000107.
19. Anderson, K.S., Schreck, K.M., Hillmyer, M.A., 2008. "Toughening polylactide." *Polymer Reviews* 48(1), 85–108. doi:10.1080/15583720701834216.
20. Ganster, J., Fink, H.P., 2010. "PLA-based bio- and nanocomposites." In: Lau, A.K.T, Hussain, F., and Lafdi, K., *Nano- and Biocomposites* (pp. 275–290). Boca Raton, FL: CRC Press. ISBN 9781138112124—CAT#.
21. Darie, R.N., Bodirlau, R., Teaca, C.A., Macyszyn, J., Kozlowski, M., Spiridon, I., 2013. "Influence of accelerated weathering on the properties of polypropylene/polylactic acid/eucalyptus wood composites." *International Journal of Polymer Analysis and Characterization* 18(4), 315–327. doi:10.1080/1023666X.2013.784936.
22. Balakrishnan, H., Hassan, A., Imran, M., Wahit, M.U., 2012. "Toughening of polylactic acid nanocomposites: A short review." *Polymer—Plastics Technology and Engineering* 51(2), 175–192. doi:10.1080/03602559.2011.618329.
23. Tajitsu, Y., Kanesaki, M., Tsukiji, M., Imoto, K., Date, M., Fukada, E., 2005. "Novel tweezers for biological cells using piezoelectric polylactic acid fibers." *Ferroelectrics* 320(1), 133–139. doi:10.1080/00150190590966982.
24. Jayaramudu, J., Guduri, B.R., Varada Rajulu, A., 2009. "Characterization of natural fabric Sterculia urens." *International Journal of Polymer Analysis and Characterization* 14(2), 115–125. doi: 10.1080/10236660802601415.
25. Jayaramudu, J., Reddy, Siva Mohan G., Varaprasad, K., Sadiku, E.R., Sinha Ray, S, Varada Rajulu, A., 2013. "Preparation and properties of biodegradable films from Sterculia urens short fiber/cellulose green composites." *Carbohydrate Polymers* 93(2), 622–627. doi:10.1016/j.carbpol.2013.01.032.
26. Jayaramudu, J., Siva Mohan Reddy, G., Varaprasad, K., Sadiku, E.R., Sinha Ray, S, Varada Rajulu, A., 2013. "Structure and properties of poly (lactic acid)/Sterculia urens uniaxial fabric biocomposites." *Carbohydrate Polymers* 94(2), 822–828. doi:10.1016/j.carbpol.2013.02.002.
27. Reddy, K.R.N., Rao, D.K.N., Rao, K.G.K., Kar, K.K., 2012. "Studies on woven century fiber polyester composites." *Journal of Composite Materials* 46(23), 2919–2933. doi: 10.1177/0021998311434789.
28. Obireddy, K., Zhang, Jinming, Zhang, Jun, Varada Rajulu, A., 2014. "Preparation and properties of self-reinforced cellulose composite films from Agave microfibrils using an ionic liquid." *Carbohydrate Polymers* 114, 537–545. doi:10.1016/j.carbpol.2014.08.054.

29. Ashok, B., Naresh, S., Reddy, K.O., Madhukar, K., Cai, J., Zhang, L., Rajulu, A.V., 2014. "Tensile and thermal properties of poly (lactic acid)/eggshell powder composite films." *International Journal of Polymer Analysis and Characterization* 19(3), 245–255. doi:10.1080/1023666X.2014.879633.
30. Zhang, X., Liu, T., Sreekumar, T.V., Kumar, S., Moore, C.V., Hauge, R.H., Smalley, R.E., 2003. "Poly (vinyl alcohol)/SWNT composite film." *Nano Letters* 3(9), 1285–1288. doi:10.1021/nl034336t.
31. Feng, Y., Ashok, B., Madhukar, K., Zhang, J., Reddy, K.O., Reddy, K.O., Rajulu, A.V., 2014. "Preparation and characterization of polypropylene carbonate bio-filler (egg shell powder) composite films." *International Journal of Polymer Analysis and Characterization* 19(7), 637–647. doi:10.1080/1023666X.2014.953747.
32. Cyras, V.P., Commisso, M.S., Mauri, A.N., Vazquez, A., 2007. "Biodegradabledouble-layer films based on biological resources: Polyhydroxybutyrate and cellulose." *Journal of Applied Polymer Science* 106(2), 749–756. doi:10.1002/app.26663.
33. Torres, F.G.P, Troncoso, O., Torres, C., Díaz, D.A., Amaya, E., 2011. "Biodegradability and mechanical properties of starch films from Andean crops." *International Journal of Biological Macromolecules* 48(4), 603–606. doi:10.1016/j.ijbiomac.2011.01.026.
34. Pereda, M., Dufresne, A., Aranguren, M.I., Marcovich, N.E., 2014. "Polyelectrolyte films based on chitosan/olive oil and reinforced with cellulose nanocrystals." *Carbohydrate Polymers* 101, 1018–1026. doi:10.1016/j.carbpol.2013.10.046.
35. Chumbhale, D.S., Upasani, C.D., 2012. "Pharmacognostic standardization stems of Thespesia lampas (Cav.) Dalz & Gibs." *Asian Pacific Journal of Tropical Biomedicine* 2(5), 357–363. doi:10.1016/S2221-1691(12)60056-2.
36. Obireddy, K., Ashok, B., Raja Narebder Reddy, K., Feng, Y.E., Jun, Zhang, Varada Rajulu, A., 2014. "Extraction and characterization of novel lignocellulosic fibers from thespesia lampas plant." *International Journal of Polymer Analysis and Characterization* 19(1), 48–61. doi:10.1080/1023666X.2014.854520.
37. John, M.J., Anandjiwala, R.D., 2008. "Recent developments in chemical modification and characterization of natural fiber-reinforced composites." *Polymer Composites* 29(2), 187–207. doi:10.1002/pc.20461.
38. Reddy, K.O., Maheswari, C.U., Shukla, M., Varada Rajulu, A., 2012. "Chemical composition and structural characterization of Napier grass fibers." *Materials Letters* 67(1), 35–38. doi:10.3390/en8053403.
39. Obi Reddy, K., Maheswari, C.U., Shukla, M., Song, J.I., Varada Rajulu, A., 2013. "Tensile and structural characterization of alkali treated Borassus fruit fine fibers." *Composites Part B: Engineering* 44(1), 433–438. doi:10.1016/j.compositesb.2012.04.075.
40. Ashok, B., Obi Reddy, K., Madhukar, K., Cai, J., Zhang, L., Varada Rajulu, A., 2015. "Properties of cellulose/Thespesia lampas short fibers bio-composite films." *Carbohydrate Polymers* 127, 110–115. doi: 10.1016/j.carbpol.2015.03.054.
41. Bureau of Indian Standards, 1998. "Determination of overall migration of constituents of plastics materials and articles intended to come in contact with food stuffs—method of analysis." IS 9845. *Bureau of Indian Standards*, IS 9845 (New Delhi).
42. Muzzy, J.D., 2000. *Thermoplastics-Properties, Incomprehensive Composite Materials*, Kelly, A., Zweben, C. (Eds.). (p. 19). Amsterdam: Elsevier Science.

8 Electrospun Nanofibers

Fundamentals, Food Packaging Technology, and Safety

Juliana Botelho Moreira, Suelen Goettems Kuntzler, Ana Luiza Machado Terra, Jorge Alberto Vieira Costa, and Michele Greque de Morais

CONTENTS

8.1 INTRODUCTION

The packaging protects foods against contamination and mechanical damage during transport, which may cause deteriorating effects on the product (Biji et al., 2015; Marsh and Bugusu, 2007; Sohail et al., 2018). Loss of food quality may be related to several intrinsic aspects such as the presence of spoilage microorganisms and physicochemical changes (moisture content, pH, and presence of preservatives) (Kalpana et al., 2019). External factors such as gas composition, relative humidity, and storage temperature also affect the shelf life of products (Abreu et al., 2012).

To ensure the safety and nutritional quality of food products, without affecting the environment, innovative technologies for packaging development are emerging. In this context, the development of active and intelligent materials using nanotechnology can be included (Maftoonazad and Ramaswamy, 2019; Restuccia et al., 2010; Vanderroost et al., 2014). Nanofibers are promising materials for application in active and intelligent food packaging due to their unique properties such as high surface area, high porosity, biocompatibility, biodegradability, and no toxicity (Francis et al., 2010; Işik et al., 2019).

Several technologies are used to synthesize nanofibers such as drawing, template synthesis, phase separation, self-assembly, and electrospinning. The electrospinning technique offers flexibility in the manipulation of process conditions, allows the use of various polymeric materials, and produces porous nanofibers with uniform diameters (Fang et al., 2011; Geoffry and Achur, 2018; Li et al., 2014). The main advantage of electrospinning over other processes is the scale-up that enables the application of nanofibers in the food industry (Petrík, 2011).

Bioactive compounds can be encapsulated into polymeric nanofibers applied in active and intelligent packaging to improve stability, promote a controlled release of the compound, and preserve antioxidant, antimicrobial, and antifungal properties (Kumar et al., 2019). In this context, the objective of this chapter is to address the electrospun nanofibers technology and the addition of biocompounds in this process for application in active and intelligent food packaging.

8.2 INTRINSIC AND EXTRINSIC FACTORS THAT INFLUENCE FOOD QUALITY

Food preservation is related to the potential for contamination by pathogenic microorganisms and the capacity for production toxins. Several factors may influence the stability of the food product. The manufacturing process, packaging material, environmental, and transport conditions may limit or inhibit the development of unwanted microorganisms (Hernández-Cortez et al., 2017). Depending on the origin of these factors, changes can be classified as extrinsic or intrinsic. Intrinsic parameters such as water activity, pH, and food composition are properties determined by factors inherent in the formulation. Extrinsic parameters are the properties related to the environment around foods such as temperature, humidity, and gas composition (Brecic et al., 2017).

8.2.1 Water Activity and Food Composition

The water activity (aw) refers to the free water within the intergranular spaces and pores of the food, directly influencing the conservation of the products (Muller et al., 2014). Aw interferes with the speed of spoilage reactions in food, as it acts on chemical and enzymatic reactions and nutrient dispersion media for the growth of microorganisms. Most fresh foods such as meat, vegetables, and fruits have aw values close to the optimum growth level of most bacteria (0.97–0.99). Aw can be controlled in foods through the addition of solutes (such as salt or sugar) or physical removal of water (by drying or cooking) (Buncic, 2006).

Another intrinsic factor that influences product quality is food composition. Microorganisms require nutrients for the growth and maintenance of their metabolic functions, like carbon, nitrogen, vitamins, and minerals (Tortora et al., 2011). Most animal foods are low in carbohydrates. Vegetables have high concentrations of different carbohydrates, and varying levels of protein, minerals, and vitamins. Information about food composition allows knowing the possible contaminating microorganism and the stability of the product (Fenster et al., 2019; Merchant and Helmann, 2012).

8.2.2 pH

pH refers to the function of the concentration of hydrogen ions, which can influence the growth rate and multiplication of the microorganisms in foods. The pH of foods such as beef (5.1–6.2) and fish (6.6–6.8) have values close to neutrality. Vegetables have a pH in the range of 3.6 to 6.5, while most fruits have a pH between 1.8 and 6.7 (Jay, 2000). The pH value of the food significantly influences the heat treatment intensity and storage conditions.

Foods of low acidity (pH >4.5) require longer thermal processes for the viability of pathogenic bacteria in vegetative form. However, there is a need for secondary treatments in these foods, such as keeping them in low-temperature conditions, to disadvantage the spore germination and toxin production. In acidic foods (pH <4.5), there is the inactivation of pathogenic and spoilage microorganisms, such as fungi and yeast, with short-term heat treatments. However, inhibition of sporulated bacteria does not occur (Cebrián et al., 2017; Martins et al., 2019). Foods with low buffering capacity change pH rapidly in response to acidic or alkaline compounds produced by microorganisms. In general, animal source foods present a greater buffering character than vegetables due to proteins in their composition (Petruzzi et al., 2017).

8.2.3 Temperature and Relative Humidity

Food storage temperature is an extrinsic factor that influences enzymatic and biochemical reactions, membrane fluidity, and folding of DNA, RNA, and ribosomes. Microorganisms have a specific multiplication temperature range between minimum, optimal, and maximum growth (Kong and Singh, 2016). At low temperatures, two factors contribute to the inhibition of microbial growth. The first one is based on the enzymatic reaction rates in the body, which becomes slower. The other refers to the reduction of cytoplasmic membrane fluidity, interfering in the nutrient transport mechanisms. However, the microbial growth rate increases with increasing temperature until the maximum temperature is reached. At high temperatures, the structural cellular components are denatured, and the inactivation of heat-sensitive enzymes occurs (Gaur et al., 2019).

Relative humidity is another extrinsic factor that influences product quality. If food with low aw is stored in an environment with high relative humidity, its aw will increase, resulting in changes in physical appearance, and microbial growth. Foods with low aw may be subject to moisture condensation on the surface due to high-temperature variation in the environment. The accumulation of surface water will result in environments favorable to the multiplication of spoilage microorganisms. Thus, for foods stored at lower temperatures, the relative humidity should be reduced (Amit et al., 2017; Hernández-Cortez et al., 2017).

8.2.4 Gaseous Composition

Unwanted enzymatic and microbial activities result in food degradation with the formation of gases that alter product quality. The gas production in food may also be linked to acidification of the medium. The main gas-producing bacteria in dairy products are coliforms and some *Bacillus* and *Clostridium* species which produce CO_2 and H_2. Propionic and heterolactic bacteria produce only CO_2 (Ekezie et al., 2017).

Spore-forming gram-positive microorganisms such as *Bacillus* and *Clostridium* are found in thermally processed foods. *Clostridium thermosaccharolyticum* spoils canned foods and, due to the production of undesirable gases, causes bulking in cans. The microorganism *Desulfotomaculum nigrificans* produces hydrogen sulfide, which causes foul-smelling and stuffing in canned goods. The presence of *Pseudomonas* in foods such as meat results in proteins degradation and production of esters and volatile sulfide compounds (Forsythe, 2010). It results in a modification in physical, chemical, or organoleptic characteristics.

8.3 FOOD PACKAGING

The packaging has a protective function that is associated with the barrier between the product and the external environment. With this, there is a prevention of the migration of chemicals into food (Koeijer et al., 2017; Mcmillin, 2017). The industry has been looking for new technologies and processes to meet the growing demand for safe food, and reduce production costs (Rolle and Enriquez, 2017). In this way, active and intelligent systems are being developed through nanotechnology techniques that can preserve the quality of the product and provide more dynamic feedback to the consumer (Abdullahi, 2018).

8.3.1 Active Packaging

Active packaging allows the protection of food against degradation and microbiological contamination, creating a barrier to external conditions. Active packaging has components that can be antimicrobials, antioxidants, and enzymes. These systems are applied for microbial growth control, inhibition of oxidative degradation reactions, and directed biocatalysis (Bastarrachea et al., 2015). Active packaging systems can be divided into active scavenging systems (absorbers) and active-releasing systems (emitters) (Yildirim et al., 2018) as can be observed in several studies (Table 8.1).

Active scavenging systems are used to remove/control environmental parameters such as moisture and odor, and unwanted concentrations of oxygen, CO_2, and ethylene. Active release systems add and release compounds in headspace inside the packaging or packed food. These compounds can be antimicrobials, antioxidants, or flavoring substances in foods. Depending on the physical form of the active systems, the packaging is of sachet, label, or film type. The sachets and labels are placed

TABLE 8.1
Potential Application of Active Packaging Systems

Active systems	Action in foods	Potential application	References
O_2 elimination	Microbial control and shelf life prolongation	Fresh fruits and vegetables	Turan et al. (2017)
	Inhibition of vitamin degradation and lipid oxidation	Fish oil-in-water emulsions	Johnson et al. (2018)
CO_2 elimination	Prevention of microbial spoilage	Sliced fresh vegetables	Kapetanakou et al. (2019)
	Prevention of volume expansion and breakage of packaging	Kimchi	Lee et al. (2019a)
Moisture absorption	Prevention of spoilage to extend storage shelf life	Strawberries	Jalali et al. (2019)
	Reduction of deterioration	Avocado	Gaona-Forero et al. (2018)
Ethylene elimination	Ethylene concentration control and moisture removal	Guava	Murmu and Mishra (2018)
	Reduction of ripening and inhibition of microbial growth	Fresh fruit	Siripatrawan and Kaewklin (2018)
CO_2 emission	Inhibition of microbial growth	Pork meat	Brodowska et al. (2019)
	Inactivation of pathogenic microorganisms	Fresh fish	Provincial et al. (2013)
Antimicrobial release	Bacterial growth reduction	Chicken meat	Tas et al. (2019)
	Control of pathogenic bacteria proliferation	Salmon and bread	Wang et al. (2019)
Antioxidant release	Free radical elimination	Cashew nut	Rambabu et al. (2019)
	Oxidative stability improvement	French fries	Oudjedi et al. (2019)

inside the packaging in contact with the food. In the film system, the active compounds are incorporated directly into the packaging matrix to which the food is coated (Yildirim et al., 2018). Thus, the addition of active compounds in this system will contribute to prolonging the shelf life of products and ensure food quality, safety, and integrity (Moreira et al., 2018b).

8.3.2 Intelligent Packaging

Intelligent packaging has a communication function, which facilitates the decision of retailers and consumers about the viability of food. According to European Regulation (EC/450/2009), intelligent packaging is an indicator system (internal or external) that monitors the condition of food during transport and storage to provide information on the quality of packaged food.

Intelligent systems are classified as devices or indicators (Table 8.2) and can be of the type temperature, pH, freshness and/or ripeness indicators, gas detectors, and radio frequency identification (RFID) systems. Devices and indicators report changes that occur in the food or its environment, such as temperature, amounts of CO_2, O_2, food components, microbial, and volatile compounds. As a result of this interaction, these intelligent systems will visually change packaging conditions if the food is not suitable for consumption (Sohail et al., 2018).

Intelligent packaging can also help improve Hazard Analysis and Critical Control Point (HACCP) systems (Heising et al., 2014). Therefore, it is possible to detect instabilities in food packaging, identify probable health risks, and establish strategies to reduce or eliminate these occurrences (Vanderroost et al., 2014).

8.4 POTENTIAL BIOCOMPOUNDS FOR FOOD PACKAGING

Quality and food safety are the main goals of the food industry. Different biocompounds can be added as coating materials for active packaging to preserve food products. Moreover, the incorporation of active molecules into intelligent systems can detect physicochemical or biochemical changes in food caused by improper transport and storage conditions.

8.4.1 Proteins

Coating materials from renewable sources such as proteins have promising properties for food packaging applications (Ahmad et al., 2016; Ahmad et al., 2012). Proteins can form stabilized three-dimensional networks reinforced by hydrogen bonds, hydrophobic interactions, and disulfide bonds, allowing the creation of intermolecular bonds and cohesive matrices (Benbettaïeb et al., 2016a; Thomas et al., 2013).

Soy protein isolate (SPI) is an amphiphilic molecule obtained as a refined byproduct from the soybean oil industry (Cao et al., 2007). Due to its non-cytotoxicity, abundance in nature, low cost, and nutritional value, SPI has attracted attention in the packaging industry to replace synthetic polymers (Denavi et al., 2009). Whey protein isolate (WPI) is an edible, biodegradable polymer that can be obtained from

TABLE 8.2
Potential Application of Intelligent Packaging Systems to Monitor the Quality of Foods

Intelligent systems	Material used for indicator development	Detection parameter of the system	Response	References
Time-temperature indicator	Label consisting of starch, iodine, and α-amylase enzyme	Change in optimal temperature of product	Loss of coloring	Brizio and Prentice (2015)
pH indicator	Cellulose nanofibers containing anthocyanins	Response behavior of the indicator to the total volatile basic nitrogen content (TVBN) after microbiological contamination	Color change	Moradi et al. (2019)
Freshness indicator	Multi-layer film composed of low density polyethylene and green bromocresol	Perception of CO_2 and TVBN production after microbial growth	Color change	Lee et al. (2019b)
Radio frequency identification device	Metal oxide plate	Measurement of the total concentration of the volatile organic compound (TVOC), CO_2, temperature, and humidity	Real-time data collection and transmission of information by computer systems	Cao and Chung (2019)
Gas sensor	Label developed from polyethylene glycol, ethylcellulose, and polylysine solution with anthocyanins	Microbiological contamination with CO_2 production	Color change	Saliu and Pergola (2018)

whey generated by the cheese industry and used as a polymeric matrix in the formation of packaging materials (Cinelli et al., 2016; Coltelli et al., 2016; Lara et al., 2019; Schmid, 2013). WPI-based film formation includes the heating of proteins, inducing their denaturation. Therefore, proteins become insoluble and aggregated. As a result, WPI films have desirable transparency and oxygen barrier properties (Schmid, 2013).

Collagen is a versatile natural polymer due to its amino acid residues in variable proportions and distributions along the macromolecular structure (Ahmad et al., 2010). Collagen has properties of barriers against oxygen and flavorings and can be used for the production of biodegradable packaging materials (Ahmad et al., 2016). However, some studies have shown that the hydrophilic nature of SPI, WPI, and collagen affects mechanical performance and thermal stability that limits their applications in the packaging industry (Benbettaïeb et al., 2016b; Zahedi et al., 2010). Thus, studies have explored the use of nanotechnology in combination with other polymers and biocompounds to improve the mechanical properties and stability of protein-based materials (Arinstein and Zussman, 2011; Huang et al., 2004; Wang et al., 2013).

The incorporation of protein hydrolysates is a promising strategy in the development of active packaging. Antioxidant films were produced with the addition of silver carp protein hydrolysate (Rostami et al., 2017). According to Filho et al. (2019), protein hydrolysates increase water vapor permeability and packaging material thickness. Besides, they observed that raising the concentration of these molecules contributes to improving the visible light barrier properties. Filho et al. also noted that the color of the film became darker, redder, and yellowish with the addition of extract protein, making these hydrolysates interesting for application in intelligent systems.

The zein is a prolamine protein obtained from corn, is water-insoluble, odorless, and tasteless. This protein is widely used in candy, nut, and fruit coating foods (Kumar et al., 2019). Moreover, the encapsulation of omega-3 fatty acids in zein fibers contributed to the stability of fish oils (Moomand and Lim, 2015). Another recently studied protein source is microalgae. The application of microalgae proteins in the production of active materials is an innovative strategy to explore new food preservation packaging through nanotechnology (Moreira et al., 2019b).

8.4.2 Phycocyanin

Phycocyanin is an accessory photosynthetic pigment extracted mainly from the microalga *Spirulina* (İlter et al., 2018; Martelli et al., 2014; Prates et al., 2018). Approximately 14% (w/w) of the pigment is found in the protein content of microalgal biomass. Phycocyanin is composed of an apoprotein and a non-protein component (corresponding to the chromophore fraction) known as phycocyanobilin (PCB). Apoprotein is connected to PCB by a thioether bond, and the protein portion of phycocyanin consists of α and β subunits that have molecular weights in the range of 18 and 20 kDa, respectively (Morais et al., 2018).

Phycocyanin is a bioactive compound that has been studied due to its antioxidant (Renugadevi et al., 2018), anti-inflammatory (Cherng et al., 2007; Talero et al., 2015), anticancer (Pan et al., 2015; Talero et al., 2015), and antibacterial (Chentir et al., 2018; El-Sheekh et al., 2014) potentials. Despite presenting water solubility and intense blue coloration, the application of phycocyanin in food has been limited due to its low stability (Falkeborg et al., 2018). Phycocyanin is unstable to light, low pH values, strong ionic forces, high temperatures, and presence of alcohol (Chaiklahan et al., 2012; Hsieh-Lo et al., 2019).

The stability of phycocyanin increases at higher temperatures combined with lower pH levels (Antelo et al., 2008). Studies have shown that the use of citric acid (Chaiklahan et al., 2012; Pan-Utai et al., 2017), glucose (Chaiklahan et al., 2012), sodium chloride (Chaiklahan et al., 2012), high pressure treatment (Zheng et al., 2020), and encapsulation into nanofibers (Braga et al., 2016; Moreira et al., 2019b; Moreira et al., 2018b) and nanoparticles (Schmatz et al., 2019) improves the stability of phycocyanin, preserving its bioactive properties during food processing or packaging applications.

However, the instability of phycocyanin is also interesting in the area of food. As pigment changes color depending on the variation of intrinsic and extrinsic parameters of food, this dye can be used to develop indicators for intelligent packaging applications to ensure food safety to the consumer (Moreira et al., 2018b).

8.4.3 Carotenoids

Carotenoids are lipophilic pigments responsible for the yellow, orange, and red coloration of fruits, vegetables, and marine sources (Roohbakhsh et al., 2017). Carotenoids are classified into carotenes, composed only of carbon and hydrogen atoms, and xanthophylls, which in addition to carbon and hydrogen, have oxygen in the molecule as part of a functional group (hydroxyl, epoxide, carbonyl) (Gong and Bassi, 2016).

Marketed carotenoids are mainly produced by chemical synthesis. However, these dyes can be extracted from natural sources such as annatto, paprika, turmeric, tomatoes, calendula, and microalgae (Rodriguez-Amaya, 2015). The chemical synthesis of carotenoids has limited use for human consumption due to safety concerns in the production process. Thus, natural sources of carotenoids have been sought given their use in the nutraceutical market (Gong and Bassi, 2016). In this context, microalgae are promising alternative sources of carotenoids in terms of economic viability (Liu et al., 2016). Consequently, food industries have produced carotenoids (astaxanthin) from microalgae (Panis and Carreon, 2016).

Carotenoids have beneficial effects in preventing various diseases, particularly cardiovascular disease (Jonasson et al., 2003; Pashkow et al., 2008) and cancer (Jain et al., 2017; Tanaka et al., 2012; Wan et al., 2014). Due to their antioxidant properties, carotenoids protect cells from reactive radicals and prevent lipid peroxidation (Grossman et al., 2004). Moreover, carotenoids, such as β-carotene and β-cryptoxanthin, have provitamin A activity, which plays a role in human health (Gurmu et al., 2014; Mercadante et al., 2017).

As with other natural pigments, the instability of carotenoids is a problem for its application in food. Stability is influenced by the nature of the carotenoid (carotene or xanthophyll, E or Z configuration, esterified or not esterified) and its interaction with the food matrix (fruits, roots, leaves, juices). Carotenoid degradation increases with high temperatures, light exposure, and extreme pH values. Thus, severe and/or improper processing and storage conditions, as well as the permeability of the packaging material to O_2, interfere with pigment stability (Rodriguez-Amaya, 2015). In this context, studies of microencapsulation (Meroni and Raikos, 2018; Mirafzali et al., 2014; Rodriguez-Amaya, 2015) and nanoencapsulation (Mehrad et al., 2018; Moreira et al., 2019a) have been developed to overcome these obstacles by improving carotenoid stability and bioavailability (Rodriguez-Amaya, 2015) for food packaging application.

8.4.4 Anthocyanin

Anthocyanins are secondary metabolites of plant products that give red, purple, and blue coloration to fruits and vegetables. The conjugated double bonds of the anthocyanidin portion constitute the chromophore, which is responsible for the characteristic anthocyanin stain (Rodriguez-Amaya, 2019). These polyphenolic pigments are water-soluble (Bunea et al., 2013) and less stable than carotenoids (Dey and Harborne, 1989). Factors such as temperature, pH, light, oxygen, enzymes, metal ions, ascorbic acid, sugars, and sulfites affect anthocyanin stability (Castañeda-Ovando et al., 2009).

In an aqueous medium as in foods, anthocyanins undergo reversible structural changes with pH, manifested in color change. At pH below 2, the red flavylium cation predominates. At pH 3 to 6, rapid hydration of the flavylium cation occurs at C-2 to form the colorless carbinol pseudobase, which may open the ring to a yellow chalcone. Under slightly acidic to neutral conditions, deprotonation of flavylium cation generates the blue quinoidal base (Rodriguez-Amaya, 2019). Because the process of food spoilage is often accompanied by pH changes, innovative packaging materials (intelligent indicators) can be produced by adding anthocyanins to monitor food freshness (Choi et al., 2017; Zhai et al., 2017).

Besides presenting potential for application in intelligent packaging systems, anthocyanins are also promising candidates for the development of bioactive packaging. Anthocyanins are used to remove free radicals and have anti-tumor and anti-inflammatory properties (Rodriguez-Amaya, 2019). Moreover, the anthocyanin structure contains phenolic hydroxyl groups, which can protect the food matrix from oxidizable components (Tsuda et al., 1996).

The use of anthocyanins in packaging development is also interesting due to their antibacterial properties. A study by Yoon et al. (2018) investigated the antimicrobial effect of anthocyanins extracted from beans and found inhibition against *Escherichia coli*. Wafa et al. (2017) extracted anthocyanins from pomegranates and found an antibacterial effect of pigment against *Salmonella* in chicken stored at 4°C. Thus, the anthocyanin addition to the packaging material can help preserve food and ensure product quality for the consumer.

Despite the great application potential that anthocyanins represent for the food industries, their use has been limited due to their relative instability and low yields in the extraction process (Castañeda-Ovando et al., 2009). Research has been looking for new technologies to improve stability (Otálora et al., 2015; Stoll et al., 2016).

Stabilization methods include co-addition, O_2 deletion, and encapsulation (Rodriguez-Amaya, 2019). Nanoencapsulation has been reported to develop natural dyes with greater stability, solubility, bioavailability, and better release control (Akhavan et al., 2018; Chi et al., 2019; Rezaei et al., 2019) than microencapsulation.

8.4.5 Quercentin and Curcumin

Different types of phenolic compounds have been incorporated into food packaging films to improve their physical property and antioxidant capacity (Arcan and Yemenicioğlu, 2011; Sun et al., 2017). Quercetin is a flavonoid found in vegetables, fruits, medicinal herbs, and red wines and has antioxidant potential due to its ability to eliminate free radicals and resist lipid peroxidation (Bozic et al., 2012). Thus, the incorporation of quercetin may increase the antioxidant capacity of materials for the development of active packaging.

Excessive Al^{3+} intake can lead to osteoporosis, Parkinson's and Alzheimer's diseases (Tang et al., 2017). In this context, Bai et al. (2019) evaluated the incorporation of quercetin in carboxymethylcellulose and chitosan films for the development of antioxidant and intelligent packaging. Bai et al. found that the addition of quercetin significantly affected the physical properties (color, thickness, moisture content, UV-vis transmission, opacity, elongation at break, and thermal stability) and structural characterization (microstructure and crystallinity) of the developed films. Furthermore, it was found that quercetin films can be used in intelligent packaging systems to be used as sensors to detect Al^{3+} in food products.

Topas is a cyclo-olefin often used in packaging materials. In a study performed by Masek et al. (2018), the color of the Topas composites containing quercetin changed under the influence of time and climatic factors. Therefore, the use of flavonoids as a natural indicator of packaging material aging time has been presented as an innovative material for the food industry. Moreover, antifungal activity was conferred to composites containing quercetin, demonstrating the relevance for application of these compounds in the development of antimicrobial packaging to guarantee quality and food safety.

Curcumin is a natural lipophilic phytochemical obtained from dried *Curcuma longa* rhizomes (Yılmaz et al., 2016). This polyphenol is widely used in the food and dye industry. Moreover, its antimicrobial properties, along with its antioxidant capacity, make curcumin a promising preservative for the food industry (Gómez-Estaca et al., 2017).

Most studies on curcumin use in food packaging focus on its combination with natural polymers. Zia et al. (2019) found that curcumin incorporation increases the thermal stability of active films. In addition, Zia et al. noted that the addition of dye improved film tensile modulus and water vapor barrier performance. These characteristics combined with the biological properties of curcumin make this pigment an excellent candidate for the coating of active food packaging.

Curcumin is also known for its antiviral, anticancer, anti-inflammatory, and antimicrobial activities (Rafiee et al., 2019). Vimala et al. (2011) produced curcumin encapsulated in chitosan–poly(vinyl alcohol)–silver nanocomposite films and they observed reduction in *E. coli* growth, demonstrating the potential of phytochemicals to be developed for food preservation uses in packaging. Also, curcumin's ability to change color through pH variations has led to the development of intelligent packaging systems from this pigment (Musso et al., 2017).

Moreover, curcumin is a promising bioactive compound for application in controlled release devices for food packaging (Baldino et al., 2017). However, use focused on its bioactivity becomes limited due to low water solubility, sensitivity to light, enzymes, heat, and oxygen. In this context, new materials such as nanofibers could be developed to improve stability (Uebel et al., 2016), solubility, and bioavailability of curcumin (Shlar et al., 2015; Yılmaz et al., 2016).

8.5 NANOTECHNOLOGY AND FOOD SAFETY

Nanotechnology faces a challenge in using low-cost processing operations to create nontoxic delivery systems and develop safe formulations for human consumption (Adabi et al., 2017). There is a compelling demand to address these issues to expand knowledge about the use of nanostructures in terms of biocompatibility, safety, and toxicity in the food sector (Bajpai et al., 2018). Guidance documents related to the potential risks associated with the use of these materials have been issued by regulatory bodies such as the United States Environmental Protection Agency, National Institute for Occupational Safety and Health, the Food and Drug Administration (FDA), the Health and Consumer Protection Directorate of the European Commission, and the International Organization for Standardization and the Organization for Economic Cooperation and Development (He and Hwang, 2016).

Most research emphasizes the regulation of nanotechnology in food packaging and processing (Chau et al., 2007). The effect of nanomaterials on humans, animals, and the environment is unpredictable due to changes in their properties over time. Nanostructures can cross biological barriers and enter cells and organs (Jovanovic, 2015; Su and Li, 2004). However, studies on migration are limited and further investigation is required before the extensive application of these materials in food and packaging can take place.

Nanoencapsulation allows direct contact of nanomaterials with humans through oral ingestion and inhalation. Toxicity will depend on the concentration used and the amount consumed or inhaled. Allergy and heavy metal release are the most discussed concerns in the literature (He and Hwang, 2016), as well as bioaccumulation of nanomaterials in food and humans. Therefore, risk assessment procedures must be strictly followed during food processing. Furthermore, future research on nanoencapsulated foods should investigate long-term toxicity (Bajpai et al., 2018).

Although the fate and potential toxicity of nanomaterials are not fully understood at the moment, it is evident that nanotechnological advances are contributing significantly to the food industries. Thus, the development of nanotechnology and its application in food science must be carried out with caution, and the benefits it confers cannot be overlooked.

8.6 ELECTROSPINNING

The electrospinning technique has been known since the 1930s and has been refined due to new applications of nanobiotechnology for nanometer-scale polymer fiber formation. Electrospinning enables producing nanofibers with unique properties such as morphology and size, high surface area, and controllable porous structure. The potential for semipermeable nanofiber formation has aroused interest in the development of active and intelligent packaging to apply it to foods with high moisture content (Sill and Von Recum, 2008).

The principle of the electrospinning technique is based on the theory that electrical charges applied to the liquid droplet exiting the capillary are capable of deforming the droplet interface. These charges generate electrostatic forces inside the droplet, known as the Coulomb force, which compete with the surface tension of the droplet, forming the Taylor cone (Bhardwaj and Kundu, 2010) (Figure 8.1). The main components of electrospinning equipment are the high potential electric source, positive detachment pump, capillary, and grounded collecting plate (static or rotary).

The Taylor cone subjected to high electrical potential causes a solution strain that tends to conical shape. The performance of the electrospinning technique depends on proper control of the parameters that contribute to the nanofibers formation (Kuntzler et al., 2018a; Reneker et al., 2000). Parameters include polymer solution properties, environmental conditions, and process. These factors also influence the morphology of the developed nanofibers (Bhardwaj and Kundu, 2010; Moreira et al., 2018b).

8.6.1 Properties of Polymeric Solution

The properties of the solution such as viscosity, electrical conductivity, and surface tension influence the morphology of the nanofibers. These parameters are related to the physicochemical properties of polymers, solvents, and polymer-solvent interactions (Kuntzler et al., 2018a). Entanglement of the polymeric chain is required to obtain uniform nanofibers without droplets. During the stretching of the polymer

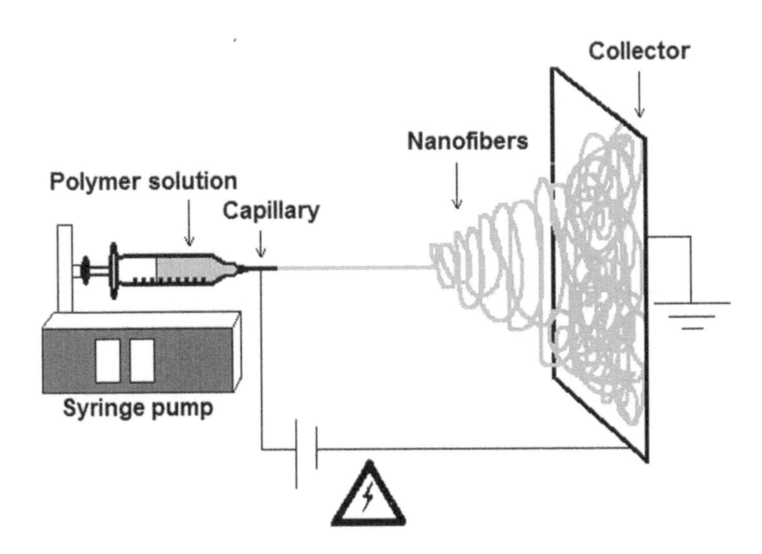

FIGURE 8.1 Illustrative representation of the electrospinning technique.

solution, the entanglement of the molecule chains keeps the jet electrically continuous (Haider et al., 2018).

When the polymeric concentration is not satisfactory, there is jet instability that can form nanoparticles or obstruct the capillary. Increasing the concentration of the solution causes an increase in viscosity, which may contribute to entanglement between polymer chains (Haider et al., 2013). Moreover, viscosity affects the diameter of nanofibers. If viscosity is higher, pores and droplet structures are less likely to form during the process. If the viscosity is too low, the jet from the Taylor cone will drop into droplets, resulting in granular structure (Khan et al., 2013).

Surface tension is directly related to Taylor cone formation. This phenomenon occurs when the applied electrical potential is high enough to make the electrostatic forces exceed the surface tension of the drop. From this electrical potential, called the critical value, the electrospinning process begins (Kuntzler et al., 2018a).

The conductivity of the solution is mainly determined by the type of polymer and solvent used, and the availability of ionizable salts. The electrospinning technique involves droplet stretching caused by repulsion of charges on the surface. Therefore, if the conductivity of the solution is high, more charges can be carried by the electrostatic jet, forming nanofibers (Ravandi et al., 2015). The conductivity of the solution affects both the Taylor cone formation and nanofiber diameter. In low conductivity solutions, the droplet surface will not be sufficiently charged to form the Taylor cone, and the electrospinning process will not occur. Raising solution conductivity to a critical value increases drop surface loading and reduces nanofiber diameter (Sun et al., 2014).

8.6.2 Process and Environment Parameters

The ideal range of applied electrical potential for a given polymer-solvent system implies the electrostatic repulsion force in the solution, which favors the jet elongation of the polymeric solution (Shi et al., 2015; Sill and Von Recum, 2008). The feed rate of the polymer solution directly influences solvent evaporation. The flow rate should be sufficient for the solvent to evaporate while elongating the solution jet. Thus, uniform nanofibers can be produced without the occurrence of obstruction in the capillary (Haider et al., 2018).

Similar to the applied electrical potential and feed rate, the distance between the capillary and the collector also affects nanofiber morphology. Therefore, a critical distance must be maintained to produce uniform nanofibers. This distance may vary depending on the polymer system used, as it directly influences the jet air time and solvent evaporation rate. The jet will have a short distance to reach the collector if the distance is reduced. As a consequence, the time may not be sufficient for the solvent to evaporate (Bhardwaj and Kundu, 2010; Matabola and Moutloali, 2013).

The environmental parameters are related to the conditions and properties of the place where the electrospinning technique occurs. Increasing the temperature occasions faster evaporation of the solvent, which may cause obstruction of the capillary. Therefore, solvents with a lower evaporation rate are preferred (Chen and Wong, 2008). Regarding humidity, lower values of this parameter may cause solvent drying when the evaporation rate is high (Bhardwaj and Kundu, 2010). On the other hand, if the humidity is high, nanofibers with larger diameters may be formed due to the charge neutralization in the jet and the elongation force that becomes small (Li and Wang, 2013).

8.7 ELECTROSPUN NANOFIBERS FOR FOOD PACKAGING

The electrospinning technique allows for producing nanofibers with several applications. It is the most recent technique in the food industry in the area of packaging for the production of innovative materials with improved properties (Wang et al., 2011; Moreira et al., 2018a). Nanofibers for food packaging applications present structural advantages, in addition to high mechanical resistance (Shen et al., 2011).

Nanofibers added into active and intelligent food packaging present functional benefits such as the encapsulation of bioactive compounds that are unstable to factors such as temperature and gases, which enables their degradation to be avoided (Bhushani and Anandharamakrishnan, 2014). Besides, the high surface area of nanofibers increases the reactivity of the material. With this increase, adsorption or release mechanisms are accelerated, increasing the number of sites for interaction or binding of reactive materials for the packaging development (Mercante et al., 2017).

Nanofibers can be used in food preservation due to the ability to immobilize or encapsulate bioactive compounds such as enzymes, vitamins, antioxidants, and antimicrobials (Kumar et al., 2019). Packaging developed with nanofibers incorporated with photo-sensible compounds can be used as sensors of volatile gas, pH, and time-temperature, which facilitate the identification of the quality of foods (Table 8.3).

8.7.1 Antimicrobial Nanofibers

The use of antimicrobial agents capable of diffusing in food and inhibiting the proliferation of microorganisms during storage is a type of active packaging system (Neo et al., 2013). Nanofibers produced by electrospinning can encapsulate antimicrobial substances to preserve the freshness of the product. Moreover, nanofibers allow the controlled release of antimicrobial agents with high transport capacity due to its porosity (Geoffry and Achur, 2018).

Nanofibers with antibacterial activity were studied by Kowsalya et al. (2019). Kowsalya et al. produced silver nanoparticles (AgNPs) through biosynthesis with grape peel and a silver nitrate solution. The AgNPs added into PVA nanofibers presented an inhibition halo of 21.47 ± 0.15 mm against *Staphylococcus aureus*. Antibacterial activity of the nanofibers was also analyzed in lemons and strawberries, in which a storage time of three times longer (ten days) was obtained over the control group. According to Martinez-Abad et al. (2012) and Rai et al. (2009), AgNPs are most commonly studied among metallic nanoparticles for application in active packaging due to their greater antimicrobial efficiency against microorganisms such as bacteria, yeasts, fungi, and viruses.

Nanofibers developed from natural compounds or bioactive plant extracts are used for food packaging as they reduce risks to health and the environment. Pan et al. (2019) produced PVA nanofibers with added cinnamon essential oil, and analyzed the antibacterial activity *in vitro* and in mushrooms. Pan et al. added 0.5, 1.0, and

TABLE 8.3
Recent Studies about the Polymeric Nanofibers with the Addition of Bioactive Compounds for Application in Food Packaging

Polymers	Bioactive substance	Potential application	References
Gelatin and zein	Curcumin	Active packaging	Alehosseini et al. (2019)
Soy protein isolate/zein/ poly (ethylene oxide) (PEO) blend	Ginger essential oil	Antibacterial packaging	Silva et al. (2018)
Chitosan/PEO blend	Phenolic compounds	Antibacterial packaging	Kuntzler et al. (2018b)
Gelatin	Moringa oil	Antibacterial packaging for cheese	Lin et al. (2019)
Gelatin	Thyme essential oil	Antibacterial packaging for chicken	Lin et al. (2018)
Polyvinyl alcohol (PVA)/whey protein blend	Thymol	Antifungal packaging for cheese	Tatlisu et al. (2019)
Protein concentrate/PEO blend	Phycocyanin	Antioxidant packaging	Moreira et al. (2019b)
Poly(caprolactone)	Biopeptides	Antioxidant packaging	Gonçalves et al. (2017)
PVA	Xylanase	Immobilization of enzyme	Santos et al. (2018)
PVA	Anthocyanin extract from red cabbage	pH biosensor in fresh date fruit	Maftoonazad and Ramaswamy (2019)
Zein	Anthocyanin extract from red cabbage	pH indicator	Prietto et al. (2018)
PEO	Ferric chloride	Sensor of volatile compounds	Bagchi et al. (2017)
Polyvinylpyrrolidone	Cholesteric liquid crystal	Thermal sensor	Guan et al. (2018)

1.5% of oil into nanofibers, and the results showed an increase in the inhibition halo of the microorganisms *Staphylococcus aureus* and *Escherichia coli*. Thus, by increasing the essential oil concentration, the nanofibers reduced the mushroom decomposition during five days of storage.

Another notable property of nanofibers in active food packaging is its antifungal property. Curcumin is a highly biocompatible and biodegradable compound, which enables it to be quickly metabolized (Mishra et al., 2005). Yilmaz et al. (2016) evaluated the antifungal activity of nanofibers produced with zein containing 2.5 or 5% of curcumin. Yilmaz et al. applied nanofibers to apple surfaces and observed inhibitory effects against *Penicillium expansum* and *Botrytis cinerea*, demonstrating the potential of curcumin to protect the fruit from these microorganisms.

8.7.2 Antioxidant Nanofibers

In the active packaging system with antioxidant compounds, it is essential to highlight that compatibility with the polymeric material, and homogeneous distribution capacity in the packaging are requirements for choosing the antioxidant (Falguera et al., 2011; Decker, 1998). Besides, the encapsulation of antioxidants by polymeric nanofibers increases the stability of the compound, since it avoids its degradation by the action of environmental factors such as temperature, light, and humidity (Schmatz et al., 2019; Shishir et al., 2018).

Galic acid is one of the phenolic compounds and natural antioxidants commonly used in the pharmaceutical, cosmetic, and food industries. However, undesirable reactions of oxidation and degradation due to light, oxygen, and temperature can be reduced by incorporating it into polymeric nanofibers (Aytac et al., 2016). In the study by Aydogdu et al. (2019), lentil flour-based nanofibers were developed with the addition of gallic acid. Aydogdu et al. evaluated the application of nanofibers as active packaging for walnut conservation. The peroxide value, a product resulting from lipid oxidation, present in the packaging of nanofibers and gallic acid, was significantly lower than the control packaging without gallic acid. Thus, nanofibers containing gallic acid can be an option for improving food preservation.

Another study performed by Fonseca et al. (2019) addressed the development of nanofibers from potato starch containing carvacrol. Carvacrol is an antioxidant agent present in oregano or thyme oils that can be used in food product packaging to avoid oxidation during storage (Altan et al., 2018; Gómez-Estaca et al., 2014). Fonseca et al. (2019) found that starch nanofibers containing 30 and 40% ($v\ v^{-1}$) of carvacrol showed the high antioxidant activity of 76.2 and 83.1% of oxidation inhibition, respectively. Therefore, it is possible to suggest that the nanofibers containing carvacrol present potential for use in active packaging.

The incorporation of bioactive peptides into nanofibers is also an option for the development of active packaging. Hosseini et al. (2019) studied the application of purified peptides from fish in nanofibers as a replacement for synthetic antioxidant compounds. In this study, nanofibers produced with chitosan and PVA blend formed the support for peptides. The antioxidant activity of nanofibers containing the compound ranged from 15.2 to 44.5% of inhibition, and it can be concluded that the peptide derived from fish maintained its activity even when incorporated into nanofibers.

8.7.3 Nanofibers as Sensors

Sensors are another application of nanofibers, presenting rapid response, higher sensitivity, and selectivity for intelligent packaging systems (Bhushani and Anandharamakrishnan, 2014; Kuswandi et al., 2011; Silvestre et al., 2011). Generally, the sensors can be added into or outside the packaging, providing qualitative information through colorimetric changes (Biji et al., 2015; Han and Scanlon, 2005). In this perspective, natural pigments are incorporated into intelligent packaging systems. Sensitive to pH changes, these dyes detect food deterioration and provide direct information on product quality (Shahid et al., 2013).

Moreira et al. (2018b) studied the use of phycocyanin in polymeric nanofibers for pH indicator development. According to Moreira et al., the indicator presented potential applicability to monitoring food, presenting a fast response of color variation between a pH range of 1–10. Silva et al. (2019) produced nanofibers loaded with anthocyanins extracted from açaí fruit. Silva et al. observed that the material is sensitive to different pH values (1–10), causing a color change.

Carbon dioxide (CO_2) is another available indicator to assess the quality and safety of packaged foods (Puligundla et al., 2012). CO_2 is used as an active gas in protective atmosphere technology to inhibit microbial metabolism (Church, 1995; Saliu and Pergola, 2018). In the study by Shang et al. (2017), photo-reticulated nanofibers were manufactured from poly(glycidyl methacrylate) containing acrylobenzophenone. Shang et al. found that the nanofibers detect dissolved and gaseous CO_2 through the display of corresponding color change. Furthermore, these nanofibers changed color rapidly (< 2s) to dissolved CO_2.

8.7.4 Nanofibers for Enzyme Immobilization

Enzymatic immobilization in nanoscale systems exerts better performance compared to conventional ones due to their high surface contact area and mass transfer rate (Fernández et al., 2008). Nanofibers have the potential to immobilize enzymes because they have several functional and structural characteristics, such as appropriate groups and high porosity, leading to improved thermal stability and recovery capacity (Cipolatti et al., 2016; El-Aassar, 2013; Song et al., 2019; Zdarta et al., 2018).

The main advantage of nanofibers compared to other nanostructured materials is the possibility of their recovery from the reaction of the medium (Rojas-Mercado et al., 2018). Enzymes have a variety of applications in the food industry. Thus, immobilization improves enzyme stability at different pH values and high temperatures. Also, immobilization provides a suitable environment for repeated use or controlled release of the enzyme (Kandimalla et al., 2006; Lopez-Rubio et al., 2006; Rodríguez-deLuna et al., 2017).

Bromelain is found in the stem and fruits of pineapple (*Ananas comosus*) (Indumathy et al., 2017; Ketnawa et al., 2011). This enzyme is used in food industries, such as in breweries and in the processing and softening of meat (Arshad et al., 2014). Cellulose acetate nanofibers produced by Brites et al. (2020) were used as a support surface for bromelain immobilization. The results showed that bromelain immobilized in nanofibers obtained 47.1% recovery from enzyme activity and 92% protein recovery.

The ficin is a protease that suffers denaturation and the autolysis process. This process occurs when the proteases use other enzymes as substrate, and this hinders the extensive and industrial use of the plant (Tavano, 2013). In this context, the ficin extracted from the latex of figs was immobilized in PVA nanofibers by Rojas-Mercado et al. (2018). Rojas-Mercado et al. found that the immobilized ficin extract had the capacity of nine cycles of reuse and maintained stability during 25 days of storage. Moreover, the optimal concentration of ficin extract in the enzyme activity test was 25%, resulting in the preservation of 92% of enzyme activity.

8.8 CONCLUSIONS

Nanotechnology research is expanding in several areas, mainly in the food industry. In this sector, nanotechnology has innovated in food packaging development from the use of nanofibers. These materials configure better food protection performance and present differentiated functions. As a result, there is growing investment in research on the use of nanofibers produced by electrospinning in the development of active and intelligent packaging. The main advantage of using nanofibers as a material for active and intelligent systems is the ability to provide thermal stability to compounds. In this way, nanotechnology serves as an essential tool for overcoming the challenges associated with packaging materials. This advance will positively affect quality, shelf life, protection, and safety of foods, benefiting both producers and consumers. However, studies on the migration behavior of nanomaterials in food and its toxicity should be further explored to clarify the impacts on health and the environment.

REFERENCES

Abdullahi, N., 2018. Advances in food packaging technology—A review. *J. Post Harvest Technol.* 6, 55–64.

Abreu, D.A.P., Cruz, J.M., Losada, P.P., 2012. Active and intelligent packaging for the food industry. *Food Rev. In.* 28(2), 146–187. doi:10.1080/87559129.2011.595022.

Adabi, M., Naghibzadeh, M., Adabi, M., Zarrinfard, M.A., Esnaashari, S.S., Seifalian, A.M., Faridi-Majidi, R., Tanimowo Aiyelabegan, H., Ghanbari, H., 2017. Biocompatibility and nanostructured materials: Applications in nanomedicine. *Artif. Cells Nanomed. Biotechnol.* 45(4), 833–842. doi:10.1080/21691401.2016.1178134.

Ahmad, M., Benjakul, S., Nalinanon, S., 2010. Compositional and physicochemical characteristics of acid solubilized collagen extracted from the skin of unicorn leatherjacket (Aluterus monoceros). *Food Hydrocol* 24(6–7), 588–594. doi:10.1016/j.foodhyd.2010.03.001.

Ahmad, M., Benjakul, S., Prodpran, T., Agustini, T.W., 2012. Physico-mechanical and antimicrobial properties of gelatin film from the skin of unicorn leatherjacket incorporated with essential oils. *Food Hydrocol* 28(1), 189–199. doi:10.1016/j.foodhyd.2011.12.003.

Ahmad, M., Nirmal, N.P., Danish, M., Chuprom, J., Jafarzedeh, S., 2016. Characterisation of composite films fabricated from collagen/chitosan and collagen/soy protein isolate for food packaging applications. *RSC Adv.* 6(85), 82191–82204. doi:10.1039/C6RA13043G.

Akhavan, S., Assadpour, E., Katouzian, I., Jafari, S.M., 2018. Lipid nano scale cargos for the protection and delivery of food bioactive ingredients and nutraceuticals. *Trends Food Sci. Technol.* 74, 132–146. doi:10.1016/j.tifs.2018.02.001.

Alehosseini, A., Gómez-Mascaraque, L.G., Martínez-Sanz, M., López-Rubio, A., 2019. Electrospun curcumin-loaded protein nanofiber mats as active/bioactive coatings for food packaging applications. *Food Hydrocol* 87, 758–771. doi:10.1016/j.foodhyd.2018.08.056.

Altan, A., Aytac, Z., Uyar, T., 2018. Carvacrol loaded electrospun fibrous films from zein and poly(lactic acid) for active food packaging. *Food Hydrocol* 81, 48–59. doi:10.1016/j.foodhyd.2018.02.028.

Amit, S.K., Uddin, M.M., Rahman, R., Islam, S.M.R., Khan, M.S., 2017. A review on mechanisms and commercial aspects of food preservation and processing. *Agric. Food Secur.* 6(1), 1–22. doi:10.1186/s40066-017-0130-8.

Antelo, F.S., Costa, J.A.V., Kalil, S.J., 2008. Thermal degradation kinetics of the phycocyanin from *Spirulina platensis*. *Biochem. Eng. J.* 41(1), 43–47. doi:10.1016/j.bej.2008.03.012.

Arcan, I., Yemenicioğlu, A., 2011. Incorporating phenolic compounds opens a new perspective to use zein films as flexible bioactive packaging materials. *Food Res. Int.* 44(2), 550–556. doi:10.1016/j.foodres.2010.11.034.

Arinstein, A., Zussman, E., 2011. Electrospun polymer nanofibers: mechanical and thermodynamic perspectives. *J. Polym. Sci. B Polym. Phys.* 49, 691–707. doi:10.1002/polb.22247.

Arshad, Z.I.M., Amid, A., Yusof, F., Jaswir, I., Ahmad, K., Loke, S.P., 2014. Bromelain: An overview of industrial application and purification strategies. *Appl. Microbiol. Biotechnol.* 98(17), 7283–7297. doi:10.1007/s00253-014-5889-y.

Aydogdu, A., Yildiz, E., Aydogdu, Y., Sumnu, G., Sahin, S., Ayhan, Z., 2019. Enhancing oxidative stability of walnuts by using gallic acid loaded lentil flour based electrospun nanofibers as active packaging material. *Food Hydrocol* 95, 245–255. doi:10.1016/j.foodhyd.2019.04.020.

Aytac, Z., Kusku, I., Durgun, E., Uyar, T., 2016. Encapsulation of gallic acid/cyclodextrin inclusion complex in electrospun polylactic acid nanofibers: Release behavior and antioxidant activity of gallic acid. *Mater. Sci. Eng. C* 63, 231–239. doi:10.1016/j.msec.2016.02.063.

Bagchi, S., Achla, R., Mondal, S.K., 2017. Electrospun polypyrrole-polyethylene oxide coated optical fiber sensor probe for detection of volatile compounds. *Sens. Actuators B* 250, 52–60. doi:10.1016/j.snb.2017.04.146.

Bai, R., Zhang, X., Yong, H., Wang, X., Liu, Y., Liu, J., 2019. Development and characterization of antioxidant active packaging and intelligent Al^{3+}-sensing films based on carboxymethyl chitosan and quercetin. *Int. J. Biol. Macromol.* 126, 1074–1084. doi:10.1016/j.ijbiomac.2018.12.264.

Bajpai, V.K., Kamle, M., Shukla, S., Mahato, D.K., Chandra, P., Hwang, S.K., Kumar, P., Huh, Y.S., Han, Y.-K., 2018. Prospects of using nanotechnology for food preservation, safety, and security. *J. Food Drug Anal.* 26(4), 1201–1214. doi:10.1016/j.jfda.2018.06.011.

Baldino, L., Cardea, S., Reverchon, E., 2017. Biodegradable membranes loaded with curcumin to be used as engineered independent devices in active packaging. *J. Taiwan Inst. Chem. Eng.* 71, 518–526. doi:10.1016/j.jtice.2016.12.020.

Bastarrachea, L.J., Wong, D.E., Roman, M.J., Lin, Z., Goddard, J.M., 2015. Active packaging coatings. *Coatings* 5(4), 771–791. doi:10.3390/coatings5040771.

Benbettaïeb, N., Gay, J.P., Karbowiak, T., Debeaufort, F., 2016b. Tuning the functional properties of polysaccharide–protein bio-based edible films by chemical, enzymatic, and physical cross-linking. Compr. *Rev. Food Sci.* 15, 739–752. doi:10.1111/1541-4337.12210.

Benbettaïeb, N., Karbowiak, T., Brachais, C.H., Debeaufort, F., 2016a. Impact of electron beam irradiation on fish gelatin film properties. *Food Chem.* 195, 11–18. doi:10.1016/j.foodchem.2015.03.034.

Bhardwaj, N., Kundu, S.C., 2010. Electrospinning: A fascinating fiber fabrication technique. *Biotechnol. Adv.* 28(3), 325–347. doi:10.1016/j.biotechadv.2010.01.004.

Bhushani, J.A., Anandharamakrishnan, C., 2014. Electrospinning and electrospraying techniques: Potential food based applications. *Trends Food Sci. Technol.* 38(1), 21–33. doi:10.1016/j.tifs.2014.03.004.

Biji, K.B., Ravishankar, C.N., Mohan, C.O., Srinivasa, G.T.K., 2015. Smart packaging systems for food applications: A review. *J. Food Sci. Technol.* 52(10), 6125–6135. doi:10.1007/s13197-015-1766-7.

Bozic, M., Gorgieva, S., Kokol, V., 2012. Homogeneous and heterogeneous methods for laccase-mediated functionalization of chitosan by tannic acid and quercetin. *Carbohydr. Polym.* 89(3), 854–864. doi:10.1016/j.carbpol.2012.04.021.

Braga, A.R.C., Figueira, F.S., Silveira, J.T., Morais, M.G., Costa, J.A., Kalil, S.J., 2016. Improvement of thermal stability of C-Phycocyanin by nanofiber and preservative agents. *J. Food Process. Preserv.* 40(6), 1264–1269. doi:10.1111/jfpp.12711.

Brecic, R., Mesic, Ž., Cerjak, M., 2017. Importance of intrinsic and extrinsic quality food characteristics by different consumer segments. *Br. Food J.* 4(4), 845–862. doi:10.1108/BFJ-06-2016-0284.

Brites, M.M., Cerón, A.A., Costa, S.M., Oliveira, R., Ferraz, H.G., Catalini, L.H., Costa, S., 2020. Bromelain immobilization in cellulose triacetate nanofiber membranes from sugarcane bagasse by electrospinning technique. *Enzyme Micro. Tech.* 132, 1–9. doi:10.1016/j.enzmictec.2019.109384.

Brizio, A.P.D.R., Prentice, C., 2015. Development of an intelligent enzyme indicator for dynamic monitoring of the shelf-life of food products. *Innov. Food Sci. Emerg. Technol.* 30, 208–217. doi:10.1016/j.ifset.2015.04.001.

Brodowska, M., Guzek, D., Jóźwik, A., Głąbska, D.,Godziszewska, J., Wojtasik-Kalinowska, I., Zarodkiewicz, M., Gantner, M., Wierzbicka, A., 2019. The effect of high-CO_2 atmosphere in packaging of pork from pigs supplemented with rapeseed oil and antioxidants on oxidation processes. *LWT Food Sci. Technol.* 99, 576–582. doi:10.1016/j.lwt.2018.09.077.

Buncic, S., 2006. *Integrated Food Safety and Veterinary Public Health.* Cromwell Press, Trowbidge.

Bunea, A., Rugină, D. Sconţa, Pop., R.M., Pintea, A., Socaciu, C., Tăbăran, F., Grootaert, C., Struijs, K., Van Camp, J., 2013. Anthocyanin determination in blueberry extracts from various cultivars and their antiproliferative and apoptotic properties in B16-F10 metastatic murine melanoma cells. *Phytochemistry* 95, 436–444. doi:10.1016/j.phytochem.2013.06.018.

Cao, N., Fu, Y., He, J., 2007. Preparation and physical properties of soy protein isolate and gelatin composite films. *Food Hydrocol* 21(7), 1153–1162. doi:10.1016/j.foodhyd.2006.09.001.

Cao, X.-T., Chung, W.-Y., 2019. Range-extended wireless food spoilage monitoring with a high energyefficient battery-free sensor tag. *Sens. Actuat. A Phys.* 299, 1–8. doi:10.1016/j.sna.2019.111632.

Castañeda-Ovando, A., Pacheco-Hernández, M.L., Páez-Hernández, M.E., Rodríguez, J.A., Galán-Vidal, C.A., 2009. Chemical studies of anthocyanins: A review. *Food Chem.* 113(4), 859–871. doi:10.1016/j.foodchem.2008.09.001.

Cebrián, G., Condón, S., Mañas, P., 2017. Physiology of the inactivation of vegetative bacteria by thermal treatments: Mode of action, influence of environmental factors and inactivation kinetics. *Foods* 6(12), 1–21. doi:10.3390/foods6120107.

Chaiklahan, R., Chirasuwan, N., Bunnag, B., 2012. Stability of phycocyanin extracted from *Spirulina* sp.: Influence of temperature, pH and preservatives. *Process Biochem.* 47(4), 659–664. doi:10.1016/j.procbio.2012.01.010.

Chau, C.F., Wu, S.H., Yen, G.C., 2007. The development of regulations for food nanotechnology. *Trends Food Sci. Technol.* 18(5), 269–280. doi:10.1016/j.tifs.2007.01.007.

Chen, T., Wong, Y.-S., 2008. In vitro antioxidant and antiproliferative activities of selenium-containing phycocyanin from selenium-enriched *Spirulina platensis. J. Agric. Food Chem.* 56(12), 4352–4358. doi:10.1021/jf073399k.

Chentir, I., Hamdi, M., Li, S., Doumandji, A., Markou, G., Nasri, M., 2018. Stability, bio-functionality and bio-activity of crude phycocyanin from a two-phase cultured Saharian *Arthrospira* sp. strain. *Algal Res.* 35, 395–406. doi:10.1016/j.algal.2018.09.013.

Cherng, S.-C., Cheng, S.-N., Tarn, A., Chou, T.-C., 2007. Anti-inflammatory activity of c-phycocyanin in lipopolysaccharide-stimulated RAW 264.7 macrophages. *Life Sci.* 81(19–20), 1431–1435. doi:10.1016/j.lfs.2007.09.009.

Chi, J., Ge, J., Yue, X., Liang, J., Sun, Y., Gao, X., Yue, P., 2019. Preparation of nanoliposomal carriers to improve the stability of anthocyanins. *LWT Food Sci. Technol.* 109, 101–107. doi:10.1016/j.lwt.2019.03.070.

Choi, I., Lee, J.Y., Lacroix, M., Han, J., 2017. Intelligent pH indicator film composed of agar/potato starch and anthocyanin extracts from purple sweet potato. *Food Chem.* 218, 122–128. doi:10.1016/j.foodchem.2016.09.050.

Church, N., 1995. Developments in modified-atmosphere packaging and related technologies. *Trends Food Sci. Technol.* 5(11), 345–352. doi:10.1016/0924-2244(94)90211-9.

Cinelli, P., Schmid, M., Bugnicourt, E., Coltelli, M.B., Lazzeri, A., 2016. Recyclability of PET/WPI/PE multilayer films by removal of whey protein isolate-based coatings with enzymatic detergents. *Materials Basel* 9(6), 1–15. doi:10.3390/ma9060473.

Cipolatti, E.P., Valério, A., Henriques, R.O., Moritz, D.E., Ninow, J.L., Freire, D.M.G., Manoel, E.A., Fernandez-Lafuente, R., Oliveira, D., 2016. Nanomaterials for biocatalyst immobilization—State of the art and future trends. *RSC Adv.* 6(106), 104675–104692. doi:10.1039/C6RA22047A.

Coltelli, M.-B., Wild, F., Bugnicourt, E., Cinelli, P., Lindner, M., Schmid, M., Weckel, V., Müller, K., Rodríguez-Turienzo, L., Lazzeri, A, 2016. State of the art in the development and properties of protein based films and coatings and their applicability to cellulose based products: An extensive Re-view. *Coatings* 6(1), 1–59. doi:10.3390/coatings6010001.

Decker, E.A., 1998. Strategies for manipulating the prooxidative/antioxidative balance of foods to maximize oxidative stability. *Trends Food Sci. Technol.* 9(6), 241–248. doi:10.1016/S0924-2244(98)00045-4.

Denavi, G.A., Pérez-Mateos, M., Añón, M.C., Montero, P., Mauri, A.N., Gómez-Guillén, M.C., 2009. Structural and functional properties of soy protein isolate and cod gelatin blend films. *Food Hydrocol* 23(8), 2094–2101. doi:10.1016/j.foodhyd.2009.03.007.

Dey, P.M., Harborne, J.B., 1989. *Methods in Plant Biochemistry, Volume 1: Plant Phenolics*, 1st ed. Academic Press, London.

Ekezie, F.G.C., Sun, D.W., Cheng, J.F., 2017. A review on recent advances in cold plasma technology for the food industry: Current applications and future trends. *Trends Food Sci. Technol.* 69, 46–58. doi:10.1016/j.tifs.2017.08.007.

El-Aassar, M.R., 2013. Functionalized electrospun nanofibers from poly (AN-co-MMA) for enzyme immobilization. *J. Mol. Catal. B Enzym.* 85–86, 140–148. doi:10.1016/j.molcatb.2012.09.002.

El-Sheekh, M.M., Daboor, S.M., Swelim, M.A., Mohamed, S., 2014. Production and characterization of antimicrobial active substance from *Spirulina platensis*. *Iran. J. Microbiol.* 6(2), 112–119.

European Commission (EC), 2009. Guidance to the Commission Regulation (EC) No 450/2009 of 29 May 2009 on active and intelligent materials and articles intended to come into the contact with food (version 1.0). Bruxelas. https://eur-lex.europa.eu/LexUriServ/LexUriServ.do?uri=OJ:L:2009:135:0003:0011:EN:PDF (accessed December 13, 2019).

Falguera, V., Quintero, J.P., Jimenez, A., Munoz, J.A., Ibarz, A., 2011. Edible films and coatings: Structures, active functions and trends in their use. *Trends Food Sci. Technol.* 22(6), 292–303. doi:10.1016/j.tifs.2011.02.004.

Falkeborg, M.F., Roda-Serrat, M.C., Burnæs, K.L., Nielsen., A.L.D., 2018. Stabilising phycocyanin by anionic micelles. *Food Chem.* 239, 771–780. doi:10.1016/j.foodchem.2017.07.007.

Fang, X., Ma, H., Xiao, S., Shen, M., Guo, R., Cao, X., Shi, X., 2011. Facile immobilization of gold nanoparticles into electrospun polyethyleneimine/polyvinyl alcohol nanofibers for catalytic applications. *J. Mater. Chem.* 21(12), 4493–4501. doi:10.1039/C0JM03987J.

Fenster, K., Freeburg, B., Hollard, C., Wong, C., Laursen, R.R., Ouwe-hand, A.C., 2019. The production and delivery of probiotics: A review of a practical approach. *Microorganisms* 7(3), 1–17. doi:10.3390/microorganisms7030083.

Fernández, A., Cava, D., Ocio, M.J., Lagaron, J.M., 2008. Perspectives for biocatalysts in food packaging. *Trends Food Sci. Technol.* 19(4), 198–206. doi:10.1016/j.tifs.2007.12.004.

Filho, J.G.O., Rodrigues, J.M., Valadares, A.C.F., Almeida, A.B., Lima, T.M., Takeuchi, K.P., Alves, C.C.F., Sousa, H.A.F., Silva, E.R., Dyszy, F.H., Egea, M.B., 2019. Active food packaging: Alginate films with cottonseed protein hydrolysates. *Food Hydrocol* 92, 267–275. doi:10.1016/j.foodhyd.2019.01.052.

Fonseca, L.M., Cruxen, C.E.S., Bruni, G.P., Fiorentini, A.M., Zavareze, E.R., Lim, L.-T., Dias, A.R.G., 2019. Development of antimicrobial and antioxidant electrospun soluble potato starch nanofibers loaded with carvacrol. *Int. J. Biol. Macromol.* 139, 1182–1190. doi:10.1016/j.ijbiomac.2019.08.096.

Forsythe, S.J., 2010. *The Microbiology of Safe Food*. Blackwell Publishing Ltd., Chichester.

Francis, L., Giunco, F., Balakrishnan, A., Marsano, E., 2010. Synthesis, characterization and mechanical properties of nylon–silver composite nanofibers prepared by electrospinning. *Curr. Appl. Phys.* 10(4), 1005–1008. doi:10.1016/j.cap.2009.12.025.

Gaona-Forero, A., Agudelo-Rodríguez, G., Herrera, A.O., Castellanos, D.A., 2018. Modeling and simulation of an active packaging system with moisture adsorption for fresh produce. Application in “Hass” avocado. *Food Packag. Shelf Life* 17, 187–195. doi:10.1016/j.fpsl.2018.07.005.

Gaur, R., Singh, A., Tripathi, A., 2019. Microbial environment of food. In: Singh, R.L., Mondal, S. (Eds.), *Food Safety and Human Health*. Academic Press, Cambridge, pp. 189–218.

Geoffry, K., Achur, R.N., 2018. Screening and production of lipase from fungal organisms. *Biocatal. Agric. Biotechnol.* 14, 241–253. doi:10.1016/j.bcab.2018.03.009.

Gómez-Estaca, J., Balaguer, M.P., López-Carballo, G., Gavara, R., Hernández-Muñoz, P., 2017. Improving antioxidant and antimicrobial properties of curcumin by means of encapsulation in gelatin through electrohydrodynamic atomization. *Food Hydrocol* 70, 313–320. doi:10.1016/j.foodhyd.2017.04.019.

Gómez-Estaca, J., López-de-Dicastillo, C., Hernández-Muñoz, P., Catalá, R., Gavara, R., 2014. Advances in antioxidant active food packaging. *Trends Food Sci. Technol.* 35(1), 42–51. doi:10.1016/j.tifs.2013.10.008.

Gonçalves, C.F., Schmatz, D.A., Uebel, L.S., Kuntzler, S.G., Costa, J.A.V., Zimmer, K.R., Morais, M.G., 2017. Microalgae biopeptides applied in nanofibers for the development of active packaging. *Polímeros* 27(4), 290–297. doi:10.1590/0104-1428.2403.

Gong, M., Bassi, A., 2016. Carotenoids from microalgae: A review of recent developments. *Biotechnol. Adv.* 34(8), 1396–1412. doi:10.1016/j.biotechadv.2016.10.005.

Grossman, A.R., Lohr, M., Im, C.S., 2004. *Chlamydomonas reinhardtii* in the landscape of pigments. *Annu. Rev. Genet.* 38, 119–173. doi:10.1146/annurev.genet.38.072902.092328.

Guan, Y., Zhang, L., Li, M., West, J.L., Fu, S., 2018. Preparation of temperature-response fibers with cholesteric liquid crystal dispersion. *Colloids Surf. A Physicochem. Eng. Asp.* 546, 212–220. doi:10.1016/j.colsurfa.2018.03.011.

Gurmu, F., Hussein, S., Laing, M., 2014. The potential of orange-fleshed sweet potato to prevent vitamin A deficiency in Africa. *Int. J. Vitam. Nutr. Res.* 84(1–2), 65–78. doi:10.1024/0300-9831/a000194.

Haider, A., Haider, S., Kang, I.K., 2018. A comprehensive review summarizing the effect of electrospinning parameters and potential applications of nanofibers in biomedical and biotechnology. *Arab. J. Chem.* 11(8), 1165–1188. doi:10.1016/j.arabjc.2015.11.015.

Haider, A., Mahmood, W., Al-Masry, M., Imran, M., Aijaz, M., 2013. Highly aligned narrow diameter chitosan electrospun nanofibers. *J. Polym. Res.* 20(4), 1–11. doi:10.1007/s10965-013-0105-9.

Han, J., Scanlon, M., 2005. Mass transfer of gas and solute through packaging materials. In: Han, J. (Ed.), *Innovations in Food Packaging*. Academic Press, Winnipeg, pp. 37–49.

He, X., Hwang, H.-M., 2016. Nanotechnology in food science: Functionality, applicability, and safety assessment. *J. Food Drug Anal.* 24(4), 671–681. doi:10.1016/j.jfda.2016.06.001.

Heising, J.K., Dekker, M., Bartels, P.V., Van Boekel, M.A., 2014. Monitoring the quality of perishable foods: Opportunities for intelligent packaging. *Crit. Rev. Food Sci. Nutr.* 54(5), 645–654. doi:10.1016/j.jfda.2016.06.001.

Hernández-Cortez, C., Palma-Martínez, I., Gonzalez-Avila, L.U., Guerrero-Mandujano, A., Solís, R.C., Castro-Escarpulli, G., 2017. Food poisoning caused by bacteria (food toxins). In: Malangu, N. (Ed.), *Poisoning—From Specific Toxic Agents to Novel Rapid and Simplified Techniques for Analysis.* Intech Open, London, pp. 33–72.

Hosseini, S.F., Nahvi, Z., Zandi, M., 2019. Antioxidant peptide-loaded electrospun chitosan/poly(vinyl alcohol) nanofibrous mat intended for food biopackaging purposes. *Food Hydrocol* 89, 637–648. doi:10.1016/j.foodhyd.2018.11.033.

Hsieh-Lo, M., Castillo, G., Ochoa-Becerra, M.A., Mojica, L., 2019. Phycocyanin and phycoerythrin: Strategies to improve production yield and chemical stability. *Algal Res.* 42, 1–11. doi:10.1016/j.algal.2019.101600.

Huang, Z.M., Zhang, Y.Z., Ramakrishna, S., Lim, C.T., 2004. Electrospinning and mechanical characterization of gelatin nanofibers. *Polymer* 45(15), 5361–5368. doi:10.1016/j.polymer.2004.04.005.

İlter, I., Akyıl, S., Demirel, Z., Koç, M., Conk-Dalay, M., Kaymak-Ertekin, F., 2018. Optimization of phycocyanin extraction from *Spirulina platensis* using different techniques. *J. Food Compost Anal.* 70, 78–88. doi:10.1016/j.jfca.2018.04.007.

Indumathy, S., Kiruthiga, K., Saraswathi, K., Arumugan, P., 2017. Extraction, Partial purification and characterization of bromelain enzyme from pineapple (*Ananas comosus*). *Indo Am. J. Pharm. Res.* 7, 566–579. doi:10.5281/zenodo.1036502.

Işik, C., Arabaci, G., Doğaç, Y.I., Deveci, İ., Teke, M., 2019. Synthesis and characterization of electrospun PVA/Zn^{2+} metal composite nanofibers for lipase immobilization with effective thermal, pH stabilities and reusability. *Mater. Sci. Eng. C* 99, 1226–1235. doi:10.1016/j.msec.2019.02.031.

Jain, A., Sharma, G., Kushwah, V., Thakur, K., Ghoshal, G., Singh, B., Jain, S., Shivhare, U.S., Katare, O.P., 2017. Fabrication and functional attributes of lipidic nanoconstructs of lycopene: An innovative endeavour for enhanced cytotoxicity in MCF-7 breast cancer cells. *Colloids Surf. B* 152, 482–491. doi:10.1016/j.colsurfb.2017.01.050.

Jalali, A., Rux, G., Linke, M., Geyer, M., Pant, A., Saengerlaub, S., Pramod, M., 2019. Application of humidity absorbing trays to fresh produce packaging: Mathematical modeling and experimental validation. *J. Food Eng.* 244, 115–125. doi:10.1016/j.jfoodeng.2018.09.006.

Jay, J.M., 2000. Indicators of food safety and quality, principles of quality control, and microbial criteria. In: Jay, J. (Ed.), *Modern Food Microbiology.* Aspen Publishers, Gaithersburg, pp. 385–422.

Johnson, D.R., Inchingolo, R., Decker, E.A., 2018. The ability of oxygen scavenging packaging to inhibit vitamin degradation and lipid oxidation in fish oil-in-water emulsions. *Innov. Food Sci. Emerg. Technol.* 47, 467–475. doi:10.1016/j.ifset.2018.04.021.

Jonasson, L., Wikby, A., Olsson, A.G., 2003. Low serum β-carotene reflects immune activation in patients with coronary artery disease. *Nutr. Metab. Cardiovasc. Dis.* 13(3), 120–125. doi:10.1016/S0939-4753(03)80170-9.

Jovanovic, B., 2015. Critical review of public health regulations of titanium dioxide, a human food additive. *Integr. Environ. Assess. Manag.* 11(1), 10–20. doi:10.1002/ieam.1571.

Kalpana, S., Priyadarshini, S.R., Leena, M.M., Moses, J.A., Anandharamakrishnan, C., 2019. Intelligent packaging: Trends and applications in food systems. *Trends Food Sci. Technol.* 93, 145–157. doi:10.1016/j.tifs.2019.09.008.

Kandimalla, V.B., Tripathi, V.S., Ju, H., 2006. Immobilization of biomolecules in sol-gels: Biological and analytical applications. *Crit. Rev. Anal. Chem.* 36(2), 73–106. doi:10.1080/10408340600713652.

Kapetanakou, A.E., Taoukis, P., Skandamis, P.N., 2019. Model development for microbial spoilage of packaged fresh–Cut salad products using temperature and in-package CO_2 levels as predictor variables. *LWT Food Sci. Technol.* 113, 1–11. doi:10.1016/j.lwt.2019.108285.

Ketnawa, S., Chaiwut, P., Rawdkuen, S., 2011. Extraction of bromelain from pineapple peels. *Food Sci. Technol. Int.* 17(4), 395–402. doi: 10.1177/1082013210387817 .

Khan, W.S., Asmatulu, R., Ceylan, M., Jabbarnia, A., 2013. Recent progress on conventional and non-conventional electrospinning processes. *Fibers Polym.* 14(8), 1235–1247. doi:10.1007/s12221-013-1235-8.

Koeijer, B., Lange, J., Wever, R., 2017. Desired, perceived, and achieved sustainability: Trade-offs in strategic and operational packaging development. *Sustainability* 9(10), 1–29. doi:10.3390/su9101923.

Kong, F., Singh, R.P., 2016. Chemical deterioration and physical instability of foods and beverages. In: Subramaniam, P. (Ed.), *The Stability and Shelf Life of Food.* Elsevier, Amsterdam, pp. 43–76.

Kowsaly, E., Christas, K.M., Balashanmugam, P., Selvi, A.T., Rani, I.J.C., 2019. Biocompatible silver nanoparticles/poly(vinyl alcohol) electrospun nanofibers for potential antimicrobial food packaging applications. *Food Packag. Shelf Life* 21, 1–8. doi:10.1016/j.fpsl.2019.100379.

Kumar, T.S.M., Kumar, K.S., Rajini, N., Siengchin, S., Ayrilmis, N., Rajulu, A.V., 2019. A comprehensive review of electrospun nanofibers: Food and packaging perspective. *Compos. B Eng.* 175, 1–11. doi:10.1016/j.compositesb.2019.107074.

Kuntzler, S.G., Costa, J.A.V., Morais, M.G., 2018b. Development of electrospun nanofibers containing chitosan/PEO blend and phenolic compounds with antibacterial activity. *Int. J. Biol. Macromol.* 117, 800–806. doi:10.1016/j.ijbiomac.2018.05.224.

Kuntzler, S.G., Uebel, L.S., Schmatz, D.A., Barcia, M.T., Costa, J.A.V., Morais, M.G., 2018a. Nanofibers and their applications. In: Nalwa, H.S. (Ed.), *Encyclopedia of Nanoscience and Nanotechnology.* American Scientific Publishers, Valencia, pp. 387–414.

Kuswandi, B., Wicaksono, Y., Abdullah, A., Heng, L.Y., Ahmad, M., Ahmad, M., 2011. Smart packaging: Sensors for monitoring of food quality and safety. *Sens. Instrumen. Food Qual.* 5(3–4), 137–146. doi:10.1007/s11694-011-9120-x.

Lara, B.R.V., Araújo, A.C.M.A., Dias, M.V., Guimarães Jr., M., Santos, T.A., Ferreira, L.F., Borges, S.V., 2019. Morphological, mechanical and physical properties of new whey protein isolate/ polyvinyl alcohol blends for food flexible packaging. *Food Packaging Shelf Life* 19, 16–23. doi:10.1016/j.fpsl.2018.11.010.

Lee, H.-G., Jeong, S., Yoo, S.R., 2019a. Development of food packaging materials containing calcium hydroxide and porous medium with carbon dioxide-adsorptive function. *Food Packag. Shelf Life* 21, 1–8. doi:10.1016/j.fpsl.2019.100352.

Lee, K., Baek, S., Kim, D., Seo, J., 2019b. A freshness indicator for monitoring chicken-breast spoilage using a Tyvek® sheet and RGB color analysis. *Food Packag. Shelf Life* 19, 40–46. doi:10.1016/j.fpsl.2018.11.016.

Li, H., Xu, Y., Xu, H., Chang, J., 2014. Electrospun membranes: Control of the structure and structure related applications in tissue regeneration and drug delivery. *J. Mater. Chem. B* 2(34), 5492–5510. doi:10.1039/C4TB00913D.

Li, Z., Wang, C., 2013. Effects of working parameters on electrospinning. In: Li, Z., Wang, C. (Eds.), *One-Dimensional Nanostructures: Electrospinning Technique and Unique Nanofibers.* Springer-Verlag, Heidelberg, pp. 15–28.

Lin, L., Gu, Y., Cui, H., 2019. Moringa oil/chitosan nanoparticles embedded gelatin nanofibers for food packaging against *Listeria monocytogenes* and *Staphylococcus aureus* on cheese. *Food Packag., Shelf Life* 19, 86–93. doi:10.1016/j.fpsl.2018.12.005.

Lin, L., Zhu, Y., Cui, H., 2018. Electrospun thyme essential oil/gelatin nanofibers for active packaging against *Campylobacter jejuni* in chicken. *LWT Food Sci. Technol.* 97, 711–718. doi:10.1016/j.lwt.2018.08.015.

Liu, J., Sun, Z., Gerken, H., 2016. *Recent Advances in Microalgal Biotechnology.* OMICS Group EBooks, Foster City.

Lopez-Rubio, A., Gavara, R., Lagaron, J.M., 2006. Bioactive packaging: Turning foods into healthier foods through biomaterials. *Trends Food Sci. Technol.* 17(10), 567–575. doi:10.1016/j.tifs.2006.04.012.

Maftoonazad, N., Ramaswamy, H., 2019. Design and testing of an electrospun nanofiber mat as a pH biosensor and monitor the pH associated quality in fresh date fruit (*Rutab*). *Polym. Test.* 75, 76–84. doi:10.1016/j.polymertesting.2019.01.011.

Marsh, K., Bugusu, B., 2007. Food packaging-roles, materials, and environmental issues. *J. Food Sci.* 72(3), R39–R55. doi:10.1111/j.1750-3841.2007.00301.x.

Martelli, G., Folli, C., Visai, L., Daglia, M., Ferrari, D., 2014. Thermal stability improvement of blue colorant C-Phycocyanin from *Spirulina platensis* for food industry applications. *Process Biochem.* 49(1), 154–159. doi:10.1016/j.procbio.2013.10.008.

Martinez-Abad, A., Lagaron, J.M., Ocio, M.J., 2012. Development and characterization of silver-based antimicrobial ethylene-vinyl alcohol copolymer (EVOH) films for food-packaging applications. *J. Agric. Food Chem.* 60(21), 5350–5359. doi:10.1021/jf300334z.

Martins, F., Sentanin, M.A., Souza, D.D., 2019. Review: Analytical methods in food additives determination: Compounds with functional applications. *Food Chem.* 272, 732–750. doi:10.1016/j.foodchem.2018.08.060.

Masek, A., Lato, M., Piotrowska, M., Zaborski, M., 2018. The potential of quercetin as an effective natural antioxidant and indicator for packaging materials. *Food Packag. Shelf Life* 16, 51–58. doi:10.1016/j.fpsl.2018.02.001.

Matabola, K.P., Moutloali, R.M., 2013. The influence of electrospinning parameters on the morphology and diameter of poly(vinyledene fluoride) nanofibers-effect of sodium chloride. *J. Mater. Sci.* 48(16), 5475–5482. doi:10.1007/s10853-013-7341-6.

McMillin, K.W., 2017. Advancements in meat packaging. *Meat Sci.* 132, 153–162. doi:10.1016/j.meatsci.2017.04.015.

Mehrad, B., Ravanfar, R., Licker, J., Regenstein, J.M., Abbaspourrad, A., 2018. Enhancing the physicochemical stability of β-carotene solid lipid nanoparticle (SLNP) using whey protein isolate. *Food Res. Int.* 105, 962–969. doi:10.1016/j.foodres.2017.12.036.

Mercadante, A.Z., Rodrigues, D.B., Petry, F.C., Mariutti, L.R.B., 2017. Carotenoid esters in foods—A review and practical directions on analysis and occurrence. *Food Res. Int.* 99(2), 830–850. doi:10.1016/j.foodres.2016.12.018.

Mercante, L.A., Scagion, V.P., Migliorini, F.L., Mattoso, L.H.C., Correa, D.S., 2017. Electrospinning-based (bio)sensors for food and agricultural applications: A review. *Trends Anal. Chem.* 91, 91–103. doi:10.1016/j.trac.2017.04.004.

Merchant, S.S., Helmann, J.D., 2012. Elemental economy: Microbial strategies for optimizing growth in the face of nutrient limitation. *Adv. Microb. Physiol.* 60, 91–210. doi:10.1016/B978-0-12-398264-3.00002-4.

Meroni, E., Raikos, V., 2018. Formulating orange oil-in-water beverage emulsions for effective delivery of bioactives: Improvements in chemical stability, antioxidant activity and gastrointestinal fate of lycopene using carrier oils. *Food Res. Int.* 106, 439–445. doi:10.1016/j.foodres.2018.01.013.

Mirafzali, Z., Thompson, C.S., Tallua, K., 2014. Application of liposomes in the food industry. In: Gaonkar, A.G., Vasisht, N., Khare, A.R., Sobel, R. (Eds.), *Microencapsulation in the Food Industry.* Elsevier, San Diego, pp. 139–150.

Mishra, S., Narain, U., Mishra, R., Misra, K., 2005. Design, development and synthesis of mixed bioconjugates of piperic acid–glycine, curcumin–glycine/alanine and curcumin–glycine–piperic acid and their antibacterial and antifungal properties. *Bioorg. Med. Chem.* 13(5), 1477–1486. doi:10.1016/j.bmc.2004.12.057.

Moomand, K., Lim, L.-T., 2015. Properties of encapsulated fish oil in electrospun zein fibres under simulated in vitro conditions. *Food Bioprocess Technol.* 8(2), 431–444. doi:10.1007/s11947-014-1414-7.

Moradi, M., Tajik, H., Almasi, H., Forough, M., Ezati, P., 2019. A novel pH-sensing indicator based on bacterial cellulose nanofibers and black carrot anthocyanins for monitoring fish freshness. *Carbohydr. Polym.* 222, 1–10. doi:10.1016/j.carbpol.2019.115030.

Morais, M.G., Prates, D.F., Moreira, J.B., Duarte, J.H., Costa, J.A.V., 2018. Phycocyanin from microalgae: Properties, extraction and purification, with some recent applications. *Ind. Biotechnol.* 14(1), 30–37. doi:10.1089/ind.2017.0009.

Moreira, J.B., Goularte, P.G., Morais, M.G., Costa, J.A.V., 2019a. Preparation of beta-carotene nanoemulsion and evaluation of stability at a long storage period. *Food Sci. Technol.* 39(3), 599–604. doi:10.1590/fst.31317.

Moreira, J.B., Lim, L.-T., Zavareze, E.R., Dias, A.R.G., Costa, J.A.V., Morais, M.G., 2019b. Antioxidant ultrafine fibers developed with microalga compounds using a free surface electrospinning. *Food Hydrocol* 93, 131–136. doi:10.1016/j.foodhyd.2019.02.015.

Moreira, J.B., Morais, M.G., Morais, E.G., Vaz, B.S., Costa, J.A.V., 2018a. Electrospun polymeric nanofibers in food packaging. In: Grumezescu, A.M., Holban, A.M. (Eds.), *Impact of Nanoscience If Food Industry.* Academic Press, London, pp. 387–417.

Moreira, J.B., Terra, A.L.M., Costa, J.A.V., Morais, M.G., 2018b. Development of pH indicator from PLA/PEO ultrafine fibers containing pigment of microalgae origin. *Int. J. Biol. Macromol.* 118(B), 1855–1862. doi:10.1016/j.ijbiomac.2018.07.028.

Muller, C., Mazel, V., Dausset, C., Busignies, V., Bornes, S., Nivoliez, A., Tchoreloff, P., 2014. Study of the *Lactobacillus rhamnosus* Lcr35(R) properties after compression and proposition of a model to predict tablet stability. *Eur. J. Pharm. Biopharm.* 88(3), 787–794. doi:10.1016/j.ejpb.2014.07.014.

Murmu, S.B., Mishra, H.N., 2018. Selection of the best active modified atmosphere packaging with ethylene and moisture scavengers to maintain quality of guava during low temperature storage. *Food Chem.* 253, 55–62. doi:10.1016/j.foodchem.2018.01.134.

Musso, Y.S., Salgado, P.R., Mauri, A.N., 2017. Smart edible films based on gelatin and curcumin. *Food Hydrocol* 66, 8–15. doi:10.1016/j.foodhyd.2016.11.007.

Neo, Y.P., Ray, S., Jin, J., Gizdavic-Nikolaidis, M., Nieuwoudt, M.K., Liu, D., Quek, S.Y., 2013. Encapsulation of food grade antioxidant in natural biopolymer by electrospinning technique: A physicochemical study based on zein–gallic acid system. *Food Chem.* 136(2), 1013–1021. doi:10.1016/j.foodchem.2012.09.010.

Otálora, M.C., Carriazo, J.G., Iturriaga, L., Nazareno, M.A., Osorio, C., 2015. Microencapsulation of betalains obtained from cactus fruit (*Opuntia fícus-indica*) by spray drying using cactus cladode mucilage and maltodextrin as encapsulating agents. *Food Chem.* 187, 174–181. doi:10.1016/j.foodchem.2015.04.090.

Oudjedi, K., Manso, S., Nerin, C., Hassissen, N., Zaidi, F., 2019. New active antioxidant multilayer food packaging films containing Algerian Sage and Bay leaves extracts and their application for oxidative stability of fried potatoes. *Food Control* 98, 216–226. doi:10.1016/j.foodcont.2018.11.018.

Pan, J., Ai, F., Shao, P., Chen, H., Gao, H., 2019. Development of polyvinyl alcohol/β-cyclodextrin antimicrobial nanofibers for fresh mushroom packaging. *Food Chem.* 300, 1–8. doi:10.1016/j.foodchem.2019.125249.

Pan, R., Lu, R., Zhang, Y., Zhu, M., Zhu, W., Yang, R., Zhang, E., Ying, J., Xu, T., Yi, H., Li, J., Shi, M., Zhou, L., Xu, Z., Li, P., Bao, Q., 2015. *Spirulina* phycocyanin induces differential protein expression and apoptosis in SKOV-3 cells. *Int. J. Biol. Macromol.* 81, 951–959. doi:10.1016/j.ijbiomac.2015.09.039.

Panis, G., Carreon, J.R., 2016. Commercial astaxanthin production derived by green alga *Haematococcus pluvialis*: A microalgae process model and a techno-economic assessment all through production line. *Algal Res.* 18, 175–190. doi:10.1016/j.algal.2016.06.007.

Pan-Utai, W., Kahapana, W., Iamtham, S., 2017. Extraction of C-phycocyanin from *Arthrospira* (*Spirulina*) and its thermal stability with citric acid. *J. Appl. Phycol.* 30(1), 1–12. doi:10.1007/s10811-017-1155-x.

Pashkow, F.J., Watumull, D.G., Campbell, C.L., 2008. Astaxanthin: A novel potential treatment for oxidative stress and inflammation in cardiovascular disease. *Am. J. Cardiol.* 101(10A), S58–S68. doi:10.1016/j.amjcard.2008.02.010.

Petrík, S., 2011. Industrial production technology for nanofibers. In: Lin, T. (Ed.), *Nanofibers Production, Properties and Functional Applications*. Intech Open, London, pp. 420–427.

Petruzzi, L., Corbo, M.R., Sinigaglia, M., Bevilacqua, A., 2017. Microbial spoilage of foods: Fundamentals. In: Bevilacqua, A., Corbo, M.R., Sinigaglia, M. (Eds.), *The Microbiological Quality of Food: Foodborne Spoilers*. Elsevier, Foggia, pp. 1–21.

Prates, D.F., Radmann, E.M., Duarte, J.H., Morais, M.G., Costa, J.A.V., 2018. *Spirulina* cultivated under different light emitting diodes: Enhanced cell growth and phycocyanin production. *Bioresour. Technol.* 256, 38–43. doi:10.1016/j.biortech.2018.01.122.

Prietto, L., Pinto, V.Z., Halal, S.L.M.E., Morais, M.G., Costa, J.A.V., Lim, L.T., Dias, A.R.G., Zavareze, E.D.R., 2018. Ultrafine fibers of zein and anthocyanins as natural pH indicator. *J. Sci. Food Agric.* 98(7), 2735–2741. doi:10.1002/jsfa.8769.

Provincial, L., Guillén, E., Gil, M., Alonso, V., Roncalés, P., Beltrán, J.A., 2013. Survival of *Listeria monocytogenes* and *Salmonella enteritidis* in sea bream (*Sparus aurata*) fillets packaged under enriched CO_2 modified atmospheres. *Int. J. Food Microbiol.* 162(3), 213–219. doi:10.1016/j.ijfoodmicro.2013.01.015.

Puligundla, P., Jung, J., Ko, S., 2012. Carbon dioxide sensors for intelligent food packaging applications. *Food Control* 25(1), 328–333. doi:10.1016/j.foodcont.2011.10.043.

Rafiee, Z., Nejatian, M., Daeihamed, M., Jafari, S.M., 2019. Application of curcumin-loaded nanocarriers for food, drug and cosmetic purposes. *Trends Food Sci. Technol.* 88, 445–458. doi:10.1016/j.tifs.2019.04.017.

Rai, M., Yadav, A., Gade, A., 2009. Silver nanoparticles as a new generation of antimicrobials. *Biotechnol. Adv.* 27(1), 76–83. doi:10.1016/j.biotechadv.2008.09.002.

Rambabu, K., Bharath, G., Banat, F., Show, P.L., Cocoletzi, H.H., 2019. Mango leaf extract incorporated chitosan antioxidant film for active food packaging. *Int. J. Biol. Macromol.* 126, 1234–1243. doi:10.1016/j.ijbiomac.2018.12.196.

Ravandi, S.A.H., Tork, R.B., Dabirian, F., Gharehaghaji, A.A., Sajjadi, A., 2015. Characteristics of yarn and fabric made out of nanofibers. *Mater. Sci. Appl.* 6(1), 103–110. doi:10.4236/msa.2015.61013.

Reneker, D.H., Yarin, A.L., Fong, H., Koombhongse, S., 2000. Bending instability of electrically charged liquid jets of polymer solutions in electrospinning. *J. Appl. Phys.* 87(9), 4531–4547. doi:10.1063/1.373532.

Renugadevi, K., Valli Nachiyar, C., Sowmiya, Sunkar, S., 2018. Antioxidant activity of phycocyanin pigment extracted from marine filamentous cyanobacteria *Geitlerinema* sp TRV57. *Biocat. Agric. Biotechnol.* 16, 237–242. doi:10.1016/j.bcab.2018.08.009.

Restuccia, D., Spizzirri, U.G., Parisi, O.I., Cirillo, G., Curcio, M., Iemma, F., Puoci, F., Vinci, G., Piccia, N., 2010. New EU regulation aspects and global market of active and intelligent packaging for food industry applications. *Food Control* 21(11), 1425–1435. doi:10.1016/j.foodcont.2010.04.028.

Rezaei, A., Fathi, M., Jafari, S.M., 2019. Nanoencapsulation of hydrophobic and low-soluble food bioactive compounds within different nanocarriers. *Food Hydrocol* 88, 146–162. doi:10.1016/j.foodhyd.2018.10.003.

Rodriguez-Amaya, D.B., 2015. *Food Carotenoids: Chemistry, Biology and Technology*, 1st ed. IFT Press-Wiley, Oxford.

Rodriguez-Amaya, D.B., 2019. Update on natural food pigments—A mini-review on carotenoids, anthocyanins, and betalains. *Food Res. In.* 124, 200–205. doi:10.1016/j.foodres.2018.05.028.

Rodríguez-deLuna, S.E., Moreno-cortez, I.E., Garza-Navarro, M.A., Lucio-Porto, R., Pavón, L.L., González-González, V.A., 2017. Thermal stability of the immobilization process of horseradish peroxidase in electrospun polymeric nanofibers. *J. Appl. Polym. Sci.* 134, 1–10. doi:10.1002/app.44811.

Rojas-Mercado, A.S., Moreno-Cortez, I.E., Lucio-Porto, R., Pavón, L.L., 2018. Encapsulation and immobilization of ficin extract in electrospun polymeric nanofibers. *Int. J. Biol. Macromol.* 118(B), 2287–2295. doi:10.1016/j.ijbiomac.2018.07.113.

Rolle, R., Enriquez, O., 2017. *Souvenir Food Packaging: A Training Resource for Small Food Processors and Artisans.* FAO (Food and Agriculture Organization of the United Nations), Rome.

Roohbakhsh, A., Karimi, G., Iranshahi, M., 2017. Carotenoids in the treatment of diabetes mellitus and its complications: A mechanistic review. *Biomed. Pharmacother.* 91, 31–42. doi:10.1016/j.biopha.2017.04.057.

Rostami, A.H., Motamedzadegan, A., Hosseini, S.E., Rezaei, M., Kamali, A., 2017. Evaluation of plasticizing and antioxidant properties of silver carp protein hydrolysates in fish gelatin film. *J. Aquat. Food Prod. Technol.* 26, 1–11. doi:10.1080/10498850.2016.1213345.

Saliu, F., Pergola, R.D., 2018. Carbon dioxide colorimetric indicators for food packaging application: Applicability of anthocyanin and poly-lysine mixtures. *Sens. Actuat. B-Chem.* 258, 1117–1124. doi:10.1016/j.snb.2017.12.007.

Santos, J.P., Zavareze, E.R., Dias, A.R.G., Vanier, N.L., 2018. Immobilization of xylanase and xylanase-β-cyclodextrin complex in polyvinyl alcohol via electrospinning improves enzyme activity at a wide pH and temperature range. *Int. J. Biol. Macromol.* 118(B), 1676–1684. doi:10.1016/j.ijbiomac.2018.07.014.

Schmatz, D.A., Costa, J.A.V., Morais, M.G., 2019. A novel nanocomposite for food packaging developed by electrospinning and electrospraying. *Food Packag. Shelf Life* 20, 1–8. doi:10.1016/j.fpsl.2019.100314.

Schmid, M., 2013. Properties of cast films made from different ratios of whey protein isolate, hydrolysed whey protein isolate and glycerol. *Materials* 6(8), 3254–3269. doi:10.3390/ma6083254.

Shahid, M., Islam, S.I., Mohammad, F., 2013. Recent advancements in natural dye applications: A review. *J. Clean. Prod.* 53, 310–331. doi:10.1016/j.jclepro.2013.03.031.

Shang, J., Lin, S., Theato, P., 2017. Fabrication of color changeable CO_2 sensitive nanofibers. *Polym. Chem.* 8(48), 7446–7451. doi:10.1039/c7py01628j.

Shen, X., Yu, D., Zhu, L., Branford-White, C., White, K., Chatterton, N.P., 2011. Electrospun diclofenac sodium loaded Eudragit L 100–55 nanofibers for colon targeted drug delivery. *Int. J. Pharm.* 408(1–2), 200–207. doi:10.1016/j.ijpharm.2011.01.058.

Shi, X., Zhou, W., Ma, D., Ma, Q., Bridges, D., Ma, Y., Hu, A., 2015. Electrospinning of nanofibers and their applications for energy devices—Review. *J. Nanomater.* 2015, 1–20. doi:10.1155/2015/140716.

Shishir, M.R.I., Xie, L., Sun, C., Zheng, X., Chen, W., 2018. Advances in micro and nano-encapsulation of bioactive compounds using biopolymer and lipid-based transporters. *Trends Food Sci. Technol.* 78, 34–60. doi:10.1016/j.tifs.2018.05.018.

Shlar, I., Poverenov, E., Vinokur, Y., Horev, B., Droby, S., Rodov, V., 2015. High-throughput screening of nanoparticle-stabilizing ligands: Application to preparing antimicrobial curcumin nanoparticles by antisolvent precipitation. *Nano Micro Lett.* 7(1), 68–79. doi:10.1007/s40820-014-0020-6.

Sill, T.J., Recum, H.A., 2008. Electrospinning: Applications in drug delivery and tissue engineering—Review. *Biomaterials* 29(13), 1989–2006. doi:10.1016/j.biomaterials.2008.01.011.

Silva, C.K., Mastrantonio, D.J.S., Costa, J.A.V., Morais, M.G., 2019. Innovative pH sensors developed from ultrafine fibers containing açaí (*Euterpe oleracea*) extract. *Food Chem.* 294, 397–404. doi:10.1016/j.foodchem.2019.05.059.

Silva, F.T., Cunha, K.F., Fonseca, L.M., Antunes, M.D., ElHalal, S.L.M., Fiorentini, A.M., Zavareze, E.R., Dias, A.R.G., 2018. Action of ginger essential oil (*Zingiber officinale*) encapsulated in proteins ultrafine fibers on the antimicrobial control *in situ*. *Int. J. Biol. Macromol.* 118(A), 107–115. doi:10.1016/j.ijbiomac.2018.06.079.

Silvestre, C., Duraccio, D., Cimmino, S., 2011. Food packaging based on polymer nanomaterials. *Progr. Polym. Sci.* 36(12), 1766–1782. doi:10.1016/j.progpolymsci.2011.02.003.

Siripatrawan, U., Kaewklin, P., 2018. Fabrication and characterization of chitosan-titanium dioxide nanocomposite film as ethylene scavenging and antimicrobial active food packaging. *Food Hydrocol* 84, 125–134. doi:10.1016/j.foodhyd.2018.04.049.

Sohail, M., Sun, D.W., Zhu, Z., 2018. Recent developments in intelligent packaging for enhancing food quality and safety. *Crit. Rev. Food Sci. Nutr.* 58(15), 1–41. doi:10.1080/10408398.2018.1449731.

Song, W., Zhao, B., Wang, C., Lu, X., 2019. Electrospun nanofibrous materials: A versatile platform for enzyme mimicking and their sensing applications. *Compos. Commun.* 12, 1–13. doi:10.1016/j.coco.2018.12.005.

Stoll, L., Costa, T.M.H., Jablonski, A., Flôres, S.H., Rios, A.O., 2016. Microencapsulation of anthocyanins with different wall materials and its application in active biodegradable films. *Food Bioprocess Tech.* 9(1), 172–181. doi:10.1007/s11947-015-1610-0.

Su, S.L., Li, Y., 2004. Quantum dot biolabeling coupled with immunomagnetic separation for detection of *Escherichia coli* O157:H7. *Anal. Chem.* 76(16), 4806–4810. doi:10.1021/ac049442+.

Sun, B., Long, Y.Z., Zhang, H.D., Li, M.M., Duvail, J.L., Jiang, X.Y., Yin, H.L., 2014. Advances in three-dimensional nanofibrous macrostructures via electrospinning. *Prog. Polym. Sci.* 39(5), 862–890. doi:10.1016/j.progpolymsci.2013.06.002.

Sun, L., Sun, J., Chen, L., Niu, P., Yang, X., Guo, Y., 2017. Preparation and characterization of chitosan film incorporated with thinned young apple polyphenols as an active packaging material. *Carbohydr. Polym.* 163, 81–91. doi:10.1016/j.carbpol.2017.01.016.

Talero, E., García-Mauriño, S., Ávila-Román, J., Rodrígues-Luna, A., Alcaide, A., Motilva, V., 2015. Bioactive compounds isolated from microalgae in chronic inflammation and cancer. *Mar. Drugs* 13(10), 6152–6209. doi:10.3390/md13106152.

Tanaka, T., Shnimizu, M., Moriwaki, H., 2012. Cancer chemoprevention by carotenoids. *Molecules* 17(3), 3202–3242. doi:10.3390/molecules17033202.

Tang, L., Ding, S., Zhong, K., Hou, S., Bian, Y., Yan, X., 2017. A new 2-(2′-hydroxyphenyl) quinazolin-4(3H)-one derived acylhydrazone for fluorescence recognition of Al^{3+}. *Spectrochim. Acta A* 174, 70–74. doi:10.1016/j.saa.2016.11.026.

Tas, B.A., Sehit, E., Tas, C.E., Unal, S., Cebeci, F.C., Menceloglu, Y.Z., Unal, H., 2019. Carvacrol loaded halloysite coatings for antimicrobial food packaging applications. *Food Packag. Shelf Life* 20, 1–6. doi:10.1016/j.fpsl.2019.01.004.

Tatlisu, N.B., Yilmaz, M.T., Aricic, M., 2019. Fabrication and characterization of thymol-loaded nanofiber mats as a novel antimould surface material for coating cheese surface. *Food Packag. Shelf Life* 21, 1–10. doi:10.1016/j.fpsl.2019.100347.

Tavano, O.L., 2013. Protein hydrolysis using proteases: An important tool for food biotechnology. *J. Mol. Catal. B Enzym.* 90, 1–11. doi:10.1016/j.molcatb.2013.01.011.

Thomas, S., Durand, D., Chassenieux, C., Jyotishkumar, P., 2013. *Handbook of biopolymer-based materials: From blends and composites to gels and complex networks.* Weinheim, Germany: Wiley-VCH Verlag GmbH & Co. KGaA.

Tortora, G.J., Funke, B.R., Case, C.L., 2011. *Microbiology: An Introduction*, 8th ed. Pearson Benjamin Cummings, San Francisco.

Tsuda, T., Shiga, K., Ohshima, K., Kawakishi, S., Osawa, T., 1996. Inhibition of lipid peroxidation and the active oxygen radical scavenging effect of anthocyanin pigments isolated from *Phaseolus vulgaris* L. *Biochem. Pharmacol.* 52(7), 1033–1039. doi:10.1007/978-4-431-67017-9_63.

Turan, D., Sängerlaub, S., Stramm, C., Gunes, G., 2017. Gas permeabilities of polyurethane films for fresh produce packaging: Response of O_2 permeability to temperature and relative humidity. *Polym. Test.* 59, 237–244. doi:10.1016/j.polymertesting.2017.02.007.

Uebel, L.S., Schmatz, D.A., Kuntzler, S.G., Dora, C.L., Muccillo-Baisch, A.L., Costa, J.A.V., Morais, M.G., 2016. Quercetin and curcumin in nanofibers of polycaprolactone and poly(hydroxybutyrate-co-hydroxyvalerate): Assessment of *in vitro* antioxidant activity. *J. Appl. Polym. Sci.* 133(30), 1–7. doi:10.1002/app.43712.

Vanderroost, M., Ragaert, P., Devlieghere, F., Meulenaer, B., 2014. Intelligent food packaging: The next generation. *Trends Food Sci. Technol.* 39(1), 47–62. doi:10.1016/j.tifs.2014.06.009.

Vimala, K., Yallapu, M.M., Varaprasad, K., Reddy, N.N., Ravindra, S., Naidu, N.S., Raju, K.M., 2011. Fabrication of curcumin encapsulated chitosan-PVA silver nanocomposite films for improved antimicrobial activity. *J. Biomater. Nanobiotechnol.* 2(1), 55–64. doi:10.4236/jbnb.2011.21008.

Wafa, B.A., Makni, M., Ammar, S., Khannous, L., Hassana, A.B., Bouaziz, M., Es-Safi, N.E., Gdoura, R., 2017. Antimicrobial effect of the Tunisian Nana variety *Punica granatum* L. extracts against *Salmonella enterica* (serovars Kentucky and Enteritidis) isolated from chicken meat and phenolic composition of its peel extract. *Int. J. Food Microbiol.* 241, 123–131. doi:10.1016/j.ijfoodmicro.2016.10.007.

Wan, L., Tan, H.-L., Thomas-Ahner, J., Pearl, D.K., Erdman Jr., J.W., Moran, N.E., Clinton, S.K., 2014. Dietary tomato and lycopene impact androgen signaling-and carcinogenesis-related gene expression during early TRAMP prostate carcinogenesis. *Cancer Prev. Res.* 7(12), 1228–1239. doi:10.1158/1940-6207.CAPR-14-0182.

Wang, K., Lim, P.N., Tong, S.Y., Thian, E.S., 2019. Development of grapefruit seed extract-loaded poly(ε-caprolactone)/chitosan films for antimicrobial food packaging. *Food Packag. Shelf Life* 22, 1–9. doi:10.1016/j.fpsl.2019.100396.

Wang, S., Marcone, M., Barbut, S., Lim, L.-T., 2013. Electrospun soy protein isolate-based fibre fortified with anthocyanin-rich red raspberry (*Rubus strigosus*) extracts. *Food Res. Int.* 52(2), 467–472. doi:10.1016/j.foodres.2012.12.036.

Wang, X., Ding, B., Yu, J., Wang, M., 2011. Engineering biomimetic superhydrophobic surfaces of electrospun nanomaterials. *Nano Today* 6(5), 510–530. doi:10.1016/j.nantod.2011.08.004.

Yildirim, S., Röcker, B., Pettersen, M.K., Nilsen-Nygaard, J., Ayhan, Z., Rutkaite, R., Radusin, T., Suminska, P., Marcos, B., Coma, V., 2018. Active packaging applications for food. *Compr. Rev. Food Sci. Food Saf.* 17(1), 165–199. doi:10.1111/1541-4337.12322.

Yilmaz, A., Bozkurt, F., Cicek, P.K., Dertli, E., Durak, M.Z., Yilmaz, M.T., 2016. A novel antifungal surface-coating application to limit postharvest decay on coated apples: Molecular, thermal and morphological properties of electrospun zein–nanofiber mats loaded with curcumin. *Innov. Food Sci. Emerg. Technol.* 37, 74–83. doi:10.1016/j.ifset.2016.08.008.

Yoon, B.I., Bae, W.J., Choi, Y.S., Kim, S.J., Ha, U.S., Hong, S.H., Sohn, D.W., Kim, S.W., 2018. Antiinflammatory and antimicrobial effects of anthocyanin extracted from black soybean on chronic bacterial prostatitis rat model. *Chin. J. Integr. Med.* 24(8), 621–626. doi:10.1007/s11655-013-1547-y.

Zahedi, Y., Ghanbarzadeh, B., Sedaghat, N., 2010. Physical properties of edible emulsified films based on pistachio globulin protein and fatty acids. *J. Food Eng.* 100(1), 102–108. doi:10.1016/j.jfoodeng.2010.03.033.

Zdarta, J., Meyer, A.S., Jesionowski, T., Pinelo, M., 2018. A general overview of support materials for enzyme immobilization: Characteristics, properties, practical utility. *Catalysts* 8(2), 1–27. doi:10.3390/catal8020092.

Zhai, X., Shi, J., Zou, X., Wang, S., Jiang, C., Zhang, J., Huang, X., Zhang, W., Holmes, M., 2017. Novel colorimetric films based on starch/polyvinyl alcohol incorporated with roselle anthocyanins for fish freshness monitoring. *Food Hydrocol* 69, 308–317. doi:10.1016/j.foodhyd.2017.02.014.

Zheng, J.-X., Yin, H., Shen, C.-C., Zhang, L., Ren, D.-F., Lu, J., 2020. Functional and structural properties of Spirulina phycocyanin modified by ultra-high-pressure composite glycation. *Food Chem.* 306, 1–8. doi:10.1016/j.foodchem.2019.125615.

Zia, J., Paul, U.C., Heredia-Guerrero, J.A., Athanassiou, A., Fragouli, D., 2019. Low-density polyethylene/curcumin melt extruded composites with enhanced water vapor barrier and antioxidant properties for active food packaging. *Polymer* 175, 137–145. doi:10.1016/j.polymer.2019.05.012.

9 Influence of Nanoparticles on the Shelf Life of Food in Packaging Materials

Malinee Sriariyanun, Atthasit Tawai, and Yu-Shen Cheng

CONTENTS

9.1 INTRODUCTION

The current trends of the food industry worldwide are targeted to meet consumer demand for safe and fresh food with high quality in terms of nutrition. Due to the development of global trading and advancement in logistics business, the management in process chains of food production has been modified from how it was. The fresh products as raw materials from agricultural fields and husbandry farms in local areas are transported across the country in shorter periods of time by using highly developed transportation systems to the consumers in modern trade channels or to the food processing industries. Fresh foods are processed to make modern forms of food products with the concept of ready to cook" and "ready to eat" or "instant meal" to provide the comfortable life the consumers want. Due to current living styles of human activities, the need of processed foods in the global market is rising, with a challenge to meet strict food safety regulations. Additionally, to increase the benefit of food business, it is necessary for both fresh foods and processed food, to maintain their food nutrition and sensory quality of product during the period of time used from when they are produced to when they are consumed. Thus, the development of food packaging materials has been motivated to find a novel, cost-effective, and safe method for food transportation, delivery, and storage from producers to sellers, and to consumers.

9.2 MECHANISMS OF SHELF LIFE DETERMINATION

Shelf life of food product is defined, according to the Institute of Food Science and Technology (IFST), as the period of time during which the product will remain safe with the desired characteristics including sensory, chemical, physical, and biological aspects and comply with the product label declaration of nutrition under the recommended storage conditions. The shelf life of the product can be affected by many factors, including initial quality, the natural properties of products, production formulation, processing methods, packaging, transportation and storage conditions, and consumer handling (Ellis, 1994). The water activity in a food product also determines the period of shelf life; the product with high water activity has a shorter shelf life. Addition of compounds with antioxidant properties or preservatives can extend the product shelf life. The numbers of populations of remaining microbes in the product is considered as the main factor to determine the quality and safety of food. The shelf life can be indicated based on governmental legislation and expected industrial standards. The determination of shelf life relies on sensory perception and customer acceptance tests by comparing aging samples with fresh samples. Sometimes, the sensory perception analysis is conducted by using expert panelists to detect the change in numbers of sensory attributes (i.e., taste, texture, color, odor, appearance, and acidity), and the tested results could be representative of the customer acceptance (Manzocco, 2016; Azad et al., 2019).

Food spoilage is defined as the reduction of food edibility, and it is related to food safety. The food spoilage can be indicated by the change in physical, chemical, and biological properties, such as texture, acidity, color, smell, and flavor (Gram et al., 2002) (see Figure 9.1). These property changes can be caused by many factors as described below, for example, pH, temperature, oxygen, moisture, etc.

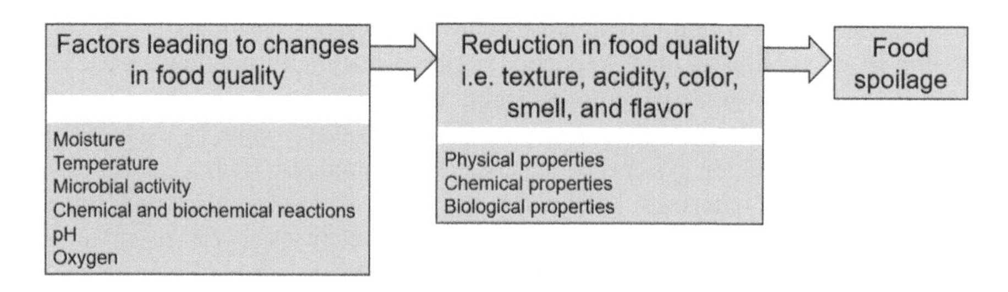

FIGURE 9.1 Mechanisms of shelf-life determination.

1. *Moisture*: The change in moisture content in food products by decreasing or increasing of water caused by water transfer phenomena and related to the water activity (a_w). Water activity (a_w) is a ratio of the partial vapor pressure of water in a product to the standard state partial vapor pressure of pure water at the same condition, especially at the same temperature. Water activity usually relates to the temperature and the microbial activities (Parra and Magan, 2004). The water transfer also leads to disaggregation of food texture and change in the chemical properties of the food product.
2. *Temperature*: This is one of the key factors in determining the shelf life of fresh products, such as fruits, vegetables, and meats. The optimal temperature, optimal moisture, and appropriate flow of air for product storage can prolong the shelf life of fresh products. Temperature is a physical parameter controlling the rates of chemical reactions and metabolism rates of microbes and flesh products (Olafsdottir et. al., 2006). In the case of food storage at low temperature, freeze damage may occur by which the ice crystals are formed within the cells and in extracellular matrixes. The sharpness of ice crystals can pierce through the cells' structures and membranes, leading to cell deaths of the products.
3. *Microbial activity*: This factor is a primary determinator of the shelf life and relates to the foodborne diseases. The rate of microbial growth can be reduced by controlling storage temperature, water activity, pH, and sometimes by using preservatives. The microbial contamination is also impacted by the original quality and cleanliness of food products (Wu et. al., 2018). Using good practice, according to the relevant safety standard in the process of food production can help to lessen the problem caused by microbial activities.
4. *Chemical and biochemical reactions*: The intrinsic chemical and biochemical properties of food result in unfavorable sensory perception to the consumers. For example, the oxidation reactions on the fatty acids, especially unsaturated fatty acids, cause rancid taste and smell in the products. The rates of reactions are associated with the physical parameters, including temperature, moisture, and flow of air. Oxidation reaction requires the presence of oxygen as a reactant. Oxygen could react with biomolecules in foods, such as fatty acid and amino acid. Amino acids in food products could be oxidized to be ammonia and organic acids, which are the source of spoilage reaction to fresh meats (Huis in't Veld, 1996). Another unpleasant chemical reaction is a Maillard reaction, which produces browning in the product from amino acid and sugar in foods. The browning of the product additionally leads to reduction in protein solubility and nutritional value, and a bitter flavor. The Maillard reaction can be observed in dairy products and processed foods.

It is to be noted that food spoilage and shelf life of the product can be controlled by physical, chemical, and biological factors. Not only do individual factors affect the shelf life; the combination of factors could also synergistically accelerate the food spoilage. Due to large numbers of these parameters, the solutions to prolong the shelf life of food products are varied. However, the selection of the solution should be conducted based on the types of foods, cost, customer acceptance, and legislation.

9.3 DIVERSITY IN NANO-PACKAGING MATERIALS

Food packaging has been produced and developed for food industries to achieve the purpose of providing ready to use, ready to eat, and ready to cook products for consumers. To fit with our conventional lifestyle and modernized trend of food trading from producer to retailers and consumers, the maintenance of fresh food quality and properties is a crucial concern to make the commercialization survive. Different types of food packaging have been produced and applied to ready-to-sell products with many purposes (see Figure 9.2). Food packaging provides a suitable environment to maintain the quality of food product and assist the transportation and distribution of the product to different areas of consumers, even across countries or continents.

Different nanoparticle additives are added into the matrixes of food packages to improve or maintain the food quality. The nano-packagings are currently categorized, depending on the purpose of customer uses, to be improved packaging, active packaging, smart packaging, and biobased packaging (Sharma et al., 2017; Rossi et al., 2017). From the point of view and perception of public customers, nanomaterials are related to remarkable antimicrobial activities. However, they exhibit improved mechanical and physical performance and promote resistance to environmental stress. Improved packaging is made from mixtures of polymeric-based compounds and nanoparticles to improve gas barrier properties, and temperature and moisture resistance of packaging. The packaging containing nanocomposite has been approved for use with food products. Currently, developed and developing countries worldwide adopt applications of nanoparticles in food packages e.g., the United States, European Union, Japan, India, Iran, China, South Korea, and Thailand (Tsagkaris et al., 2018;

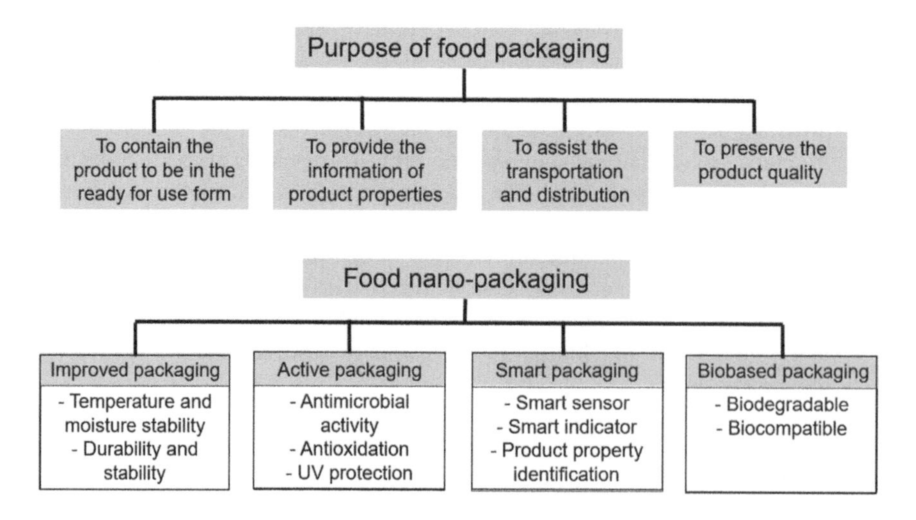

FIGURE 9.2 Purpose of food packaging and nano-packaging.

Thiruvengadam et. al., 2018; Singh et al., 2017). Active packaging uses nanomaterials that react with chemicals in foods or the environment to enhance the efficiency to maintain food product qualities. The types of nanoparticles used in active packaging include metal oxide compounds, such as nanosilver oxide, nanotitanium oxide, nanocopper oxide, and nanomagnesium oxide, and they possess antimicrobial activities to slow food spoilage. Smart packaging is developed to have a special property to sense the change of chemicals profiles in foods as well as growth of microbes as a signal of food spoilage. The smart packaging can report the food quality to consumers in the form of color change. Biobased packagings are made of natural polymers which are obtained from extraction of natural biomass, chemical synthesis of biobased monomers, and microbial synthesis.

9.3.1 Active Packaging

The active package contains nanoparticles that react with chemicals in foods or environment and perform the function of extending shelf life, maintaining food quality and food safety. The active nanoparticles in packaging can release active compounds to food content or protect the unwanted compounds from the environment contacting food products. Based on the targeted functions, active packages can (1) capture or scavenge CO_2, O_2, and ethylene (by using zeolite, ferrous oxide, activated carbon, catechol or ascorbate salts), (2) release active compounds (i.e., antimicrobial compound, antioxidant compound, insect repellent compound, and flavor), and (3) regulate or absorb moisture in the package (by using sachets or pads) (Sharma et al., 2017; Thiruvengadam et al., 2018).

One promising type of active packages is antimicrobial packaging systems with properties that prevent microbial growth in food products and to extend food product shelf life. Antimicrobial packages are developed to be able to act either by killing or preventing the microbial population growth and their activities. The common antimicrobial agents in active packages are metallic salts, enzymes, organic acids, chitins, chitosans, natural extracts, and essential oils (Huang et al., 2018). The classic scenario to explain how metallic salts or metal oxides express antimicrobial activities is presented in Figure 9.3. The metal ions released from suspension of metallic salts or metal oxides in liquid or gas bulks move to be in contact with cell walls or cell membranes of targeted microbes, i.e., bacteria. The reactions between bacterial cells and nanometal ions generate reactive oxygen species, which subsequently (1) inactivate protein, (2) damage DNA, (3) impair electro transport chain in mitochondria, and (4) interfere in the arrangement of phospholipid in cell membrane. Then, once the metabolic mechanism in bacterial cell fails and the cell containment collapses, cell cytoplasm leaks out from the cell leading to cell death (Bumbudsanpharoke et al., 2015; Vasile, 2018). This metallic nanopackage has been developed in many research projects and applied in industrial uses (e.g., Philadelphia cheese, Nestle milk product, etc.) as listed in Table 9.1.

In addition to metallic salts or metal oxides, the natural antimicrobial agents obtained from plants, animals, and microbes, for example, bacteriocins, enzymes, and phenolic compounds are gaining more interest from consumers due to their environmental impact, and safety and regulation issues. Much research has demonstrated

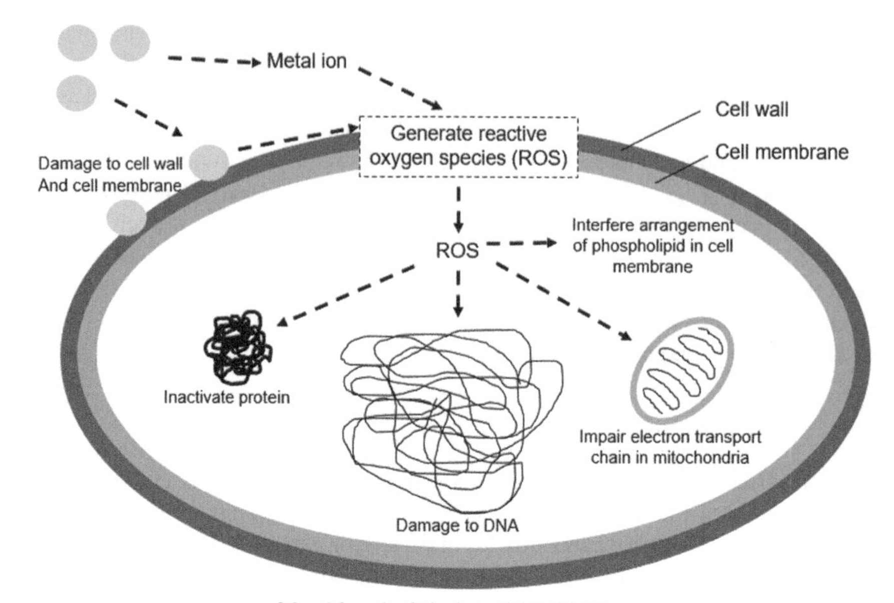

FIGURE 9.3 Multi-actions of metallic salts or metal oxides in antimicrobial activities.

the productions of these natural antimicrobial agents and their mechanistic effect as listed in Table 9.2. Vasile et. al., 2019 developed a multifunctional poly(lactic acid) (PLA) and polyethylene glycol (PEG)-based active food packaging to be used for medical and food product purposes. The PLA bioplastic was mixed during melting with natural active nanocompounds, e.g., chitosan and rosemary extract. This biocomposite was expected to be biodegradable and thermoplastic with heat-resistant and mechanical-tolerant properties and possessed good antimicrobial and antioxidative activity from chitosan and rosemary extract, respectively. The optimal formulation of PEG-plasticized PLA containing 6 wt % chitosan and 0.5 wt % rosemary extract (so-called PLA/PEG/6CS/0.5R) demonstrated high hydrophilicity and good *in vitro* and *in vivo* biocompatibility tested by observing the changes in hematological, biochemical, and immunological phenotypes obtained from subcutaneous implantation. The PLA/PEG/6CS/0.5R expressed antioxidative activities to the level of about 40–50% of vitamin E compound, and it has excellent antimicrobial activity at 100%, 90%, and 82% inhibition of *Bacillus cereus*, *Salmonella Enteritidis*, and *Escherichia coli,* respectively. To show the application of active packaging to extend food shelf life, Heras-Mozos et al. (2019) produced composite films of polyethylene (PE), ethylene-vinyl alcohol copolymer and zein (so-called PE/EVOH/Zein) and mixed it with garlic extract for preservation of free-slice pan loaf. PE/EVOH/Zein film containing 0.25 and 0.5% w/w of garlic extract was challenged *in vitro* with *Penicillium expansum* and *in vivo* with natural sliced bread. The result clearly showed that this active film can protect bread product from mold infection for 30 days.

TABLE 9.1
Applications of Active Package Containing Metallic Salts or Metal Oxides

Polymer matrix	Nanoparticles	Tested food	Application	References
Polyethylene	Silver nanoparticles	Pasteurized orange juice without pulp, fresh gala apples, kingsmill white soft bread, prepackaged butter, fresh carrot, prepackaged Philadelphia soft cheese, fresh Aberdeen Angus, Hereford cross ground beef, Nestle MAP milk powder, and water	Packaging material: nanosilver containers and coated films (Fresh Box, Blue Moon Goods, the United States)	Metak et al., 2015
Poly(vinyl alcohol)/ graphene oxide/ starch	Silver nanoparticles	Chicken sausages	Packaging pouches	Mathew et al., 2019
Poly(vinyl alcohol)	Silver nanoparticles	Lemon and strawberry	Nanofiber	Kowsalya et al., 2019
Gelatin nanofiber	Moringa oil-loaded chitosan nanoparticles	Cheese	Nanofiber	Lin et al., 2019
Corn starch	Talc nanoparticles	Cherry tomatoes	Packaging bags	López et al., 2015
Carboxymethyl cellulose	Zinc oxide nanoparticles	Chicken breast meat	Nanocomposite film	Mohammadi et al., 2019
Polylactide/polyethylene glycol/ polycaprolactone	Zinc oxide nanoparticles	Scrambled egg	Composite film	Ahmed et al., 2019
Low density polyethylene	Copper nanoparticles	Peda (Indian sweet dairy product)	Packing film	Lomate et al., 2018
Liposomes	Silica (SiO_2) nanoparticles	Beef	Nanofibrous membranes	Cui et al., 2017

TABLE 9.2
Active Package Containing Natural Antimicrobial Agents

Nanoparticles	Sources	Targeted microbes	References
Zein film contains natural phenolic compounds, e.g., phenolic acids (gallic, vanillic, cinnamic acids), essential oils (carvacrol, thymol, eugenol, citral), phenolic extracts	Clove, oregano, artichoke stem, and walnut shells	*Erwinia amylovora*, *Erwinia carotovora*, *Xanthomonas vesicatoria*, and *Pseudomonas syringae*	Alkan and Yemenicioğlu, 2016
Electrospun polyvinyl alcohol nanofibrous film containing cinnamon essential oil/β-cyclodextrin	Cinnamon	*Staphylococcus aureus* and *Escherichia coli*	Wen et al., 2016
Coating with a carnauba wax with grapefruit seed extract	Grapefruit	*Penicillium digitatum*	Choi et al., 2019
Poly(lactic acid) (PLA) and polyethylene glycol (PEG) -based materials containing chitosan and rosemary extract	Chitosan and rosemary	*Escherichia coli*, *Salmonella enteritidis*, and *Bacillus cereus*	Vasile et al., 2019
Edible-coating chitosan and nanochitosan films with pomegranate peel extract	Chitosan from *Aspergillus niger* mycelia, and pomegranate	*Aspergillus flavus*, *Aspergillus ochraceus*, and *Fusarium moniliforme*	Alotaibi et al., 2019
Polyethylene aqueous emulsion and ethylene-vinyl alcohol copolymer and zein hydroalcoholic solutions containing garlic extract	Garlic	*Penicillium expansum*	Heras-Mozos et al., 2019
Poly(lactic acid) (PLA) containing powdered rosemary ethanolic extract	Rosemary	*Bacillus cereus*, *Salmonella typhymurium*, and *Escherichia coli*	
Polypropylene (PP) films containing oregano essential oil and Proallium® (an Allium extract)	Oregano, allium	*Brochothrix thermosphacta*	Llana-Ruiz-Cabello et al., 2018
Nanomontmorillonite-chitosan and nanomontmorillonite-carboxymethyl cellulose containing *Ziziphora clinopodioides* essential oil and *Ficus carica* extract	*Ziziphora clinopodioides* and *Ficus carica*	*Listeria monocytogenes* and *Escherichia coli* O157:H7	Khezrian et al., 2018
Methylcellulose films containing natural extracts from the stems of Ginja cherry	Ginja cherry	*Listeria innocua*, methicillin-sensitive *Staphylococcus aureus*, methicillin-resistant *S. aureus*, *Salmonella enteritidis*, *Escherichia coli*	Campos et al., 2014

9.3.2 Smart Packaging

Smart packaging promotes the advancement of food industries by adding information about food quality and condition on the shelf to let customers make decisions before purchasing and allow retailers to effectively manage the stocks. The food conditions in the packages, e.g., time, temperature, freshness, leakage, or spoilage, could be monitored and reported by different methods, especially nanosensor (see Figure 9.4). The reactivities of nanosensors added in the packages to the targeted parameters are designed to indicate changes to the physical, chemical, and biological properties of food products in the packages. For example, gas sensors detect targeted gases, e.g., hydrogen, sulfide, CO_2, and volatile amines, released from spoiled food and report as a luminescent or fluorescent signal that can be visualized by a specific camera or color changes that can be detected by the naked eye.

9.3.2.1 Detection of Leakages in Food Package

To extend the shelf life of food products, the packages are vacuumed to remove oxygen and to prevent air ventilation, and occasionally nitrogen gas or other inert gas are applied in the sealed food containers. Based on this scenario, the presence of oxygen in the food package could be used as an indicator of package leakage. In 2005, Lee et al. developed and characterized a UV-activated colorimetric oxygen indicator made of TiO_2 nanoparticles. The TiO_2 nanoparticles are activated upon UVA light irradiation and reduce the methylene blue dye in a polymer encapsulation medium. UV irradiation leads to bleaching of indicator or dye and remains colorless in dark conditions, and once the indicator is exposed to oxygen, the original color of dye is restored. The detection rate of this O_2 sensor was demonstrated to be dependent on the level of O_2. Mills and Hazafy (2009) synthesized nanocrystalline SnO_2 as a photosensitizer to detect O_2 presence by using glycerol as an electron donor, methylene blue as a redox dye, and hydroxyethyl cellulose as an encapsulating polymer. Exposure of UVB light activated photobleaching of methylene blue. Without oxygen, the composite film showed colorless due to bleaching, but the original color

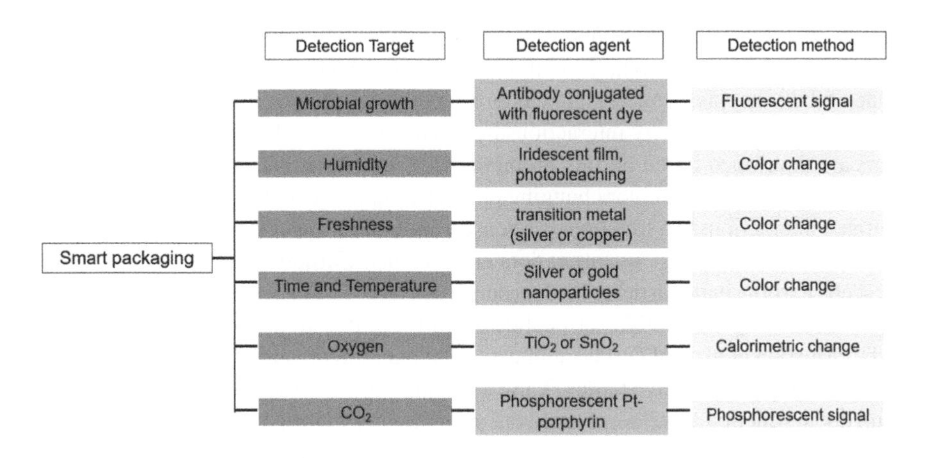

FIGURE 9.4 Targeted applications of smart packaging for food products.

was restored when the package leaked. López-Carballo et al. (2019) targeted to use halogen lamp irradiation to activate photosensitization of TiO_2 nanoparticles embedded in ethylene-vinyl alcohol copolymer (EVOH) to activate reduction reaction of methylene blue. This O_2 nanosensor was fabricated as a wafer to place in the headspace of food package without contact with the food product. Zhu et al. (2017), reported the use of a wireless oxygen sensor, namely O_2-p-CARD, made of a mixture of Fe(II)-poly(4-vinylpyridine) in carbon nanotube on a commercial passive near-field communication tag picked up the signal of O_2 detection in the nitrogen-filled vegetable package and could be monitored by using a smartphone.

Humidity or moisture changes in a package could be used as another indicator for package leakage. The monitoring of this parameter could ensure the integrity of the package reflecting the quality and food safety of food products in the packages. Different designs of humidity nanosensors were developed by using different indicators and methods. Bridgeman et al. (2014) synthesized a humidity sensor by coating a low-toxicity ionic-liquid composite material onto a porous substrate. This low-cost sensor reported the change of humidity in the range of 5–100% based on the colorimetric change, from yellow to blue, of DenimBlu30 dye as a redox compound. Another humidity sensor was developed by modification of psychrometer and tested in dry-cured meat products (Zhang et al., 2016). The methodology of signal receiving and reporting of psychrometer was improved to monitor the relative humidity with high accuracy at ±0.6%. Ghadiry et al. (2016) developed the humidity sensors based on the principle that the cladding refractive index (RI) increases due to cladding water absorption. The graphene oxide layers were fabricated as a support for nano-anatase TiO_2. The response to the relative humidity in the range of 35 to 98% was reported as transmitted optical signals with a response time of only ~0.7 sec.

9.3.2.2 Detection of Microbial Activities and Food Spoilage

The main causal agent of food spoilage is microbes contamination in food packages. The importance of this concern gets more serious if the contaminated microbes are pathogenic species that lead to foodborne diseases such as *Salmonella* spp., *Clostridium perfringens*, *E. coli* O157:H7, and *Listeria monocytogenes*. Due to human health concerns, precise, reliable, repeatable, rapid, and cost-effective detection sensors are necessary to prevent the outbreak of disease and loss of spoiled food products. One of the concepts of microbial detection method relied on is immunological assays, which are high-specificity interactions between antibody and antigen. Microbial sensors are developed based on this immunological principle, but incorporated with the reporting system using color changing or electric signal generated from antibody and antigen interactions. To function as a sensor, antibodies must be tagged or conjugated with nanomaterials, e.g., quantum dots and aptamer. Carbon quantum dots are luminescent carbon nanoparticles with reactive surface properties. Aptamers are short synthesized oligonucleotides or peptides with the ability to impinge in carbon nanotube (Mustafa et al., 2017). The application of quantum dots in microbial sensors has been developed by many studies. Yang and Li (2006) modified semiconductor quantum dot to function as a fluorescence tag on antibodies in immunoassays to simultaneously detect two foodborne pathogens, *E. coli* O157:H7 and *Salmonella typhimurium*. Irradiation with a specific wavelength light on the quantum dot-conjugated antibody to *E. coli* O157 and *Salmonella* caused the shift of emission wavelengths that could be detected by using a fluorescent microscope. Quantum dot-conjugated antibodies were applied to other foodborne pathogenic bacteria, such as *Shigella flexneri*, with a combination of magnetic gamma-Fe(II)O_3 microparticles impregnated on silica by using a water-in-oil reverse microemulsions method. This microbial sensor was demonstrated to detect the low concentration of pathogen in a food matrix at 10^{-3} cfu/mL within 2 h (Zhao et al., 2009). Different attempts to find conjugation agents, such as streptavidin with quantum dots conjugated with nanobeads were conducted to detect three major pathogenic bacteria, *Salmonella typhimurium, E. coli* O157:H7, and *Listeria monocytogenes*, in food products (Wang et al., 2011). Recently, Yousefi et al. (2018) developed a transparent, flexible, and realtime sensor suitable for food packaging based on interactions between *E. coli*-RNA-cleaving fluorogenic DNAzyme probe to a cyclo-olefin polymeric film arranged in microarrays. This DNAzyme sensor was demonstrated to remain stable for at least 14 days to detect *E. coli* (minimum concentration at 10^3 CFU/mL) in meat and apple juice. The presence of microbial growth could be reported as a fluorescent signal without the need to take the food sample out of the package.

Another approach to be relied on to detect microbial growth could be the monitoring of microbial metabolic activities, which directly result in changes of food appearance and off-taste perception. During the activities of contaminated microbes in foods, volatile gases, such as sulfide and amine derivatives, etc. are produced (Mihindukulasuriya and Lim, 2014). Based on this concept, Diaz et al. (2017) applied the use of Meldrum's activated furan (MAF) as an indicator of primary and secondary amines with ppm levels released from fish samples and the interactions of these compounds resulted in color change that could be observed by the naked eye. Schaude et al. (2017) invented a colorimetric layer sensor for detection of ammonia and biogenic amines by a pH indicator dye immobilized onto cellulose microparticles and subsequently embedded into non cytotoxic food-grade silicone. The ammonia derivatives putrescine and cadaverine were also tested as food spoilage indicators with wireless sensors, with which the signal was transmitted to smartphone via Near-field communication. The polymer-based gas nanosensor specific to putrescine and cadaverine was tested with minimum 5 ppm concentration and it showed that the readout signal was successfully activating the Near-field communication circuit to monitor meat spoilage (Ma et al., 2018). Oh et al. (2019) invented field-effect transistor (FET) biosensors to detect liquid and gaseous cadaverine as an indicator of food spoilage by using nanodisc bioelectronic nose. The reactions between a nickel (Ni)-decorated carboxylated polypyrrole nanoparticle (cPPyNP)-FET with cadaverine generated signal on the transistor in the readable format by the device. For H_2S detection, Shu et al. (2018) created a paper-based nanosensor made from Cu xO-polypyrrole conductive aerogel loaded on graphene oxide framework. The electrical signal generated from the reactions between H2S and Cu xO-polypyrrole was reported by light emitting diode (LED), and monitored in real time by smartphone. To be integrated into a food monitoring system and food packaging, a real-time monitoring H_2S sensor made of silver nanoparticles that was encapsulated in gellan gum for meat products was developed (Zhai et al., 2019). The reactions between silver nanoparticles with H_2S yielded Ag_2S with colorimetric change from yellow to colorless. This sensor was also tested in chicken breast and silver carp, and it showed high selectivity to H_2S, but not to other volatile compounds in tested packaged foods.

9.3.2.3 Detection of Physical Parameters of Food Packaging

The physical parameters, especially temperature, are prime determinators of food shelf life. During transportation and distribution of food products from food producer to retailers and consumers, the temperature conditions may fluctuate and this scenario could accelerate the microbial activities and food deterioration. Unfortunately, these temperature fluctuations are unknown to customers. Although perfect food packages with un-expired date may be in consumer hands, no one is able to guarantee that the quality of the food product is still acceptable. The irreversible temperature sensor is consequently useful in monitoring the thermal history during food storage, handling, and distribution and to provide information to consumers and retailers to help them make decisions during purchasing and stock management. A time-temperature indicator (TTI) is a sensor device, such as a temperature data logger, used in the monitoring of time-temperature history of the product during the transportation and distribution process. This smart logger is important to control the quality of food, especially frozen food or cold chain products. The application of this sensor device is not limited to only food industries, but also pharmaceutical, medical, and chemical industries as well. Zhang et al. (2013) developed a TTI device from Ag nanocubes with 30 to 70 nm-size from a mixture of CH3COOAg, sodium hydrosulfide, and hydrochloric acid. The detection of time-temperature profile is reported as a colorimetric change and could be visualized by the naked eye. Youn et al. (2018) reported the production of silver nanoparticle-based sensor for blood samples in the form of silver nanowires and colorless polyimide film integrated with a wireless data transmission circuit. The change of time and temperature activated the resistance of transmission circuit and the signal was transmitted in real time to the receivers. In addition to the silver nanoparticle-based sensor, other types of sensors were developed. For example, a glucose biosensor based on glucose oxidase was applied as a TTI sensor for smart food packaging (Rahman et al., 2018). The glucose sensor was integrated with a three-electrode potentiostat to report the time and temperature profile as an electrical signal and color development at the same time. The reaction kinetics of this sensor monitoring by the colorimetric change was studied based on Arrhenius assumption under isothermal condition. Anbukarasu et al. (2017) invented a TTI based on the function of depolymerase enzyme that reported the signal as the color change. This smart sensor was made from a dye-loaded polyhydroxybutyrate (PHB) film mixed with a depolymerase enzyme. Therefore, during the progress of time and temperature, the hydrolysis reaction proceeded to cause the release of dye. The dye release kinetics were studied at different temperatures ranging from 4 to 37°C, and the sensor can function to report the color change up to 168 h.

9.3.3 Biobased Packaging

Biobased materials are a focus of worldwide research and industrial sectors due to awareness of global warming and petroleum-based plastic contamination of the environment. The biobased packages to be used as matrix could be obtained from extraction of biomass of plants, animals, and microbes. The organic or inorganic active nanoparticles, as a filler, with desired properties to differentially function, such as active package or smart sensor could be added in the biocomposite matrix. The inventions and characterizations of biobased packaging with nanoparticles have been demonstrated in many studies as listed in Table 9.3. The major groups of matrix materials of biobased packages are (1) bioplastic, e.g., polylactic acid and polyhydroxybutyrate, (2) gelatin and alginate, and (3) natural polymers, e.g., chitosan, cellulose, and, starch. The fillers for biobased food packages are widely varied to fit the various purposes of food producers and could be categorized as (1) metal oxide nanoparticles, (2) herbal extract and essential oil, and (3) natural active polymers, e.g., chitosan. To fulfill the functions of food packaging, these bio-composites were invented with antimicrobial, antioxidant activities, as well as some mechanical and physical properties, e.g., water vapor protection and UV protection. Also, a nanosensor device could be integrated into biobased composites to make smart packaging as discussed in the previous section.

9.4 TOXICITY OF NANOMATERIALS AND SAFETY ISSUES

The applications of nanomaterials in food packaging with the main purpose to inhibit the growth of contaminated microorganisms in food products consequently extend food shelf life. Additional benefits of nanomaterials in food packaging are demonstrated as functioning antioxidants, improvement of food quality, and the development of smart sensors. Altogether, the commercial use of these nanomaterials in food industries has been expanded worldwide in developed and developing countries, in a wide range of products. The antimicrobial mechanism of nanoparticles can be explained by the release of reactive oxygen species (ROS) from nanoparticles and lead to the biological damage on the cellular level. This principle raises the awareness of safety issues to conduct studies to evaluate the nanoparticle's impacts before the use of products by consumers. Many research groups have focused on the negative effects, e.g., human-toxicity and ecotoxicity of nanoparticles. The toxicity levels of nanoparticles have been demonstrated *in vivo* to be dependent on the particle size and concentration in different organisms, including bacterial, yeast, algal cells, crustaceans, and mammalian cells (Ivask et al., 2014). The *in vivo* dose-dependent toxicity assessment of silver nanoparticles was conducted in Wistar rat to find the safe level of nanoparticles by monitoring the function of liver enzymes, histological phenotypes, and hematological parameters., The results showed that the safe concentration of silver nanoparticles was less than 10 mg/kg (Tiwari et al., 2011).

Nanoparticles in food packages could enter humans during food production processes, waste treatment, and consumer consumption. Therefore, the portal of nanoparticle exposure could be relevant to inhalation, skin penetration, and ingestion. Through food ingestion, nanoparticles in food packages could possibly contaminate food products and be delivered into the gastrointestinal tract. Through inhalation, un-immobilized nanoparticles could possibly free-flow into the respiratory system and be trapped in respiratory organs. Hernandez et al. (2019) investigated the contamination of microplastics and/or nanoplastics of plastic tea bags

TABLE 9.3
Biobased Packaging Containing Nanoparticles for Applications in Food Packaging

Filler	Metrix	Application	References
Silver-montmorillonite nanoparticles	Sodium alginic acid solution	Modified-atmosphere packaging (MAP) for Fior di latte cheese to preserve quality of and extend shelf life	Gammariello et al., 2011
Chitosan nanoparticles and *Origanum vulgare* L. essential oil	Fish gelatin	Bio-based nanocomposite films with antimicrobial activities against food pathogens, namely *Staphylococcus aureus*, *Listeria monocytogenes*, *Salmonella enteritidis*, and *Escherichia coli*	Hosseini et al., 2016
Ag/TiO2-SiO2, Ag/N-TiO2 or Au/TiO2	Paper	Paper packages for white bread storage to extend shelf life	Peter et al., 2016
Cellulose nanocrystals from sugarcane bagasse	Polyvinyl alcohol/carboxymethyl cellulose	Eco-friendly bio-nanocomposite films with transparency and water vapor permeability protection	El Achaby et al., 2017
Chitin nanoparticles	Gelatin	Bio-nanocomposite film with antifungal properties against *Aspergillus niger*	Sahraee et al., 2017
Silver nanoparticles	Tragacanth/hydroxypropyl methylcellulose/beeswax	Edible and biocomposite film with antimicrobial properties	Bahrami et al., 2019
Mahua oil and zinc oxide nanoparticles	Polyurethane and chitosan	Biodegradable composite films with properties of antibacterial activities, extending of shelf life, auto-degrading within 28 days	K et al., 2019
Cinnamon essential oil and chitosan nanoparticles	Zein film	Biodegradable zein composites film with antimicrobial properties against *Escherichia coli* and *Staphylococcus aureus*	Vahedikia et al., 2019
Carboxymethyl chitosan and ZnO nanoparticles	Sodium alginate and chitosan	Biodegradable films with water vapor resistance and antibacterial activity against *S. aureus* and *E. coli*	Wang et al., 2019
Chitosan nanofiber and ZnO nanoparticles	Gelatin	Biodegradable active packaging with water barrier properties and antibacterial activity	Amjadi et al., 2019

(Continued)

TABLE 9.3 (CONTINUED)
Biobased Packaging Containing Nanoparticles for Applications in Food Packaging

Filler	Metrix	Application	References
Miswak (Salvadora persica L.) extract and titanium dioxide nanoparticles	Carboxymethyl cellulose bio-nanocomposites	Bio-nanocomposite film with antimicrobial against both Staphylococcus aureus and *Escherichia coli*	Ahmadi et al., 2019
Alcoholic extracts of red propolis	Starch, glycerol, and cellulose nanocrystals from licuri leaves	Active packaging films for cheese curds and butter packaging with water vapor permeability, antimicrobial and antioxidant efficacy	Costa et al., 2014
Three essential oils cinnamon, garlic, and clove	Polylactide and polyethylene glycol	Biobased and biodegradable film with antimicrobial properties against *Staphylococcus aureus* and *Campylobacter jejuni*	Ahmed et al., 2016
Cellulose nanocrystals and lignin nanoparticles	Poly(lactic acid)	Ternary nanocomposite films with antioxidant and antimicrobial activities against plant pathogens, i.e. *Xanthomonas axonopodis* pv. *vesicatoria* and *Xanthomonas arboricola* pv. *pruni*	Yang et al., 2016
SiO_2 nanoparticles	Agar/sodium alginate	Nanocomposite films with water resistance and thermal stability	Hou et al., 2019
Chitosan nanoparticles	Chitosan	Edible biobased and unplasticized film with antimicrobial activity against both Gram-positive and Gram-negative bacteria	Gomes et al., 2018

(made of nylon and polyethylene terephthalate) during a typical steeping process. The brewing temperature at 95°C released microplastics and nanoplastics of several orders of magnitude higher than that reported in other foods. Zhao et al. (2019) reported especially on the discovery of the foodborne nanoparticles generated during food processing, in this case in roasted pork. The released rate of fluorescent nanoparticles was selected to be monitored for this work at different temperatures. The tracking of released nanoparticle in tested mice showed that the accumulations were found in liver, kidney, and testis, as well as penetrating the blood-brain barrier to enter into the brain. This study showed that oral ingestion of packed roast pork in mice at a dose of 2 g/kg did not cause obvious toxicity, but the toxicity effects were clearly found in *Caenorhabditis elegans.*

The toxicity of nanoparticles was assessed in many studies, especially to internal organs of different organisms. Osman et al. (2019) conducted a clinical evaluation to assess the *in vivo* interaction between airborne nanoparticles and pulmonary tracts. Pavičić et al. (2019) evaluated the effect of silver nanoparticles on neuronal precursor cells *in vivo.* The toxicity of nanoparticles was determined based on cellular viability, apoptosis, induction, oxidative stress response, and, cellular and mitochondrial membrane damage, DNA damage, inflammation response, and neural stem cell regulation. The safe dose of nanoparticles to neuronal precursor cells was shown to be less than 5 mg Ag/L. Hu et al. (2019) studied the impact of using silicon dioxide nanoparticles (SiO_2 NPs), which are mostly used in food industrial sectors, on the endocrine in mice. Through observation of responses in several signaling pathways and endoplasmic reticulum, SiO_2 NPs induce resistance to insulin administration. This finding suggested the potential of SiO_2 NPs to cause negative effects to glucose level in blood over a long term. Another systematical study looked at the growing of lettuce in a cerium- (IV), copper- (II), and zinc oxide nanoparticles-contaminated environment (Li et al., 2019). The digested lettuce containing nanoparticle was applied to human intestine and human liver cells *in vitro.* The exposure to nanoparticle showed a detrimental effect on intestine cells, but not on liver cells. Additionally, different types of nanoparticle (comparing between ZnO, CuO, and CeO2) showed different levels of negative effects. This work suggested the possible pathway of contaminated nanoparticle in the environment that could be delivered to humans through food consumption. Interestingly, the benefit of using nanoparticle for treatment of illness was demonstrated as well. Hamza et al. (2019) applied the zinc oxide nanoparticles/green tea extract (ZnO NPs/GTE), by oral administration, to monosodium glutamate (MSG)-treated rats. This led to a reduction of superoxide dismutase, catalase, and glutathione peroxidase activities as well as of the levels of brain-derived neurotrophic factor (BDNF) and glutathione (GSH) in the cerebral cortex of rats. Co-treatment with ZnO NPs/GTE inhibited the negative effects induced by MSG suggesting the potential of nanoparticles against oxidative stress and neuronal necrosis. Regarding these scenarios however, it is necessary to conduct further research and studies on the effect of different physical and biological parameters, such as temperature, pressure, pH, size of nanoparticles, exposure time, and cross-reactivity of food to nanoparticles on the release rate kinetics of nanoparticles. In addition, the mode of translocation of nanoparticles and their derivatives, e.g., ionic form, in environment and in sensitive human organs and parts, e.g., stomach, placenta, breast, and brain, should be studied for potential side effects.

9.5 CONCLUSIONS

The integrations of nanotechnology into food packaging has been gaining a more important role in the development of food industries and is being extended to other industries, such as medical, pharmaceutical, and chemical sectors. Applications of nanoparticles in food packaging can benefit food producers, retailers, and consumers to achieve high-quality food products, to reduce product loss during transportation and distribution, and to assist the decision making of consumers. The primary function of nanomaterials known to the public is their broad-spectrum antimicrobial activities with no clear evidence that they affect human health. Therefore, they can be used for production of food packaging to prevent the growth of contaminated microorganisms in food products and to prolong food shelf life. The extended purposes of nanomaterial packaging are developed for active packaging, smart packaging, and biobased packaging to provide numerous choices for cost-effective, eco-friendly, degradable, and renewable packaging materials. It is important to note that the toxicity and ecotoxicity effects of nanomaterials in food packaging applications cannot be ignored. Therefore, during the introduction phase of nanomaterials in food industries, it is necessary to develop public awareness to understand the risk and safety protocol about the use of nanomaterials, to achieve the goal of delivering fresh and safe quality food to our world.

CONFLICT OF INTEREST

The authors declare that they have no conflict of interest.

REFERENCES

Ahmadi R, Tanomand A, Kazeminava F, Kamounah FS, Ayaseh A, Ganbarov K, Yousefi M, Katourani A, Yousefi B, Kafil HS. Fabrication and characterization of a titanium dioxide (TiO(2)) nanoparticles reinforced bio-nanocomposite containing Miswak (Salvadora persica L.) extract—The antimicrobial, thermo-physical and barrier properties. *Int J Nanomed* 2019;14:3439–3454.

Ahmed J, Hiremath N, Jacob H. Antimicrobial, rheological, and thermal properties of Plasticized Polylactide Films Incorporated with essential oils to inhibit Staphylococcus aureus and Campylobacter jejuni. *J Food Sci* 2016 Feb;81(2):E419–429.

Ahmed J, Mulla M, Jacob H, Luciano G, Bini TB, Almusallam A. Polylactide/poly (ε-caprolactone)/zinc oxide/clove essential oil composite antimicrobial films for scrambled egg packaging. *Food Packag Shelf Life* 2019;21:100355.

Alkan D, Yemenicioğlu A. Potential application of natural phenolic antimicrobials and edible film technology against bacterial plant pathogens. *Food Hydrocoll* 2016;55:1–10.

Alotaibi MA, Tayel AA, Zidan NS, El Rabey HA. Bioactive coatings from nano-biopolymers/plant extract composites for complete protection from mycotoxigenic fungi in dates. *J Sci Food Agric* 2019 Jul;99(9):4338–4343.

Amjadi S, Emaminia S, Heyat Davudian S, Pourmohammad S, Hamishehkar H, Roufegarinejad L. Preparation and characterization of gelatin-based nanocomposite containing chitosan nanofiber and ZnO nanoparticles. *Carbohydr Polym* 2019 Jul 15;216:376–384.

Anbukarasu P, Sauvageau D, Elias AL. Time-temperature indicator based on enzymatic degradation of dye-loaded polyhydroxybutyrate. *Biotechnol J* 2017 Sep;12(9):11.

Azad ZRAA, Ahmad MF, Siddiqui WA (2019). Food spoilage and food contamination. In: A. Malik, Erginkaya Z., Erten H. (eds) *Health and Safety Aspects of Food Processing Technologies*, Springer, Cham.

Bahrami A, Rezaei Mokarram R, Sowti Khiabani M, Ghanbarzadeh B, Salehi R. Physico-mechanical and antimicrobial properties of tragacanth/hydroxypropyl methylcellulose/beeswax edible films reinforced with silver nanoparticles. *Int J Biol Macromol* 2019 May 15;129:1103–1112.

Bridgeman D, Corral J, Quach A, Xian X, Forzani E. Colorimetric humidity sensor based on liquid composite materials for the monitoring of food and pharmaceuticals. *Langmuir* 2014 Sep 9;30(35):10785–10791.

Bumbudsanpharoke N, Choi J, Ko S. Applications of nanomaterials in food packaging. *J Nanosci Nanotechnol* 2015 Sep;15(9):6357–6372.

Campos D, Piccirillo C, Pullar RC, Castro PM, Pintado MM. Characterization and antimicrobial properties of food packaging methylcellulose films containing stem extract of Ginja cherry. *J Sci Food Agric* 2014 Aug;94(10):2097–2103.

Choi HY, Bang IH, Kang JH, Min SC. Development of a microbial decontamination system combining washing with highly activated calcium oxide solution and antimicrobial coating for improvement of mandarin storability. *J Food Sci* 2019 Aug;84(8):2190–2198.

Costa SS, Druzian JI, Machado BA, de Souza CO, Guimarães AG. Bi-functional biobased packing of the cassava starch, glycerol, licuri nanocellulose and red propolis. *PLoS One* 2014 Nov 10;9(11):e112554.

Cui H, Yuan L, Li W, Lin L. Antioxidant property of SiO2-eugenol liposome loaded nanofibrous membranes on beef. *Food Packag Shelf Life* 2017;11:49–57.

Diaz YJ, Page ZA, Knight AS, Treat NJ, Hemmer JR, Hawker CJ, Read de Alaniz J. A versatile and highly selective colorimetric sensor for the detection of amines. *Chemistry* 2017 Mar 13;23(15):3562–3566.

El Achaby M, El Miri N, Aboulkas A, Zahouily M, Bilal E, Barakat A, Solhy A. Processing and properties of eco-friendly bio-nanocomposite films filled with cellulose nanocrystals from sugarcane bagasse. *Int J Biol Macromol* 2017 Mar;96:340–352.

Ellis MJ (1994). The methodology of shelf life determination. In: C.M.D., Man, A.A. Jones (eds). *Shelf Life Evaluation of Foods*, Springer, Boston, MA.

Gammariello D, Conte A, Buonocore GG, Del Nobile MA. Bio-based nanocomposite coating to preserve quality of Fior di latte cheese. *J Dairy Sci* 2011 Nov;94(11):5298–5304.

Ghadiry M, Gholami M, Lai CK, Ahmad H, Chong WY. Ultra-sensitive humidity sensor based on optical properties of graphene oxide and nano-anatase TiO2. *PLoS One* 2016 Apr 21;11(4):e0153949.

Gomes LP, Souza HKS, Campiña JM, Andrade CT, Silva AF, Gonçalves MP, Paschoalin VMF. Edible chitosan films and their nanosized counterparts exhibit antimicrobial activity and enhanced mechanical and barrier properties. *Molecules* 2018 Dec 31;24(1): pii:E127.

Gram L, Ravn L, Rasch M, Bruhn JB, Christensen AB, Givskov M. Food spoilage—Interactions between food spoilage bacteria. *Int J Food Microbiol* 2002 Sep 15;78(1–2):79–97.

Hamza RZ, Al-Salmi FA, El-Shenawy NS. Evaluation of the effects of the green nanoparticles zinc oxide on monosodium glutamate-induced toxicity in the brain of rats. *PeerJ* 2019 Sep 23;7:e7460.

Heras-Mozos R, Muriel-Galet V, López-Carballo G, Catalá R, Hernández-Muñoz P, Gavara R. Development and optimization of antifungal packaging for sliced pan loaf based on garlic as active agent and bread aroma as aroma corrector. *Int J Food Microbiol* 2019 Feb 2;290:42–48.

Hernandez LM, Xu EG, Larsson HCE, Tahara R, Maisuria VB, Tufenkji N. Plastic teabags release billions of microparticles and nanoparticles into tea. *Environ Sci Technol* 2019 Nov 5;53(21):12300–12310.

Hosseini SF, Rezaei M, Zandi M, Farahmandghavi F. Development of bioactive fish gelatin/chitosan nanoparticles composite films with antimicrobial properties. *Food Chem* 2016 Mar 1;194:1266–1274.

Hou X, Xue Z, Xia Y, Qin Y, Zhang G, Liu H, Li K. Effect of SiO(2) nanoparticle on the physical and chemical properties of eco-friendly agar/sodium alginate nanocomposite film. *Int J Biol Macromol* 2019 Mar 15;125:1289–1298.

Hu H, Fan X, Guo Q, Wei X, Yang D, Zhang B, Liu J, Wu Q, Oh Y, Feng Y, Chen K, Hou L, Gu N. Silicon dioxide nanoparticles induce insulin resistance through endoplasmic reticulum stress and generation of reactive oxygen species. *Part Fibre Toxicol* 2019 Nov 7;16(1):41.

Huang Y, Mei L, Chen X, Wang Q. Recent developments in food packaging based on nanomaterials. *Nanomaterials (Basel)* 2018 Oct 13;8(10):5.

Huis in 't Veld JH. Microbial and biochemical spoilage of foods: An overview. *Int J Food Microbiol* 1996 Nov;33(1):1–18.

Ivask A, Kurvet I, Kasemets K, Blinova I, Aruoja V, Suppi S, Vija H, Käkinen A, Titma T, Heinlaan M, Visnapuu M, Koller D, Kisand V, Kahru A. Size-dependent toxicity of silver nanoparticles to bacteria, yeast, algae, crustaceans and mammalian cells in vitro. *PLoS One* 2014 Jul 21;9(7):e102108.

Indumathi MP, Rajarajeswari GR. Mahua oil-based polyurethane/chitosan/nano ZnO composite films for biodegradable food packaging applications. *Int J Biol Macromol* 2019 Mar 1;124:163–174.

Khezrian A, Shahbazi Y. Application of nanocomposite chitosan and carboxymethyl cellulose films containing natural preservative compounds in minced camel's meat. *Int J Biol Macromol* 2018 Jan;106:1146–1158.

Kowsalya E, MosaChristas K, Balashanmugam P, Rani JC, I JCR. Biocompatible silver nanoparticles/poly (vinyl alcohol) electrospun nanofibers for potential antimicrobial food packaging applications. *Food Packag Shelf Life* 2019;21:100379.

Li J, Song Y, Vogt RD, Liu Y, Luo J, Li T. Bioavailability and cytotoxicity of Cerium- (IV), Copper- (II), and Zinc oxide nanoparticles to human intestinal and liver cells through food. *Sci Total Environ* 2019 Nov 3;702:134700.

Lin L, Gu Y, Cui H. Moringa oil/chitosan nanoparticles embedded gelatin nanofibers for food packaging against Listeria monocytogenes and Staphylococcus aureus on cheese. *Food Packag Shelf Life* 2019;19:86–93.

Llana-Ruiz-Cabello M, Pichardo S, Bermudez JM, Baños A, Ariza JJ, Guillamón E, Aucejo S, Cameán AM. Characterisation and antimicrobial activity of active polypropylene films containing oregano essential oil and Allium extract to be used in packaging for meat products. *Food Addit Contam A* 2018 Apr;35(4):782–791.

Lomate GB, Dandi B, Mishra S. Development of antimicrobial LDPE/Cu nanocomposite food packaging film for extended shelf life of peda. *Food Packag Shelf Life* 2018;16:211–219.

López OV, Castillo LA, Garcia MA, Villar MA, Barbosa SE. Food packaging bags based on thermoplastic corn starch reinforced with talc nanoparticles. *Food Hydrocoll* 2015;43:18–24.

López-Carballo G, Muriel-Galet V, Hernández-Muñoz P, Gavara R. Chromatic sensor to determine oxygen presence for applications in intelligent packaging. *Sensors (Basel)* 2019 Oct 28;19(21):pii: E4684.

Ma Z, Chen P, Cheng W, Yan K, Pan L, Shi Y, Yu G. Highly sensitive, printable nanostructured conductive polymer wireless sensor for food spoilage detection. *Nano Lett* 2018 Jul 11;18(7):4570–4575.

Manzocco L. The acceptability limit in food shelf life studies. *Crit Rev Food Sci Nutr* 2016 Jul 26;56(10):1640–1646.

Mathew S, Snigdha S, Mathew J, Radhakrishnan EK. Biodegradable and active nanocomposite pouches reinforced with silver nanoparticles for improved packaging of chicken sausages. *Food Packag Shelf Life* 2019;19:155–166.

Metak AM, Nabhani F, Connolly SN. Migration of engineered nanoparticles from packaging into food products. *LWT Food Sci Technol* 2015;64(2):781–787.

Mihindukulasuriya SDF, Lim L-T. Nanotechnology development in food packaging: A review. *Trends Food Sci Technol* 2014;40(2):149–167.

Rahman ATM, Kim DH, Jang HD, Yang JH, Lee SJ. Preliminary study on biosensor-type time-temperature integrator for intelligent food packaging. *Sensors (Basel)* 2018 Jun 15;18(6):11.

Mills A, Hazafy D. Nanocrystalline SnO2-based, UVB-activated, colourimetric oxygen indicator. *Sens Actuators B* 2009;136(2):344–349.

Mohammadi H, Kamkar A, Misaghi A, Zunabovic-Pichler M, Fatehi S. Nanocomposite films with CMC, okra mucilage, and ZnO nanoparticles: Extending the shelf-life of chicken breast meat. *Food Packag Shelf Life* 2019;21:100330.

Mustafa F, Hassan R, Andreescu S. Multifunctional nanotechnology-enabled sensors for rapid capture and detection of pathogens. *Sensors (Basel, Switzerland)* 2017;17(9):2121. doi:10.3390/s17092121.

Oh J, Yang H, Jeong GE, Moon D, Kwon OS, Phyo S, Lee J, Song HS, Park TH, Jang J. Ultrasensitive, selective, and highly stable bioelectronic nose that detects the liquid and gaseous cadaverine. *Anal Chem* 2019 Oct 1;91(19):12181–12190.

Olafsdottir G, Lauzon HL, Martinsdottir E, Kristbergsson K. Influence of storage temperature on microbial spoilage characteristics of haddock fillets (Melanogrammus aeglefinus) evaluated by multivariate quality prediction. *Int J Food Microbiol* 2006 Sep 1;111(2):112–125.

Osman NM, Sexton DW, Saleem IY. Toxicological assessment of nanoparticle interactions with the pulmonary system. *Nanotoxicology* 2020;14:21–58.

Parra R, Magan N. Modelling the effect of temperature and water activity on growth of Aspergillus niger strains and applications for food spoilage moulds. *J Appl Microbiol* 2004;97(2):429–438.

Pavičić I, Milić M, Pongrac IM, Brkić Ahmed L, Matijević Glavan T, Ilić K, Zapletal E, Ćurlin M, Mitrečić D, Vinković Vrček I. Neurotoxicity of silver nanoparticles stabilized with different coating agents: In vitro response of neuronal precursor cells. *Food Chem Toxicol* 2019 Nov 3(136):110935.

Peter A, Mihaly-Cozmuta L, Mihaly-Cozmuta A, Nicula C, Ziemkowska W, Basiak D, Danciu V, Vulpoi A, Baia L, Falup A, Craciun G, Ciric A, Begea M, Kiss C, Vatuiu D. Changes in the microbiological and chemical characteristics of white bread during storage in paper packages modified with Ag/TiO2-SiO2, Ag/N-TiO2 or Au/TiO2. *Food Chem* 2016 Apr 15;197(A):790–798.

Rossi M, Passeri D, Sinibaldi A, Angjellari M, Tamburri E, Sorbo A, Carata E, Dini L. Nanotechnology for food packaging and food quality assessment. *Adv Food Nutr Res* 2017;82:149–204.

Sahraee S, Milani JM, Ghanbarzadeh B, Hamishehkar H. Physicochemical and antifungal properties of bio-nanocomposite film based on gelatin-chitin nanoparticles. *Int J Biol Macromol* 2017 Apr;97:373–381.

Schaude C, Meindl C, Fröhlich E, Attard J, Mohr GJ. Developing a sensor layer for the optical detection of amines during food spoilage. *Talanta* 2017 Aug 1;170:481–487.

Sharma C, Dhiman R, Rokana N, Panwar H. Nanotechnology: An untapped resource for food packaging. *Front Microbiol* 2017 Sep 12;8:1735.

Shu J, Qiu Z, Tang D. Self-Referenced Smartphone Imaging for Visual Screening of H(2) S Using Cu (x)O-Polypyrrole Conductive Aerogel Doped with Graphene Oxide Framework. *Anal Chem* 2018 Aug 21;90(16):9691–9694.

Singh T, Shukla S, Kumar P, Wahla V, Bajpai VK. Application of nanotechnology in food science: Perception and overview. *Front Microbiol* 2017 Aug 7;8:1501.

Lee Soo-Keun, Sheridan Martin, Mills A, Novel UV-activated colorimetric oxygen indicator. *Chem Mater* 2005;17(10):2744–2751.

Thiruvengadam M, Rajakumar G, Chung IM. Nanotechnology: Current uses and future applications in the food industry. *3 Biotech* 2018 Jan;8(1):74.

Tiwari DK, Jin T, Behari J. Dose-dependent in-vivo toxicity assessment of silver nanoparticle in Wistar rats. *Toxicol Mech Methods* 2011 Jan;21(1):13–24.

Tsagkaris AS, Tzegkas SG, Danezis GP. Nanomaterials in food packaging: State of the art and analysis. *J Food Sci Technol* 2018 Aug;55(8):2862–2870.

Vahedikia N, Garavand F, Tajeddin B, Cacciotti I, Jafari SM, Omidi T, Zahedi Z. Biodegradable zein film composites reinforced with chitosan nanoparticles and cinnamon essential oil: Physical, mechanical, structural and antimicrobial attributes. *Colloids Surf B Biointerfaces* 2019 May 1;177:25–32.

Vasile C. Polymeric nanocomposites and nanocoatings for food packaging: A review. *Materials (Basel)* 2018 Sep 26;11(10):1834.

Vasile C, Stoleru E, Darie-Niţă RN, Dumitriu RP, Pamfil D, Tarţău L. Biocompatible Materials Based on Plasticized poly(lactic acid), chitosan and Rosemary ethanolic Extract I. *Eff Chitosan Prop Plasticized Poly (Lactic Acid) Mater Polym (Basel)* 2019 May 30;11(6):pii:E941.

Wang H, Gong X, Miao Y, Guo X, Liu C, Fan YY, Zhang J, Niu B, Li W. Preparation and characterization of multilayer films composed of chitosan, sodium alginate and carboxymethyl chitosan-ZnO nanoparticles. *Food Chem* 2019 Jun 15;283:397–403.

Wang H, Li Y, Wang A, Slavik M. Rapid, sensitive, and simultaneous detection of three foodborne pathogens using magnetic nanobead-based immunoseparation and quantum dot-based multiplex immunoassay. *J Food Prot* 2011 Dec;74(12):2039–2047.

Wen P, Zhu D-H, Wu H, Zong M-H, Jing Y-R, Han S-Y. Encapsulation of cinnamon essential oil in electrospun nanofibrous film for active food packaging. *Food Control* 2016;59:366–376.

Wu S, Xu S, Chen X, Sun H, Hu M, Bai Z, Zhuang G, Zhuang X. Bacterial communities changes during food waste spoilage. *Sci Rep* 2018 May 29;8(1):8220.

Yang W, Fortunati E, Dominici F, Giovanale G, Mazzaglia A, Balestra GM, Kenny JM, Puglia D. Effect of cellulose and lignin on disintegration, antimicrobial and antioxidant properties of PLA active films. *Int J Biol Macromol* 2016 Aug;89:360–368.

Yang L, Li Y. Simultaneous detection of Escherichia coli O157:H7 and Salmonella Typhimurium using quantum dots as fluorescence labels. *Analyst* 2006 Mar;131(3):394–401.

Youn DY, Jung U, Naqi M, Choi SJ, Lee MG, Lee S, Park HJ, Kim ID, Kim S. Wireless real-time temperature monitoring of blood packages: Silver nanowire-embedded flexible temperature sensors. *ACS Appl Mater Interfaces* 2018 Dec 26;10(51):44678–44685.

Yousefi H, Ali MM, Su HM, Filipe CDM, Didar TF. Sentinel wraps: Real monitoring of food contamination by printing DNAzyme probes on food packaging. *ACS Nano*. 2018 Apr 24;12(4):3287–3294.

Zhai X, Li Z, Shi J, Huang X, Sun Z, Zhang D, Zou X, Sun Y, Zhang J, Holmes M, Gong Y, Povey M, Wang S. A colorimetric hydrogen sulfide sensor based on gellan gum-silver nanoparticles bionanocomposite for monitoring of meat spoilage in intelligent packaging. *Food Chem* 2019 Aug 30;290:135–143.

Zhang C, Yin AX, Jiang R, Rong J, Dong L, Zhao T, Sun LD, Wang J, Chen X, Yan CH. Time--Temperature indicator for perishable products based on kinetically programmable Ag overgrowth on Au nanorods. *ACS Nano* 2013 May 28;7(5):4561–4568.

Zhang W, Ma H, Yang SX. An inexpensive, stable, and accurate relative humidity measurement method for challenging environments. *Sensors (Basel)* 2016 Mar 18;16(3):8.

Zhao X, Shan S, Li J, Cao L, Lv J, Tan M. Assessment of potential toxicity of foodborne fluorescent nanoparticles from roasted pork. *Nanotoxicology* 2019 Dec;13(10):1310–1323.

Zhu R, Desroches M, Yoon B, Swager TM. Wireless oxygen sensors enabled by Fe(II)-polymer wrapped carbon nanotubes. *ACS Sens* 2017 Jul 28;2(7):1044–1050.

10 Optoelectronic and Electronic Packaging Materials and Their Properties

Theivasanthi Thirugnanasambandan and Karthikeyan Subramaniam

CONTENTS

10.1 INTRODUCTION

Packaging can protect food from the external environment and can communicate information about the food to consumers with labels. Intelligent food packaging needs the integration of many disciplines such as printed electronics, nanotechnology, silicon photonics, and biotechnology [1]. Intelligent food packaging can monitor the condition of food; whether the food is in good condition or spoiled. Parameters like presence of gases, humidity, temperature, and available microorganisms in the food environment are analyzed using sensors and indicators. Any change in these parameters can cause changes in the electrical properties like resistance and capacitance and can be measured with the respective instruments. Electrical circuits can be printed on polymers and papers and these labels are used to check the food condition. The use of this kind of intelligent food packaging technology can avoid the wastage of food in large amounts. Printed electronic labels can give a quantitative measure of the parameters. They are a good alternative to indication using colors because indication using colors may confuse the consumers. Even a small amount of color change will lead to wasting of food.

When foods are spoiled, the sensors on the packaging can detect gases produced by the microorganisms. This kind of "smart packaging" can give more accurate information than the expiry date available on the package. This can give a digital display to warn about spoiling of food, for example wine labels that can alert when white wine is at its optimum temperature. Active food packaging may release some components like antimicrobial agents to increase the shelf life of food. But intelligent packaging materials need to get released into the food.

Researchers are working to increase the shelf life of food with a technology called modified atmosphere packaging (MAP). In this method, an increased amount of carbon dioxide is maintained, which can reduce the respiration of microorganisms inside packaging. Introducing essential oils inside packaging can also fight against the bacteria that assist in the spoiling of food. Smart labels and tags are made using radio frequency identification (RFID) tags, sensors, and printed electronics to monitor food. This technology is more suitable for perishable food items such as meat, dairy, and fish products. Sensors integrated into food packages will be more helpful to consumers to check the freshness and quality of food stocks.

Sensors can be printed on smart radio-frequency labels to measure temperature, humidity, gases, and chemicals. They can be made through high-quality screen printing and lamination technologies on low-cost foils in combination with pick and place technology. For fish, trimethylamine is used as a marker in low concentrations in the order of parts per million which is to be accurately and reliably measured. Development of low-cost electronics and sensors leads to many solutions to the intelligent food packaging industry. Compared to silicon electronics, printed electronic devices can be produced on a large scale at low cost. Inkjet printing can allow for the rapid prototyping of electronic components, particularly sensors which can offer a direct printing feature, high spatial resolution, and compatibility with many substrates. Conducting polymer materials such as PEDOT:PSS- poly(3,4-e thylenedioxythiophene):polystyrene sulfonate and PANI (polyaniline) and other conducting materials like silver nanoparticles are compatible with inkjet printing technology.

E-nose consists of multiple sensors as an array to make a pattern employed for sensing. The development of sensor technology and material science can support the food packaging industry in many ways. Intelligent packaging technologies are available as data carriers, indicators, and sensors. Monitoring environmental conditions can be carried out using time temperature indicators, gas leakage indicators, and relative humidity sensors and can be placed outside or inside the packaging. Biosensors and freshness indicators measure the quality of the food and are placed inside the packaging. Indicators, sensors, and RFID tags are the available components in smart packaging.

The data like packaging date, batch number, packaging weight, nutritional information, or preparation instructions can be stored in one-dimensional (barcode) or two-dimensional (QR code) barcodes. Figure 10.1 shows the bar code and QR code which can be scanned to get the coded information. Also, it shows the RFID tagging system.

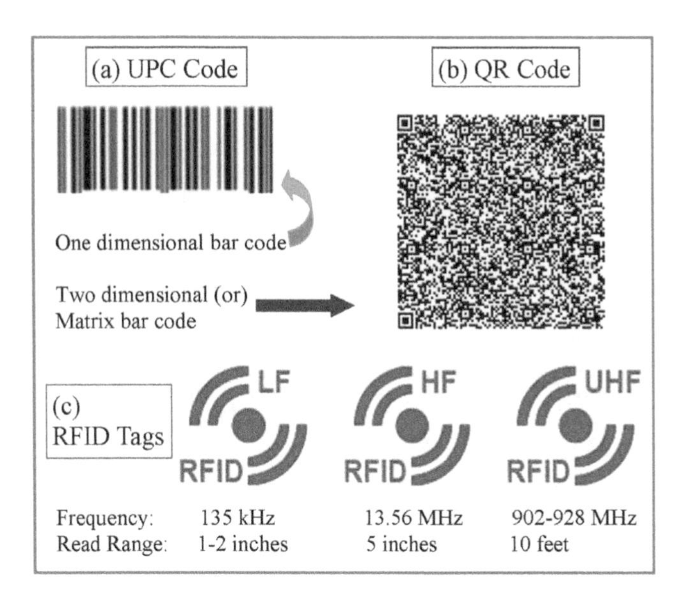

FIGURE 10.1 Coded information (optoelectronic and electronic) packaging. (a) UPC (Universal Product Code) is a one-dimensional barcode. (b) QR code (quick response code) is a two-dimensional or matrix-type barcode. (c) RFID (radio-frequency identification) tagging system. Low frequency (LF), high frequency (HF), and ultra high frequency (UHF) types of RFID tags are shown with their frequency and read range.

10.2 INDICATORS

Indicators change their color in accordance with the freshness, temperature, and spoilage of the food. Figure 10.2 displays the product Fresh-Check from Lifeline technologies, which is a time temperature indicator which works on the principle of color change due to polymerization reaction [2].

The product CoolVu is an indicator and has an etchant material which can give visual information such as "use"/"do not use." The company Flex Alert launched an RFID sensor for the detection of *Escherichia coli* (*E. coli*) and *Salmonella* in packaged foods. RipeSense® is a product used as an indicator for the ripening of fruits, based on the fact that ethylene is used as a ripening indicator which is released during ripening [3].

Because of the surface plasmon resonance property of gold nanoparticles, they are employed as efficient thermal history indicator. The color of the indicator becomes more intense with increased temperature. The color changes from gray to red at high temperature (40°C) [4].

10.3 SENSORS

Supermarkets are using bar codes that can store data about the product in the label. The cost of packaging is sometimes up to 30–40% of the cost of the products. Sensors can be printed on paper or polymers that can monitor the environment of the

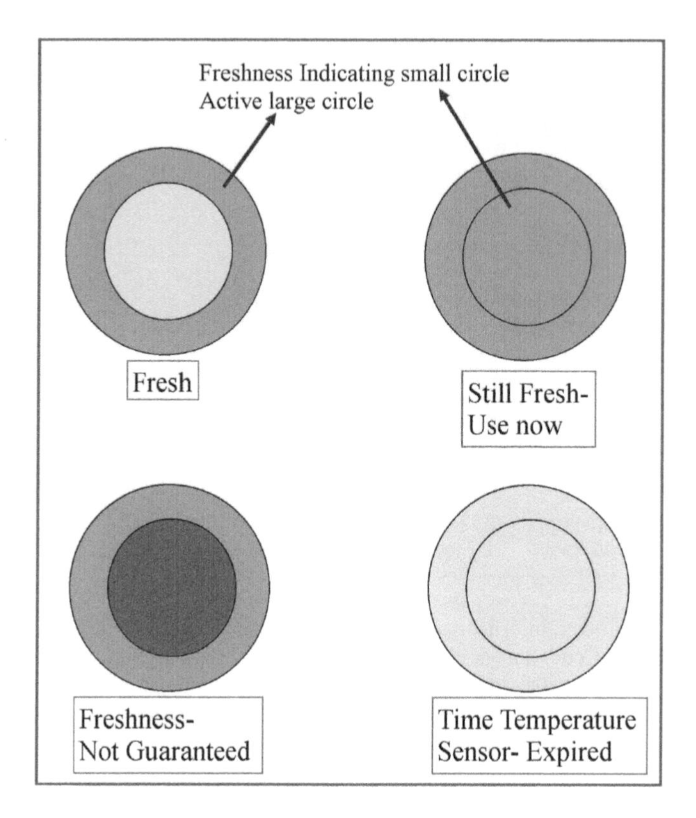

FIGURE 10.2 Freshness checking time temperature sensing indicators. The packed food product should be utilized before the color of the small circle is darker than the color of large circle. The light color of the large circle (like the light color of the small circle) indicates an expired sensor.

food [5]. The sensors can monitor the freshness of food which can reduce the wastage of food and can replace use-by dates that are not reliable.

The sensing ability of carbon nanotubes can be enhanced by chemically modifying them with metalloporphyrins, which contain a central metal atom bound to several nitrogen-containing rings with cobalt at its center. Metalloporphyrins can bind to nitrogen-containing compounds called amines, particularly biogenic amines such as putrescine and cadaverine produced by decaying meat. When the cobalt-mixed porphyrin bind to these amines, the electrical resistance of the carbon nanotube increases which can be measured easily. These kinds of sensors can be made to be portable and require less power for operation [6].

Paper-based electrical gas sensors have been explored to detect gases like ammonia and trimethylamine in meat and fish products. Smartphones can be used to read the sensor data. Paper-based electrical gas sensors are developed by printing carbon electrodes onto readily available cellulose paper and are combined with near field communication (NFC) tags. These sensors are able to pick-up trace amounts of gases quickly and more accurately than existing sensors. They can also function effectively even at 100% humidity where most sensors struggle above 90%. The sensors can work at room temperature and consume very low amounts of energy. More specifically, they are sensitive only to the gases involved in food spoilage whereas the other sensors can be triggered by non-spoilage gases [7].

A gas sensor made of carbon nanotubes has been fabricated to check the freshness of food in food packaging. The sensor can detect gases like ammonia, carbon dioxide, and nitrogen oxide. The amount of carbon dioxide inside the packaging can provide the shelf life of meat. The electrical resistivity of the films is changed by the adsorption of gas molecules which is in turn measured [8].

10.4 MATERIALS WITH HIGH SENSITIVITY FOR THE DETECTION OF GASES AND MICROORGANISMS

A composite of Na_2CO_3 and $NaNO_2$ is reported to measure the concentration of CO_2. The interdigitated electrodes of a field-effect-transistor (FET)-type gas sensor are formed by coating a layer of (3-aminopropyl) triethoxysilane (APTES), and then the sensing material is printed by inkjet printing method [9].

Thin films are prepared by depositing highly sensitive nanomaterials on substrates and then sintering up to high temperatures. These thin films are used to detect various gases when adsorbed on the surface. Palladium-sensitized La_2O_3 thin films sintered at a temperature of 523 K with maximum sensitivity of 64% can measure CO_2 gas at a concentration of ppm level [10].

A thickness of 40 nm thin film made of nanostructured ZnO is deposited on glass substrates using a direct current (DC) reactive magnetron sputtering technique. A high sensitivity for the detection of CO_2 is achieved by annealing the substrate at 450°C [11]. Silver-doped $BaTiO_3$-CuO composite is fabricated for making carbon dioxide gas sensors. The composite shows a high electrical conductivity and a high sensitivity toward CO_2 [12]. $LaFeO_3$ nanocrystalline powders are prepared by sol–gel method and annealed at 800°C for 2 h. The sensing response of the nanomaterial to CO_2 gas is confirmed when the resistance of the $LaFeO_3$ sensor increases with increasing concentration of CO_2 [13]. Conducting polymers like polythiophene, polyaniline, and polypyrrole are identified as suitable materials for sensing gases and biological molecules.

A composite film made of PEDOT and BPEI branched polyethylenimine (BPEI) is prepared by in situ polymerization to measure CO_2 gas. The sensor performance is highly appreciated due to the synergistic effect of BPEI with PEDOT rather than of PEDOT alone [14]. Conducting polymers (PEDOT and PSS) and carbon material (grapheme) are inkjet printed on interdigited electrodes printed onto a polyethylene terephthalate (PET) substrate. A high sensitivity of CO_2 gas is achieved because of the high electrical conductivity of graphene and more adsorption of gas molecules [15].

A resistive sensor made of polypyrrole and ZnO nanosheets is used to detect NH_3 gas with sensing range (0.5–200 ppm), good repeatability, and selectivity [16]. ZnO nanorods are grown on substrates over which polyaniline is sprayed. The addition of ZnO nanorods enhances the sensitivity of polyaniline toward ammonia gas [17]. NH_3 gas sensing performance at room temperature has been studied with SnO_2/polyaniline composite in a polyethylene terephthalate substrate [18, 19]. Trace level detection of ammonia (NH_3) gas is possible with carboxylated multiwalled carbon nanotubes and polyaniline composite [20].

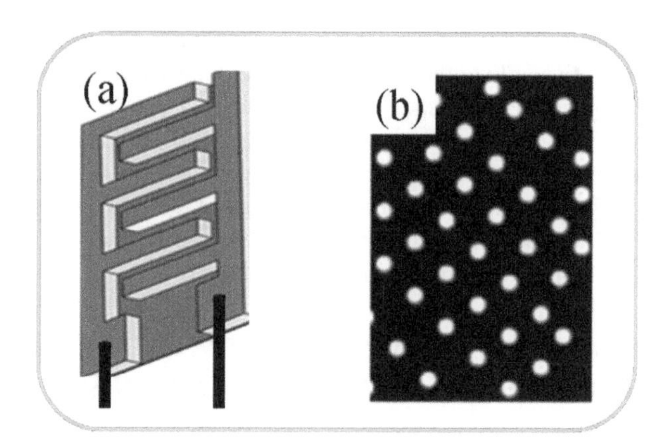

FIGURE 10.3 Schematic diagrams of second generation sensors. These types of sensors are fabricated by screen printing of nanoparticles on flexible polymer substrate. (a) Lateral capacitor. (b) Perforated flat plate capacitor.

A composite of silver and poly(m-aminobenzene sulfonic acid) functionalized single-walled carbon nanotubes (SWNT-PABS) can be inkjet printed on paper to make an ammonia gas sensor. Silver is first printed to prepare the electrodes and then SWNT-PABS dispersion is then printed on silver electrode. This paper-based flexible sensor can be made to be simple and cost effective [21].

Capacitive-based humidity sensors are developed by making inter-digitated capacitors (IDC) printed with silver nanoparticle-based ink on a flexible polyethylene terephthalate substrate using a gravure printing process. Figure 10.3 shows the second generation lateral and perforated flat plate capacitors. These sensors can also be prepared by inkjet printing silver on polyimide film. Flat plate capacitors can be fabricated by screen printing silver ink on both sides of the substrates like paper and cardboard [22].

Optical on-pack sensors can be used to avoid destructive sampling of packages. To monitor the modified atmosphere packaging, oxygen sensing is performed by measuring the degree of quenching of a fluorescent ruthenium complex entrapped in a sol–gel matrix [23].

Biosensors can also be included in food packaging. Biosensors include the sensing material and some bio-recognition elements or probes (like antibodies, enzymes and aptamers or nucleic acid sequences). They can measure any biological molecule. The transducers may be optical, electrochemical, and electrical as in highly sensitive FET transistors. Even simple paper-based colorimetric sensors are available for glucose sensing. These sensors are working in such a way that the intensity of the color increases on the cellulose paper as the concentration of the glucose increases. Food Sectinel System® from SIRA Technologies Inc. is a biosensor that can detect food pathogens. Specific antibodies are attached with the barcode acting as bio-recognition element. If the food is contaminated with more pathogens, the bar code becomes dark and the information in the barcode becomes unreadable. Another product, ToxinGuard® from ToxinAlert, Canada, is able to detect pathogens such as *Salmonella sp.*, *Campylobacter sp.*, *E coli.*, and *Listeria sp.* This product is made of polyethylene plastic coated with antibodies [24]. Titanium dioxide nanotubes are well known for their gas-sensing applications. For preparing the indicator solution, the TiO_2 is combined with glycerol, methylene blue, and hydroxyethyl cellulose which is then printed on polyethylene terephthalate film by screen printing [25].

Due to aging of food, toxins and volatile organic compounds are produced that can be analyzed with chemical sensors. Metal oxide nanomaterials are researched more for making chemical sensors. The biodegradation of fish and meat releases trimethylamine gas. So, the detection of trimethylamine can give information about the freshness of food. Metal oxide nanomaterials like TiO_2, WO_3, MoO_3, and ZnO nanoparticles with different morphology are reported for the detection of trimethylamine. Gold nanoparticles have very good catalytic activity. So, the metal oxide nanomaterials can be functionalized with Au NPs. ZnO nanowires are suitable for the detection of dimethyamine and ammonia. Hydrogen sulfide gas produced from eggs and egg products during heating can also be detected using metal oxide nanomaterials. Metal oxide nanomaterials like TiO_2, NiO, and CuO are the materials reported to be usable for H_2S detection [26].

10.5 RFID TAGS

RFID tags use radio waves to transmit data and can track objects. Radio frequency identification tags are attached with the label. They transmit the information to a mobile phone. A tag can be read from several feet away and does not need to be within direct line of sight of the reader. These tags do not require a battery and will emit a weak electrical signal when the gas inside the package changes. Sensors attached with RFID tags can provide remote and non-destructive monitoring.

RFID sensors require lower installation costs without the need for extensive wiring. RFID sensor antenna generates the electric field which is affected by the ambient environment. The equivalent circuit of the developed sensors forms an inductor-capacitor-resistor (LCR) circuit [27]. A coating of gelatin and sodium caseinate for a thickness of 38 μm and 125 μm is required and can be used as temperature sensor in RFID tags [28]. As the food decomposes, the gelatin breaks down and the microorganisms will release carbon dioxide, ethanol, and aromatic molecules. This is a plant-based sensor that can detect the freshness of the food [29].

Each product can be labeled individually using RFID tags and it is possible to track the single product rather than the shipping container level [30]. RFID tags have been developed with thin film transistors which are made of carbon nanotubes with increased charge-carrier mobility and decreased on/off ratio [31]. The labels are prepared by screen printing and lamination technologies on low-cost foils. Current RFID tags have silicon chips containing memory that can be wirelessly read out to uniquely identify products [32].

A near field communication tag is a label with small microchips that can be read by mobile phones. Information is stored in these microchips. Gases like ammonia, hydrogen peroxide, and cyclohexanone can be detected using a simple sensor that can be read by a smartphone. This is possible with some components known as chemiresistors that can change their resistance when exposed to chemicals. The sensors are made from modified near field communication tags. As sensing materials, carbon nanotubes are coated over the tags. Coating over the tags is done by mechanical drawing using pencil [33].

Electrical components like resistors, inductors, and capacitors can be printed by 3D printers and can be used as wireless electrical sensing systems. A wireless "smart cap" embedded with sensors in a milk carton has been designed to detect the signs of spoilage. A system is built using polymers and wax. Selective removal of wax will create hollow tubes which are filled with liquid metal, usually silver, and then cured. If the milk contains more bacteria, then there is a change in the electrical properties of the smart cap which can be checked with a smart phone [34].

Microsensors created by encapsulating a superfine, tightly wound electrical filament made of magnesium, silicon dioxide, and nitride in a compostable polymer are used for the detection of poisonous gases. The polymer is produced from corn and potato starch [35]. The development of RFID sensor tags related to smart food packaging have been reported in many articles. Smits et al. have explained a method for the preparation of RFID sensor tags. As per this report, silver paste with low resistance and good mechanical stability is used to make the circuit board. The antenna is screen printed on low-cost poly ethylene naphthalate (PEN) foils. Interconnections are made by stencil printing conductive adhesives. Microprocessors (MSP430G) are used for logical purposes. The sensor resistances are converted to a voltage suitable for readout by a voltage divider circuit with a 10-M Ω counter resistor. The data is then given to a digital link to a radio chip located on the same foil [36]. A RFID tag embedded with pH sensor has been reported to monitor fish products. RFID tags with sensors are used to measure temperature, humidity, milk freshness, gases, and the presence of volatile amine compounds [37]. Paper can be a valuable substrate for making RFID tags. Commercial paper samples have dielectric constant values between 3 and 3.20 and loss-tangent values between 0.02 and 0.05 for frequencies up to 7 GHz [38].

10.6 PRINTED ELECTRONICS

Printed electronics can provide lightweight, bendable, and foldable sensors to the food packaging industry. Fluids with conducting nanoparticles like silver nanoparticles and grapheme and sensing nanomaterials can be printed on polymers like polyimide, polyethylene terephthalate, and polyethylene naphthalate. This new technology is often referred to as printed electronics. With the help of this new technology, people will be able to read out the results from a paper- or polymer-based sensor very much similar to the barcodes as labels on the product. Polymer polymethyl methacrylate (PMMA) is used as a substrate in printed electronics. It has an advantage, i.e. enabling of transistors and integrated circuits to operate at low voltages and high speeds [39].

Printing the semiconductor channels of transistors at low temperatures and in open air is a requirement for low-cost manufacturing in printed electronics as an alternative to silicon technology. In this technology, common liquid-deposition techniques like spin coating, screen or stencil printing, offset or inkjet printing are used. Conducting polymers like polythiophene semiconductor can be used because they not only possess air stability but also exhibit excellent self-assembly behavior. They are processed into ordered semiconductor nanoparticles to make a nanoparticle ink.

The ink is used to produce organic transistor channel layers under ambient conditions by inkjet printing [7].

Circuits can be printed on plastics, films, and textiles by using conductive inks. If the ink is available at low cost, that can lead to more opportunities for flexible electronics. Materials with more electrical conductivity like silver nanoparticles and graphene are mostly employed for these kinds of applications. The melting temperature of silver ink should be lower than the melting temperature of the plastic substrate. A new silver ink that can melt at 140°C has been identified because the melting temperature of plastic is at 150°C. With this feature, low-cost radio frequency identification tags can be constructed [40].

A low-cost plastic sensor circuit has been developed which consists of an analog–digital converter (ADC). A sensor circuit consists of the sensor, an amplifier, an ADC to digitize the signal, and a radio transmitter [41]. Two-dimensional (2D) materials like graphene can be used to produce water-based inks for printed electronics that can be printed by simple techniques such as inkjet printing. These inks are found to be biocompatible and can be used for printing on consumer goods [42]. Graphene inks can be printed by roll-to-roll printing process like that used to print newspapers for making printed electronics, intelligent packaging, and disposable sensors. Graphene possesses high flexibility, good optical transparency, and a very high electrical conductivity, which makes it suitable for printed electronics. Graphene particles are suspended in a solvent and then added to conductive water-based ink formulations. The ingredients can be adjusted to control the liquid's properties to reduce the resistance. This graphene-based ink formulation may be cheaper when compared to the silver ink formulation [43].

Graphene ink can be used to print circuits and other electronic components. A radio frequency antenna is printed using graphene ink. This antenna is found to be flexible and environmentally friendly and mass-production is possible at low cost. Graphene ink is prepared by dispersing graphene flakes and a binder ethyl cellulose in a solvent. The ink possesses high electrical conductivity with the binder when the binder is broken down during annealing. After printing on the substrate, the ink is compressed into a graphene laminate which enhances its electrical conductivity. Graphene-based RFID tags can be produced at low cost when compared to RFID tags made from metals like aluminum and copper [44].

An example of the Internet of Things (IoT) coming into effect is when milk cartons can send messages to mobile phones about when the milk will be out of date. Conducting, semiconducting, and insulating 2D nanomaterials (graphene nanosheets, tungsten diselenide, and boron nitride) are combined together to print transistors. Liquid processing is usually performed to prepare high-quality 2D materials which are then used to process into inks [45]. The ink for printed electronics can be formulated by using solvent, nanoparticles, binders, and additives. In this formulation, properties such as surface tension, viscosity, and particle size of nanoparticles are to be considered. After printing on the substrate, treatment like sintering of the ink is required. Sintering can be performed by exposure to heat or light.

Conductive inks are used to print transistors and circuits by screen printing and gravure printing. Sensors can be printed on the package. The key component in a printed sensor is an analog-to-digital converter. This circuit converts the sensor output in digital form and is transmitted to near field communication (NFC) found in the smartphone. The phone will read out the information on conservation temperature from a sensor embedded in the packaging around meat, thus making possible a more precise calculation of the meat's expiration date [46].

The conductive inks for printed electronics are carbon-based materials (graphite and carbon nanotubes) and metal nanoparticles (like silver and gold). Cellulose nanocrystals can be used in the conductive inks to create transparent films and can bring rheological properties and stability. Cellulose can be used for both the substrates and inks. Cellulose fibrils can provide desired mechanical properties, control porosity, and enable recycling for substrates.

The sintering process is more important in printed electronics. It removes excess solvent and increases the bond between the deposited material and substrate. The conductivity of the conductive ink is also varied based on the curing temperature and the sintering time. Sintering process temperature is usually between 150°C and 300°C. The sintered inks possess high electrical conductivity [47].

10.7 CONCLUSION

The next generation printed electronics can provide support for wearable sensors, printable solar cells, and food packaging technology. In the future, these things can be mass-produced like copying in xerox shops. The technology can be commercialized into small-scale industries where these devices can be fabricated by people who do not have any technical knowledge. A high sensitivity and high electrical conductivity can be achieved with advanced nanomaterials. Flexible substrates can be formed easily. Also, they are foldable. The role of printed electronics in making intelligent food packaging has been clearly elucidated in this chapter. The usage of RFID tags in labeling food products with advanced technology has been introduced. The use of nanomaterials as sensors and in the preparation of conductive ink has been described in detail. The possibility of using paper-based and polymer-based substrates in electronic labels has been well analyzed.

ACKNOWLEDGMENT

The authors acknowledge the assistance of International Research Center, Kalasalingam Academy of Research and Education (Deemed University), Krishnankoil—626 126, (India) for providing necessary support and facilities.

REFERENCES

1. Vanderroost, Mike, Peter Ragaert, Frank Devlieghere, and Bruno De Meulenaer, 2014. Intelligent food packaging: The next generation. *Trends in Food Science and Technology*, 39(1), 47–62.
2. Müller, Patricia, and Markus Schmid, 2019. Intelligent packaging in the food sector: A brief overview. *Foods*, 8(1), 16.
3. Mustafa, Fatima, and Silvana Andreescu, 2018. Chemical and biological sensors for food-quality monitoring and smart packaging. *Foods*, 7(10), 168.

4. Wang, Yi-Cheng, Lin Lu, and Sundaram Gunasekaran, 2017. Biopolymer/gold nanoparticles composite plasmonic thermal history indicator to monitor quality and safety of perishable bioproducts. *Biosensors and Bioelectronics*, 92, 109–116.
5. Forbes Media, L. L. C., 2015. The smart labels that will power the internet of things. https://www.forbes.com/sites/mikekavis/2015/02/17/the-smart-labels-that-will-power-the-internet-of-things/#226d7d362ba6.
6. Liu, Sophie F., Alexander R. Petty, Graham T. Sazama, and Timothy M. Swager, 2015. Single-walled carbon nanotube/metalloporphyrin composites for the chemiresistive detection of amines and meat spoilage. *Angewandte Chemie International Edition*, 54(22), 6554–6557. DOI: 10.1002/anie.201501434.
7. Barandun, Giandrin, Matteo Soprani, Sina Naficy, Max Grell, Michael Kasimatis, Chiu Kwan Lun, Andrea Ponzoni, and Firat Güder, 2019. Cellulose fibers enable near zero-cost electrical sensing of water-soluble gases. *ACS Sensors*, 4(6), 1662–1669. DOI: 10.1021/acssensors.9b00555.
8. Abdellah, Alaa, Ahmed Abdelhalim, Markus Horn, Giuseppe Scarpa, and Paolo Lugli, 2013. Scalable spray deposition process for high performance carbon nanotube gas sensors. *IEEE Transactions on Nanotechnology*, 12(2), 174–181. DOI: 10.1109/TNANO.2013.2238248.
9. Wu, Meile, Jongmin Shin, Yoonki Hong, Dongkyu Jang, Xiaoshi Jin, Hyuck-In Kwon, and Jong-Ho Lee, 2018. An FET-type gas sensor with a sodium ion conducting solid electrolyte for CO2 detection. *Sensors and Actuators B: Chemical*, 259, 1058–1065.
10. Yadav, A. A., A. C. Lokhande, J. H. Kim, and C. D. Lokhande, 2017. Enhanced sensitivity and selectivity of CO2 gas sensor based on modified La2O3 nanorods. *Journal of Alloys and Compounds*, 723, 880–886.
11. Kannan, Padmanathan Karthick, Ramiah Saraswathi, and John Bosco Balaguru Rayappan, 2014. CO2 gas sensing properties of DC reactive magnetron sputtered ZnO thin film. *Ceramics International*, 40(8), 13115–13122.
12. El-Sayet, A. M., F. M. Ismail, and S. M. Yakout, 2011. Electrical conductivity and sensitive characteristics of Ag-added BaTiO3-CuO mixed oxide for CO2 gas sensing. *Journal of Materials Science and Technology*, 27(1), 35–40.
13. Wang, Xiaofeng, Hongwei Qin, Lihui Sun, and Jifan Hu, 2013. CO2 sensing properties and mechanism of nanocrystalline LaFeO3 sensor. *Sensors and Actuators B: Chemical*, 188, 965–971.
14. Chiang, Chi-Ju, Kang-Ting Tsai, Yi-Huan Lee, Hung-Wei Lin, Yi-Lung Yang, Chien-Chung Shih, Chia-Yu Lin, H. Jeng, Y. Weng, Y. Cheng, K. Ho, and C. Dai, 2013. In situ fabrication of conducting polymer composite film as a chemical resistive CO2 gas sensor. *Microelectronic Engineering*, 111, 409–415.
15. Andò, B., S. Baglio, G. Di Pasquale, A. Pollicino, S. D'Agata, C. Gugliuzzo, C. Lombardo, and G. Re, 2015. An inkjet printed CO2 gas sensor. *Procedia Engineering*, 120, 628–631.
16. Li, Yang, Mingfei Jiao, and Mujie Yang, 2017. In-situ grown nanostructured ZnO via a green approach and gas sensing properties of polypyrrole/ZnO nanohybrids. *Sensors and Actuators B: Chemical*, 238, 596–604.
17. Zhu, Guotao, Qiuping Zhang, Guangzhong Xie, Yuanjie Su, Kang Zhao, Hongfei Du, and Yadong Jiang, 2016. Gas sensors based on polyaniline/zinc oxide hybrid film for ammonia detection at room temperature. *Chemical Physics Letters*, 665, 147–152.
18. Bai, Shouli, Yanli Tian, Meng Cui, Jianhua Sun, Ye Tian, Ruixian Luo, Aifan Chen, and Dianqing Li, 2016. Polyaniline@ SnO2 heterojunction loading on flexible PET thin film for detection of NH3 at room temperature. *Sensors and Actuators B: Chemical*, 226, 540–547.
19. Bera, Susanta, Susmita Kundu, Hasmat Khan, and Sunirmal Jana, 2018. "Polyaniline coated graphene hybridized SnO2 nanocomposite: Low temperature solution synthesis, structural property and room temperature ammonia gas sensing. *Journal of Alloys and Compounds*, 744, 260–270.

20. Abdulla, Sukhananazerin, Thalakkotur Lazar Mathew, and Biji Pullithadathil, 2015. Highly sensitive, room temperature gas sensor based on polyaniline-multiwalled carbon nanotubes (PANI/MWCNTs) nanocomposite for trace-level ammonia detection. *Sensors and Actuators B: Chemical*, 221, 1523–1534.
21. Huang, Lianghui, Peng Jiang, Dong Wang, Yuanfang Luo, Mufang Li, Hoseon Lee, and Rosario A. Gerhardt, 2014. A novel paper-based flexible ammonia gas sensor via silver and SWNT-PABS inkjet printing. *Sensors and Actuators B: Chemical*, 197, 308–313.
22. Yousefi, Hanie, Hsuan-Ming Su, Sara M. Imani, Kais Alkhaldi, Carlos D. M. Filipe, and Tohid F. Didar, 2019. Intelligent food packaging: A review of smart sensing technologies for monitoring food quality. *ACS Sensors*, 4(4), 808–821.
23. McEvoy, Aisling K., Christoph Von Bueltzingsloewen, Colette M. McDonagh, Brian D. MacCraith, Ingo Klimant, and Otto S. Wolfbeis, 2003. Optical sensors for application in intelligent food-packaging technology. *Opto-Ireland 2002: Optics and Photonics Technologies and Applications*, 4876, 806–815.
24. Biji, K. B., C. N. Ravishankar, C. O. Mohan, and T. K. Srinivasa Gopal, 2015. Smart packaging systems for food applications: A review. *Journal of Food Science and Technology*, 52(10), 6125–6135.
25. Wen, Junwei, Shuting Huang, Yu Sun, Zhengjie Chen, Yixiang Wang, Houbin Li, and Xinghai Liu, 2018. Titanium dioxide nanotube-based oxygen indicator for modified atmosphere packaging: Efficiency and accuracy. *Materials*, 11(12), 2410.
26. Galstyan, Vardan, Manohar P. Bhandari, Veronica Sberveglieri, Giorgio Sberveglieri, and Elisabetta Comini, 2018. Metal oxide nanostructures in food applications: Quality control and packaging. *Chemosensors*, 6(2), 16.
27. Potyrailo, Radislav A., Nandini Nagraj, Zhexiong Tang, Frank J. Mondello, Cheryl Surman, and William Morris, 2012. Battery-free radio frequency identification (RFID) sensors for food quality and safety. *Journal of Agricultural and Food Chemistry*, 60(35), 8535–8543.
28. Silva, Fernando Teixeira, 2017. Intelligent packaging: Feasibility of using a biosensor coupled to a UHF RFID tag for temperature monitoring. *Food Engineering*. Université Montpellier; Universidade Federal de Rio de Janeiro. Escola de Química. https://tel.archives-ouvertes.fr/tel-01815980/document.
29. Phys.org, 2018. To reduce food waste, scientists are making labels that track produce as it spoils. https://phys.org/news/2018-11-food-scientists-track.html.
30. Forbes Media LLC, 2015. The smart labels that will power the internet of things. https://www.forbes.com/sites/mikekavis/2015/02/17/the-smart-labels-that-will-power-the-internet-of-things/#a23e8f42ba61.
31. Sun, Dong-ming, Marina Y. Timmermans, Ying Tian, Albert G. Nasibulin, Esko I. Kauppinen, Shigeru Kishimoto, Takashi Mizutani, and Yutaka Ohno, 2011. Flexible high-performance carbon nanotube integrated circuits. *Nature Nanotechnology*, 6(3), 156–161. DOI: 10.1038/NNANO.2011.1.
32. Smits, Edsger, Jeroen Schram, Matthijs Nagelkerke, Roel Kusters, Gert van Heck, Victor van Acht, Marc Koetse, Jeroen van den Brand, and G. Gerlinck, 2012. Development of printed RFID sensor tags for smart food packaging. In: *Proceedings of the 14th International Meeting on Chemical Sensors*, Nuremberg, Germany, 20–23.
33. Azzarelli, Joseph M., Katherine A. Mirica, Jens B. Ravnsbæk, and Timothy M. Swager, 2014. Wireless gas detection with a smartphone via rf communication. *Proceedings of the National Academy of Sciences of the United States of America*, 111(51), 18162–18166.
34. Phys.org, 2015. 3D-Printed smart cap uses electronics to sense spoiled food. https://phys.org/news/2015-07-3d-printed-smart-cap-electronics-food.html.
35. Printed Electronics World, 2017. *Biodegradable Microsensors for Food Monitoring*. https://www.printedelectronicsworld.com/articles/12877/biodegradable-microsensors-for-food-monitoring.

36. Smits, Edsger, Jeroen Schram, Matthijs Nagelkerke, Roel Kusters, Gert van Heck, Victor van Acht, Marc Koetse, Jeroen van den Brand, and G. Gerlinck, 2012. Development of printed RFID sensor tags for smart food packaging. In: *Proceedings of the 14th International Meeting on Chemical Sensors*, Nuremberg, Germany, 20–23.
37. Kuswandi, B., 2017. Freshness sensors for food packaging. *Reference Module in Food Science*. DOI: 10.1016/b978-0-08-100596-5.21876-3.
38. Lakafosis, Vasileios, Amin Rida, Rushi Vyas, Li Yang, Symeon Nikolaou, and Manos M. Tentzeris, 2010. Progress towards the first wireless sensor networks consisting of inkjet-printed, paper-based RFID-enabled sensor tags. *Proceedings of the IEEE*, 98(9), 1601–1609.
39. Dong-Ming Sun, W., Marina Y. Timmermans, Antti Kaskela, Albert G. Nasibulin, Shigeru Kishimoto, Takashi Mizutani, Esko I. Kauppine, and Yutaka Ohno, 2013. Mouldable all-carbon integrated circuits. *Nature Communications*, 4(1). DOI: 10.1038/ncomms3302.
40. Phys.org, 2009. Xerox develops silver ink for cheap printable electronics. https://phys.org/news/2009-10-xerox-silver-ink-cheap-printable.html.
41. Phys.org, 2013. Invention opens the way to packaging that monitors food freshness. https://phys.org/news/2013-02-packaging-food-freshness.html.
42. McManus, Daryl, Sandra Vranic, Freddie Withers, Veronica Sanchez-Romaguera, Massimo Macucci, Huafeng Yang, Roberto Sorrentino, Khaled Parvez, Seok-Kyun Son, Giuseppe Iannaccone, Kostas Kostarelos, Gianluca Fiori, and Cinzia Casiraghi, 2017. Water-based and biocompatible 2D crystal inks for all-inkjet-printed heterostructures. *Nature Nanotechnology*, 12(4), 343–350. DOI: 10.1038/nnano.2016.281.
43. Phys.org, 2015. New graphene-based inks for high-speed manufacturing of printed electronics. https://phys.org/news/2015-10-graphene-based-inks-high-speed-electronics.html.
44. Huang, Xianjun, Ting Leng, Xiao Zhang, Jia Cing Chen, Kuo Hsin Chang, Andre K. Geim, Kostya S. Novoselov, and Zhirun Hu, 2015. Binder-free highly conductive graphene laminate for low cost printed radio frequency applications. *Applied Physics Letters*, 106(20). DOI: 10.1063/1.4919935.
45. Kelly, Adam G., Toby Hallam Claudia Backes, Andrew Harvey, Amir Sajad Esmaeily, Ian Godwin, João Coelho Valeria Nicolosi, Jannika Lauth, Aditya Kulkarni, Sachin Kinge, Laurens D. A. Siebbeles, Georg S. Duesberg, Jonathan N. Coleman, G. S. Duesberg, and J. N. Coleman, 2017. All-printed thin-film transistors from networks of liquid-exfoliated nanosheets. *Science*, 356(6333), 69–73.
46. SPIE, The International Society for Optics and Photonics, 2015. Printed electronics for food packaging sensors. https://spie.org/news/6173-printed-electronics-for-food-packaging-sensors?SSO=1.
47. Liao, Yu, Rui Zhang, and Jun Qian, 2019. Printed electronics based on inorganic conductive nanomaterials and their applications in intelligent food packaging. *RSC Advances*, 9(50), 29154–29172.
48. Huang, Xianjun, Ting Leng, Thanasis Georgiou, Abraham Jijo, Rahul Raveendran Nair, Kostya S. Novoselov, and Zhirun Hu, 2018. Graphene oxide dielectric permittivity at GHz and its applications for wireless humidity sensing. *Scientific Reports*, 8(1), 43.
49. Mo, Lixin, Zhenxin Guo, Zhenguo Wang, Li Yang, Yi Fang, Zhiqing Xin, Xiu Lim, Y. Chen, M. Cao, Q. Zhang, and L. Li, 2019. Nano-silver ink of high conductivity and low sintering temperature for paper electronics. *Nanoscale Research Letters*, 14(1), 197.
50. Bülbül, Gonca, Akhtar Hayat, and Silvana Andreescu, 2015. Portable nanoparticle-based sensors for food safety assessment. *Sensors*, 15(12), 30736–30758.
51. Wen, Junwei, Shuting Huang, Yu Sun, Zhengjie Chen, Yixiang Wang, Houbin Li, and Xinghai Liu, 2018. Titanium dioxide nanotube-based oxygen indicator for modified atmosphere packaging: Efficiency and accuracy. *Materials*, 11(12), 2410.

11 Processing and Properties of Chitosan and/or Chitin Biocomposites for Food Packaging

Maria Râpă and Cornelia Vasile

CONTENTS

11.1 INTRODUCTION: GENERAL REQUIREMENTS FOR FOOD PACKAGING BIOCOMPOSITES

Chitin/chitosan-based food packaging materials are known in the form of bags, sachets, pouches and heat-sealable flexible lidding materials. They show some particularities in terms of their properties, existent practices, limitations, and technologies. Knowledge of them is useful both for scientists and manufacturers of food packaging (Agulló et al. 2003; Srinivasa and Tharanathan, 2007). Food product quality and freshness during commercialization and consumption should be checked by the correct choice of materials.

In the European market, polyethylene (PE) is the plastic in highest demand (56%), followed by polypropylene (PP), polyethylene terephthalate (PET), polystyrene (PS) and polyvinyl chloride (PVC) (Coles and Kirwan, 2011). Flexible food packaging films based on conventional polymers are used for fresh, refrigerated or frozen foods such as desserts, salads and snacks, and as a barrier layer for packaged foods to be stored for extended periods of time at ambient temperature. These conventional materials show good mechanical and barrier properties, are low cost and are easily manufactured (Kim et al. 2013).

The methods to extend the shelf life refer to modified atmosphere, packaging refrigeration, and the use of synthetic antioxidants and natural antibacterial compounds. However, the use of synthetic antioxidants incorporated into food products may result in excessive amounts of the antioxidant agent which may change the taste of the food, cause inactivation or evaporation of active agents and rapid migration into the bulk of the food (Nguyen et al., 2016). Once these additives are consumed in reaction, protection ceases and oxidation reactions increase rapidly. There are known antibacterial nanocomposites based on polyethylene and copper nanofibers that exhibit enhanced tensile strength, Young's modulus, and oxygen barrier as well as antibacterial properties (Bikiaris and Triantafyllidis, 2013). Increasing the levels of additives in the formulation is not feasible due to Food and Drug Administration (FDA) regulations and also because higher concentrations could cause pro-oxidation reactions in lipids.

However, conventional food packaging materials represent a huge problem because of petrochemical resources that are limited and their durability. Another disadvantage of conventional food packaging is related to its short lifetime and environmental pollution.

Because of these reasons, in the last few years serious efforts have been made to develop biodegradable materials from renewable resources for food packaging that ensures full quality of food.

Like conventional packaging, the food packaging biocomposites must fulfill some important functions, including containment and protection of food and maintaining its sensory quality and safety and consumer acceptance (Coma, 2013; Goyal and Goyal, 2012). Also, when designing flexible bioplastic packaging systems, the following characteristics are desired to be accomplished simultaneously (López et al., 2015): good thermal properties of materials, such as glass transition temperature, melting temperature, crystallization temperature, chemical resistance, permeability (solubility and diffusivity), mechanical properties, such as tensile strength, elongation, viscosity, elasticity, plasticity, modulus, and characteristic morphological features. A high oxygen barrier property is required for most food packages to maintain qualities and shelf life of foods since oxygen permeated through packaging materials may cause rancidity and other oxidative degradation reactions in lipid-containing foods resulting in quality losses. Another aspect to when the consumers decide to purchase food products is the aspect of packaging, and for this the high transparency of food packaging is required. Colored films may affect the consumer acceptance of the products when they are applied to light-colored foods. Overall migration of some components from packaging formulations tested into simulant media should be below the overall migration limit of 10 mg dm^{-2} of food contact surface as required by the Regulation (EU) No. 10/2011. At the same time, the materials used for packaging of food must have and maintain antibacterial activity for a long period of time (Appendini and Hotchkiss, 2002; Hanušová et al., 2009).

11.2 STRUCTURE, PROPERTIES AND LIMITS OF CHITIN/CHITOSAN

Chitin is present in nature as ordered microfibrils and is the major structural component in the cuticles of insects, arthropods such as crustacean shells, beaks of cephalopods and the cell walls of fungi and yeast. It is presented in three different structural forms (α, β and γ), and they are obtained by chemical or biological methods, such as by microbial fermentation and enzymatic reaction (Gamal et al., 2016; Musarrat et al., 2013; Theruvathil et al., 2007). The main commercial sources of chitin are crab and shrimp shells, which are abundantly supplied as waste products of the seafood industry.

The structure of chitin, a polysaccharide composed of β (1→4)-N-acetyl D-glucosamine units, is as shown in Figure 11.1.

Chitin itself is insoluble in aqueous and organic solvents and it exhibits a high degree of crystallinity. Nano-chitin exists in the form of chitin whiskers and/or nanocrystals (CHNC) with rod morphology and chitin nanofibers (CHNF) which are fibrilar (Salaberria et al., 2015a). Chitin whiskers have been prepared by acid hydrolysis and successfully used as reinforcing fillers in polymeric matrices due to their high aspect ratios and highly crystalline nature. The main problems associated with nano-chitins are a strong tendency to aggregate and the fact that they are difficult to redisperse due to the strong hydrogen bonds formed when dried (Herrera et al., 2016).

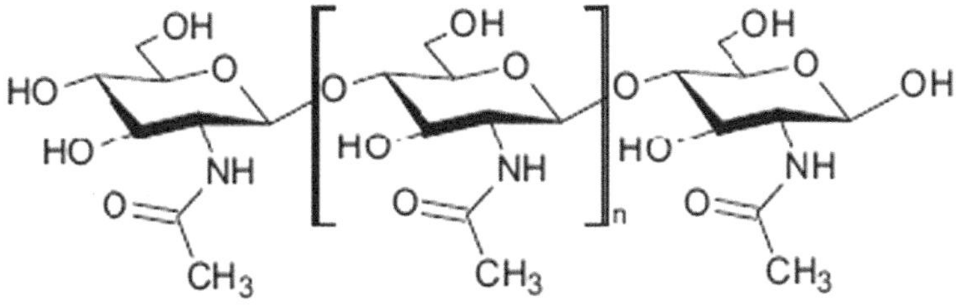

FIGURE 11.1 Structure of chitin.

Chitosan (CS) biopolymer structure (see Figure 11.2) consists of glucosamine and *N*-acetylglucosamine units linked by β-1,4 glycosidic bonds. Due to its antibacterial activity, CS has received a considerable amount of interest for commercial food packaging applications. Although CS has been approved as a food additive in some countries, for instance in Japan and Korea, in Europe, it has not yet been accepted as a food contact material. Because most CS products are originated from crustacean chitin, it should be properly labeled as a hypoallergenic material when used for food products (Corrales et al., 2014).

The CS is obtained by the deacetylation of the chitin. The main source of commercial CS is the deacetylation of its parent polymer chitin. It is present in green algae, the cell walls of fungi and in the exoskeleton of crustaceans. There are also known procedures for preparing chitosan directly from shrimp shells (De Queiroz et al., 2017). It is soluble under acidic conditions.

Due to the extensive hydrogen bonding, the CS degrades prior to melting. To overcome this drawback, usually CS is dissolved in an appropriate solvent. Under acidic conditions, CS can be dissolved in water after the amino group protonation to confer positive charges, gelations, and membrane-forming properties.

The physical and chemical properties of CS depend mainly on its molecular weight and degree of deacetylation. Based on molecular weight, the CS can be classified into high molecular weight, low molecular weight, and oligochitosans. For example, if the degree of deacetylation is low, the tensile strength and elongation can be high (Muzzarelli et al., 2012). The use of CS with high degree of deacetylation and the use of glycerol as a plasticizer resulted in films with higher crystallinity. Low-molecular-weight CS molecules could penetrate the nuclei of microorganisms and extend shelf life and improve the microbial safety of packaged foods. Poor water barrier properties, fairly brittle nature, as well as the differences in the morphology/crystallinity of the films upon preparation/processing are considered to be the key deficiencies of CS use in food active packaging (Vlacha et al., 2016).

The antibacterial effects of CS are reported to be dependent on its molecular weight (Cruz-Romero et al., 2013), its degree of deacetylation, the type and concentration of the organic solvent employed, the pH of the medium, and the type of bacterium (Fernandez-Saiz et. al., 2008). Chitosan as powder has no antibacterial activity; only in viscous acid solution forms and as chitosonium-acetate films does it show optimum biocide properties. Because CS contains amino and hydroxyl reactive groups, its chemical modifications can be applied in order to improve the antimicrobial activity and physicochemical properties (Belalia et al., 2008; Khwaldia et al., 2010; Sashiwa and Aiba, 2004; Vroman and Tighzert, 2009).

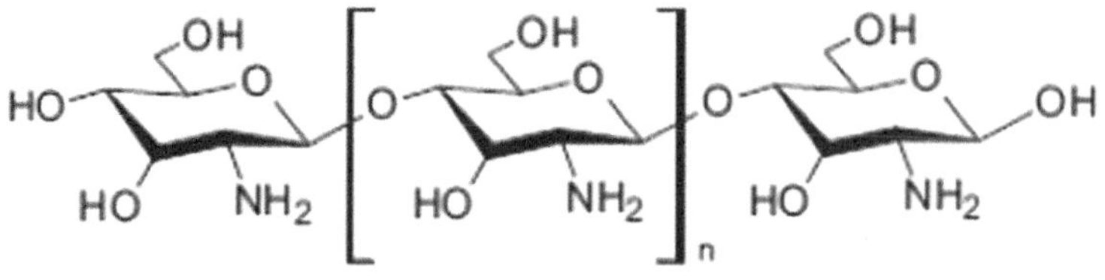

FIGURE 11.2 Structure of chitosan.

However, due to the fascinating chitin and CS structures, there are multiple possibilities for modifying them by chemical or mechanical treatments or using nanotechnology to generate novel properties desired for food packaging film application.

11.3 PROCESSING AND PROPERTIES OF CHITOSAN/CHITIN BIOCOMPOSITES

Numerous strategies such as direct solution casting (solvent evaporation), coating, freeze drying, dipping, layer-by-layer assembly, co-precipitation, complex coacervation, polyelectrolyte complexes, electrospinning, chemical modification and common processing techniques such as compression molding, thermal molding, melt blending, and twin-screw extrusion, have been employed to prepare chitosan/chitin-based films and other materials with multiple functionalities (Hooda et al., 2018; Moura et al., 2016; Thomas et al., 2019; Wang et al., 2018).

11.3.1 Solvent Casting (Evaporation Technique)

Usually, the coating technique is required to decrease or avoid moisture transfer through the food and the neighboring atmosphere. Casting solution is the most useful technique for preparing active food packaging at laboratory scale. Generally, in this technique 1 g of CS was added to the 1% acetic acid solution and stirred constantly until it was dissolved. Utilization of formic acid or propionic acid, however, has created higher-quality films than those films produced using acetic acid as solvent. After elimination of impurities and insoluble CS by filtration, various plasticizers, emulsifiers, nanomaterials, etc. are added into dissolved CS depending on the desired properties. Adjusting the pH and/or heating the solutions may be necessary to facilitate the solubility of some biopolymers. Finally, the solution is cast onto the Petri-plates and is dried at 50°C in an oven overnight. The casted CS films are removed from the glass Petri-dish either with 5 M NaOH or without NaOH.

Overall, it has been shown that glycerol is a more effective plasticizer than oleic acid in terms of processing, elongation at break and water sorption. Vlacha et al. (2016) found that oleic acid offers a better barrier to water vapor permeability with up to three times lower permeability rates and that it improves the antimicrobial activity of chitosan with up to 75% lower relative bacterial growth as compared with glycerol when they prepared CS/clay nanocomposites films.

Different chitin/CS systems and their related properties for food packaging application are summarized in Table 11.1.

According to data shown in Table 11.1, the CS/chitin active packaging contains antimicrobial and antioxidant compounds as bioactive agents, as nanoparticles (NPs) or nanofillers, which enhance their activity. It is worth highlighting that the enhancement of antioxidant activity of CS films' bionanocomposites has led to the prolongation of shelf life of packaged food. An increase of antioxidant activity of functionalized chitosan fibers with flavonoids (flavanols, flavonols, flavone, flavanone, isoflavone) was investigated by Sousa et al. (2009) who found that some flavonoids used increased antimicrobial activity of chitosan against *Bacillus subtillis* and *Pseudomonas aeruginosa*.

TABLE 11.1
Chitosan/Chitin-Based Coating Materials and Their Properties

Composition	Improved properties	References
Chitin nanofibers (N-chitin)/corn oil/gelatin nanocomposite films	Improved mechanical, thermal and antifungal properties	Sahraee et al. (2017a)
CS/carboxymethyl cellulose/ ZnO NPs	Increase the shelf life of soft white cheese stored at 7°C for 30 days	Youssef et al. (2016)
CS/carboxymethyl cellulose/ oleic acid/(ZnO NPs)	Increased the shelf life (microbial and staling) of sliced wheat bread to 35 days of storage	Noshirvani et al. (2017)
CS/ZnO NPs (up to 0.5%)/ neem oil	Tensile strength max. 60 MPa, elongation 16% and improved film thickness and film transparency, the decreased water solubility (from 80% in the case of CS to 30%), water vapor permeability properties (from 1.57 mg/cm^{2}xh in the case of CS to 0.41 mg/cm^{2}xh) were achieved. The antibacterial activity against *E. coli* proved the potential use of this composition as food packaging coating system for carrot	Sanuja et al. (2015)
CS NPs-CuO NPs with different CuO percentage	Synergistic antibacterial activity against *E. coli*, *Salmonella typhi*, *Pseudomonas aeruginosa*, *S. aureus*	Syame et al. (2017)
CS/Mahua oil-based polyurethane (PU)/ZnO NPs films	The shelf life period of carrot pieces wrapped with the composite film was extended up to nine days	Saral Sarojini et al. (2019)
CS/reduced graphene oxide (rGO) with caffeic acid	Great enhancement of antioxidant activity of bionanocomposites with incorporation of rGO	Barra et al. (2019)
CS/gelatin/ polyethylene glycol/ silver nanoparticles (AgNPs)	Shelf life of the red grapes packaged with hybrid films extended for additional two weeks	Kumar et al. (2018)
CS nanofibers/CS film	Effectiveness to prevent moisture transfer between the layered fresh beef patties	Deng et al. (2017)
CS NPs/Carum copticum (CEO) prepared by emulsion-ionic gelation method	The DPPH scavenging activities of CEO, before and after its encapsulation in CS NPs were in the range of 21.8%–63.9% and 10.6%–60.6%, respectively at concentrations (50–800 g mL^{-1}). Highest antibacterial activities against *S. aureus*, *S. epidermidis*, *B. cereus*, *E. coli*, *S. typhimurium* and *P. vulgaris*, respectively	Esmaeili and Asgari (2015)

(*Continued*)

TABLE 11.1 (CONTINUED)
Chitosan/Chitin-Based Coating Materials and Their Properties

Composition	Improved properties	References
CS/Carum copticum essential oil/ cellulose nanofibers or lignocelluloses nanofibers	Antimicrobial effects against *E. coli* O157:H7 (ATCC 25922), and *B. cereus* (PTCC 1154). Solubility of the bionanocomposite films reduced in the range of 22.81–22.84%	Jahed et al. (2017a)
CS/nanocrystalline cellulose (NCC)/styrylquinoxalin films	Antibacterial activity against *Pseudomonas Aeruginosa* (ATCC 27853) comparable to ampicillin antibiotic	Fardioui et al. (2018)
CS NPs/cinnamon essential oil (CE-NPs)	Sensory evaluation of chilled pork coated with 527 nm CE-NPs during storage at 4°C for 12 days revealed the following attributes: red color 2.67±0.49, discoloration 2.92±0.51, and off odor 2.83±0.41. 2-thiobarbituric acid (TBA) value was 2.09 mg malonaldehyde kg^{-1} meat sample	Hu et al. (2015)
CS NPs/ summer savory (*Satureja hortensis* L.) essential oil	Strong antibacterial (against *Staphylococcus aureus*, *Listeria monocytogenes* and *E. coli* O157:H7	Cansu Feyzioglu and Tornuk (2016)
CS NPs/lime essential oil and CS nanocapsules/lime essential oil were synthesized by nanoprecipitation and nanoencapsulation methods, respectively	The antibacterial activity was tested against four foodborne pathogens (*Shigella dysenteriae*) being higher for NPs than for nanocapsules	Sotelo-Boyás et al. (2017)
CS NPs/*Schinus molle* EO were prepared by ionotropic gelation	At a concentration of 500 µg mL^{-1}, the *Schinus molle* L. essential oil-loaded CS NPs inhibited up to 59% of the aflatoxin production stored grains	López-Meneses et al. (2018)
CS NPs/ clove essential oil	Antifungal activity against *Aspergillus niger*	Hasheminejad et al. (2019)
CS NPs/cinnamomum zeylanicum essential oil (CEO) coatings	Extended the shelf life of cucumbers up to 21 days at 10±1°C while uncoated fruit was unmarketable in less than 15 days	Mohammadi et al. (2015)
Cellulose nanocrystal (CNC)/ chitosan/nisin disodium ethylene diamine tetraacetate (EDTA) by using genipin as a cross-linking agent	The films restricted the growth of psychrotrophs, mesophiles and *Lactobacillus spp.* in fresh pork loin meats and increased the microbiological shelf life of meat sample by more than five weeks	Khan et al. (2016)

(Continued)

TABLE 11.1 (CONTINUED)
Chitosan/Chitin-Based Coating Materials and Their Properties

Composition	Improved properties	References
CS/nanoclay (MMT)/*Silybum marianum* L. extract	DPPH radical scavenging activity 84.13%	Ghelejlu et al. (2016)
CS/MMT or Cloisite 30B by an exchange reaction or chitosan acetate/MMT/glycerol or oleic acid by casting method	Thermal stability is remarkably improved. Antimicrobial against *E. coli* and *S. Aureus*. Increased stiffness and strength	Giannakas and Leontiou (2017)
CS/nanoclay/rosemary essential oil	Incorporation of rosemary led to the highest value of total phenol content (2.23 mg gallic acid/g sample) and the best antibacterial activity in the disk diffusion test on Gram-positive bacteria (*L. monocytogenes*, *S. agalactiae*)	Abdollahi et al. (2012)
CS/rosehip seed oil/ montmorillonite nanoclay	Inhibit at 24 h the *Escherichia coli* (72% reduction), *Salmonella typhymurium* (65% reduction), and *Bacillus cereus* (82% reduction) microorganisms Introduction of rosehip seed oil led to increase up to threefold of the radical scavenging activity	Butnaru et al. (2019)
CS/(Cloisite®Na+ and Cloisite®Ca^{+2})/rosemary essential oil (REO) or ginger essential oil (GEO)	Nanoclays reduced lipid oxidation by half and microbiological contamination by 6%–16%, but incorporation with essential oils (EOs) only improves the barrier to oxidation	Pires et al. (2018)
Moringa oil/CS NPs/gelatin	Coating of cheese surfaces for ten days at 4°C exhibited excellent antibacterial activity against *L. monocytogenes* and *S. aureus*	Lin et al. (2019)
CS/gelatin/β-carotene/starch nanocrystals films	91.5±0.3% of radical scavenging activity	Hari et al. (2018)
Hydrophobized and antimicrobial papers by coating with CS/palmitic acid emulsions or a blend of CS and O,O′-dipalmitoylchitosan	Improved water-resistance, antimicrobial, environmentally friendly food packaging materials. All CS-coated materials exhibited over 98% inhibition of *Salmonella typhimurium* and *Listeria monocytogenes*	Bordenave et al. (2010)
CS and casein films	The protein-polysaccharide complexes exhibit better functional properties than proteins and polysaccharides alone.	Ponce et al. (2016)

(*Continued*)

TABLE 11.1 (CONTINUED)
Chitosan/Chitin-Based Coating Materials and Their Properties

Composition	Improved properties	References
CS/other polysaccharides (starch, pectin or alginate)	Polyelectrolyte complexes, improved mechanical and barrier properties, better performance in terms of water vapor permeability and lower water solubility	Ferreira et al. (2016)
CS/halloysite nanotubes/TiO_2 composite coatings by electrophoretic deposition	Hydrophobicity, good mechanical properties	Darie –Niţă and Vasile (2018); Jackoub Raddaha et al. (2015)
Crab shell CS/graphene oxide (GO) composite films (GO in content of 0.5, 1 and 2% w/w)	Tensile strength increased from 6.99 MPa for chitosan to 5.32 MPa for CS/GO 2%, glass transition temperature (T_g) increased with ~26°C; water vapor transmission rate and oxygen permeability of film were reduced significantly with increasing the GO loading, from 22.4 gm/m^{2}day to 9.6 gm/m^{2}day and from 8.6 (cm^3 mm/m^2 d atm) to 2.99 (cm^3 mm/m^2 d atm), respectively	Ahmed et al. (2017)
CS/paper packaging (CS coating level of 5.41%)	Addition of CS allowed an improvement of the fat barrier of Kraft paper for pet food packaging but treatment cost remained high compared with fluorinated resins	Ham-Pichavant et al. (2005)
Chitin nanocrystals (CHNC)/thermoplastic starch Chitin nanofibers/thermoplastic starch	Negative effect on the thermal stability of the starch/CHNC nanocomposite films and increases of Young's modulus from 25 to 75 MPa and of tensile strength from 2 to 3 MPa were recorded. Increased thermal stability with 5°C, Young's modulus to 425 MPa and tensile strength of 11 MPa were obtained in the case of CHNF use	Salaberria et al. (2015b)
CS whisker/rectorite ternary composite films	Tensile strength of the composite films increased from 9 to 17 MPa, while the percentage of elongation increased from 60 to 129%; the water uptake of the chitosan/chitin whisker film decreased from 214 to 153% with an increase of the whisker content from 0 to 229 mg/g. The composite exhibited a stronger antibacterial effect against *Escherichia coli* and *Staphylococcus aureus*	Li et al. (2011)

(*Continued*)

TABLE 11.1 (CONTINUED)
Chitosan/Chitin-Based Coating Materials and Their Properties

Composition	Improved properties	References
CS/olive residue flour (10 wt%, 20 wt% and 30 wt% into CS)	10% microparticles of olive residue flour significantly increased with ~ 34% tensile strength and with ~43% elongation at break of films compared to the control film. The antioxidant capacity of the films was proportionally increased with the concentration of flour or microparticles added to the film. The films with 30% of flour or microparticles were effective as protective packaging against the oxidation of nuts for 31 days	Crizel et al. (2018)
CS/ellagic acid films (0.5, 1.0, 2.5 and 5.0% w/w ellagic acid relative to CS)	The obtained results revealed Young's modulus of 3.21–3.57 GPa, thermal stability up to 215–220°C, UVA- and UVB-barrier properties, moderate water vapor permeability and totally inhibited the growth of both Gram-positive (*S. aureus*) and Gram-negative (*P. aeruginosa*) food pathogenic bacteria	Vilela et al. (2017)
CS/propolis extract films (containing 0, 2.5, 5, 10 and 20% wt propolis extract)	Tensile strength increased to fourfold that of the CS film control for 20% propolis extract and elongation at break increased approximately threefold in the case of 10% propolis extract content incorporated into CS, while water vapor permeability and oxygen permeability decreased with ~20% and ~50%, respectively, with increasing propolis extract concentration; antimicrobial activity against *Staphylococcus aureus*, *Salmonella Enteritidis*, *Escherichia coli* and *Pseudomonas aeruginosa* by using agar diffusion technique was enhanced	Siripatrawan and Vitchayakitti (2016)
Chitosonium-acetate/ ethylene–vinyl alcohol copolymers blend	This film showed enhanced phase morphology, transparency, up to 86% water permeability reduction compared to pure chitosonium-acetate films, as well as excellent antimicrobial activity	Fernandez-Saiz et al. (2010)
CS/polyvinyl alcohol (PVA) coating; Glutaraldehyde was used as crosslinker	Microbiological analyses against food pathogenic bacteria viz. *Escherichia coli*, *Staphylococcus aureus*, and *Bacillus subtilis* revealed the improved property performed on tomatoes	Tripathi et al. (2009)

Another way to improve not only the properties of CS/chitin but also to add antibacterial, antioxidant and nutraceutical capacities to food packaging film consists in the application of nanotechnology (Malathi et al. 2014). Nanoparticles (NPs) can be added into CS matrix as reinforcement or as active agents. For this purpose, Salaberria et al. (2015b) evidenced the role of chitin nanocrystals (CHNC) and chitin nanofibers reinforcements on the structural and functional properties of thermoplastic starch-based films prepared by a casting evaporation approach. By adding nanofillers together with antioxidant agents into CS films a diminishing of antioxidant properties was recorded. Thus, by adding lignocelluloses nanofibers (LCNF) and cellulose nanofibers (CNF) in CS matrix together with 5% (w/w) Carum copticum essential oil, as active agent, a decreased antioxidant activity than that of CS films was noted (Jahed et al., 2017a). Nanofillers can act as a release controller of antioxidant ingredients from active films. Jahed et al. have explained the reduced antioxidant property in the presence of nanofillers and antioxidant compounds by (1) the creation of more tortuous paths in the CS matrix for antioxidant molecules to pass through, (2) decreasing the porosity of the CS matrix and thus decreasing the segmental mobility of essential oil molecules, (3) effect of nanofibers nucleation that increases the crystallization of CS molecular chains, and (4) interaction between nanofibers and essential oil compounds (Jahed et al., 2017a). Similar findings are reported by Ghelejlu et al. (2016), that investigated the antioxidant activity of nanocomposites films containing CS, sodium montmorillonite (MMT) (1, 3 and 5% w/w based on CS) and milk thistle plant extract (0.5, 1 and 1.5% v/v). Ghelejlu et al. have observed that the antioxidant activity of nanocomposites films is strongly dependent on antioxidant agent and is not dependent on the amounts of MMT. It is possible that MTT could be used to reduce the antioxidant activity of active agent due to the fixing and stabilization of phenolic compounds at the gallery space of MMT layers. Ghelejlu et al. considered that MTT could act as a release controlling device in bionanocomposites films. The incorporation of nanoclays into packaging material may represent a food safety hazard once those nanoparticles can migrate to package food. In a study performed by (Pires et al., 2018) to develop bionanocomposites based on chitosan reinforced with two different montmorillonites (MMTs) (Cloisite®Na+ and Cloisite®Ca^{+2}) incorporated with rosemary essential oil or ginger essential oil, it was established that there was no migration of MMT nanoparticles from films to the meat.

Inorganic nanoparticles, as Cu NPs, ZnO NPs have been incorporated into polymer matrix as reinforcement to improve the mechanical properties of biocomposites and also to add functional properties, such as antimicrobial activity (Noshirvani et al., 2017; Sanuja et al., 2015; Syame et al., 2017; Youssef et al., 2016). Several parameters related to NPs incorporation may be controlled in order to tune the composite's properties, including particle size, chemical composition, crystallinity and shape. However, the NPs have the potential to migrate to the foodstuff packaged. The toxicity and ecotoxicity of inorganic NPs exposed via food packaging films exposure are not yet clear, so many toxicological studies are still necessary (Duncan, 2011; Souza and Fernando, 2016).

Data presented in Table 11.1 evidenced that reinforcements at nano scale and active compounds are necessary for preparation of CS-based nano(biocomposites) films with potential as active food packaging due to the synergistic interactions of components. In order to find the optimization of the carboxymethyl cellulose (CMC) (4–16 wt%), chitin nanofiber (CHNF) (2–5 wt%) and Trachyspermum ammi (Ajowan) essential oil (AJEO) (0.26–1 v/v%) components on the antimicrobial and physical properties of the gelatin-based nanocomposite films, a study by response surface methodology (RSM) was performed (Azarifar et al., 2019). It was found the optimal values of CMC, CHNF and AJEO for composition of the gelatin film were: 15.83 wt%, 3 wt% and 0.74 v/v%, respectively.

A new trend in obtaining of novel bioactive food packaging systems is to incorporate bioactive compounds obtained from waste materials, such as elagic acid (Vilela et al., 2017), olive pomace flour (Crizel et al., 2018), fish gelatin and crustacean shells (De la Caba et al., 2019) in line with the main goal of the circular economy. Bionanocomposite films, including chitin nanofibers and chitosan from α-chitin powder extracted from lobster wastes, demonstrated an inhibitory effect (>80%) against *A. Niger* (Salaberria et al., 2015a). Natural chitin films from insects are a new approach to obtain chitin-based films with resistance to bacterial biofilm formation and antifungal activity (Kaya et al., 2017a). This is a potential solution to producing chitin films without using any chemical additives.

11.3.2 Edible Coating/Edible Film

In recent years, edible films-based natural resources have emerged as the most attractive food packaging, as an alternative to those obtained from synthetic polymers. CS has been documented to possess a film-forming property for use as edible film or coating, to decrease water vapor and oxygen transmission, diminish respiration rate and increase shelf life of fruit. The coating methods involve dipping and spraying of the food products (Dutta et al., 2009). Edible coatings are applied in liquid form, meanwhile edible films are obtained as solid sheets and then used to wrap food products. The materials used to obtain edible coatings with improved properties of food packaging are made from chitin/CS, protein, polysaccharides (Altieri et al., 2005; Higueras et al., 2013). Among the materials derived from agro-food industry, wastes are preferred due to their low cost. In most cases, the addition of plasticizers is required in order to obtain protein- and polysaccharides-based films. Some examples of CS compounds used as edible coating/edible films to improve the preservation of a variety of food by providing a relevant antimicrobial activity are shown in Table 11.2.

In a recent paper (Uranga et al., 2019) fish gelatin/CS composite films were prepared by solution casting. Introduction of 20 wt% citric acid in fish gelatin/CS composite led to reducing the *E. coli* growth, proving the potential use of this film as active food packaging.

Also, the use of fungal-derived CS is a new trend for the development of naturally activated edible films (Gomaa et al., 2018).

Incorporation of several polyphenols such as fruit extracts (Genskowsky et al., 2015) and blue-green algae (Balti et al., 2017) into CS have been demonstrated in recent years to lead to the high quality and extended shelf life of food. Other researches are devoted to the enhancing of durability, hydrophobicity (as related by

TABLE 11.2
Composition and Properties of the Edible Films-Based Chitosan/Chitin Materials

Packaging material/Composition	Improved active properties	References
CS/15 wt% cellulose nanofibers/18 wt% glycerol	Edible films or coatings; cellulose nanofibers can improve the mechanical and water vapor barrier properties of chitosan films; environmentally friendly properties	Azeredo et al. (2010)
CS/essential oil edible coating CS:oil ratio of 2% w/w: 1% w/v	Increased antioxidant and antimicrobial activity of fish and meat	Yuan et al. (2016)
Quinoa protein/ CS thymol nanoparticles edible films CS solution 2% w/v; citric acid 2% w/v; quinoa solution (0.72 g of protein per 100mL; CS-thymol NPs were prepared by diluting 1.9 g of citric acid in 100 mL of 1 mg/mL of thymol in water	Extension of postharvest life of blueberries and tomato cherries. CS formed a water barrier between the fruit and external environment, reducing the external transfer of water	Medina et al. (2019)
CS/5% w/w *Origanum vulgare* ssp. gracile essential oil/4% w/w cellulose nanofibers (CNF) or lignocellulose nanofibers (LCNF)	Highest antioxidant and antimicrobial activities were obtained in CS-EO film. Addition of organic nanofibers (CNF or LCNF) to CS-EO film led to decreased DPPH scavenging and antimicrobial activity of bionanocomposites, showing these nanofibers act as a release controller of antioxidant and antimicrobial compounds	Jahed et al. (2017b)
CS/oregano essential oil	Antimicrobial packaging. CS edible films still suffer from high water vapor permeation; oregano essential oil increases its hydrophobicity and improves its water vapor permeation. Films show a high level of antimicrobial activity against different strains of Gram-positive and Gram-negative bacteria and antioxidant activities	Elsabee et al. (2016)
Galactomannans (tara gum)/CS NPs edible films Sodium tripolyphosphate (TPP) used as crosslinker	CSNPs were found to be less effective against *E. coli* (Gram-negative) compared to *S. aureus* (Gram-positive) probably due to agglomeration of NPs	Antoniou et al. (2015)

(Continued)

TABLE 11.2 (CONTINUED)
Composition and Properties of the Edible Films-Based Chitosan/Chitin Materials

Packaging material/Composition	Improved active properties	References
CS/maqui berry (MB) edible films. MB in content of 0, 0.5; 1%; glycerol in content of 2.5 mL/g CS was used as plasticizer	Total phenol content (TPC) was: 4.75 mg GAE/g for CH + 0.5% MB and 8.44 mg GAE/g for CH + 1% MB Antioxidant activity as DPPH was 2.06 mg Trolox equivalent/g film for CH + 0.5% MB and 2.8 mg Trolox equivalent/g film for CH + 1% MB with values of IC_{50} comprised between 1.62 and 20.10 µg/mL Edible films were effective against *Serratia marcescens*, *Aeromonas hydrophila*, *Achromobacter denitrificans*, *Alcaligenes faecalis*, *Pseudomonas fluorescens*, *Citrobacter freundii* and *Shewanella putrefaciens* test bacteria by disk diffusion method	Genskowsky et al. (2015)
Sporopollenin/CS edible film, Sporopollenin in content of 10, 20 and 40 mg in 100 mL chitosan gel	Young modulus values for CS, CS10, CS20 and CS40 were recorded as 1.69±0.12; 5.08±0.23;11.37±0.40 and 16.29±1.72 MPa respectively; elongation at break had the values 35.07±4.09 for CS, 40.09±1.73 for CS10 and 42.42±1.26% for CS20; tensile strength values of blend film compared to control film were: 0.68±0.05 for CS, 2.04±0.12 for CS10, 2.91±0.13 for CS20 and 2.92±0.15 MPa for CS40; the overall antifungal activity against *A. niger* of the films and the antioxidant activity (inhibition %) of the films were found to be 16.57±0.41 for CS, 38.90±0.61 for CS10, 47.48±2.14 for CS20 and 47.53±1.68 for CS40	Kaya et al. (2017b)
CS/*Spirulina* extract (SE) edible films. SE concentrations of 0, 2.5, 5, 10, 15 and 20% (w/v); glycerol, as plasticizer, was used at the constant concentration of 30% w/w	Tensile strength significantly increased from 22.45±1.17 to 29.65±1.43 MPa when SE concentration increased from 2.5 to 20% The elongation at break of the films increased when *Spirulina* extract concentration increased from 2.5 to 10% up to 39.5% as compared with control (26.1%) The total phenolic content ranged from 5.62 to 21.85 mg GAE/g film and the highest value was recorded with the film formulated with 20% SE The CS/SE films were more effective ($p < 0.05$) against Gram-positive bacteria (*L. monocytogenes*) than Gram-negative bacteria (*E. coli*, *P. aeruginosa* and *S. typhimurium*)	Balti et al. (2017)

(Continued)

TABLE 11.2 (CONTINUED)
Composition and Properties of the Edible Films-Based Chitosan/Chitin Materials

Packaging material/Composition	Improved active properties	References
Gelatin/nano chitin/CS NPs nanocomposite films N-Chitin in concentration of 3, 5 and 10% wt on dry gelatin basis was used 30 wt.% glycerol (on dry gelatin) was added as a plasticizer 50%wt glutaraldehyde aqueous solution (1% wt on dry gelatin basis) used as crosslinker	The reduced water vapor permeability (WVP) and solubility and higher surface hydrophobicity of the nanocomposite films were obtained by enhancing N-chitin concentration in film formulation The use of N-chitin up to 5% concentration led to a tensile strength at break of 119.08MPa and elongation at break of 5.82% The results of differential scanning calorimetry (DSC) and thermogravimetric analysis (TGA) confirmed improved stability of nanocomposite films against melting and degradation at high temperatures in comparison to neat gelatin film The gelatin-based nanocomposite films had antifungal properties against *Aspergillus niger* in the contact surface zone	Sahraee et al. (2017b); Zafar et al. (2016)
Gelatin/CS/eugenol and ginger essential oils blend	The best antioxidant activity based on the Trolox-equivalent-antioxidant-capacity test	Bonilla et al. (2018)
Turmeric (TEE)/CS film 20% sol. Na_2SO_4 used as crosslinker	The incorporation of TEE caused a significant increase in tensile strength from 32.2 MPa to 47.9 MPa However, turmeric incorporation did not modify the water vapor barrier property and thermal behavior of the films, significantly. Films have also good light barrier properties but are transparent enough to fulfill consumer demands Better antimicrobial activity was observed against *Salmonella* and *Staphylococcus aureus*	Kalaycıoglu et al. (2017)
Diatomite/CS edible films 50 mg and 100 mg of diatomite were used; Glutaraldehyde used as crosslinker	Film wettability, elongation at break and thermal stability (264–277°C) were enhanced The incorporation of diatomite did not influence the overall antioxidant activity of the composite films, but revealed a notable enhancement in the antimicrobial activity	Akyuz et al. (2017)

(Continued)

TABLE 11.2 (CONTINUED)
Composition and Properties of the Edible Films-Based Chitosan/Chitin Materials

Packaging material/Composition	Improved active properties	References
CS/poly(vinyl alcohol)/pectin ternary film 1 wt% clear CS solution; 1 wt%, 1.5 wt%, and 2 wt% pectin solution 1% clear PVA solution	The thermogravimetry analysis (TG) depicted the weight losses at 200–300°C resulting from ternary film for degradation of CS molecule Antimicrobial activity was demonstrated against *Escherichia coli*, *Staphylococcus aureus*, *Bacillus subtilis*, *Pseudomonas*, and *Candida albicans* bacteria	Tripathi et al. (2010)
CS/glycerol/apple peel polyphenols (APP. CS solution (2%, w/v) in acetic acid aqueous solution (1.0%, v/v) Glycerol (30%, w/w) APP: 0.25, 0.50, 0.75 and 1.0%	Scavenging activities on DPPH and ABTS radicals significantly ($p < 0.05$) increased with the addition of APP. Films showed antimicrobial activity against *E. coli*, *B. cereus*, *S. aureus* and *S. typhimurium* The free amino groups in CS might bind to cell surface, disturbing the cell membrane, and thus causing death of the cell by inducing leakage of intracellular components Antimicrobial activities of CS-APP films might be the outcome of both functions of CS and polyphenols	Riaz et al. (2018)
Zein/CS/gallic acid edible film Dicarboxylic acids (adipic acid or succinic acid) used as crosslinker	Phenolic compounds led to decrease of the brightness while the mechanical properties significantly increased 22% elongation at break was recorded for composites containing CS, gallic acid and dicarboxylic acid	Cheng et al. (2015)
Zein/CS edible films 25–75%, 50–50% and 75–25% zein–CS, weight basis; ethanol/zein 10:1 w/v, without glycerol	The composite films presented better barrier and mechanical properties than single ingredient films and increasing roughness values as the zein concentration increased	Escamilla-Garcia et al. (2013)

(Continued)

TABLE 11.2 (CONTINUED)
Composition and Properties of the Edible Films-Based Chitosan/Chitin Materials

Packaging material/Composition	Improved active properties	References
Galactomannans (tara gum)/CS NPs edible films TPP used as crosslinker	When 10% w/w of CS NPs was added, the tensile strength was 58.44±3.23 MPa, WVP value was 0.357 g mm m^{-2} h^{-1} kPa^{-1} and storage modulus (E′) increased. The decreased moisture content of edible tara gum film was explained by authors due to the compact structure of CS NPs allowing them to occupy more free volume in the polymer matrix. Although the roughness surface of film increased, CS NPs were found to be less effective against *E. coli* (Gram-negative) compared to *S. aureus* (Gram-positive) probably due to agglomeration of NPs	Antoniou et al. (2015)
Starch/CS-based composite blending at 13500 rpm for 5 minutes. CS (1%) in dilute acetic acid (1%) dissolving with 5 g starch in 95 mL water and heating (90°C/15 min) and addition of glycerol (3 g). The thickness of the composite edible film was 206.25 µm	Antimicrobial edible composite film was effective against bacteria, yeast and molds and the antimicrobial efficacy varied with the species: *Saccharomyces cerevisiae* (NCDC-186), *Aspergillus flavus* (NCDC-268) and *Lactobacillus acidophilus* (NCDC-14)	Priyanka and Ganguly (2015)

contact angle), and mechanical properties of CS edible films by incorporation of sporopollenin (Kaya et al., 2017b), carrageenan (Zia et al., 2017), diatomite (Akyuz et al., 2017). It was found that incorporation of natural additives into CS matrix successfully increased the hydrophobicity of blend films and improved mechanical properties and antimicrobial activity as compared with unmodified CS film.

A modern trend for developing active edible films is to combine different base materials and to incorporate multiple functional ingredients. From this perspective, Cheng et al. (2015) prepared composite edible films based on zein and CS as base materials and supplemented them with phenolic compounds (ferulic acid or gallic acid) and 0.5% dicarboxylic acids (adipic acid or succinic acid). With the addition of phenolic compounds, the highest brightness significantly decreased while the mechanical properties significantly increased. It was assumed that phenolic compounds serving as plasticizers contribute to increasing the break value and break strength. In the case of maximum elongation, a combination of gallic acid and dicarboxylic acid conferred the highest values for the composite films. It is interesting to note that the inhibitory activity of zein/chitosan films against *S. aureus* was higher than that against *E. coli* due to the contribution of phenolic compounds in edible films.

Investigated edible film formulations revealed a significant effect of additives on apparent color and transparency of the CS films and a certain improvement of mechanical and antimicrobial properties. This improvement of tensile strength may be due to the interaction between functional groups of additive and $-NH_2$ groups in chitosan.

11.3.3 Electrospinning

Electrospinning is a novel technique, which has been applied to synthetic polymers for a while, but as the technology is getting better, it is also possible to electrospun biopolymers. What makes electrospinning of biopolymers difficult is that they in general cannot handle as much mechanical stress as synthetic polymers. Electrospun nanofibers show improved physicochemical properties compared to macroscale fibers and are therefore increasingly investigated for use in novel food packaging systems. Due to the abundance of CS resources in nature and its excellent biocompatibility, this cationic polysaccharide is a very promising polymer for producing functional nanofibers. Although the material has good physicochemical properties, the electrospinning of the polymer is far from being easy. Torres-Giner et al. (2008) investigated the obtainment of CS-based nanoporous structures starting from blends of CS and PLA (polylactic acid). Even though there are very interesting possibilities in this technology and research going on, there are currently very few applications of this in the food sector (Kriegel et al., 2008).

Recently, Munteanu et al. (2018) have shown how to obtain PLA nanocoating by simultaneously using encapsulation of clove and argan oils into chitosan and their deposit onto PLA surface by coaxial electrospinning. The advantage of this method consists of the distribution of bioactive compounds only to the surface contact of film with the food. It is expected that the low quantity of bioactive compounds is necessary to provide antioxidant and antimicrobial activities. By encapsulation, the bioactive agents are protected from environmental media. The higher antibacterial activity was obtained in the case of nanocoating that contained clove oil. This is due to the composition of compounds responsible for antimicrobial activity. Clove oil is rich in eugenol (~80%) responsible for high antimicrobial and antifungal activities, in comparison to argan oil, with contains a low amount of phenolic compounds.

11.3.4 Extrusion Process

11.3.4.1 Polyolefin/Chitosan Films

The research interest in proving antimicrobial activity of CS in combination with synthetic polymers, because of the low cost production/performance ratio of the resulting film food packaging, has increased remarkably during the past decade.

For obtaining foils at industrial scale, the CS has a disadvantage, because it is not thermoplastic as it degrades before the melting. Therefore, unlike conventional thermoplastic polymers, CS cannot be extruded or molded and the films cannot be heat-sealed (Cazón et al., 2017). The manufacturers of food packaging should take into account the properties of chitin/CS material including, for example, flexibility, transmission characteristics, orientation of the polymer, crystallinity and incorporation of additives, etc. The main disadvantage of extrusion is the use of high temperatures and shearing forces that can reduce antimicrobial activities. High viscosity of the composite materials with high CS content normally requires the addition of plasticizers in order to increase the molecular mobility and consequently enhance the material flexibility (Ansari and Fatma, 2014). The most used plasticizer used to improve the processability of chitosan is glycerol. CS is hygroscopic; thus, before melt processing, usually it is dried at 110°C for 24 h. Although the data on the processing of CS films by thermoforming technology at industrial scale are still scarce, active packaging obtained by extrusion processing of recyclable and biodegradable polymers have been reported (Del Nobile et al., 2009; Nam et al., 2007).

The compatibilizers are also used in order to improve the compatibility between incompatible blends, and the mechanical properties. Also, a decrease of the tensile strength of approximately 13% and of the elongation at break of 60% with increased CS content up to 20 wt% was obtained by Quiroz-Castillo et al. (2014) when they prepared extruded poly(ethylene) (PE)/CS blends compatibilized with PE-graft-maleic anhydride in the presence of glycerol, at 130°C and 140°C and 40 rpm. Similarly, Reesha et al. (2015) used a compatibilizer to increase the ductility of the LDPE/CS blends, 5% weight of MA-g-LDPE which has both hydrophilic and hydrophobic molecules. These molecules react with hydrophobic non polar LDPE, and the other polar end groups react with hydrophilic CS. Firstly, Reesha et al. obtained compounds at the temperatures maintained in the compression zones 1 and 2, using 103°C and 132°C for 1%, 107°C and 129°C for 3%, and 99°C and 138°C for 5% LDPE/CS blends. The LDPE/CS tubular films were obtained by processing compounded pellets in a blown extrusion at a temperature of 184°C and a melt pressure of 20 bar. It was found that the transparency, tensile strength and elongation at break and heat sealing strength decreased through incorporation of chitosan, while oxygen permeability showed good values and the overall migration rate (OMR) of LDPE/CS films was within the stipulated upper limit of 60 mg L^{-1} (Reesha et al., 2015).

Investigation of antimicrobial properties of LDPE/CS films showed 85–100% inhibition of *Escherichia coli*. Virgin LDPE and 1%, 3% and 5% LDPE/CS films were also tested as packaging films for chill-stored tilapia, and showed that samples packed in LDPE films were rejected by the seventh day, whereas fish packed in 1% and 3% remained acceptable up to 15 days. Similar results have been obtained by other authors (Vasile et al., 2014).

Martínez-Camacho et al. (2013) obtained CS extruded films by using LDPE as a matrix polymer and ethylene-acrylic acid copolymer as an adhesive, in order to ensure adhesion in the interphase of the immiscible polymers.

11.3.4.2 Processing and Properties of PLA/Chitosan Films

An approach to overcome the use of compatibilizer has been evidenced by Akopova et al. (2012), who have been developing the solid-state reactive blending technique. Accordingly, PLA/CS blends were processed in a semi-industrial co-rotating twin-screw extruder, at different temperatures chosen within the range of 50–150°C and the screw rotation speed of 100 rpm, based on activation of the substrate and the reactant during their intensive intermixing by applying external mechanical energy. This method shows the advantages because the entire modification process proceeds in the solid state and does not require the components melting or employment of any solvents as the reaction medium.

Herrera et al. (2016) successfully obtained green films nanocomposites-based plasticized PLA and chitin nanocrystals prepared by using extrusion and compression molding with two cooling rates. It is recommended not to dry the chitin nanocrystals prior to extrusion in order to avoid the aggregate of nanomaterial into polymeric matrix. Triethyl citrate (TEC) was used both as plasticizer and processing aid for assuring the dispersion of chitin nanocrystals into PLA. The extrusion using liquid feeding was carried out at a screw speed of 300 rpm and with a temperature profile ranging from 170°C at the feeding zone to 200°C at the die. It was demonstrated that the fast cooling rate led to the most transparent PLA/chitin nanocrystals film with appropriated tensile strength and elongation at break, comparable with tensile properties recorded for polyethylene, than in the case of slow cooling rate.

Bonilla et al. (2013) successfully prepared PLA/CS (content of CS in composites of 5 and 10 wt%) films by extrusion technique. In another study, melt blending of PLA with tributyl o-acetyl citrate (ATBC) and various contents of CS was carried out in a Brabender Plastograph at a temperature of 170°C, at a fixed screw speed of 60 rpm for 6 min by Râpă et al. (2016). The polymeric blends were prepared in two stages: firstly, the mixing of chitosan with the plasticizer in order to obtain a homogeneous mass and then, the melt mixing of PLA with plasticizer/CS mixture on the Brabender Plastograph. Once the melted materials were obtained, the thin films, with maximum thickness of 100 mm, were prepared by compression molding using a hydraulic press at a pressure of 150 atm/125 atm at 175°C and pressing time of 3 min. CS incorporated into plasticized PLA led to a decrease in the tensile strength and elongation at break. (see Table 11.3). The decrease in tensile strength with the increase in CS content could be explained by the lack of compatibility between PLA and CS particles and the generation of discontinuities in the PLA matrix which decreased the overall cohesion forces in the matrix and favor its

TABLE 11.3
Tensile Properties and Water Vapor Transmission Rate (WVTR) for PLA/CS Films Compared with Plasticized PLA (Adapted from Râpăet al. 2016)

Sample	Tensile strength at break (MPa)	Elongation at break (%)	Young's modulus (MPa)	WVTR ($g/m^2/24$ h)
Plasticized PLA	33±1.5	253±2.6	23±1.2	42.50
PLA/CS1	22±3.1	207±19	17±3	43.91
PLA/CS3	20±0.1	195±5	16±0.9	53.67
PLA/CS5	14±2.4	138±15	846±41	48.30

flow and break (Bonilla et al., 2013; Quiroz-Castillo et al., 2014). Water vapor transmission rate (WVTR) was higher than that of plasticized PLA, proved by the greater water affinity of CS. All samples in contact with 10% (v/v) ethanol (Simulant A) for ten days at a temperature of 20°C showed very low migration, far below the overall migration limit of 10 mg dm^{-2} of food contact surface required by the "Regulation (EU) No. 10/2011." Also, prepared PLA/CS films showed good antifungal activity against *Aspergillus brasiliensis ATCC 16404, Penicillium corylophilum CBMF1* and *Fusarium graminearum G87* and a significant reduction in S. *aureus* and *E. coli* on the contact surfaces.

11.3.4.3 Other Chitosan Film Formulations

Using the extrusion technology to manufacture active packaging materials containing eugenol-loaded CS NPs incorporated into thermoplastic flour at temperatures above 150°C was reported (Woranuch and Yoksan, 2013).

Correlo et al. (2005) showed that adding CS to the blends based on poly-ε-caprolactone (PCL), poly(butylene succinate) (PBS), PLA, poly(butylene terephthalate adipate) (PBTA), and poly(butylene succinate adipate) (PBSA) led to a decreased tensile strength and a significant decrease in the elongation. This behavior was explained due to the reduced adhesive interaction between the chitosan and the polyesters.

11.3.5 Thermoforming

Thermoforming is one of the fastest growing segments of the plastics processing sector due to the major advantages it has such as the cost-effectiveness of thermoforming tools, reasonably priced thermoforming machines and the possibility for processing even multi-layered materials (Ashter, 2014). By thermoforming, the thermoplastic film is heated to a temperature above its softening temperature for amorphous polymers, or slightly below the melting point, for semi-crystalline materials and then formed/stretched. During thermoforming, the increase of packaging area and thinning of the film may be a problem leading to increase of oxygen transmission rate (OTR) (Ashter, 2014).

Although CS-based films have proven to be very effective in food preservation, there are limited data on antimicrobial packaging based on chitin/chitosan materials produced in line with existing infrastructures and logistics systems. A single polymer is often unable to provide a suitable barrier; thus, most food packaging materials are multilayer constructions obtained by thermoforming. In multilayer constructions, polymers with different barrier properties are combined, where at least one layer acts as an oxygen barrier and other layers act as water barriers and polymers with sealing properties (Pettersen et al., 2005).

11.3.6 Intelligent Packaging

Thanks to the safe use of CS in food packaging, intelligent packaging has been reported (Jung et al., 2013; Pereira Jr. et al., 2014) with the purpose to monitor the conditions of the food in real time, to inform consumers about the conditions of transport and storage of these products and to establish if the actual parameters of food quality are safe before consumption.

Li et al. (2019) developed a novel pH-sensitive intelligent packaging by using purple potato extractions, CS and surface-deacetylated chitin nanofibers. This intelligent packaging has antioxidant properties higher than that of CS, assigned to the purple potato extract.

The grafting to CS of some functional molecules, like phenolic compounds or essential oils, adds to its antioxidant, antimicrobial and others properties, while the addition of nanofillers to CS and other biopolymers improves the required properties for food applications and also can add attributes of electrical conductivity and magnetic properties for active and intelligent packaging. Electrical conductivity is a required property for the processing of food at low temperature using electric fields or for sensors application (Nunes et al., 2018).

11.3.7 Packaging under Modified Atmosphere

Modified atmosphere packaging (MAP) is traditionally used to preserve the freshness of products, meats and fish by controlling their biochemical metabolism, such as respiration. It was found that the combined effect of dipping in a CS solution (1 g/100 mL) and packaging under modified atmosphere (MAP, 70% CO_2, 30% N_2) on shelf life extension of refrigerated chicken fillets showed the best effect on microbiological (total viable counts of *Pseudomonas* spp., lactic acid bacteria and Enterobacteriaceae), physicochemical (headspace gas composition, pH, color and thiobarbituric acid test) and sensory (odor and taste) parameters during monitoring for up to 14 days (Latou et al., 2014).

The effect of chitosan coating containing antibrowning agents and modified atmosphere packaging on the browning and shelf life of fresh-cut lotus root stored at 4°C for ten days was studied by Xing et al. (2010). It was found that at the end of storage, the coated and MAP samples exhibited the lowest polyphenol oxidase (PPO) activity and malondialdehyde (MDA) content. Its highest overall visual quality (OVQ) scores (>7) demonstrated that CS-based coating and MAP treatment could provide a better inhibitory effect on the browning and extend the shelf life of fresh-cut lotus root. In another study performed by Duan et al. (2010), CS solutions (3%) incorporating 20% krill oil (w/w CS) with or without the addition of 0.1 μL/mL cinnamon leaf essential oil were tested on fresh lingcod (*Ophiodon elongates*) fillets vacuum-impregnated with the coating solutions, vacuum or modified atmosphere (MAP) (60% CO_2 + 40% N_2) packaged, and then stored at 2°C for up to 21 days for physicochemical and microbial quality evaluation. It was reported that CS-krill oil coating increased total lipid and omega-3 fatty acid contents of the lingcod by about twofold. The combined CS coating and vacuum or MAP packaging reduced lipid oxidation as represented in thiobarbituric acid reactive substances (TBARS), chemical spoilage as reflected in total volatile base nitrogen (TVB-N), and microbiological spoilage as reported in total plate count (2.22–4.25 log reductions during storage). CS-krill oil coating did not change the color of the fresh fillets, nor affect consumers' acceptance of both raw and cooked fish samples. Consumers preferred the overall quality of CS-coated, cooked lingcod samples over the control, based on their firm texture and less fishy aroma.

11.4 QUALITY ANALYSES FOR FOOD PACKAGED IN ACTIVE PACKAGING FILMS

11.4.1 Biochemical Analysis

The mean pH measurements of barracuda steaks stored at 2°C for 20 days were measured in films prepared from CS incorporated with ginger (*Zingiber officinale*) essential oil at different concentrations (0.1, 0.2 and 0.3% v/v) compared with control, EVOH film wrapped (Remya et al., 2016). On the 12th day of the chilled storage study, a pH value of 6.15 was found in the case of CS/ginger essential oil film, very close with that of the initial pH of the fish steak (5.94). Instead, EVOH film-wrapped fish sample showed an increase in pH value to 7.3 due to accumulation of alkaline compounds derived mainly from microbial action in the fish muscle.

Another index of spoilage of the food is the total volatile base nitrogen associated with the activity of spoilage bacteria and endogenous enzymes. The limit of acceptability for TVB-N value is 30–35 mg N_2 100 g^{-1} (Remya et al., 2016). The TVB-N value of fish steak stored in CS/ginger essential oil film was 32.4±1.5 mg N_2 100 g^{-1} on the 20th day of storage, lower than the control and EVOH film-wrapped sample.

Extending the shelf life of the food is proven by the retardation of lipid oxidation. Lipid oxidation is a key factor related to the deterioration of food quality, such as undesirable rancid off-flavors and poisoning. In addition, lipids are easily oxidized in the presence of light, heat, and enzymes. The lipid oxidation takes place by the attack of oxygen at the double bond in fatty acids. Therefore, hydroperoxides, aldehyde, ketone and alcohol compounds appear as result of lipid oxidation. Crizel et al. (2018) developed CS/olive residue flour (10 wt%, 20 wt% and 30 wt% into CS) biocomposites. Crizel et al. found the antioxidant capacity of the films increased proportionally with the concentration of flour or microparticles added to the film. The films with 30% of flour or microparticles were effective as protective packaging against the oxidation of nuts for 31 days.

2-thiobarbituric acid (TBA) is a good index for evaluation of secondary lipid oxidation products.

In a study performed on minced beef samples coated with CS/kombucha tea extract KT at 3% active film, Ashrafi et al. (2018) found a TBA value of

0.31 malondialdehyde mg kg^{-1} after six days of storage at 4°C, demonstrating that this active biocomposite inhibited lipid oxidation. A TBA value of 1.5 is considered as an unacceptable off odor of meat.

TBA was used as an indicator to determine the content of the secondary lipid oxidation products closely related to the sensory quality of meat. A TBA value of 2 mg malonaldehyde kg^{-1} meat sample was the limit (Ashrafi et al., 2018). In another study performed by Duan et al. (2010), chitosan solutions (3%) incorporating 20 % krill oil (w/w chitosan) with or without the addition of 0.1 μL mL^{-1} cinnamon leaf essential oil were tested on fresh lingcod (*Ophiodon elongates*) fillets vacuum-impregnated with the coating solutions, vacuum or modified atmosphere (MAP) (60% CO_2 + 40% N_2) packaged, and then stored at 2°C for up to 21 days for physicochemical and microbial quality evaluation. It was reported that chitosan-krill oil coating increased total lipid and omega-3 fatty acid contents of the lingcod by about twofold. The combined chitosan coating and vacuum or MAP packaging reduced lipid oxidation as represented in TBARS, chemical spoilage as reflected in TVB-N, and microbiological spoilage as reported in total plate count (2.22–4.25 log reductions during storage). CS-krill oil coating did not change the color of the fresh fillets, nor did it affect consumers' acceptance of both raw and cooked fish samples. Consumers preferred the overall quality of CS-coated, cooked lingcod samples over the control, based on their firm texture and less fishy aroma.

The effect of CS coating containing antibrowning agents and modified atmosphere packaging (MAP) on the browning and shelf life of fresh-cut lotus root stored at 4°C for ten days was studied by Xing et al. (2010). It was found that at the end of storage, the coated together with MAP samples exhibited the lowest polyphenol oxidase (PPO) activity and malondialdehyde (MDA) content. Its highest overall visual quality scores (>7) demonstrated that CS-based coating together with MAP treatment could provide a better inhibitory effect on the browning and extend the shelf life of fresh-cut lotus root.

In another study, Schreiber et al. (2013) found that the gallic acid-grafted CS film-forming solution reduced level of thiobarbituric acid reactive substances, peroxide, and conjugated trienes formation for 15 weeks at 50°C under low humidity, as compared to PE bags during storage of ground peanuts.

11.4.2 Antimicrobial Analysis

Beyki et al. (2014) encapsulated *Mentha piperita* essential oils in CS-cinnamic acid nanogel in order to enhance antimicrobial activity and stability of the oils against *A. flavus*. The encapsulated oils preserved the tomato fruit well at 1000 ppm concentration during the 1-month storage period. In the case of oil concentrations of 800 and 900 ppm, the decay process was delayed to 10 and 14 days, respectively.

A comparison of the microbial assay of cheese (10 g) packaged in CS, CS containing natamycin and methylcellulose/natamycin films, then sealed in polyethylene bags was performed under accelerated conditions at 20°C for seven days (Santonicola et al., 2017). The mold/yeast counts values obtained in coated cheese were 8.30 log CFU/g for CS film, 7.91 log CFU/g for CS/natamycin film and 8.25 log CFU/g for methylcellulose/natamycin film.

Bionanocomposite films including chitin nanofibers and CS from α-chitin powder, extracted from lobster wastes demonstrated an inhibitory effect (>80%) against *A. Niger* (Salaberria et al., 2015a).

Zemljič et al. (2013) reported the antimicrobial activity of CS in combination with polyethylene terephtalate against *Salmonella enterica, Campylobacter spp., Escherichia coli, Listeria monocytogenes* and *Candida albicans.* In other studies, CS/(polyvinyl alcohol (PVA) film was found to effectively protect meat (Agostino et al., 2012; Liang et al., 2009; Zhuang et al., 2012; Wang et al., 2018), and by blending with LDPE, a potential active packaging material to increase shelf life of poultry meat was obtained (Vasile et al., 2013; Stoleru et al., 2016).

CS/polyvinyl alcohol (PVA) film was used as coating on tomatoes (Tripathi et al., 2009). The improved microbiological analysis against food pathogenic bacteria (*Escherichia coli, Staphylococcus aureus,* and *Bacillus subtilis*) was obtained. Excellent antimicrobial properties were obtained in the case of chitosonium-acetate/ethylene–vinyl alcohol (EVOH) copolymers blend (Fernandez-Saiz et al. 2010).

In another paper, Siripatrawan and Vitchayakitti (2016) studied the antimicrobial activity of CS/propolis extract films (containing 0, 2.5, 5, 10 and 20% wt propolis extract). In this case, the antimicrobial activity against *Staphylococcus aureus, Salmonella Enteritidis, Escherichia coli* and *Pseudomonas aeruginosa* was enhanced by using agar diffusion technique. Similar results were found by Li et al. (2011) that studied the antibacterial effect of CS/chitin whisker/rectorite ternary composite films against *Escherichia coli* and *Staphylococcus aureus.*

Total inhibition of the growth of both Gram-positive (*S. aureus*) and Gram-negative (*P. aeruginosa*) food pathogenic bacteria in the case of CS/ellagic acid biocomposites films (0.5, 1.0, 2.5 and 5.0% wt ellagic acid relative to CS) was demonstrated by Vilela et al. (2017).

Ashrafi et al. (2018) investigated the shelf life of the minced beef meat coated with CS/kombucha tea (1–3% w/w) active films. The results obtained showed antimicrobial effect of the film containing 3% KT was significantly higher toward *E. coli* than toward *S. aureus*, retardation of lipid oxidation and microbial growth from 5.36 to 2.11 log CFU g^{-1} in four days storage and extension of the shelf life of the minced beef meat up to three days.

A viable method to prolong the shelf life of food products consists of coating PE films by spraying with a CS/ZnO nanocomposite solution (Al-Naamani et al., 2016). Oxygen plasma treatment was used to obtain a better attachment of bioactive solution on the PE surface. Using this coating system, 99% inactivation of viable pathogenic bacteria of 99% was found.

One innovative method to design potential active food packaging was developed by Khan et al. (2016) that prepared antimicrobial nanocomposite films by covering the surface of the nanocomposite films based on CS with nisin–ethylene diamine tetraacetate (EDTA) antimicrobial solution cross-linked with genipin. Nisin is the only bacteriocin that has generally considered as safe (GRAS) status from the Food and Drug Administration. After that, the films were γ-irradiated at doses 0.5 and 1.5 kGy. Khan et al. evidenced the role of gamma irradiated cross-linked films to increase the microbiological shelf life of pork meats by more than 5 weeks.

All data reveal the great importance of active food packaging films based on CS to inhibit the development of microorganisms and thus to contribute to prolonging the shelf life of food.

11.4.3 Sensory Evaluation

For sensory evaluation, panelists were considered on the basis of their previous experience in consuming food. Four attributes are established to describe the samples: odor appearance (color and slime), flavor and overall acceptability. The acceptance of the attributes (red color, discoloration and off odor) was if the scores are below 3 using an intensity 5-point scale (Hu et al., 2015). The best effect on the preservation of pork meat sample was due to the synergistic effects of compounds of cinnamon essential oil, as reported by Hu et al. (2015).

Lin et al. (2019) obtained a bioactive coating based on CS NPs, moringa oil and gelatin, by electrospinning, with excellent antimicrobial properties against *L. monocytogenes* and *S. aureus* on cheese. At the optimal concentrations of moringa oil and CS NPs of 20 mg mL^{-1} and 3.0 mg mL^{-1}, respectively, no effect was observed on the sensory quality of cheese.

A panel of six regular members assessed the appearance, color, odor, taste, texture and flavor of the fish steak wrapped in control, EVOH and CS/ginger essential oil film, for 20 days using a nine point hedonic scale. They found the overall acceptability scope of the fish steak wrapped in CS/ginger essential oil film to be below 5, as the lower limit of acceptability (Remya et al., 2016). The shelf life of fish steak stored in active food packaging was prolonged by 12 days compared with the control sample.

In another study by Kumar et al. (2018), packaging of red grapes with hybrid nanomaterials films based on CS, gelatin and silver nanoparticles (AgNPs) 0.1% led to extend the shelf life of food by up to 18 days, before mildew occurred.

In a study carried out on the effect of mango leaf extract (MLE) incorporated into CS on the preservation of cashew nuts for 28 days storage, an increase of 56% oxidation resistance was attained in comparison with what was recorded with a commercial polyamide/PE film (Rambabu et al., 2019).

The results showed that the prepared food active packaging films based on CS and active agents could be a promising food packaging material to protect packaged food from microbial infection and extend its shelf life.

11.5 BIODEGRADABILITY

Cinelli et al. (2017) studied the degradability of nanocomposites based on PLA and chitin nano fibrils in aerobic conditions. A significant degradation of the films buried in synthetic compost was reported after 30 days. Reinforcing this biopolymer with nanoclays can lead to novel composites with enhanced physical properties, such as water resistance, without loss in biodegradability (Fernandez–Saiz, 2011). However, the behavior of NPs in soil should be evaluated.

An extensive study on the PLA/CS biocomposites sheets degradation on the soil burial was done by Vasile et al. (2018). The weight loss measurements, changes in average molecular weight and its distribution, mechanical, thermal and surface properties correlated with structural and morphological modifications suggested that CS as hydrophilic compound promotes the soil burial degradation of PLA, but longer time is necessary for entire degradation.

11.6 INDUSTRIAL DEVELOPMENT

The manufacturers and suppliers of CS and chitin products worldwide commercialize high-grade CS and chitin for food application (Ferreira et al., 2016). Some of these manufacturers and suppliers are Primex (Siglufjordur, Iceland) with ChitoClear®, based on the purest CS possible, Norwegian Chitosan (Kløfta, Norway) under brand names NorLife and Kitoflokk™, for food and beverages; G.T.C. Bio Corporation (Qingdao, China) different grades of chitin and CS with a price of around 20 €/kg for chitin and between 18 and 45 €/kg for CS (depending on required purity grade). Chitin has as some of its main applications: coffee capsules, food bags, and packaging films while those of CS are edible membranes and coatings (strawberries, cherries, mango, guava, among others) and packaging membranes for vegetables and fruit.

Nanoencapsulated CS biopolymer nanofilms considerably decreased water vapor permeability, while the introduction of the dispersed CS into protein significantly improved the mechanical strength of nanofilms, making the use of these films in the food industry practicable (Manigandan et al., 2018). CS as coatings protects fresh vegetables and fruits from fungal degradation and acts as a barrier between the produce and microorganisms (Cuq et al., 1995; Rhoades and Roller, 2000). The degraded and native CS showed antimicrobial activity against spoilage microorganisms. Mild hydrolysis of CS resulted in improved inhibitory activity in saline and greater inhibition of growth of spoilage yeasts, whereas highly degraded products of CS exhibited no antimicrobial activity.

11.7 CONCLUSIONS AND FUTURE TRENDS

The research interest in proving antimicrobial activity of CS/chitin in combination with both synthetic polymers and bioactive agents or NPs, as well as improved mechanical properties of the resulting food packaging films has increased remarkably during the past decade.

The most studied properties of chitin/CS food packaging films are related to tensile properties, water vapor permeability and oxygen barrier, solubility, swelling, color properties and morphological and structural characteristics, as well as antibacterial activity and efficacity to extend the shelf life of packaged food.

Much research deals with the use of CS in solution or blending with other polymers, as active packaging for preserving food products, assuring its quality as well as a prolonged shelf life. Chitin nanofibers are an attractive ingredient for film production since they enhance the mechanical properties of composite materials. In the majority of the studies, solvent casting is the most frequently used technique to prepare chitin/CS-based films. However, further studies are necessary to improve the functional and water barrier properties of chitin/CS films and also to validate their applications on real food products in order to scale-up their use on the industrial level.

Besides exploring the improving of properties, the researchers should investigate the possibilities to reduce the cost of chitin/CS-based materials to be comparable with commercial food packaging.

REFERENCES

Abdollahi, M., Rezaei, M., Farzi, G., 2012. A novel active bionanocomposite film incorporating rosemary essential oil and nanoclay into chitosan. *J. Food. Eng.* 111(2), 343–350. doi:10.1016/j.jfoodeng.2012.02.012.

Agostino, A.D., Gatta, A.L., Busico, T., Rosa, M.D., Schiraldi, C., 2012. Semi-interpenetrated hydrogels composed of PVA and hyaluronan or chondroitin sulphate: Chemico-physical and biological characterization. *J. Biotechnol. Biomater.* 2, 140. doi:10.4172/2155-952X.100014.

Agulló, E., Rodríguez, M.S., Ramos, V., Albertengo, L., 2003. Present and future role of chitin and chitosan in food. Review. *Macromol. Biosci.* 3(10), 521–530. doi:10.1002/mabi.200300010.

Ahmed, J., Mulla, M., Arfat, Y.A., Thai, L.A., 2017. Mechanical, thermal, structural and barrier properties of crab shell chitosan/graphene oxide composite films. *Food Hydrocoll.* 71, 141–148. doi:10.1016/j.foodhyd.2017.05.013.

Akopova, T.A., Demina, T.S., Shchegolikhin, A.N. et al., 2012. A novel approach to design chitosan-polyester materials for biomedical applications. *Int. J. Polym. Sci.* 2012, Article ID 827967. doi:10.1155/2012/827967.

Akyuz, L., Kaya, M., Koc, B. et al., 2017. Diatomite as a novel composite ingredient for chitosan film with enhanced physicochemical properties. *Int. J. Biol. Macromol.* 105(2), 1401–1411. doi:10.1016/j.ijbiomac.2017.08.161.

Al-Naamani, L., Dobretsov, S., Dutta, J., 2016. Chitosan-zinc oxide nanoparticle composite coating for active food packaging applications. *Innov. Food Sci. Emerg. Technol.* 38, 231–237. doi:10.1016/j.ifset.2016.10.010.

Altieri, C., Scrocco, C., Sinigaglia, M., Del Nobile, M.A., 2005. Use of chitosan to prolong mozzarella cheese shelf life. *J. Dairy Sci.* 88(8), 2683–2688. doi:10.3168/jds.S0022-0302(05)72946-5.

Ansari, S., Fatma, T., 2014. Polyhydroxybutyrate—A biodegradable plastic and its various formulations. *Int. J. Innov. Res. Sci. Eng. Technol.* 3(2), 9494–9499.

Antoniou, J., Liu, F., Majeed, H., Zhong, F., 2015. Characterization of tara gum edible films incorporated with bulk chitosan and chitosan nanoparticles: A comparative study. *Food Hydrocoll.* 44, 309–319. doi:10.1016/j.foodhyd.2014.09.023.

Appendini, P., Hotchkiss, J.H., 2002. Review of antimicrobial food packaging. *Innov. Food. Sci. Emerg. Technol.* 3(2), 113–126. doi:10.1016/s1466-8564(02)00012-7.

Ashrafi, A., Jokar, M., Nafchi, A.M., 2018. Preparation and characterization of biocomposite film based on chitosan and kombucha tea as active food packaging. *Int. J. Biol. Macromol.* 108, 444–454. doi:10.1016/j.ijbiomac.2017.12.028.

Ashter, S.A., 2014. Introduction to thermoforming. In: *Thermoforming Single Multilayer Laminates*, Elsevier, pp. 1–12. doi:10.1016/b978-1-4557-3172-5.00001-3.

Azarifar, M., Ghanbarzadeh, B., Khiabani, M.S., Basti, A.A., Abdulkhani, A., Noshirvani, N., Hosseini, M., 2019. The optimization of gelatin-CMC based active films containing chitin nanofiber and Trachyspermum ammi essential oil by response surface methodology. *Carbohydr. Polym.* 208, 457–468. doi:10.1016/j.carbpol.2019.01.005.

Azeredo, H.M.C., Mattoso, L.H.C., Avena-Bustillos, R.J., Filho, G.C., Munford, M.L., Wood, D., McHugh, T.H., 2010. Nanocellulose reinforced chitosan composite films as affected by nanofiller loading and plasticizer content. *J. Food Sci.* 75(1), 1–7. doi:10.1111/j.1750-3841.2009.01386.x.

Balti, R., Mansour, M.B., Sayari, N., Yacoubi, L., Rabaoui, L., Brodu, N., Massé, A., 2017. Development and characterization of bioactive edible films from spider crab (*Maja crispata*) chitosan incorporated with *Spirulina* extract. *Int. J. Biol. Macromol.* 105(2), 1464–1472. doi:10.1016/j.ijbiomac.2017.07.046.

Barra, A., Ferreira, N.M., Martins, M.A. et al., 2019. Eco-friendly preparation of electrically conductive chitosan—Reduced graphene oxide flexible bionanocomposites for food packaging and biological applications. *Compos. Sci. Technol.* 173, 53–60. doi:10.1016/j.compscitech.2019.01.027.

Belalia, R., Grelier, S., Benaissa, M., Coma, V., 2008. New bioactive biomaterials based on quaternized chitosan. *J. Agric. Food. Chem.* 56(5), 1582–1588. doi:10.1021/jf071717+.

Beyki, M., Zhaveh, S., Khalili, S.T. et al., 2014. Encapsulation of *Mentha piperita* essential oils in chitosan–cinnamicacid nanogel with enhanced antimicrobial activity against *Aspergillus flavus*. *Ind. Crops Prod.* 54, 310–319. doi:10.1016/j.indcrop.2014.01.033.

Bikiaris, D.N., Triantafyllidis, K.S., 2013. HDPE/Cu-nanofiber nanocomposites with enhanced antibacterial and oxygen barrier properties appropriate for food packaging applications. *Mater. Lett.* 93, 1–4. doi:10.1016/j.matlet.2012.10.128.

Bonilla, J., Fortunati, E., Vargas, M., Chiralt, A., Kenny, J.M., 2013. Effects of chitosan on the physicochemical and antimicrobial properties of PLA films. *J. Food Eng.* 119(2), 236–243. doi:10.1016/j.jfoodeng.2013.05.026.

Bonilla, J., Poloni, T., Lourenço, R.V., Sobral, P.J.A., 2018. Antioxidant potential of eugenol and ginger essential oils with gelatin/chitosan films. *Food Biosci.* 23, 107–114. doi:10.1016/j.fbio.2018.03.007.

Bordenave, N., Grelier, S., Coma, V., 2010. Hydrophobization and antimicrobial activity of chitosan and paper-based packaging material. *Biomacromolecules* 11(1), 88–96. doi:10.1021/bm9009528.

Butnaru, E., Stoleru, E., Brebu, M.A., Darie-Nita, R.N., Bargan, A., Vasile, C., 2019. Chitosan-based bionanocomposite films prepared by emulsion technique for food preservation. *Materials* 12(3), 373. doi:10.3390/ma12030373.

Cansu Feyzioglu, G., Tornuk, F., 2016. Development of chitosan nanoparticles loaded with summer savory (*Satureja hortensis* L.) essential oil for antimicrobial and antioxidant delivery applications. *LWT Food Sci. Technol.* 70, 104–110.

Cazón, P., Velazquez, G., Ramírez, J.A., Vázquez, M., 2017. Polysaccharide-based films and coatings for food packaging: A review. *Food Hydrocoll.* 68, 136–148. doi:10.1016/j.foodhyd.2016.09.009.

Cheng, S.Y., Wang, B.J., Weng, Y.M., 2015. Antioxidant and antimicrobial edible zein/chitosan composite films fabricated by incorporation of phenolic compounds and dicarboxylic acids. *LWT Food Sci. Technol.* 63(1), 115–121. doi:10.1016/j.lwt.2015.03.030.

Cinelli, P., Coltelli, M.B., Mallegni, N., Morganti, P., Lazzeri, A., 2017. Degradability and sustainability of nanocomposites based on polylactic acid and chitin nano fibrils. *Chem. Eng. Trans.* 60, 115–120.

Coles, R., Kirwan, M.J., 2011. *Food and Beverage Packaging Technology*, Ed. Coles, R., Kirwan, M., Oxford: Willey-Blackwell Publishing Ltd.

Correlo, V.M., Boesel, L.F., Bhattacharya, M., Mano, J.F., Neves, N.M., Reis, R.L., 2005. Properties of melt processed chitosan and aliphatic polyester blends. *Mat. Sci. Eng. A* 403(1–2), 57–68. doi:10.1016/j.msea.2005.04.055.

Coma, V., 2013. Polysaccharide-based biomaterials with antimicrobial and antioxidant properties. *Polímeros Cienc. Tecnol.* 20(2), 287–297. doi:10.4322/polimeros020ov002.

Corrales, M., Fernández, A., Han, J.H., 2014. Antimicrobial packaging systems. In: *Innovations in Food Packaging*, Elsevier, pp. 134–170. doi:10.1016/B978-0-12-394601-0.00007-2.
Crizel, T.M., Rios, A.O., Alves, V.D., Bandarra, N., Moldão-Martins, M., Flôres, S.H., 2018. Active food packaging prepared with chitosan and olive pomace. *Food Hydrocoll.* 74, 139–150. doi:10.1016/j.foodhyd.2017.08.007.
Cruz-Romero, M.C., Murphy, T., Morris, M., Cummins, E., Kerry, J.P., 2013. Antimicrobial activity of chitosan, organic acids and nano-sized solubilisates for potential use in smart antimicrobially-active packaging for potential food applications. *Food Control* 34(2), 393–397. doi:10.1016/j.foodcont.2013.04.042.
Cuq, B., Contard, N., Guilbert, S., 1995. Edible films and coatings as active layers. In: Blackie Academic & Professional, Rooney, M.L. (Eds.), *Active Food Packaging*, Glasgow, UK: Chapman and Hall, pp. 111–142.
Darie–Niţă, R.N., Vasile, C., 2018. Halloysite containing composites for food packaging applications. In: Cirillo, G., Kozlowski, M.A., Spizzirri, U.G. (Eds.), *Composites Materials for Food Packaging*, Scrivener Publishing LLC, Beverly, MA, pp. 73–122.
De la Caba, K., Guerrero, P., Trung, T.S. et al., 2019. From seafood waste to active seafood packaging: An emerging opportunity of the circular economy. *J. Clean. Prod.* 208, 86–98. doi:10.1016/j.jclepro.2018.09.164.
Del Nobile, M.A., Conte, A., Cannarsi, M., Sinigaglia, M., 2009. Strategies for prolonging the shelf life of minced beef patties. *J. Food Saf.* 29(1), 14–25. doi:10.1111/j.1745-4565.2008.00145.x.
Deng, Z., Jung, J., Zhao, Y., 2017. Development, characterization, and validation of chitosan adsorbed cellulose nanofiber (CNF) films as water resistant and antibacterial food contact packaging. *LWT Food Sci. Technol.* 83, 132–140.
De Queiroz Antonino, R.S.C.M., Lia Fook, B.R.P., de Oliveira, V.A. et al., 2017. Preparation and characterization of chitosan obtained from shells of shrimp (*Litopenaeus vannamei Boone*). *Mar. Drugs* 15(5), pii: E141. doi:10.3390/md15050141.
Duan, J., Jiang, Y., Cherian, G., Zhao, Y., 2010. Effect of combined chitosan-krill oil coating and modified atmosphere packaging on the storability of cold-stored lingcod (*Ophiodon elongates*) fillets. *Food Chem.* 122(4), 1035–1042.
Duncan, T.V., 2011. Applications of nanotechnology in food packaging and food safety: Barrier materials, antimicrobials and sensors. *J. Colloid Interface Sci.* 363(1), 1–24.
Dutta, P.K., Tripathi, S., Mehrotra, G.K., Dutta, J., 2009. Perspectives for chitosan based antimicrobial films in food applications. *Food Chem.* 114(4), 1173–1182.
Elsabee, M.Z., Morsi, R.E., Fathy, M., 2016. Chitosan-oregano essential oil blends use as antimicrobial packaging material. In: Barros-Velázquez, J. (Ed.), *Antimicrobial Food Packaging*, Elsevier Inc., Amsterdam, Netherlands, pp. 539–551.
Escamilla-Garcia, M., Calderon-Dominguez, G., Chanona-Perez, J.J. et al., 2013. Physical and structural characterisation of zein and chitosan edible films using nanotechnology tools. *Int. J. Biol. Macromol.* 61, 196–203. doi:10.1016/j.ijbiomac.2013.06.051.
Esmaeili, A., Asgari, A., 2015. In vitro release and biological activities of *Carum copticum* essential oil (CEO) loaded chitosan nanoparticles. *Int. J. Biol. Macromol.* 81, 283–290.
Fardioui, M., Kadmiri, I.M., Qaiss, A.E.K., Bouhfid, R., 2018. Bio-active nanocomposite films based on nanocrystalline cellulose reinforced styrylquinoxalin-grafted-chitosan: Antibacterial and mechanical properties. *Int. J. Biol. Macromol.* 114, 733–740.
Fernandez-Saiz, P., 2011. Chitosan polysaccharide in food packaging applications. In: Lagarón, J.M. (Ed.), *Multifunctional and Nanoreinforced Polymers for Food Packaging*, Woodhead Publishing Limited, Amsterdam, Netherlands, pp. 571–593.

Fernandez-Saiz, P., Lagaron, J.M., Hernandez-Muñoz, P., Ocio, M.J., 2008. Characterization of antimicrobial properties on the growth of *S. aureus* of novel renewable blends of gliadins and chitosan of interest in food packaging and coating applications. *Int. J. Food Microbiol.* 124(1), 13–20. doi:10.1016/j.ijfoodmicro.2007.12.019.

Fernandez-Saiz, P., Ocio, M.J., Lagaron, J.M., 2010. Antibacterial chitosan-based blends with ethylene–vinyl alcohol copolymer. *Carbohydr. Polym.* 80(3), 874–884.

Ferreira, A.R.V., Alves, V.D., Coelhoso, I.M., 2016. Polysaccharide-based membranes in food packaging applications. *Membranes (Basel)* 6(2), 22. doi:10.3390/membranes6020022.

Gamal, R.F., El-Tayeb, T.S., Raffat, E.I., Ibrahim, H.M.M., Bashandy, A.S., 2016. Optimization of chitin yield from shrimp shell waste by *Bacillus subtilis* and impact of gamma irradiation on production of low molecular weight chitosan. *Int. J. Biol. Macromol.* 91, 598–608. doi:10.1016/j.ijbiomac.2016.06.008.

Genskowsky, E., Puente, L.A., Pérez-Álvarez, J.A., Fernandez-Lopez, J., Muñoz, L.A., Viuda-Martos, M., 2015. Assessment of antibacterial and antioxidant properties of chitosan edible films incorporated with maqui berry (*Aristotelia chilensis*). *LWT Food Sci. Technol.* 64(2), 1057–1062.

Ghelejlu, S.B., Esmaiili, M., Almasi, H., 2016. Characterization of chitosan–nanoclay bionanocomposite active films containing milk thistle extract. *Int. J. Biol. Macromol.* 86, 613–621.

Giannakas, A.E., Leontiou, A.A., 2017. Montmorillonite composite materials and food packaging. In: Cirillo, G., Kozlowski, M.A., Spizzirri, U.G. (Eds.), *Composites Materials for Food Packaging*, Scrivener Publishing LLC, Wiley Blackwell, pp. 1–71.

Gomaa, M., Hifney, A.F., Fawzy, M.A., Abdel-Gawad, K.M., 2018. Use of seaweed and filamentous fungus derived polysaccharides in the development of alginate-chitosan edible films containing fucoidan: Study of moisture sorption, polyphenol release and antioxidant properties. *Food Hydrocoll.* 82, 239–247.

Goyal, S., Goyal, G.K., 2012. Nanotechnology in food packaging—A critical review. *Russian J. Agric. Socio-Econ.* 10(10), 14–24.

Ham-Pichavant, F., Sèbe, G., Pardon, P., Coma, V., 2005. Fat resistance properties of chitosan-based paper packaging for food applications. *Carbohydr. Polym.* 61(3), 259–265. doi:10.1016/j.carbpol.2005.01.020.

Hanušová, K., Dobiáš, J., Klaudisová, K., 2009. Effect of packaging films releasing antimicrobial agents on stability of food products. *Czech J. Food Sci.* 27, 347–349.

Hari, N., Francis, S., Nair, A.G.R., Nair, A.J., 2018. Synthesis, characterization and biological evaluation of chitosan film incorporated with β-carotene loaded starch nanocrystals. *Food Pack. Shelf Life* 16, 69–76.

Hasheminejad, N., Khodaiyan, F., Safari, M.., 2019. Improving the antifungal activity of clove essential oil encapsulated by chitosan nanoparticles. *Food Chem.* 275, 113–122.

Herrera, N., Salaberria, A.M., Mathew, A.P., Oksman, K., 2016. Plasticized polylactic acid nanocomposite films with cellulose and chitin nanocrystals prepared using extrusion and compression molding with two cooling rates: Effects on mechanical, thermal and optical properties. *Compos. A* 83, 89–97. doi:10.1016/j.compositesa.2015.05.024.

Higueras, L., López-Carballo, G., Hernández-Muñoz, P., Gavara, R., Rollini, M., 2013. Development of a novel antimicrobial film based on chitosan with LAE (ethyl-Nα-dodecanoyl-*l*-arginate) and its application to fresh chicken. *Int. J. Biol. Macromol.* 165, 339–345.

Hooda, R., Batra, B., Kalra, V., Rana, J.S., Sharma, M., 2018. Chitosan-based nanocomposites in food packaging. In: Ahmed, S. (Ed.), *Green and Sustainable Advanced Packaging Materials*, Springer Nature Singapore Pts. Ltd., Singapore, pp. 269–285.

Hu, J., Wang, X., Xiao, Z., Bi, W., 2015. Effect of chitosan nanoparticles loaded with cinnamon essential oil on the quality of chilled pork. *LWT Food Sci. Technol.* 63(1), 519–526.

Jahed, E., Khaledabad, M.A., Almasi, H., Hasanzadeh, R., 2017a. Physicochemical properties of *Carum copticum* essential oil loaded chitosan films containing organic nanoreinforcements. *Carbohydr. Polym.* 164, 325–338.

Jahed, E., Khaledabad, M.A., Bari, M.R., Almasi, H., 2017b. Effect of cellulose and lignocellulose nanofibers on the properties of *Origanum vulgare* ssp. gracile essential oil-loaded chitosan films. *React. Funct. Polym.* 117, 70–80.

Jung, J., Lee, K., Puligundla, P., Ko, S., 2013. Chitosan-based carbon dioxide indicator to communicate the onset of kimchi ripening. *LWT Food Sci. Technol.* 54(1), 101–106.

Kalaycıoglu, Z., Torlak, E., Akın-Evingür, G., Özen, I., Erim, F.B., 2017. Antimicrobial and physical properties of chitosan films incorporated with turmeric extract. *Int. J. Biol. Macromol.* 101, 882–888.

Kaya, M., Sargin, I., Sabeckis, I. et al., 2017a. Biological, mechanical, optical and physicochemical properties of natural chitin films obtained from the dorsal pronotum and the wing of cockroach. *Carbohydr. Polym.* 163, 162–169.

Kaya, M., Akyuz, L., Sargin, I. et al., 2017b. Incorporation of sporopollenin enhances acid–base durability, hydrophobicity, and mechanical, antifungal and antioxidant properties of chitosan films. *J. Ind. Eng. Chem.* 47, 236–245.

Khan, A., Gallah, H., Riedl, B., Bouchard, J., Safrany, A., Lacroix, M., 2016. Genipin cross-linked antimicrobial nanocomposite films and gamma irradiation to prevent the surface growth of bacteria in fresh meats. *Innov. Food. Sci. Emerg. Technol.* 35, 96–102.

Khwaldia, K., Arab-Tehrany, E., Desobry, S., 2010. Biopolymer coatings on paper packaging materials. *Compr. Rev. Food Sci. Food* 9(1), 82–91. doi:10.1111/j.1541-4337.2009.00095.x.

Kim, Y.T., Min, B., Kim, K.W., 2013. General characteristics of packaging materials for food system. In: Han, J.H. (Ed.), *Innovations in Food Packaging*, Elsevier, Netherlands, pp. 13–35.

Kriegel, C., Arrechi, A., Kit, K., McClements, D.J., Weiss, J., 2008. Fabrication, functionalization, and application of electrospun biopolymer nanofibers. *Crit. Rev. Food Sci. Nutr.* 48(8), 775–797. doi:10.1080/10408390802241325.

Kumar, S., Shukla, A., Baul, P.P., Mitra, A., Halder, D., 2018. Biodegradable hybrid nanocomposites of chitosan/gelatin and silver nanoparticles for active food packaging applications. *Food Pack. Shelf Life* 16, 178–184.

Latou, E., Mexis, S.F., Badeka, A.V., Kontakos, S., Kontominas, M.G., 2014. Combined effect of chitosan and modified atmosphere packaging for shelf life extension of chicken breast fillets. *LWT Food Sci. Technol.* 55(1), 263–268.

Li, X., Li, X., Ke, B., Shiand, X., Du, Y., 2011. Cooperative performance of chitin whisker and rectorite fillers on chitosan films. *Carbohydr. Polym.* 85(4), 747–852.

Li, Y., Ying, Y., Zhou, Y., Ge, Y., Yuan, C., Wu, C., Hu, Y., 2019. A pH-indicating intelligent packaging composed of chitosan-purple potato extractions strength by surface-deacetylated chitin nanofibers. *Int. J. Biol. Macromol.* 127, 376–384.

Lin, L., Gu, Y., Cui, H., 2019. Moringa oil/chitosan nanoparticles embedded gelatin nanofibers for food packaging against Listeria monocytogenes and Staphylococcus aureus on cheese. *Food Pack. Shelf Life* 19, 86–93.

López, O.V., Castillo, L.A., García, M.A., Villar, M.A., Barbosa, S.E., 2015. Food packaging bags based on thermoplastic corn starch reinforced with talc nanoparticles. *Food Hydrocoll.* 43, 18–24. doi:10.1016/j.foodhyd.2014.04.021.

López-Meneses, A.K., Plascencia-Jatomea, M., Lizardi-Mendoza, J., Fernández-Quiroz, D., Rodríguez-Félix, F., Mouriño-Pérez, R.R., Cortez-Rocha, M.O., 2018. *Schinus molle* L. essential oil-loaded chitosan nanoparticles: Preparation, characterization, antifungal and anti-aflatoxigenic properties. *LWT Food Sci. Technol.* 96, 597–603.

Malathi, A.N., Santhosh, K.S., Udaykumar, N., 2014. Recent trends of biodegradable polymer: Biodegradable films for food packaging and application of nanotechnology in biodegradable food packaging. *Curr. Trends Technol. Sci.* 3, 73–79.

Manigandan, V., Karthik, R., Ramachandran, S., Rajagopal, S., 2018. Chitosan applications in food industry. In: Grumezescu, A.M., Holban, A.M. (Eds.), *Handbook of Food Bioengineering, Biopolymers for Food Design*, Academic Press, Chennai, India, pp. 469–491.

Martínez-Camacho, A.P., Cortez-Rocha, M.O., Graciano-Verdugo, A.Z. et al., 2013. Extruded films of blended chitosan, low density polyethylene and ethylene acrylic acid. *Carbohydr. Polym.* 91(2), 666–674. doi:10.1016/j.carbpol.2012.08.076.

Medina, E., Caro, N., Abugoch, L., Gamboa, A., Díaz-Dosque, M., Tapia, C., 2019. Chitosan thymol nanoparticles improve the antimicrobial effect and the water vapour barrier of chitosan-quinoa protein films. *J. Food Eng.* 240, 191–198.

Mohammadi, A., Hashemi, M., Hosseini, S., 2015. Chitosan nanoparticles loaded with *Cinnamomum zeylanicum* essential oil enhance the shelf life of cucumber during cold storage. *Postharvest Biol. Technol.* 110, 203–213.

Munteanu, B.S., Sacarescu, L., Vasiliu, A.L. et al., 2018. Antioxidant/antibacterial electrospun nanocoatings applied onto PLA films. *Materials* 11(10), 1973. doi:10.3390/ma11101973.

Musarrat, H.M., Williams, P.A., Tverezovskaya, O., 2013. Extraction of chitin from prawn shells and conversion to low molecular mass chitosan. *Food Hydrocoll.* 31(2), 166–171. doi:10.1016/j.foodhyd.2012.10.021.

Muzzarelli, R.A.A., Boudrant, J., Meyer, D., Manno, N., DeMarchis, M., Paoletti, M.G., 2012. Current views on fungal chitin/chitosan, human chitinases, food preservation, glucans, pectins and inulin: A tribute to Henri Braconnot, precursor of the Carbohydr. Polym. science, on the chitin bicentennial. *Carbohydr. Polym.* 87(2), 995– 1012. doi:10.1016/j.carbpol.2011.09.063.

Nam, S., Scanlon, M.G., Han, J.H., Izydorczyk, M.S., 2007. Extrusion of pea starch containing lysozyme and determination of antimicrobial activity. *J. Food Sci.* 72(9), E477–E484. doi:10.1111/j.1750-3841.2007.00513.x.

Nguyen, N., Long, V., Joly, C., Dantigny, P., 2016. Active packaging with antifungal activities. *Int. J. Biol. Macromol.* 220, 73–90.

Noshirvani, N., Ghanbarzadeh, B., Mokarram, R.R., Hashemi, M., 2017. Novel active packaging based on carboxymethyl cellulose-chitosan—ZnO NPs nanocomposite for increasing the shelf life of bread. *Food Pack. Shelf Life* 11, 106–114.

Nunes, C., Coimbra, M.A., Ferreira, P., 2018. Tailoring functional chitosan-based composites for food applications. *Chem. Rec.* 18(7–8), 1138–1149.

Pereira Jr., V.A., de Arruda, I.N.Q., Stefani, R., 2014. Active chitosan/PVA films with anthocyanins from *Brassica oleraceae* (Red Cabbage) as time-temperature indicators for application in intelligent food packaging. *Food Hydrocoll.* 43, 180–188.

Pettersen, M.K., Eie, T., Nilsson, A., 2005. Oxidative stability of cream cheese stored in thermoformed trays as affected by packaging material, drawing depth and light. *Int. Dairy J.* 15(4), 355–362. doi:10.1016/j.idairyj.2004.08.006.

Pires, J.R.A., de Souza, V.G.L., Fernando, A.L., 2018. Chitosan/montmorillonite bionanocomposites incorporated with rosemary and ginger essential oil as packaging for fresh poultry meat. *Food Pack. Shelf Life* 17, 142–149.

Ponce, A., Roura, S.I., Moreira, M.R., 2016. Casein and chitosan polymers: Use in antimicrobial packaging. In: Barros-Velázquez, J. (Ed.), *Antimicrobial Food Packaging*, Elsevier Inc., London, UK, pp. 455–466.

Priyanka, P. Narender Raju, Ganguly, S., 2015. Starch-chitosan based composite edible antimicrobial film: modelling the growth of selected food spoilage microbiota. *Indian J. Dairy Sci.* 68, 316–320.

Quiroz-Castillo, J.M., Rodríguez-Félix, D.E., Grijalva-Monteverde, H., del Castillo-Castro, T., Plascencia-Jatomea, M., Rodríguez-Félix, F., Herrera-Franco, P.J., 2014. Preparation of extruded polyethylene/chitosan blends compatibilized with polyethylene-graft-maleic anhydride. *Carbohydr. Polym.* 101, 1094–1100. doi:10.1016/j.carbpol.2013.10.052.

Raddaha, Jackoub, Seuss, N.S.S., Boccaccini, A.R., 2015. Study of the electrophoretic deposition of chitosan/halloysite nanotubes/titanium dioxide composite coatings using Taguchi experimental design approach. *Key Eng. Mater.* 654, 230–239.

Rambabu, K., Bharath, G., Banat, F., Show, P.L., Cocoletzi, H.H., 2019. Mango leaf extract incorporated chitosan antioxidant film for active food packaging. *Int. J. Biol. Macromol.* 126, 1234–1243.

Râpă, M., Mitelut, A.C., Tanase, E.E. et al., 2016. Influence of chitosan on mechanical, thermal, barrier and antimicrobial properties of PLA-biocomposites for food packaging. *Compos. B Eng.* 102, 112–121.

Reesha, K.V., Panda, S.K., Bindu, J., Varghese, T.O., 2015. Development and characterization of an LDPE/chitosan composite antimicrobial film for chilled fish storage. *Int. J. Biol. Macromol.* 79, 934–942.

Remya, S., Mohan, C.O., Bindu, J., Sivaraman, G.K., Venkateshwarlu, G., Ravishankar, C.N., 2016. Effect of chitosan based active packaging film on the keeping quality of chilled stored barracuda fish. *J. Food Sci. Technol.* 53(1), 685–693.

Rhoades, J., Roller, S., 2000. Antimicrobial actions of degraded and native chitosan against spoilage organisms in laboratory media and foods. *Appl. Environ. Microbiol.* 66(1), 80–86.

Riaz, A., Lei, S., Akhtar, H.M.S. et al., 2018. Preparation and characterization of chitosan-based antimicrobial active food packaging film incorporated with apple peel polyphenols. *Int. J. Biol. Macromol.* 114, 547–555.

Sahraee, S., Milani, J.M., Ghanbarzadeh, B., Hamishehkar, H., 2017a. Effect of corn oil on physical, thermal, and antifungal properties of gelatin-based nanocomposite films containing nano chitin. *LWT Food Sci. Technol.* 76, 33–39.

Sahraee, S., Milani, J.M., Ghanbarzadeh, B., Hamishehkar, H., 2017b. Physicochemical and antifungal properties of bio-nanocomposite film based on gelatin-chitin nanoparticles. *Int. J. Biol. Macromol.* 97, 373–381.

Salaberria, A.M., Diaz, R.H., Labidi, J., Fernandes, S.C.M., 2015a. Preparing valuable renewable nanocomposite films based exclusively on oceanic biomass—Chitin nanofillers and chitosan. *React. Funct. Polym.* 89, 31–39.

Salaberria, A.M., Diaz, R.H., Labidi, J., Fernandes, S.C.M., 2015b. Role of chitin nanocrystals and nanofibers on physical, mechanical and functional properties in thermoplastic starch films. *Food Hydrocoll.* 46, 93–102.

Santonicola, S., García Ibarra, V., Sendón, R., Mercogliano, R., Rodríguez-Bernaldo de Quirós, A., 2017. Antimicrobial films based on chitosan and methylcellulose containing natamycin for active packaging applications. *Coatings* 7(10), 1–10.

Sanuja, S., Agalya, A., Umapathy, M.J., 2015. Synthesis and characterization of zinc oxide–neem oil–chitosan bionanocomposite for food packaging application. *Int. J. Biol. Macromol.* 74, 76–84.

Saral Sarojini, K., Indumathi, M.P., Rajarajeswari, G.R., 2019. Mahua oil-based polyurethane/chitosan/nano ZnO composite films for biodegradable food packaging applications. *Int. J. Biol. Macromol.* 124, 163–174.

Sashiwa, H., Aiba, S.I., 2004. Chemically modified chitin and chitosan as biomaterials. *Prog. Polym. Sci.* 29(9), 887–908. doi:10.1016/j.progpolymsci.2004.04.001.

Schreiber, S.B., Bozell, J.J., Hayes, D.G., Zivanovic, S., 2013. Introduction of primary antioxidant activity to chitosan for application as a multifunctional food packaging material. *Food Hydrocoll.* 33(2), 207–214.

Siripatrawan, U., Vitchayakitti, W., 2016. Improving functional properties of chitosan films as active food packaging by incorporating with propolis. *Food Hydrocoll.* 61, 695–702.

Sotelo-Boyás, M.E., Correa-Pacheco, Z.N., Bautista-Baños, S., Corona-Rangel, M.L., 2017. Physicochemical characterization of chitosan nanoparticles and nanocapsules incorporated with lime essential oil and their antibacterial activity against food-borne pathogens. *LWT Food Sci. Technol.* 77, 15–20.

Sousa, F., Guebitz, G.M., Kokol., V., 2009. Antimicrobial and antioxidant properties of chitosan enzymatically functionalized with flavonoids. *Process. Biochem.* 44(7), 749–756.

Souza, V.G.L., Fernando, A.L., 2016. Nanoparticles in food packaging: Biodegradability and potential migration to food—A review. *Food Pack. Shelf Life* 8, 63–70.

Srinivasa, P.C., Tharanathan, R.N., 2007. Chitin/chitosan—Safe, ecofriendly packaging materials with multiple potential uses. *Food Rev. Int.* 23(1), 53–72.

Stoleru, E., Munteanu, S.B., Dumitriu, R.P. et al., 2016. Polyethylene materials with multi-functional surface properties by electrospraying chitosan/vitamin E formulation destined to biomedical and food packaging applications. *Iran. Polym. J.* 25(4), 295–307. doi:10.1007/s13726-016-0421-0.

Syame, S.M., Mohamed, W.S., Mahmoud, R.K., Omara, S.T., 2017. Synthesis of copper-chitosan nanocomposites and its application in treatment of local pathogenic isolates bacteria. *Orient. J. Chem.* 33(6), 2959–2969. doi:10.13005/ojc/330632.

Theruvathil, K.S., Sethumadhavan, S., Paruthapara, T.M., 2007. Study on the production of chitin and chitosan from shrimp shell by using Bacillus subtilis fermentation. *Carbohydr. Res.* 342(16), 2423–2429. doi:10.1016/j.carres.2007.06.028.

Thomas, M.S., Koshy, R.R., Mary, S.K., Thomas, S., Pothan, L.A., 2019. Applications of polysaccharide based composites. In: *Starch, Chitin and Chitosan Based Composites and Nanocomposites, SpringerBriefs in Molecular Science*, Cham: Springer, pp. 43–55. doi:10.1007/978-3-030-03158-9_4.

Torres-Giner, S., Ocio, M.J., Lagaron, J.M., 2008. Development of active antimicrobial fiber-based chitosan polysaccharide nanostructures using electrospinning. *Eng. Life Sci.* 8(3), 303–314. doi:10.1002/elsc.200700066.

Tripathi, S., Mehrotra, G.K., Dutta, P.K., 2009. Physicochemical and bioactivity of cross-linked chitosan–PVA film for food packaging applications. *Int. J. Biol. Macromol.* 45(4), 372–376.

Tripathi, S., Mehrotra, G.K., Dutta, P.K., 2010. Preparation and physicochemical evaluation of chitosan/poly(vinyl alcohol)/pectin ternary film for food-packaging applications. *Carbohydr. Polym.* 79(3), 711–716.

Uranga, J., Puertas, A.I., Etxabide, A., Dueñas, M.T., Guerrero, P., de la Caba, K., 2019. Citric acid-incorporated fish gelatin/chitosan composite films. *Food Hydrocoll.* 86, 95–103.

Vasile, C., Darie, R.N., Cheaburu-Yilmaz, C.N. et al., 2013. Low density polyethylene—Chitosan composites. *Compos. B Eng.* 55, 314–323.

Vasile, C., Darie, R.N., Sdrobis, A. et al., 2014. Effectiveness of chitosan as antimicrobial agent in LDPE/CS composite films as minced poultry meat packaging materials. *Cell. Chem. Technol.* 48(3–4), 325–336.

Vasile, C., Pamfil, D., Râpă, M., et al., 2018. Study of the soil burial degradation of some PLA/CS biocomposites. *Compos. B Eng.* 142, 251–262.

Vilela, C., Pinto, R.J.B., Coelho, J. et al., 2017. Bioactive chitosan/ellagic acid films with UV-light protection for active food packaging. *Food Hydrocoll.* 73, 120–128.

Vlacha, M., Giannakas, A., Katapodis, P., Stamatis, H., Ladavos, A., Barkoula, N.M., 2016. On the efficiency of oleic acid as plasticizer of chitosan/clay nanocomposites and its role on thermo-mechanical, barrier and antimicrobial properties—Comparison with glycerol. *Food Hydrocoll.* 57, 10–19. doi:10.1016/j.foodhyd.2016.01.003.

Vroman I., Tighzert, L., 2009. Biodegradable polymers. *Materials.* 2(2), 307–344. doi:10.3390/ma2020307.

Wang, H., Qian, J., Ding, F., 2018. Emerging chitosan-based films for food packaging applications. *J. Agric. Food Chem.* 66(2), 395–413.

Xing, Y., Li, X., Xu, Q., Jiang, Y., Yun, J., Li, W., 2010. Effects of chitosan-based coating and modified atmosphere packaging (MAP) on browning and shelf life of fresh-cut lotus root (*Nelumbo nucifera* Gaerth). *Innov. Food. Sci. Emerg. Technol.* 11(4), 684–689.

Youssef, A.M., EL-Sayed, S.M., EL-Sayed, H.S., Salama, H.H., Dufresne, A., 2016. Enhancement of Egyptian soft white cheese shelf life using a novel chitosan/carboxymethyl cellulose/zinc oxide bionanocomposite film. *Carbohydr. Polym.* 151, 9–19.

Yuan, G., Chen, X., Li, D., 2016. Chitosan films and coatings containing essential oils: The antioxidant and antimicrobial activity, and application in food systems. *Food Res. Int.* 89(1), 117–128.

Zafar, R., Zia, K.M., Tabasum, S., Jabeen, F., Noreen, A., Zuber, M., 2016. Polysaccharide based bionanocomposites, properties and applications: A review. *Int. J. Biol. Macromol.* 92, 1012–1024.

Zemljič, L.F., Tkavc, T., Vesel, A., Šauperl, O., 2013. Chitosan coatings onto polyethylene terephthalate for the development of potential active packaging material. *Appl. Surf. Sci.* 265, 697–703.

Zhuang, P.Y., Li, Y.L., Fan, L., Lin, J., Hu, Q.L., 2012. Modification of chitosan membrane with poly(vinyl alcohol) and biocompatibility evaluation. *Int. J. Biol. Macromol.* 50(3), 658–663.

Zia, K.M., Tabasum, S., Nasif, M., Sultan, N., Aslam, N., Noreen, A., Zuber, M., 2017. A review on synthesis, properties and applications of natural polymer based carrageenan blends and composites. *Int. J. Biol. Macromol.* 96, 282–301.

Wang, H., Qian, J., Ding, F., 2018. Emerging chitosan-based films for food packaging applications. *J. Agric. Food Chem.* 17(66), 395–413.

Woranuch, S., Yoksan, R., 2013. Eugenol-loaded chitosan nanoparticles: II. Application in bio-based plastics for active packaging. *Carbohydr. Polym.* 96(2), 586– 592.

12 Chitosan-Based Hybrid Nanocomposites for Food Packaging Applications

M. Chandrasekar, K. Senthilkumar, T. Senthil Muthu Kumar, Sabarish Radoor, Jyotishkumar Parameswaranpillai, and Suchart Siengchin

CONTENTS

12.1 INTRODUCTION

Commercially available polymer-based materials presently used in food packaging applications are derived from fossil fuel resources. The ability to maintain the food quality, freshness, and shelf life of the food product has made this synthetic polymers-based packaging material inevitable for food packaging applications. However, non-biodegradability of the synthetic plastics after usage has affected the ecosystem by injuring animals, birds, marine life, etc. and has led to environmental issues such as landfills due to waste accumulation. These factors have pushed the need for environmentally friendly bio-based packaging materials from renewable resources. Biodegradable packaging films from bioplastics such as polylactic acid (PLA), starch, chitosan (CS), etc., are preferred over synthetic polymer films. The use of such environmentally friendly material in food packaging applications is limited due to their poor mechanical and gas barrier properties. Studies have shown that these limitations could be overcome with the incorporation of nanofillers. This chapter addresses the influence of various nanofillers on the chitosan-based packaging films.

12.1.1 Chitosan

Chitosan is a polysaccharide derived from chitin present in the exoskeleton of crustacean shells, cell wall of fungi, and cuticles of insects, mollusks, and annelids (Ruiz and Corrales, 2017). It is a biodegradable polymer which has additional advantages such as antimicrobial resistance, functional compatibility with the polymers, inherent film forming ability, etc. as shown in Figure 12.1.

Chitosan could be used in food packing applications in the form of (i) packaging films and/or (ii) coatings, which can be directly applied to the food materials, especially on meat products, fruits, and vegetables. The chitosan packaging films are

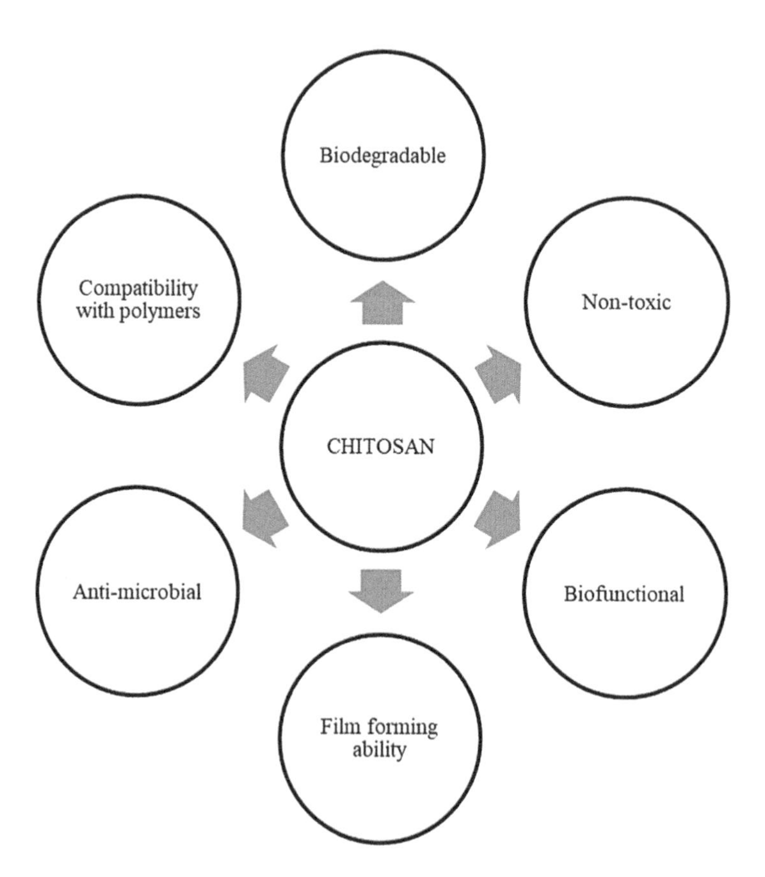

FIGURE 12.1 Characteristics of chitosan making them suitable for food packaging applications.

available in edible or inedible form. However, the coatings would be edible in most of the cases since they are applied directly on the surfaces of the food materials.

Nevertheless, the chitosan-based food packing applications are limited owing to their (1) high sensitivity to humidity and (2) moisture. Besides, the chitosan films are prone to degradation before attaining their melting temperatures. Hence, the chitosan films (1) cannot be processed by using molding and extrusion methods and (2) are hard to stretch. Therefore, the researchers incorporated many additives with chitosan films and formulated different types of materials to overcome these difficulties and improve their functionality. The different types of formulated composites consist of (1) chitosan films with nanomaterials, (2) chitosan films with clays, (3) chitosan films with fibers, whiskers, and polysaccharide particles, (4) chitosan films with natural-based oils and their extracts, (5) chitosan films with cross-linkers and plasticizers, (6) chitosan with antimicrobial agents, and (7) chitosan films blended with biopolymers or synthetic polymers.

12.1.2 Preparation and Fabrication of Chitosan-Based Packaging Films

12.1.2.1 Preparation and Fabrication of Chitosan-Based Packaging Films

The solution casting method is the commonly used technique for the fabrication of chitosan-based nanocomposite films for packaging applications. Once the film is prepared, it is subjected to various characterizations such as tensile, thermal stability, antimicrobial, etc., as shown in Figure 12.2, to aid understanding of the properties of the composites films. In the upcoming sections, factors influencing the crystallinity, thermal properties, tensile properties, water-solubility, swelling and contact angle measurement, and antimicrobial and antifungal activity of the composite films are discussed in detail.

12.2 CHARACTERIZATION OF CHITOSAN-BASED HYBRID NANOCOMPOSITES

12.2.1 Crystallinity from X-Ray Diffraction (XRD)

Chitosan is a semi-crystalline polymer and displayed characteristic reflection at $2\theta = 10°$ and $20°$ in the XRD profile which can form crystalline regions due to the presence of an –OH and –NH2 functional group. An additional broad peak that extended from 15.5 to 24° showed the amorphous region of the film (Javaid et al., 2018; Koosha et al., 2015). Owing to its biodegradable and eco-friendly nature, it has often been used to fabricate food packaging materials (Hosseini et al., 2019; Tripathi et al., 2010). The properties of the chitosan-based membranes can be further improved with the addition of nanoparticles such as nanoclay, graphene oxide (GO), silver nanoparticles (Ag NPs), zinc oxide nanoparticles, titanium dioxide nanoparticles, etc. Montmorillonite (MMT) is one of the most efficient nanoclays and is widely used to reinforce polymer composite. Abdollahi et al. (2012) investigated the influence of MMT or nanoclay on the properties of chitosan film containing essential oil. It was reported that with the addition of MMT to chitosan, the intensity of characteristic peak of chitosan at $2\theta = 10.18°$ slightly diminishes. Abdollahi et al.

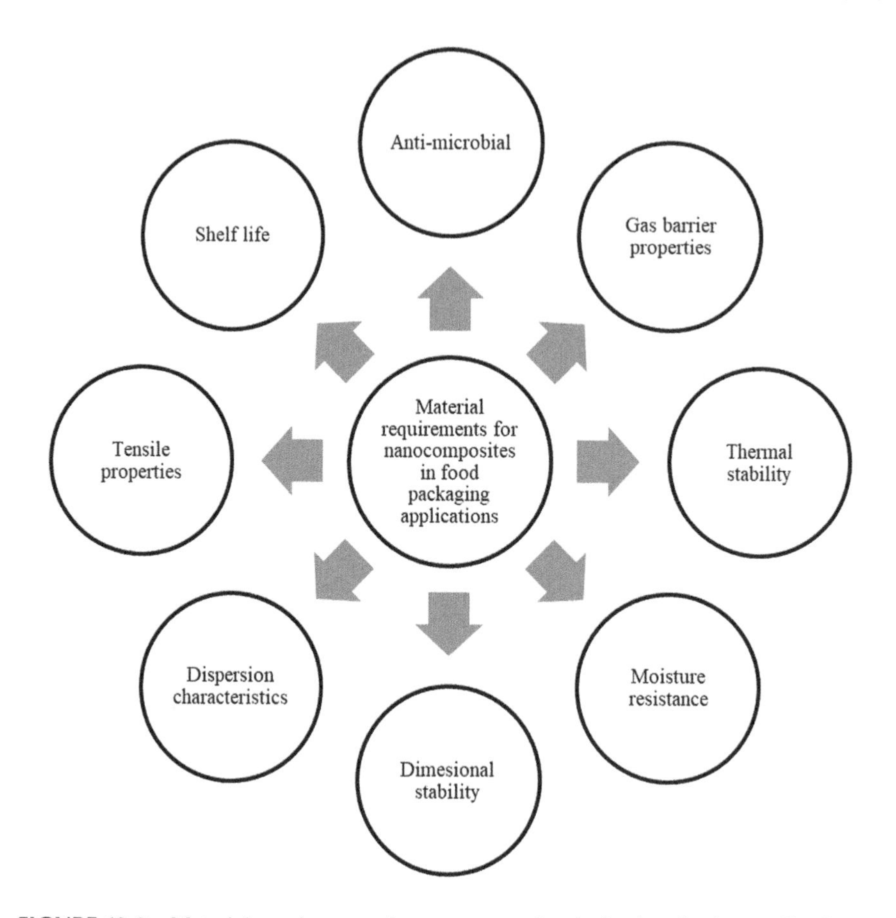

FIGURE 12.2 Material requirements for nanocomposites in food packaging applications.

varied the concentration of MMT from 1 to 5 wt.% and they observed that with increase in concentration, the intensity of the diffraction peak of MMT ($2\theta = 8.56°$) decreases, probably due to the exfoliation of clay. Meanwhile, at 5 wt.% the peaks of MMTe shifted to a lower angle of 5.32° attributed to the intercalation of clay (see Figure 12.3). A similar observation was noticed in chitosan-nanoclay by other researchers (Enescu et al., 2019; Ghelejlu et al., 2016; Xu, Ren, and Hanna, 2006).

Graphene oxide, a derivative of graphene, is used to fabricate biomaterials which find application in various fields. GO-based polymer composites possess superior properties such as large surface area, high mechanical property, high thermal conductivity, high crystallinity, etc. (Gong et al., 2019; Smith et al., 2019; Wang et al., 2013). Gong and co-workers studied the properties of graphene oxide/chitosan aerogel. Their studies revealed that the composite possesses high mechanical property and crystallinity. The XRD result shows that on incorporating GO into chitosan, there is remarkable improvement in the crystallinity of the system. This could be due to the interaction of GO with the amino group of chitosan (Gong et al., 2019). Recently, Yadav et al. (2015) developed MMT/GO/CS composite and studied their

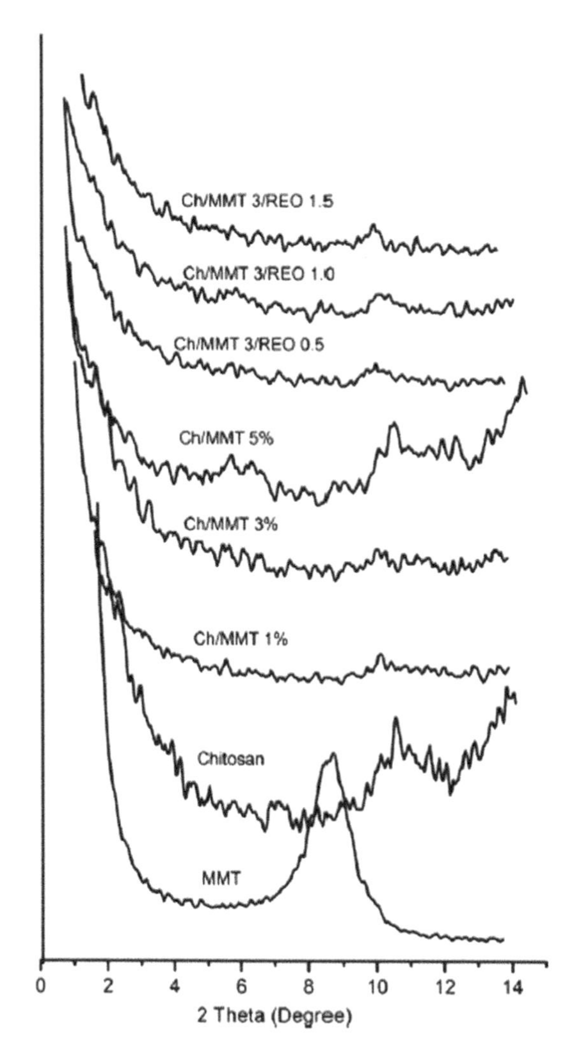

FIGURE 12.3 XRD patterns of pure chitosan, MMT, and different concentrations of MMT and rosemary essential oil-based nanocomposite films. (Source: Abdollahi et al., 2012. Used with permission.)

properties. XRD analysis of the composite indicated that the addition of MMT and GO improve the crystallinity of the composite, attributed to the closer packing in the composite system. From the XRD results, it was obvious that after incorporation, the characteristic reflection peak of GO disappears, thereby indicating the complete exfoliations of GO into the chitosan polymer (see Figure 12.4). The result is in accordance with the results of Yang and co-workers (Yang et al., 2010).

Due to their antibacterial property, silver nanoparticle/chitosan composites could be a potential candidate for fabricating the food packaging composite films (Nasef et al., 2019; Shao et al., 2019). Salari et al. (2018) developed an active food packaging material from chitosan, bacterial cellulose nanocrystal (BCNC), and Ag NPs.

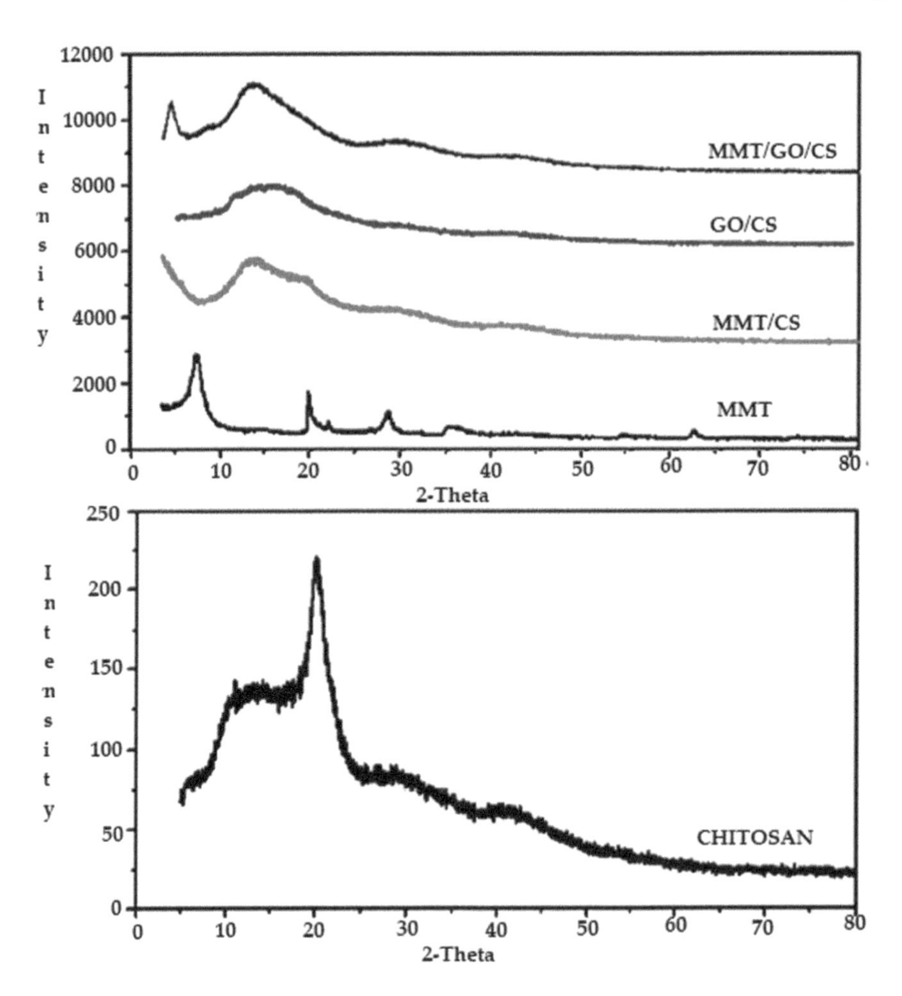

FIGURE 12.4 XRD patterns of chitosan, MMT and MMT/CS, MMT/CS, (1 wt.%) GO/CS and (5 wt.%) MMT/(1 wt.%) GO/CS. (Source: Yadav and Ahmad, 2015. Used with permission.)

It is evident from the XRD spectra that the presence of both silver and cellulose nanoparticle leads to the enhancement in the intensity of the characteristic peak of chitosan. Consequently, the overall crystallinity of the composite is also improved (see Figure 12.5). XRD results further support the successful incorporation of silver nanoparticle in the composites, which was in agreement with the previous reports (Kalaivani et al., 2018).

Noshirvani and co-workers (Noshirvani et al., 2017) observed that the high concentration of zinc oxide nanoparticle (ZnO NPs) enhances the overall crystallinity of chitosan-carboxymethyl cellulose-chitosan-oleic acid (CMC-CH-OL) composite. Moreover, the absence of any shift in the reflection peak of chitosan after incorporation of ZnO NPs suggests that the crystalline structure of polymer is not disturbed after the addition of ZnO NPs (see Figure 12.6). Similar results were reported by Abdelhady and Khorasani (AbdElhady, 2012; Khorasani et al., 2018).

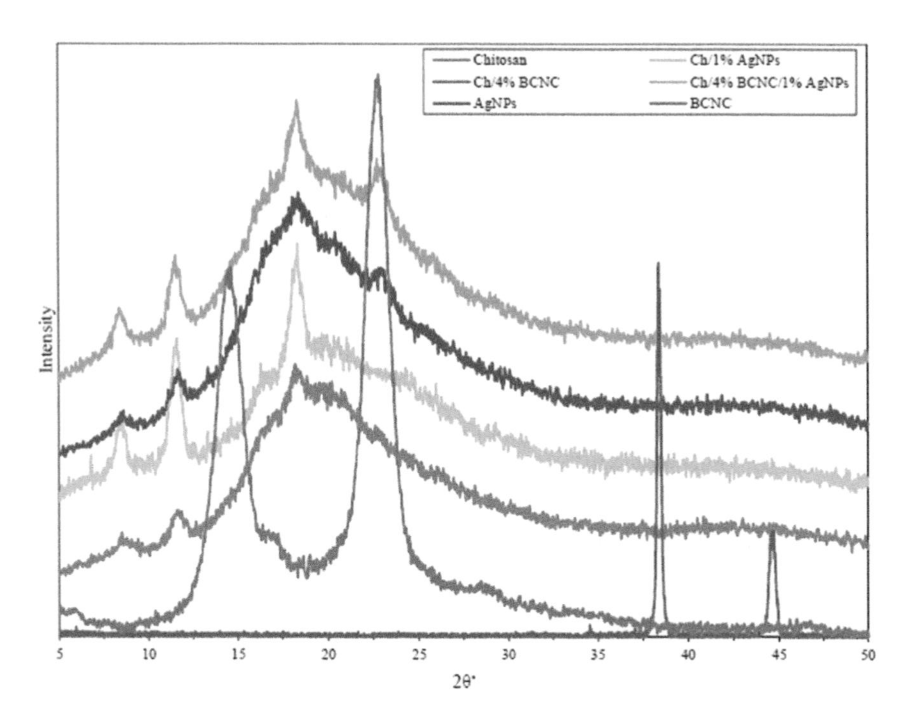

FIGURE 12.5 X-ray diffractometry of chitosan and chitosan incorporated Ag NPs and BCNC nanocomposite films. (Source: Salari et al., 2018. Used with permission.)

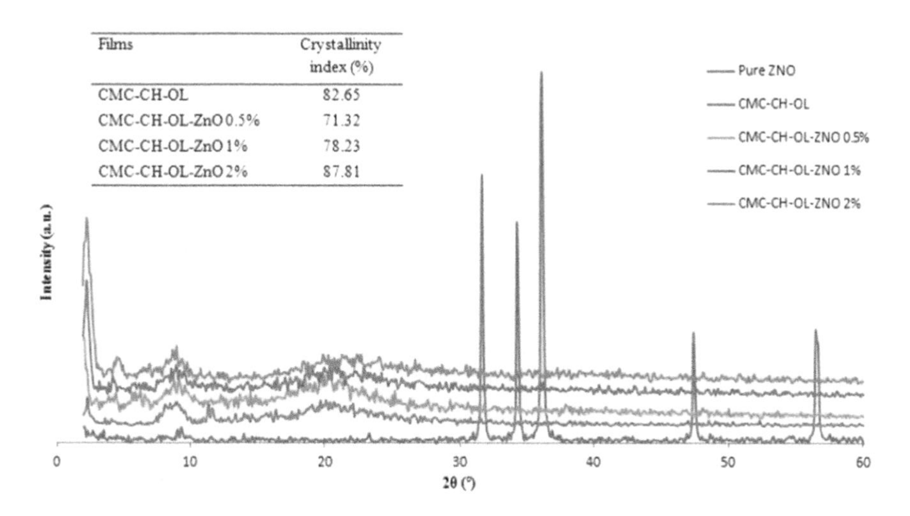

FIGURE 12.6 X-ray diffraction of ZnO NPs, CMC-CH-OL and different percentages of ZnO NPs-incorporated CMC-CH-OL. (Source: Noshirvani et al., 2017. Used with permission.)

Titanium dioxide (TiO_2) nanoparticle is usually incorporated in polymer composite to improve its photocatalytic activity and antibacterial property. Haldorai et al. (Haldorai and Shim, 2014) synthesized novel chitosan-TiO2 nanohybrid and monitored its properties. XRD analysis of the composite displayed the characteristic peak of chitosan as well as TiO_2. On comparing the diffraction peak of pure CS and its composite, it is evident that the composite has intense crystalline peak, indicating TiO2 nanoparticle induced crystallinity in the system. In another study, Ali and coworkers (Ali et al., 2018) monitored the effect of TiO_2 content on the diffraction profile of chitosan. They observed that on varying the concentration of $TiO_{2,}$ the intensity of the crystalline peak increases which is in agreement with Yang and coworkers (Yang et al., 2009). Norranattrakul et al. (2013) observed a decrease in the intensity of characteristic reflection peak of CS with the incorporation of TiO2 content, probably due to the increase of amorphous region in chitosan matrix occurring during the component blending.

12.2.2 Water Barrier Properties

Surface properties (hydrophobicity and hydrophilicity) of the chitosan-based nanocomposites films is usually assessed by water contact angle measurement (WCA) while barrier properties of the nanocomposite film to moisture and water vapor were determined from the water uptake, film solubility, and water vapor permeability (WVP) test. WVP indicates the resistance to diffusion of water vapor into the nanocomposite film. Water barrier properties from the literature on chitosan-based nanocomposites films infused with various nanoparticles are presented in Table 12.1.

Al-Naamani et al. (2016) coated polyethylene (PE) films with chitosan and zinc oxide nanoparticles (ZnO) were immersed in distilled water for 24h. The WCA, water uptake, and solubility were determined and compared to the PE film without coating (control specimen). The water contact angle measured between the film substrate and the tangent line of the liquid droplet on the substrate surface was found to be higher for the PE/chitosan-ZnO-coated film indicating hydrophobicity or lower surface wettability compared to lower angle for the control film. The hydrophobic characteristic of the coated film was also evident from the reduced water uptake. Though chitosan-ZnO coating helped in improving hydrophobicity, it dissolved easier than the control film because of the poor film solubility. The cross-linking between the chitosan and ZnO was believed to reduce the mobility of chitosan chains and decreased the water uptake into the film (Al-Naamani et al., 2016). A similar trend of decreased water uptake by the chitosan film with the introduction of MMT-AgNP and increasing concentration of MMT-Ag was observed due to the cross-linking between chitosan chains and MMT-AgNP (Lavorgna et al., 2014). In a recent study, Nouri et al. (2018) showed that water uptake of chitosan-based nanocomposite film was considerably lower due to the addition of MMT-copper oxide nanoparticles (MMT-CuO). According to Nouri et al., MMT-CuO protected the chitosan film against water uptake by cross-linking with hydroxyl groups of the chitosan. Salari et al. investigated moisture barrier properties of chitosan films reinforced with 1 wt.% AgNP and 4, 6, and 8 wt.% of bacterial cellulose nanocrystal (BCNC). Their findings indicate that WCA increased while the water uptake, WVP and film solubility declined with

TABLE 12.1
Water Barrier Properties of the Chitosan-Based Nanocomposites with Various Nanofiller Reinforcements

Material	Barrier properties				References
	Water uptake (%)	Film solubility (%)	WVP ($\times 10^{-10}$ g/msPa)	WCA (○)	
CS	–	~13	2.26	–	(Nouri et al., 2018)
CS-glycerol-1MMT-20CuO		~18	3.32		
CS-glycerol-5MMT-20CuO		~16	2.90		
CS-glycerol-5MMT-90CuO		~16	1.08		
CS	~18	~22	3.65	–	(Salari et al., 2018)
CS/1% AgNP/2% BCNC	~12	~17	2.18		
CS/1% AgNP/4% BCNC	~12	~17	2.16		
CS/1% AgNP/6% BCNC	~11	~16	2.02		
CS	~1.5	~0.4		–	(Al-Naamani et al., 2016)
PE coated with CS-ZnO	~0.3	~0.7		95	
CS	~31	~18	6.65	–	(Zhang et al., 2019)
CS-TiO_2-BPPE	~29	~23	5.12	–	

the increasing concentration of BCNC. Presence of AgNP and BCNC improved the barrier properties of nanocomposites film, slowing down the water diffusion process over time by increasing tortuosity, crystallinity, and cohesiveness of the chitosan matrix (Salari et al., 2018). Water barrier properties of chitosan films were found to be improved by the addition of clove essential oil and magnesium oxide nanoparticles (MgO). The reason behind improved water barrier properties is the hydrophobic nature of the clove oil and cross-linking between MgO and chitosan which reduces the interaction of hydroxyl groups of chitosan with the water molecules (Sanuja et al., 2014). According to Zhang et al. (2019), presence of TiO_2 and black plum peel extract (BPPE) improved the WVP as shown in Table 12.1. This is because the micro-path for water vapor diffusion in chitosan films was blocked by TiO2 and aromatic rings in the phenolic skeleton of BPPE.

12.2.3 Antimicrobial Activity

Chitosan has natural resistance to bacteria; however, it diminishes over time (Leceta et al., 2013). Various works in the literature show that the addition of nanoparticle can inhibit the microorganism and significantly improve the antimicrobial (both gram-positive and gram-negative bacteria and antifungal) activity of the chitosan-based film. Agar diffusion method is the commonly used technique from which antimicrobial and antifungal activity is identified from the formation of inhibition zones or microorganism colonies represented as colony-forming units (CFU).

Antibacterial activity of chitosan/polyvinyl alcohol films reinforced with MMT films against gram-negative bacteria such as *E. coli* and *S. typhymurium* and gram-positive bacteria (*L. monocytogenes*) was studied by Butnaru et al. (2016). The percentage inhibition of microbial growth was proportional to the concentration of MMT in the composite film. Antibacterial activity of chitosan/polyvinyl alcohol film with MMT filler followed the order 5 wt.% >3 wt.% >1 wt.%. Bourakadi et al. (2019) studied the antimicrobial activity of chitosan/polyvinyl alcohol films infused with thiabendazoluim-MMT salts at 8–16 wt.%. All the nanocomposites films showed antibacterial activity against gram-negative *E. coli* and *P. aeruginosa* as well as against the gram-positive bacteria such as *S. aureus*. The carbon chain length and the positively charged amino groups at N3 in thiabendazoluim salts are responsible for high antibacterial activity. Sanuja et al. (2014) reported increased antimicrobial activity against *S. aureus* due to the addition of clove oil-MgO nanoparticles into the chitosan matrix. According to them, phenol-reactive group such as eugenol in the clove oil, chitosan functional group, and MgO together disrupts the cell membrane of the bacteria and inhibits further growth of the microorganisms. Antimicrobial activity against gram-positive and gram-negative bacteria due to the addition of essential oils such as thyme, clove, and cinnamon oil in the chitosan was reported by Hosseini et al. (2009). They found that chitosan films mixed with essential oils were more resistant to gram-positive bacteria than the gram-negative bacteria. Variation in antibacterial activity against gram-positive and gram-negative bacteria was basically due to the differences in thickness and constituents of the outer cell wall.

The antifungal activity of chitosan nanocomposites was reported in the literature. In a study by Dananjaya et al. (2018), chitosan films incorporated with ZnO nanoparticles exhibited superior fungal resistance against *Candida albicans*. Zn^{2+} cations released from the surface of ZnO NPs penetrate the cell membrane and generate reactive oxygen which causes severe damage to the cell membrane and the intracellular components of the microbial cells. Salaberia et al. investigated antifungal activity of chitosan films reinforced with the chitosan nanocrystals (CHNC) and chitosan nanofibers (CHNF) against *A. niger* that grows on foods, fruits, and vegetables. CFU was comparatively lower for chitosan films with CHNC and CHNF fillers (Salaberria et al., 2015).

12.2.4 Tensile Properties

Butler et al. (1996) reported the tensile strength and elongation at break of chitosan films were comparable to high-density polyethylene (HDPE) and low-density polyethylene (LDPE). Kumar et al. (2018) reported a noticeable increase in elongation at break of the chitosan films obtained by incorporating gelatin, polyethylene glycol, and silver nanoparticles. Bano et al. (2014) indicated that chitosan/ polyvinyl alcohol (PVA) composites exhibited good film-forming and antimicrobial properties. However, the films were highly brittle. This problem was solved by incorporating the silver nanoparticles in PVA; hence the chitosan/PVA/silver nanoparticles hybrid composites showed an increment in the percentage of elongation (George et al., 2012). Similar to the previous study, Li et al. (Li et al., 2015) significantly improved the elongation at the break of PVA/chitosan films by adding cellulose nanowhiskers. This may be attributed to the enhanced interactions between the PVA and the cellulose nanowhiskers through the hydrogen bonding. However, the tensile strength of these hybrid composites had a slight reduction when compared to the PVA/chitosan composites.

Many researchers reported that the optimal content of nanoclay should not exceed 5% in polymer matrix composites. If the nanoclay content is more than 5%, there may be a tendency for the nanoclay to agglomerate, and as a result the composites will be highly brittle with poor mechanical properties (Pavlidou and Papaspyrides, 2008; Butnaru et al., 2016; Gierszewska et al., 2019). In an interesting work, Bourakadi et al. (2019) varied the chain length of organoclay (thiabendazolium–MMT) in chitosan/PVA mixtures, whereby the overall organoclay content was maintained as 5 wt%. From their results, it was observed that there was a substantial improvement in tensile strength and Young's modulus by increasing the chain length of the organoclay. These observations were attributed to (1) the increased value of aspect ratio and (2) the improved dispersion ability of nanofillers into the matrix. However, a limited enhancement in elongation at break was observed for the nanocomposites. In an another study, an increment of maximum stress, tensile modulus, and elongation at break of chitosan/PVA mixtures was improved by incorporating the silver nanoparticles (1–3%) owing to their enhanced dispersion of nanoparticles (Vimala et al., 2011).

Generally, the cross-linking could be used to bond the two polymer chains and helps to improve several properties such as (1) solvent resistance, (2) dimensional stability, (3) lower creep rates, and (4) improved glass transition temperature (T_g). These enhancements are obtained by achieving the cross-linked structure in polymers (Nielsen, 1969). For example, Jahan et al. (Jahan et al., 2016) established the cross-linking between the chitosan and polyvinyl alcohol polymers using potassium nitrate. It was observed that the tensile strength and elongation at break of the cross-linked chitosan/ polyvinyl alcohol blends improved linearly by increasing the content of potassium nitrate ranged from 0.1 to 0.5 g. In another study, the mechanical properties were analyzed by varying the contents of (1) chitosan, (2) chitin, and (3) tannic acid (cross-linker). The cross-linked matrix, i.e., chitosan/tannic acid, exhibited a better tensile strength and Young's modulus. However, the pure chitosan matrix showed a higher elongation at break than the rest of the composites (Rubentheren et al., 2015).

12.2.5 Thermal Properties

Thermal stability plays a significant part in defining both processing and applications of polymer nanocomposites. Thermal effects such as heating, cooling, or phase transition are inevitable in almost all food processes in the food industry. Hence, the thermal properties have considerable significance in the food packing industry. The thermal properties are vital in designing the food storage or packing material as are the mechanical, optical, and barrier properties (Berk, 2018). Further, it is also to be noted that the environmental concerns created by the petroleum-based packaging materials have created interest in using biodegradable packaging materials for food applications (Kumar et al., 2017; Senthil Muthu Kumar et al., 2018, 2019; Thiagamani et al., 2019). Hence it is necessary to find potential environmentally friendly food packing materials with better functional properties, especially thermal stability. Many researchers have reported that the most common approach to enhancing the properties of packing materials to meet standards in food packing is by reinforcing polymers with nanofillers (Giannakas et al., 2014). In this regard nano-scaled materials such as silver (Indira Devi et al., 2019; Thiagamani et al., 2019), copper (Devi et al., 2019), MMT (Bourakadi et al., 2019), zinc (Dananjaya et al., 2018), magnesium (De Silva et al., 2017), titanium (Zhang et al., 2017), etc. has been used as reinforcement materials. Apart from using nano-scaled fillers, the blending of synthetic and natural polymers to serve this purpose has gained considerable attention (2019).

The thermal stability of chitosan has been well recognized and the rate at which the polymer degrades was found to accelerate with rising temperature and duration of heat involved. The thermal degradation of the chitosan structure is a complex reaction, which involves three degradation stages. The first stage occurs at temperatures between 30 and 110°C and is attributed to the evaporation of the residual water present in chitosan. The second event was due to the polymer decomposition which could be evidenced by a broad range of temperature from 180 to 340°C. The last stage degradation above 450°C is due to the degradation of carbonaceous residue formed during the second stage (Szymańska and Winnicka, 2015).

When the chitosan is modified with reinforcements, the thermal stability may be varied based on the constituents of the reinforced materials. Table 12.2 provides an overall thermal degradation trend of chitosan-based materials.

TABLE 12.2
Overall Thermal Degradation Trend of Chitosan-Based Composites

Stage	Temperature	Significance
Stage 1	30–200°C	Evaporation of absorbed and bound water in chitosan
Stage 2	200–450°C	Chemical degradation and deacetylation of chitosan
Stage 3	450–600°C	Oxidative degradation of the carbonaceous residue formed during the second stage

AgilAbrahama and Rejinib showed that the addition of formaldehyde as cross-linking agent and glycerol as plasticizer in chitosan/polyvinyl alcohol composite films improved the thermal stability and mechanical behavior of the polymer blends (AgilAbrahama and Rejinib, 2016). Thermal stability of chitosan-based films with MMT in varying concentrations were studied by some researchers (Kasirga et al., 2012). It was found that the incorporation of nano MMT improved the thermal stability of nanocomposite films when compared with the pure chitosan. This is attributed to the better interaction between the CS and MMT; furthermore, the MMT clay acted as a heat barrier. Kasirga et al. also observed an increase in char residue with increasing MMT content.

Chitosan-based nanocomposite films with 1% w/v of silver oxides (Ag_2O) were fabricated and their thermal degradation studies were performed (Tripathi et al., 2011). It was found that the thermal degradation of the nanocomposites took place in multiple stages. The first stage was at 50–150°C, which was due to the moisture removal; the second stage was at 200–300°C corresponding to the degradation of chitosan molecules and the silver oxide compounds. Further, it was noted that the addition of Ag_2O particles improved the thermal degradation temperature of the nanocomposite when compared to the pure chitosan films. Rahman and Muraleedharan (2017) fabricated new material for enhancing the shelf life of raw meat. In their research they made chitosan/zinc oxide (ZnO) nanocomposites with varying concentrations of ZnO nanoparticles and studied the functional properties of the nanocomposites as an effect of the reinforcement of ZnO particles. They found that the thermal properties of the nanocomposites substantially enhanced with an increase in the ZnO filler.

The addition of MMT and MMT/CuO in chitosan enhanced the thermal stability of the nanocomposite films. It was reported that after pyrolysis, the nanocomposite films formed char with multilayered carbonaceous silicate structure, which enhanced the insulating property of the material, thus improving its stability. It was also noted that when glycerol was used as a plasticizer, a reduction in thermal stability was observed (Nouri et al., 2018).

Nanocomposite films of chitosan reinforced with varying loadings of cellulose nano whiskers (5, 10, 15, 20, 25, and 30 wt.%) were fabricated by Li et al. (Erukhimovich and de la Cruz, 2004). The films were characterized to find the thermal stability. The nanocomposite films presented improved thermal stability compared to the pure cellulose whiskers. The temperature at maximum decomposition rate of the nanocomposite films hardly changed with an increase of the whisker content in the matrix which suggested that the infusion of cellulose whiskers retained the thermal stability due to the strong interactions between whiskers and chitosan.

The chitosan-based composite films were fabricated by adding lactic acid oligomer grafted-chitosan as nanofiller. The addition of nanofiller improved properties such as the tensile and thermal properties and was verified as a potential food packing material (Bie et al., 2013).

From the studies, it is understandable that the thermal properties of the chitosan-based composites can be enhanced with the incorporation of fillers. However, the mechanism or phenomenon responsible for the improvement in thermal properties due to the addition of nanofiller has never been discussed. Hence, it is concluded that there is great scope for identifying the mechanism or phenomenon responsible for the improvement in the thermal properties of chitosan-based nanocomposite through *in-situ* observation.

12.3 CONCLUSION

The increase in production and consumption of processed foods over the years has increased the demand for packaging films. Despite the success of synthetic polymers in packaging films, excessive use of these materials leads to full landfills and their inability to degrade at a faster rate has put an emphasis on environmentally friendly packaging film made with renewable sources. Chitosan extracted from the exoskeleton of crustacean shells, molluscs, and insect shells has received greater interest among researchers as a viable alternative in food packaging films due to their innate antimicrobial characteristic and film forming ability. However, the tensile, crystalline, water barrier, and thermal properties were inferior compared to the synthetic packaging materials. Hence, attempts were made by researchers to improve the properties of chitosan-based composite films with the introduction of one or more nanoparticles. The following observations were made from the existing literature and recently published works on chitosan-based nanocomposites films:

- Metal oxide fillers such as titanium oxide, zinc oxide, copper oxide, silver oxide, and magnesium oxide have been proved to be effective as filler by enhancing thermal, crystalline, water barrier, and antimicrobial activity against bacteria and fungi.
- Addition of nanofillers enhanced the crystalline properties of nanocomposites films as reflected by the change in intensity of the characteristics peak and shift in 2θ peaks from the XRD spectra.
- Introduction of nanoparticle in the chitosan-based nanocomposite films increased the thermal stability by forming char and shifted the thermal degradation to slightly higher temperatures than the control film without nanofillers.
- Chitosan-based nanocomposites films possessed better antimicrobial activity against gram-positive, gram-negative, and fungal organisms as represented by the inhibition zones and colony forming units. Use of one or more nanofillers helped the chitosan-based nanocomposites films to have prolonged resistance to pathogens as visible from the inhibition zones over a longer period of

time and less CFU compared to the neat chitosan or the film without nanofillers. In general, nanoparticles disrupt the bacterial cell by penetrating the cell wall and through the generation of reactive oxygen species.

- Nanofillers such as MMT-CuO, TiO2-BPPE, thiabendazolium–MMT, etc., when introduced into the chitosan improve the water barrier properties by slowing down the water vapor permeability and moisture uptake. Hydrophobicity of nanocomposite films with such fillers was visible from the larger water contact angle indicating poor wettability of liquid with film substrate. However, the film solubility in water could be slightly higher than the neat chitosan. Combined nanofillers block the water diffusion into the film by cross-linking with the hydroxyl groups of the chitosan and reduce the interaction of chitosan with water molecules.
- Hybrid nanocomposite films with nanofillers possessed better tensile properties such as tensile strength, Young's modulus, and elongation at break. However, tensile properties can decline beyond an optimum concentration of the nanofillers. Uniform dispersion of nanofillers is critical for superior tensile properties and at higher concentration; agglomeration of nanoparticles acts as a stress concentration site and could lead to inferior properties.

REFERENCES

AbdElhady, M. M. 2012. "Preparation and Characterization of Chitosan/Zinc Oxide Nanoparticles for Imparting Antimicrobial and UV Protection to Cotton Fabric." *International Journal of Carbohydrate Chemistry* 2012, pp. 1–6.

Abdollahi, Mehdi, Masoud Rezaei, and Gholamali Farzi. 2012. "A Novel Active Bionanocomposite Film Incorporating Rosemary Essential Oil and Nanoclay into Chitosan." *Journal of Food Engineering* 111(2):343–50.

AgilAbraham, P. A. and V. O. Rejinib. 2016. "Preparation of Chitosan-Polyvinyl Alcohol Blends and Studies on Thermal and Mechanical Properties." *Procedia Technology* 24:741–48.

Al-Naamani, Laila, Sergey Dobretsov, and Joydeep Dutta. 2016. "Chitosan-Zinc Oxide Nanoparticle Composite Coating for Active Food Packaging Applications." *Innovative Food Science and Emerging Technologies* 38:231–37.

Ali, Fayaz, Sher Bahadar Khan, Tahseen Kamal, Khalid A. Alamry, and Abdullah M. Asiri. 2018. "Chitosan-Titanium Oxide Fibers Supported Zero-Valent Nanoparticles: Highly Efficient and Easily Retrievable Catalyst for the Removal of Organic Pollutants." *Scientific Reports* 8(1):1–18.

Bano, Ijaz, Muhammad Afzal Ghauri, Tariq Yasin, Qingrong Huang, and Annie D'Souza Palaparthi. 2014. "Characterization and Potential Applications of Gamma Irradiated Chitosan and Its Blends with Poly (Vinyl Alcohol)." *International Journal of Biological Macromolecules* 65:81–8.

Berk, Zeki. 2018. *Food Process Engineering and Technology*. Academic Press, UK.

Bie, Pingping, Peng Liu, Long Yu, Xiaoxi Li, Ling Chen, and Fengwei Xie. 2013. "The Properties of Antimicrobial Films Derived from Poly (Lactic Acid)/Starch/Chitosan Blended Matrix." *Carbohydrate Polymers* 98(1):959–66.

Bourakadi, Khadija El, Nawal Merghoub, Meriem Fardioui, Mohamed El Mehdi Mekhzoum, Issam Meftah Kadmiri, El Mokhtar Essassi, Abou el Kacem Qaiss, and Rachid Bouhfid. 2019. "Chitosan/Polyvinyl Alcohol/Thiabendazoluim-Montmorillonite Bio-Nanocomposite Films: Mechanical, Morphological and Antimicrobial Properties." *Composites Part B: Engineering* 172(April):103–10.

Butler, B. L., P. J. Vergano, R. F. Testin, J. M. Bunn, and J. L. Wiles. 1996. "Mechanical and Barrier Properties of Edible Chitosan Films as Affected by Composition and Storage." *Journal of Food Science* 61(5):953–56.

Butnaru, Elena, Catalina Natalia Cheaburu, Onur Yilmaz, Gina Mihaela Pricope, and Cornelia Vasile. 2016. "Poly(Vinyl Alcohol)/Chitosan/Montmorillonite Nanocomposites for Food Packaging Applications: Influence of Montmorillonite Content." *High Performance Polymers* 28(10):1124–38.

Dananjaya, S. H. S., R. Saravana Kumar, Minyang Yang, Chamilani Nikapitiya, Jehee Lee, and Mahanama De Zoysa. 2018. "Synthesis, Characterization of ZnO-Chitosan Nanocomposites and Evaluation of Its Antifungal Activity against Pathogenic Candida albicans." *International Journal of Biological Macromolecules* 108:1281–88.

Devi, M. P. Indira, N. Nallamuthu, N. Rajini, T. Senthil Muthu Kumar, Suchart Siengchin, A. Varada Rajulu, and Nadir Ayrilmis. 2019. "Biodegradable Poly (Propylene) Carbonate Using in-Situ Generated CuNPs Coated Tamarindus indica Filler for Biomedical Applications." *Materials Today Communications* 19:106–13.

Enescu, Daniela, Christian Gardrat, Henri Cramail, Cédric Le Coz, Gilles Sèbe, and Véronique Coma. 2019. "Bio-Inspired Films Based on Chitosan, Nanoclays and Cellulose Nanocrystals: Structuring and Properties Improvement by Using Water-Evaporation-Induced Self-Assembly." *Cellulose* 26(4):2389–401.

Erukhimovich, Igor and Monica Olvera de la Cruz. 2004. "Phase Equilibria and Charge Fractionation in Polydisperse Polyelectrolyte Solutions" (March):15–7.

George, Johnsy, Vallayil Appukuttan Sajeevkumar, Karna Venkata Ramana, and Shanmugam Nadana Sabapathy. 2012. "Augmented Properties of PVA Hybrid Nanocomposites Containing Cellulose Nanocrystals and Silver Nanoparticles." *Journal of Materials Chemistry* 22(42):22433–39.

Ghelejlu, Sara Beigzadeh, Mohsen Esmaiili, and Hadi Almasi. 2016. "Characterization of Chitosan–Nanoclay Bionanocomposite Active Films Containing Milk Thistle Extract." *International Journal of Biological Macromolecules* 86:613–21.

Giannakas, Aris, Kalouda Grigoriadi, Areti Leontiou, Nektaria Marianthi Barkoula, and Athanasios Ladavos. 2014. "Preparation, Characterization, Mechanical and Barrier Properties Investigation of Chitosan-Clay Nanocomposites." *Carbohydrate Polymers* 108(1):103–11.

Gierszewska, Magdalena, Ewelina Jakubowska, and Ewa Olewnik-Kruszkowska. 2019. "Effect of Chemical Crosslinking on Properties of Chitosan-Montmorillonite Composites." *Polymer Testing* 77(April):105872.

Gong, Yang, Yu Yingchun, Huixuan Kang, Xiaohong Chen, Hao Liu, Yue Zhang, Yimeng Sun, and Song Huaihe. 2019. "Synthesis and Characterization of Graphene Oxide/Chitosan Composite Aerogels with High Mechanical Performance." *Polymers* 11(5):777.

Haldorai, Yuvaraj and Jae-Jin Shim. 2014. "Novel Chitosan-TiO2 Nanohybrid: Preparation, Characterization, Antibacterial, and Photocatalytic Properties." *Polymer Composites* 35(2):327–33.

Hosseini, M. H., S. H. Razavi, and M. A. Mousavi. 2009. "Antimicrobial, Physical and Mechanical Properties of Chitosan-Based Films Incorporated with Thyme, Clove and Cinnamon Essential Oils." *Journal of Food Processing and Preservation* 33(6):727–43.

Hosseini, Seyed Fakhreddin, Zahra Nahvi, and Mojgan Zandi. 2019. "Antioxidant Peptide-Loaded Electrospun Chitosan/Poly (Vinyl Alcohol) Nanofibrous Mat Intended for Food Biopackaging Purposes." *Food Hydrocolloids* 89:637–48.

Indira Devi, M. P., N. Nallamuthu, N. Rajini, T. Senthil Muthu Kumar, Varada Rajulu A. Suchart Siengchin, and N. Hariram. 2019. "Antimicrobial Properties of Poly(Propylene) Carbonate/Ag Nanoparticle-Modified Tamarind Seed Polysaccharide with Composite Films." *Ionics*25:3461–71.

Jahan, Firdos, R. D. Mathad, and Shazia Farheen. 2016. "Effect of Mechanical Strength on Chitosan-Pva Blend through Ionic Crosslinking for Food Packaging Application." *Materials Today: Proceedings* 3(10):3689–96.

Javaid, Muhammad Asif, Muhammad Rizwan, Rasheed Ahmad Khera, Khalid Mahmood Zia, Kei Saito, Muhammad Zuber, Javed Iqbal, and Peter Langer. 2018. "Thermal Degradation Behavior and X-Ray Diffraction Studies of Chitosan Based Polyurethane Bio-Nanocomposites Using Different Diisocyanates." *International Journal of Biological Macromolecules* 117:762–72.

Kalaivani, R., M. Maruthupandy, T. Muneeswaran, A. Hameedha Beevi, M. Anand, C. M. Ramakritinan, and A. K. Kumaraguru. 2018. "Synthesis of Chitosan Mediated Silver Nanoparticles (Ag NPs) for Potential Antimicrobial Applications." *Frontiers in Laboratory Medicine* 2(1):30–5.

Kasirga, Yasemin, Ayhan Oral, and Cengiz Caner. 2012. "Preparation and Characterization of Chitosan/Montmorillonite-K10 Nanocomposites Films for Food Packaging Applications." *Polymer Composites* 33(11):1874–82.

Khorasani, Mohammad Taghi, Alireza Joorabloo, Armaghan Moghaddam, Hamidreza Shamsi, and Zohreh MansooriMoghadam. 2018. "Incorporation of ZnO Nanoparticles into Heparinised Polyvinyl Alcohol/Chitosan Hydrogels for Wound Dressing Application." *International Journal of Biological Macromolecules* 114: 1203–15.

Koosha, Mojtaba, Hamid Mirzadeh, Mohammad Ali Shokrgozar, and Mehdi Farokhi. 2015. "Nanoclay-Reinforced Electrospun Chitosan/PVA Nanocomposite Nanofibers for Biomedical Applications." *RSC Advances* 5(14):10479–87.

Kumar, Santosh, Ankita Shukla, Partha Pratim Baul, Atanu Mitra, and Dipankar Halder. 2018. "Biodegradable Hybrid Nanocomposites of Chitosan/Gelatin and Silver Nanoparticles for Active Food Packaging Applications." *Food Packaging and Shelf Life* 16(March):178–84.

Kumar, T., Senthil Muthu, N. Rajini, Huafeng Tian, A. Varada Rajulu, J. T. Winowlin Jappes, and Suchart Siengchin. 2017. "Development and Analysis of Biodegradable Poly (Propylene Carbonate)/Tamarind Nut Powder Composite Films." *International Journal of Polymer Analysis and Characterization* 22(5):415–23.

Lavorgna, M., I. Attianese, G. G. Buonocore, Amalia Conte, Matteo Alessandro Del Nobile, F. Tescione, and E. Amendola. 2014. "MMT-Supported Ag Nanoparticles for Chitosan Nanocomposites: Structural Properties and Antibacterial Activity." *Carbohydrate Polymers* 102:385–92.

Leceta, I., P. Guerrero, I. Ibarburu, M. T. Dueñas, and K. De la Caba. 2013. "Characterization and Antimicrobial Analysis of Chitosan-Based Films." *Journal of Food Engineering* 116(4):889–99.

Li, Hong Zhen, Si Chong Chen, and Yu. Zhong Wang. 2015. "Preparation and Characterization of Nanocomposites of Polyvinyl Alcohol/Cellulose Nanowhiskers/Chitosan." *Composites Science and Technology* 115:60–5.

Nasef, Shaimaa M., Ehab E. Khozemy, Elbadawy A. Kamoun, and H. El-Gendi. 2019. "Gamma Radiation-Induced Crosslinked Composite Membranes Based on Polyvinyl Alcohol/Chitosan/AgNO3/Vitamin E for Biomedical Applications." *International Journal of Biological Macromolecules* 137:878–85.

Nielsen, Lawrence E. 1969. "Cross-Linking–Effect on Physical Properties of Polymers." *Journal of Macromolecular Science, Part C* 3(1):69–103.

Norranattrakul, Parichat, Krisana Siralertmukul, and Roongkan Nuisin. 2013. "Fabrication of Chitosan/Titanium Dioxide Composites Film for the Photocatalytic Degradation of Dye." *Journal of Metals, Materials and Minerals* 23(2):9–22.

Noshirvani, Nooshin, Babak Ghanbarzadeh, Reza Rezaei Mokarram, Mahdi Hashemi, and Véronique Coma. 2017. "Preparation and Characterization of Active Emulsified Films Based on Chitosan-Carboxymethyl Cellulose Containing Zinc Oxide Nano Particles." *International Journal of Biological Macromolecules* 99:530–38.

Nouri, Afsaneh, Mohammad Tavakkoli Yaraki, Mohammad Ghorbanpour, Shilpi Agarwal, and Vinod Kumar Gupta. 2018. "Enhanced Antibacterial Effect of Chitosan Film Using Montmorillonite/CuO Nanocomposite." *International Journal of Biological Macromolecules* 109:1219–31.

Pavlidou, S. and C. D. Papaspyrides. 2008. "A Review on Polymer–Layered Silicate Nanocomposites." *Progress in Polymer Science* 33(12):1119–98.

Rahman, P. Mujeeb, V. M. Abdu. Mujeeb, and K. Muraleedharan. 2017. "Flexible Chitosan-Nano ZnO Antimicrobial Pouches as a New Material for Extending the Shelf Life of Raw Meat." *International Journal of Biological Macromolecules* 97:382–91.

Rubentheren, Viyapuri, Thomas A. Ward, Ching Yern Chee, and Chee Kuang Tang. 2015. "Processing and Analysis of Chitosan Nanocomposites Reinforced with Chitin Whiskers and Tannic Acid as a Crosslinker." *Carbohydrate Polymers* 115:379–87.

Ruiz, Gustavo Adolfo Muñoz and Hector Fabio Zuluaga Corrales. 2017. "Chitosan, Chitosan Derivatives and Their Biomedical Applications." In: Emad A. Shalby (ed.), *Biological Activities and Application of Marine Polysaccharides* 87. IntechOpen, DOI: 10.5772/66527.

Salaberria, Asier M., Rene Herrera Diaz, Jalel Labidi, and Susana C. M. Fernandes. 2015. "Preparing Valuable Renewable Nanocomposite Films Based Exclusively on Oceanic Biomass–Chitin Nanofillers and Chitosan." *Reactive and Functional Polymers* 89:31–9.

Salari, Mahdieh, Mahmod Sowti Khiabani, Reza Rezaei Mokarram, Babak Ghanbarzadeh, and Hossein Samadi Kafil. 2018. "Development and Evaluation of Chitosan Based Active Nanocomposite Films Containing Bacterial Cellulose Nanocrystals and Silver Nanoparticles." *Food Hydrocolloids* 84:414–23.

Sanuja, S., A. Agalya, and M. J. Umapathy. 2014. "Studies on Magnesium Oxide Reinforced Chitosan Bionanocomposite Incorporated with Clove Oil for Active Food Packaging Application." *International Journal of Polymeric Materials and Polymeric Biomaterials* 63(14):733–40.

Senthil Muthu Kumar, T., N. Rajini, M. Jawaid, A. Varada Rajulu, and J. T. Winowlin Jappes. 2018. "Preparation and Properties of Cellulose/Tamarind Nut Powder Green Composites: (Green Composite Using Agricultural Waste Reinforcement)." *Journal of Natural Fibers* 15(1):11–20.

Senthil Muthu Kumar, T., K. Senthil Kumar, N. Rajini, Suchart Siengchin, Nadir Ayrilmis, and A. Varada Rajulu. 2019. "A Comprehensive Review of Electrospun Nanofibers: Food and Packaging Perspective." *Composites Part B: Engineering* 175(July):107074.

Shao, Jinlong, Bing Wang, Jinmeng Li, John A. Jansen, X. Frank Walboomers, and Fang Yang. 2019. "Antibacterial Effect and Wound Healing Ability of Silver Nanoparticles Incorporation into Chitosan-Based Nanofibrous Membranes." *Materials Science and Engineering: Part C* 98:1053–63.

De Silva, R. T., M. M. M. G. P. G. Mantilaka, S. P. Ratnayake, G. A. J. Amaratunga, and K. M. Nali. de Silva. 2017. "Nano-MgO Reinforced Chitosan Nanocomposites for High Performance Packaging Applications with Improved Mechanical, Thermal and Barrier Properties." *Carbohydrate Polymers* 157:739–47.

Smith, Andrew T., Anna Marie LaChance, Songshan Zeng, Bin Liu, and Luyi Sun. 2019. "Synthesis, Properties, and Applications of Graphene Oxide/Reduced Graphene Oxide and Their Nanocomposites." *NANO Materials Science* 1(1):31–47.

Szymańska, Emilia and Katarzyna Winnicka. 2015. "Stability of Chitosan—A Challenge for Pharmaceutical and Biomedical Applications." *Marine Drugs* 13(4):1819–46.

Thiagamani, Senthil Muthu Kumar, Nagarajan Rajini, Suchart Siengchin, A. Varada Rajulu, Natarajan Hariram, and Nadir Ayrilmis. 2019. "Influence of Silver Nanoparticles on the Mechanical, Thermal and Antimicrobial Properties of Cellulose-Based Hybrid Nanocomposites." *Composites Part B: Engineering* 165:516–25.

Tripathi, S., G. K. Mehrotra, and P. K. Dutta. 2010. "Preparation and Physicochemical Evaluation of Chitosan/Poly (Vinyl Alcohol)/Pectin Ternary Film for Food-Packaging Applications." *Carbohydrate Polymers* 79(3):711–16.

Tripathi, Shipra, G. K. Mehrotra, and P. K. Dutta. 2011. "Chitosan-Silver Oxide Nanocomposite Film: Preparation and Antimicrobial Activity." *Bulletin of Materials Science* 34(1):29–35.

Vimala, K., Murali Mohan Yallapu, K. Varaprasad, N. Narayana Reddy, S. Ravindra, N. Sudhakar Naidu, and K. Mohana Raju. 2011. "Fabrication of Curcumin Encapsulated Chitosan-PVA Silver Nanocomposite Films for Improved Antimicrobial Activity." *Journal of Biomaterials and Nanobiotechnology* 02(01):55–64.

Wang, Guang-Shuo, Zhi-Yong Wei, Lin Sang, Guang-Yi Chen, Wan-Xi Zhang, Xu-Feng Dong, and Min Qi. 2013. "Morphology, Crystallization and Mechanical Properties of Poly (Composite Films for Improved Antimicrobial Activity." *Chinese Journal of Polymer Science* 31(8):1148–60.

Xu, Yixiang, Xi Ren, and Milford A. Hanna. 2006. "Chitosan/Clay Nanocomposite Film Preparation and Characterization." *Journal of Applied Polymer Science* 99(4):1684–91.

Yadav, Mithilesh and Sharif Ahmad. 2015. "Montmorillonite/Graphene Oxide/Chitosan Composite: Synthesis, Characterization and Properties." *International Journal of Biological Macromolecules* 79:923–33.

Yang, Dong, Jie Li, Zhongyi Jiang, Lianyu Lu, and Xue Chen. 2009. "Chitosan/TiO2 Nanocomposite Pervaporation Membranes for Ethanol Dehydration." *Chemical Engineering Science* 64(13):3130–37.

Yang, Xiaoming, Yingfeng Tu, Liang Li, Songmin Shang, and Xiao-Ming Tao. 2010. "Well-Dispersed Chitosan/Graphene Oxide Nanocomposites." *ACS Applied Materials and Interfaces* 2(6):1707–13.

Zhang, Xiaodong, Gang Xiao, Yaoqiang Wang, Yan Zhao, Haijia Su, and Tianwei Tan. 2017. "Preparation of Chitosan-TiO2 Composite Film with Efficient Antimicrobial Activities under Visible Light for Food Packaging Applications." *Carbohydrate Polymers* 169:101–7.

Zhang, Xin, Yunpeng Liu, Huimin Yong, Yan Qin, Jing Liu, and Jun Liu. 2019. "Development of Multifunctional Food Packaging Films Based on Chitosan, TiO2 Nanoparticles and Anthocyanin-Rich Black Plum Peel Extract." *Food Hydrocolloids* 94:80–92.

13 Medium Density Fiberboard as Food Contact Material

A Proposed Methodology for Safety Evaluation from the Food Contact Point of View

P. Vazquez-Loureiro, F. Salgado, A. Rodríguez Bernaldo de Quirós, and R. Sendón

CONTENTS

13.1 INTRODUCTION: MDF TECHNICAL WOOD

Medium density fiberboard (MDF) is the most recent addition of wood-based panels. In Europe it was first manufactured in the early 1970s.

This product arose as a substitute for solid wood because of several advantages: homogeneity, properties that could be adjusted to be suitable for different requirements and a better use and exploitation of forest resources, which implies economic and ecological advantages. Due to these advantages, the global production and consumption of this type of material have been growing steadily since 1980 (Kim, 2019).

Wood is the main component of MDF (approximately 90%). The rest is made up of the adhesive used to join the fibers together (Kartala and Green III, 2003). Additionally, other products can be added to provide water-resistance, protective properties, etc. Urea-formaldehyde and urea-melamine-formaldehyde resins are the most common glues used to produce MDF. Installed production capacity in the world is more than 100 million m^3/year. The main producers are China, Europe (Germany, Poland, Spain, Portugal, France and Hungary), South and North America, Australia and New Zealand.

Being made of wood, MDF is a carbon sink. It has been calculated that one m^3 of MDF removes up to two tons of CO_2 from the atmosphere (Wilson, 2010), while the manufacture of other materials implies an emission of this compound (CO_2) which has negative impacts which contribute to climate change. For example, one ton of plastic produces three tons of CO_2, or around 1.5 tons when recycled plastic is produced.

Like all wood-based panels, MDF was developed looking for a better way to use the forest and wood. Several advantages can be highlighted (see Figure 13.1):

- MDF allows sustainable use of forest resources. Only about 50% of the tree is used to manufacture sawn timber or plywood, but the production of MDF uses almost the whole tree.
- The by-products of plywood and sawn wood manufacturing can be used as raw material to produce MDF while other processes use the by-products as fuel.
- Due to the wide range of sizes of boards available in the market, the right choice of MDF allows the minimization of waste material.
- MDF provides an excellent surface. The characteristics of the wood fibers which make up the MDF ensure better quality surface than other types of wood-based panels.
- MDF has excellent physical properties. There are several qualities of MDF, as well as sizes and thicknesses available on the market, which makes it easy to find the right type of MDF to satisfy the requirements of the product that is going to be made.
- It is important to underline that MDF maintains most of its physical properties even in low-temperature and high-moisture-content environments (standard storage conditions for fruits and vegetables).
- MDF is very easy to mechanize. Due to the structure and size of the fibers which make up the MDF, this board is easy to cut and shape with a wide range of mechanical tools, laser, etc. It is also shatterproof and has a superb quality of finish even when parts have been machined.
- Another advantage MDF has over other materials on contact with food is the time it takes to degrade. Table 13.1 shows the degradation time of some materials used as food contact materials (FCMs).

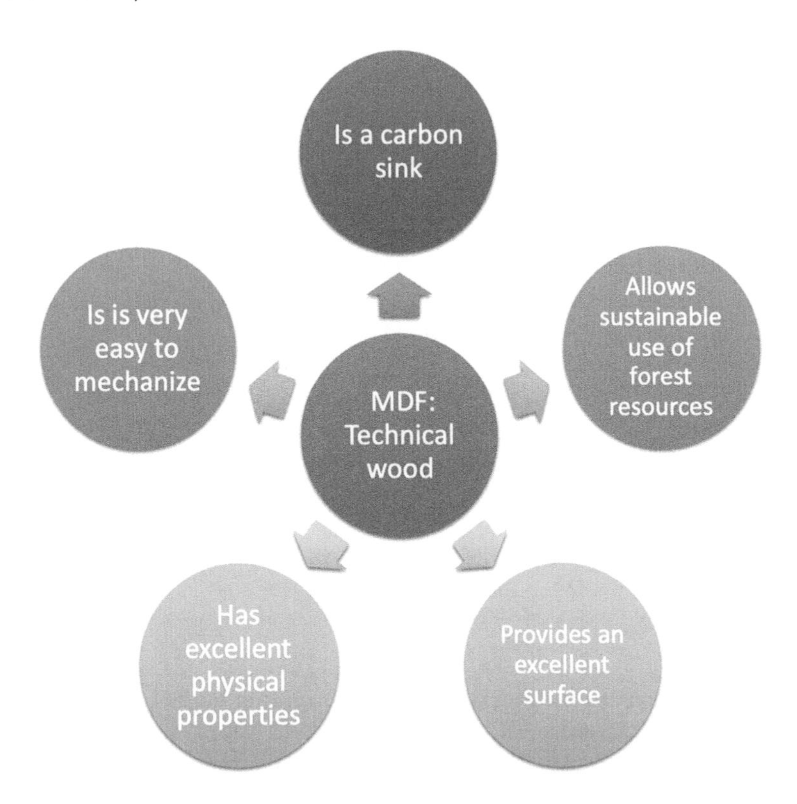

FIGURE 13.1 Main advantages of MDF.

- Last but not least, MDF is currently used as a food contact material. MDF was initially used as a solid wood substitute in construction and furniture production but, because of the above-mentioned advantages, nowadays, MDF is being used as raw material to produce packaging for whole (neither sliced nor peeled) fresh fruits and vegetables.

The type of MDF that is used for the manufacture of food packaging must comply with European food contact current regulations (EC Regulation, 1935/2004) to guarantee that no substance migrates to the food in sufficient quantity that may represent a health hazard, that causes unacceptable changes of the composition of food or alteration of the organoleptic characteristics of foods; hence to consider MDF a suitable material for food contact. To evaluate MDF safety as an FCM (for fresh fruits and vegetables, neither sliced nor peeled), a methodology is presented in this chapter.

13.2 MDF AS FOOD CONTACT MATERIAL IN THE FRUIT AND VEGETABLE SECTOR

Fruit and vegetable production in Spain is an important economic activity sector in terms of export and employment. The Spanish export of fresh fruits and

TABLE 13.1
Degradation Time of Other Materials Used as Food Contact Materials (FEDEMCO, ITENE, Plastics Europe)

Product		Time of degradation
Paper		1 to 5 months
Paperboard/cardboard		1 to 2 years
MDF, wood, boards		1 to 3 years (1 to 3 months with pretreatment)
Multilayer (paperboard/aluminium/and plastic		~ 30 years
Plastic	Bio (PLA)	2 to 3 months
	Bag (PE)	~ 150 years
	Package (PP)	450 to 1000 years
	Bottle (PET)	> 1000 years
Glass		~ 4000 years
Metal	Boat (Al)	~ 10 years
	Steel sheets, aerosols	~ 30 years
	Iron, steel	1 to 1000 years

vegetables stood at 9.2 million tons and 9.363 million euros for the first eight months of 2019. It increased 9% in volume and 4% in value compared to the same period in 2018, according to the data updated by the Custom Authorities. (MAPA, FEPEX).

The use of MDF in the packaging fruits and fresh vegetables is growing significantly. Due to the advantages mentioned above, MDF is a good alternative to plastic packaging. The production and consumption of MDF packaging has been increasing in recent years, according to data from the Spanish Federation of the Wooden Container and its Components (FEDEMCO, n.d.).

Spain is the eighth largest world producer of MDF (in 2011, it was the eleventh largest), the first producer in the EU and, since at least 2006, the first EU exporter to which 60% of the production is destined (FEPEX) while Brazil is one of the largest producers of MDF in the world (Piekarski et al., 2017). The world's major consumer and producer of forest products including wood-based panels is China (wood-based panels: 47% of global consumption and 50% of global production; sawn wood: 26% of global consumption and 18% of global production).

According to the Food and Agriculture Organization (FAO), China in 2018 was also a major exporter of wood-based panels in the world (16% of global exports).

The main materials used in Europe to package agricultural products are wood and boards, cardboard and plastic. Within the wood and wood-based panels section, MDF shows continuous growth (~ 25%). This is due to all the above-mentioned characteristics and advantages.

Because of its features, the use of MDF as food packaging is usually limited to fresh fruits and fresh or chilled vegetables, without peeling or cutting. It is a single-use container for storage of this type of food, although it can be used as storage at home, but not as raw material to manufacture new MDF as FCM.

Also, the inner part of the package, which comes into contact with the fruit or vegetable, is always MDF unsanded, uncoated and without any treatment.

The manufacturing process of the MDF package is very simple since only the pieces are cut in the appropriate dimensions and a subsequent assembly to configure the final container. These packages have a completely homogenous edges surface obtained in some cases by laser cutting, without using any other material to build the container, facilitating their recycling. An easy assembly system is used in most cases to produce a great resistance box. In other cases, to assemble these MDF boxes it is necessary to employ small wood pieces in the corners to join the different MDF parts using staples.

13.3 COMPOSITION OF MDF

MDF is mainly made of wood (approximately 88–90%), which is mostly coniferous wood although it is possible to use a small percentage of some leafy wood. An adhesive is used to bond the fibers, complementing the natural action of the lignin contained in the wood. The adhesive (glue) is the second component of the MDF (approximately 7%).

This adhesive is usually an amino resin obtained by condensation polymerization of formaldehyde, urea and sometimes a small percentage of melamine. The proportion of these three monomers and the manufacturing conditions are optimized to obtain a completely crosslinked polymer. This resin applied to the wood fibers, completes the function of lignin providing the resulting MDF with the characteristic properties of this type of board such as enough moisture resistance that makes the addition of water repellent agents unnecessary.

Wood is basically a lignocellulosic material consisting of cellulose chains, surrounded by hemicellulose fibers, all of them bound by lignin which is the natural glue that holds all the wood components together.

13.4 MDF MANUFACTURING PROCESS

The MDF for food contact is manufactured in a similar process as the MDF for other applications but with some differences. This includes different stages; briefly, the production process includes the production of wood chips, its transformation into fiber, the addition of the adhesive and finally the hot pressing of the fiber (see Figure 13.2).

As mentioned, the most commonly used wood is the conifer family. For the manufacture of the MDF that is to be employed for food contact, only first-use wood can be used, material from the recycling of wood, MDF or other wooden boards being unacceptable.

- *Wood chips production*: Wood logs without bark become wood chips by means of a knife chipper (see Figure 13.3). The wood chips are sieved to select the right size. Then, wood chips are washed with water to remove foreign matter, such as sand, that may be attached to the wood. The water resulting from the washing is processed in a treatment station for later reuse in the process.
- *Fiber production*: Sifted and washed wood chips are placed in a digester where they undergo a steam cooking process. At a temperature higher than

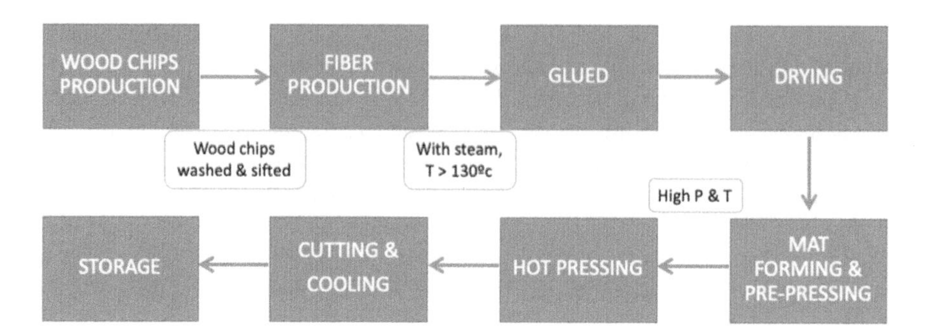

FIGURE 13.2 Scheme of MDF production.

FIGURE 13.3 Wood chips obtained during the first stage of MDF production.

130°C, the lignin of the wood softens, facilitating the subsequent mechanical separation of the cellulose and hemicellulose fibers, that takes place in a defibrator. This device consists of two disks, usually one static and one rotating, separated by a few tenths of a millimeter. The wood chips are introduced axially between the disks and are converted into fibers.

- *Gluing*: In the blow line of the defibrator the adhesive is added. This adhesive is needed to reinforce the action of the lignin to improve the characteristics of the MDF.

As described above, and because of the press temperature (higher than 100°C) and the composition and subsequent reactivity of the adhesive, this resin will be converted during the pressing step in a stable polymer.

- *Drying*: Excess moisture from the glued fiber (see Figure 13.4) is removed in the dryer by pneumatic transport with hot air.
- *Mat forming and pre-pressing*: The fiber, glued and with adequate moisture, is distributed evenly and homogeneously on a conveyor belt forming a mattress. This fiber mat is compacted by pressure at room temperature to reduce thickness and facilitate the subsequent hot-pressing process.
- *Hot pressing*: The pre-compacted glued fiber is subjected to the high pressure and temperature action to achieve the targeted thickness and the complete polymerization of the added adhesive.

  The temperature (>100°C) and pressing time employed should ensure that the adhesive, which reinforces the action of lignin, will be converted into an intercross polymer.

  In this pressing phase, the excess water contained in the pre-compacted glued fiber is released as well as the free formaldehyde that the adhesive could contain.
- *Cutting and cooling*: After this stage, the boards are cut to their final size. The final step of the process is cooling, which allows the boards to be stored correctly without damaging their dimensional stability. Normally, it is not necessary to sand thin MDF.

FIGURE 13.4 Different samples of wood chips and glued fiber employed in the manufacture of MDF.

Wood-based panels manufacturers have certified quality systems in place including inspection, quality control, traceability and in-line and product control devices such as continuous thickness measurement, internal bag detector, etc.

13.5 CHARACTERISTICS

The specifications of fiberboards are defined in the standard EN 622-5: Requirements for dry process boards (MDF), some of these parameters, from the technological point of view, are shown in Table 13.2.

Some features must be highlighted such as the density profile, and the impact of the environmental humidity on some properties of MDF.

Regarding the density profile, it is important to point out that the faces of the board have a density in the range of 1000 kg/m^3 compared to the MDF average density (850 kg/m^3), which represents an advantage in terms of barrier and surface quality.

The equilibrium humidity of the MDF depends on the relative humidity (RH). For example, MDF moisture content will be 3 to 7% in an environment of 25% RH, 7 to 12% if the RH is 65% and between 13 and 17% in cases of high humidity, of the order of 90%.

Moreover, there is a correlation between the humidity of the board and its dimensions so that when the humidity of the board increases, both the thickness, the length and the width increase.

In terms of internal bond, (the resistance of the board to delamination) must be measured according to the EN 319 standard as indicated in Table 13.2. The use of

TABLE 13.2
Specifications of Medium Density Fiberboard (MDF)

Unsanded MDF for food packaging. thickness ≤3.0 mm

Parameter	Test	Units	Value
Dimensions			
Thickness	EN 324-1	mm	± 0.2
Length and width	EN 324-1	mm/m	± 2.0
Squareness	EN 324-2	mm/m	2.0
Edge straightness	EN 324-2	mm/m	± 1.5
Dimensional stability			
Length and width	EN 318	%	0.4
Thickness	EN 318	%	10
Humidity	EN 322	%	5 to 10
Mechanical properties			
Average density	EN 323	Kg/ m^3	870 ± 30
Thick swelling, 24 h in water	EN 317	%	<45
Internal bond	EN 319	N/mm^2	> 0.90
Flexural strength	EN 310	N/mm^2	> 23

MDF in an environment with high relative humidity reduces the internal bond of the MDF (up to 40%). This is an important issue to be taken into account in designing the package.

13.6 SAFETY EVALUATION OF MDF AS FCM

In Europe, the framework regulation for food contact, Regulation (EC) No 1935/2004, establishes the legal basis for food contact materials. It sets out the general principles of safety and inertness for all food contact materials. However, among the 17 specific materials referred to in the framework regulation, only the group of plastic materials is broadly regulated. In the case of these plastic materials, experimental conditions for testing the fulfillment with legislation of these materials are detailed in Regulation (EC) N° 10/2011. Other materials are also regulated but, in less extension, as is shown in Figure 13.5.

As it can be observed, several materials widely used as food contact materials have not, at the moment, harmonized regulations at European level. Some examples are paper and board, metals and alloys, cork, glass, varnishes and coatings, waxes and wood among others. In some cases, some Member States have developed their own national regulations, as it is with the case of adhesives or paper that are regulated in Germany. Besides, in the absence of specific regulations, some other documents are used for reference, e.g., the Council of Europe (CoE) resolutions, recognized international standards or other guideline documents.

A deep report about the situation in Europe of the non-harmonized food contact materials was elaborated by Joint Research Center of the European Commission; the food contact materials included in this document that are more similar to MDF are solid wood and plywood, for which there are specific measures in France, Croatia and the Netherlands besides some national standards and good manufacturing practices (Simoneau et al., 2016; FEDEMCO, n.d.).

Currently, for the particular case of the use of MDF for the packaging of fresh fruits and vegetables, there does not exist any standard that can serve as a reference to evaluate the safety of MDF as FCM. With this purpose, and taking into account the knowledge of the raw materials employed in the manufacture of MDF, Table 13.3 includes chemical substances that could be present in the MDF and consequently could reach the food that it is in contact with the material, i.e., that could migrate. As far as possible, the Regulation (EC) N° 10/2011 (plastic regulation) has been taken as a reference. So, from the food safety point of view, the following substances should be considered (see Table 13.3).

Most of the specific migration limits (SMLs) shown in Table 13.2 are included in the plastic regulation, but for those compounds that were not included, other reference documents were used. For example, for the SML of mineral oils, mineral oil aromatic hydrocarbons with compounds between C16 and C35 (MOAHs), the fourth draft elaborated by German Federal Ministry of Food and Agriculture (BfR) for recycled cardboard for food contact was used (BfR, 2007); for the SML of some metals (As, Pb, Cd, Hg, Ni, V and Cr) the Council of Europe's "Resolution on Metals and Alloys Used in Food Contact Materials and Articles" was employed (CoE, 2013).

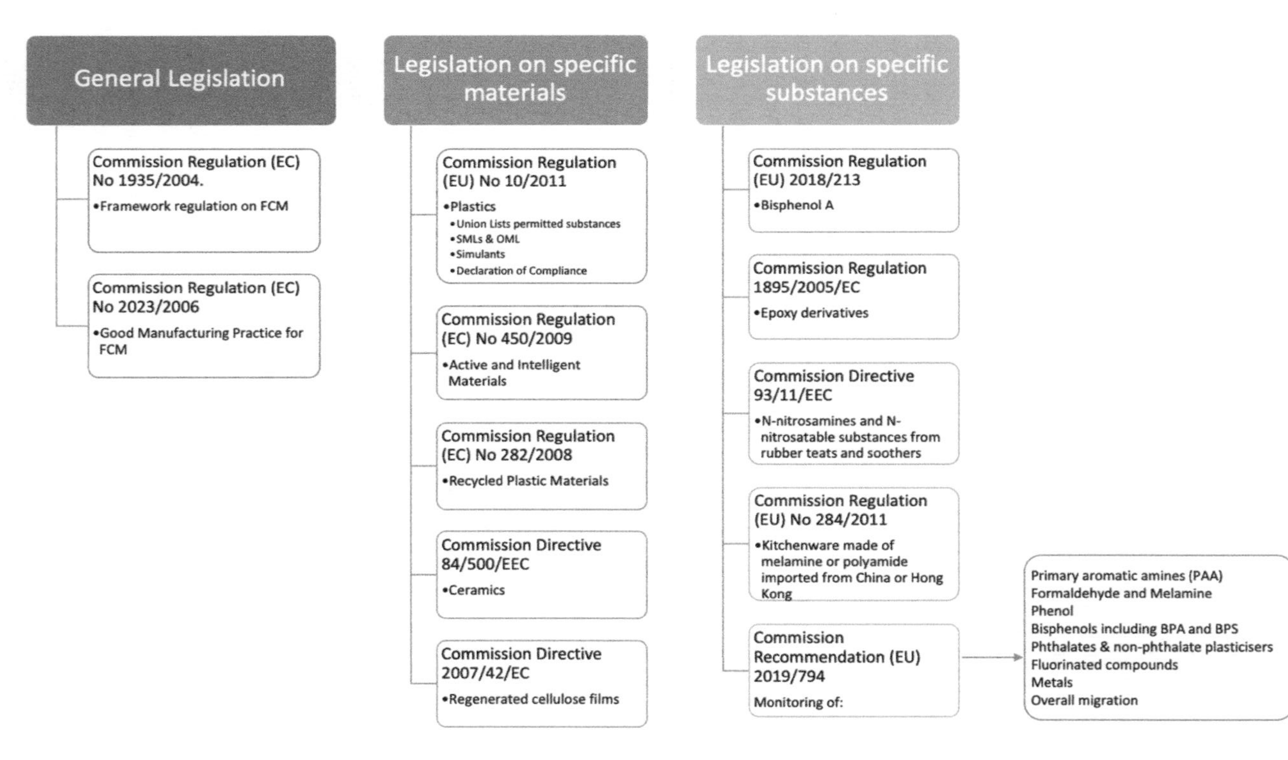

FIGURE 13.5 European legislation regarding food contact materials.

TABLE 13.3
Specifications of Medium Density Fiberboard from the Point of View of Food Safety

Parameter	Test	Units	Limit value
Unsanded MDF for food contact use; Thickness ≤3.0 mm			
Formaldehyde content	EN ISO 12460-5 Perforator	mg/100 g dry board	<8
Food safety-specific migration			
Formaldehyde	Proposed	mg/kg simulant or food	<15
Mineral oils MOAHs	Proposed		<0.5
Melamine	Proposed		<2.5
Bisphenol A	Proposed		<0.05
Pentachlorophenol	Proposed		<0.15*
Primary aromatic amines	Proposed		<0.01
Metals			
Al	Proposed	mg/kg simulant or food	<1.0
Ba			<1.0
Co			<0.05
Cu			<5.0
Fe			<48
Li			<0.6
Mn			<0.6
Zn			<5.0
As			<0.002
Pb			<0.01
Cd			<0.005
Hg			<0.003
Ni			<0.14
V			<0.010
Cr			<0.25
Organoleptic impact	ISO 13302:2008		ok

*This value refers to mg/kg of MDF.

Regarding the organoleptic impact, the UNE-ISO 13302:2008 standard refers to methods based on sensory analysis for assessing modifications to the flavor of foodstuffs due to packaging, and that can be applied to MDF.

13.6.1 Formaldehyde

One important parameter that should be controlled is the content and emission of formaldehyde (CH_2O), which is an extremely volatile chemical that is present in nature and in many foods, is generated in many natural processes and is degraded by sunlight. Formaldehyde is toxic by inhalation and is associated with possible health hazards, such as irritation of the upper respiratory tract and eyes. As a result, the European and Northern American governments have imposed regulations limiting the emission of formaldehyde from building materials and from the materials used in the manufacture of furniture and fittings (Kawouras et al., 1998).

Formaldehyde is present in non-treated wood (around 10 mg/kg) and also in the adhesive used to glue the fibers, although once the polymerization occurs its concentration in the end material should not be above the established limit. As Table 13.4 shows, several international standards detail different methods to evaluate the formaldehyde content or emission and allow the classification into several types of MDF according to formaldehyde content or emission.

The ISO 12460-5 standard (perforator method) establishes the method to determine the content of the formaldehyde extracted with boiling toluene (110.6°C) for 3 hours. In the case of MDF used for the manufacture of packaging for fruits and vegetables, which is E1 grade or equivalent, the extracted formaldehyde will be less than 8 mg/100 g of dry board.

The EN 717-1 (chamber method) standard establishes the method of determining the emission of formaldehyde by the boards. The amount of formaldehyde emitted (released) when circulating air at 23 ± 0.5°C and RH 45 ± 3% is measured at a speed between 0.1 and 0.3 m/s for a period of time between 1 and 4 weeks, more specifically until reaching a sustained value, it will be less than 0.124 mg of formaldehyde/m^3 of MDF.

Regarding the other substances listed in Table 13.3, they were selected taking into account the knowledge of the MDF manufacturing process but also the concern that exists about other substances that a priori are not expected to be in the MDF (e.g., bisphenol A).

13.6.2 Methodology for Food Safety Evaluation: Specific Migration Limits

From the food safety point of view, it is important to consider, as it has been explained above, that no specific measures exist that can be applied to MDF, so as far as possible the Regulation (EC) N° 10/2011 and amendments have been taking as a reference. This document includes a positive list of substances that can be used in the manufacture of plastics for food contact, and each substance can have several restrictions. One is the "specific migration limit" that is defined as "the maximum permitted amount of a given substance released from a material or article into food or food simulants." Briefly, these values are set usually after a risk assessment of the substance and most of them have been evaluated and a tolerable daily intake (TDI, expressed in mg/kg body weight) established; the SML is then calculated on the basis of the assumption that a person with a bodyweight of 60 kg consumes 1 kg of food containing the substance daily. So, the SMLs should be expressed as mg substance/kg of food, but if the material is not yet in contact with the food, it is necessary to know the real surface material to volume of food ratio. In some cases, as for sheets not yet in contact with food, this real ratio is not known and afterward a surface:volume ratio of 6 dm^2 per kg of food should be applied. This standard ratio

TABLE 13.4
Types of MDF According to the Formaldehyde Content or Emission

	EUROPE EN 13986			JAPAN JIS 5905			USA ANSI A 208.1 1&2	
	EN 120							
	ISO 12460-5	EN 717-1	EN 717-2		JIS A1460; JAS 233; ISO 12460-4			ASTM 1133
Method	Perforator	Chamber	Gas analysis	Method	Dessicator		Method	Big chamber
Type	mg HCHO/100 g	ppm (mg/m³)	mg/h.m²	Type	mg HCHO/l		Type	ppm
E 2	≤ 30	> 0.1	> 3.5; < 8	F*	≤ 5 (average)	≤ 7 (maximum)	After CARB II	
E 1	≤ 8	< 0.1	>2; < 3.5	F**	≤ 1.5 (average)	≤ 2.1 (maximum)	PB	≤ 0.09
E 05	≤ 4	< 0.05	< 2	F***	≤ 0.5 (average)	≤ 0.7 (maximum)	MDF	≤ 0.11
				F****	≤ 0.3 (average)	≤ 0.4 (maximum)	TMDF	≤ 0.13

results from the assumption that a cube of 6 dm^2 of total surface (each side of the cube is 1 dm^2) contains 1 kg of food (Robertson, 2012).

To check the compliance with these SMLs, it is necessary to carry out the migration assays. These migration tests should be undertaken under standardized conditions, as in plastic regulation, and to select the test conditions (temperature, time or food simulant) the real use conditions of the MDF package should be known. Taking into account that this packaging is used for fresh fruits and vegetables and that these foods are maintained at low temperatures (always lower than 12°C) for different periods and under controlled RH conditions, the migration test should be carried out at 20°C, for 10 days and at a 75% RH. It is important to point out that plastic regulation does not refer to RH, but this is an important parameter in the case of MDF because almost always the food contained in MDF packaging is kept refrigerated.

Once the temperature and time of migration test are set, a food simulant should be chosen to carry out the assays. From the food simulants proposed in plastic regulation (see Table 13.5) only simulant E (a solid polymer) can be used in contact with MDF, due to all other simulants being liquids and containing water (except the D2 that is specific for fatty foods) that affect the integrity of MDF that it swells and loses its physical properties. If the FCM is modified in a different way than during its real use, the migration test cannot be considered since it does not mimic what occurs during the real contact. In plastic regulation, simulant E is the one selected to check compliance with SMLs in the case of plastics that are intended to be in contact with fruit and vegetables, fresh or chilled, unpeeled and uncut, but applying a correcting factor of 10 (X/10) after obtaining the experimental result and before the comparison with the established SML.

Tenax® (polyoxide of 2,6-diphenyl-p-phenylene, MPPO) is a polymer of high porosity and specific surface, which is commonly used as an adsorbent of volatile compounds and as food simulant. It has own characteristics such as particle size: 60–80 mesh, pore 200 nm and high price. Moreover, it is difficult to handle, and after the migration tests, substances must be extracted prior to its analysis. In order to carry out the migration assays with Tenax, at the conditions cited above, the following experimental conditions are proposed: approx. 1 g of Tenax is homogenously distributed over a 5 × 5 cm MDF sample using a glass ring; this sample is placed in a Petri dish and at the same time it is placed inside a container where a solution is at the bottom for RH control. Figure 13.6 shows an example. After 10 days, all selected compounds trapped into Tenax need to be extracted before its determination.

TABLE 13.5
List of Food Simulants Included in Regulation (EC) N° 10/2011

Simulant	
Simulant A	10% ethanol: for aqueous foods pH > 4.5
Simulant B	3% acetic acid: for acidic foods pH <4.5
Simulant C	20% ethanol: alcoholic foods
Simulant D1	50% ethanol: alcoholic foods > 20% alcohol & oil in water emulsions including milk
Simulant D2	Vegetable oil: foods with free fat at the surface
Simulant E	Poly (2,6-diphenyl-p-phenylene oxide) (Tenax®): dry foods

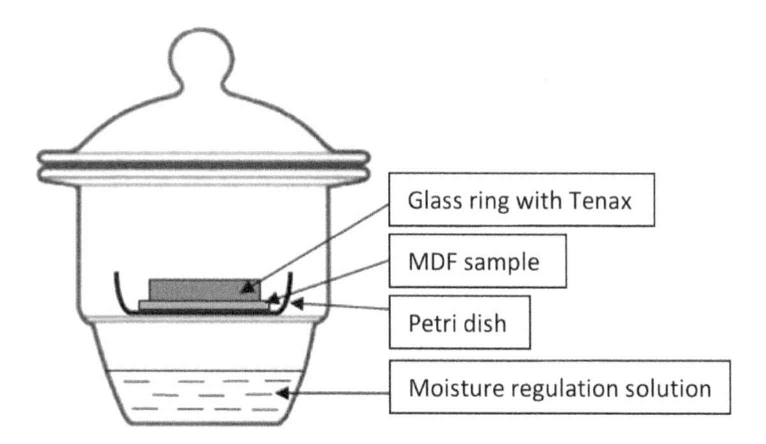

FIGURE 13.6 Scheme illustrating migration test with Tenax.

Extraction conditions will depend on the substance and also on the final analytical technique selected.

As a well-defined strategy and to minimize the number of tests to be carried out to evaluate the MDF as FCM, a screening method has been designed and it is shown in the following decision diagram (see Figure 13.7).

Three main decision steps are included in Figure 13.7.

1. *Content*: This is the determination of the content and the expression of the result in mg/kg of material (MDF), for each and every one of the selected analytes. The aim of the content determination is to know the total amount of substance that is present in the MDF and that could be able to migrate. The extraction conditions employed should ensure that all of the substance is extracted. Some extraction conditions are shown in Table 13.6.
2. *Screening*: As it has been explained before, 6 dm^2 of contact surface is associated with 1 kg of food or simulant. As worst-case scenario, it is assumed that the whole amount of analyte contained in a 6 dm^2 MDF piece migrates only through one of the faces. By means of this calculation, the maximum migration that could occur in this worst-case scenario is determined.

The calculation formula to be applied is:

$$X/10\left(\text{mg of analyte/kg of food or simulant}\right) = C \times D \times T \times 6 \times 10^{-6}$$

where

C = Content, mg analyte/kg of MDF
D = Density of MDF, kg/m^3
T = Thickness, mm
6 = ratio 6 dm^2 (material contact surface) / 1 kg of food or simulant

If $X/10$ (calculated maximum specific migration) is ≤ SML, it is not necessary to carry out the specific migration test.

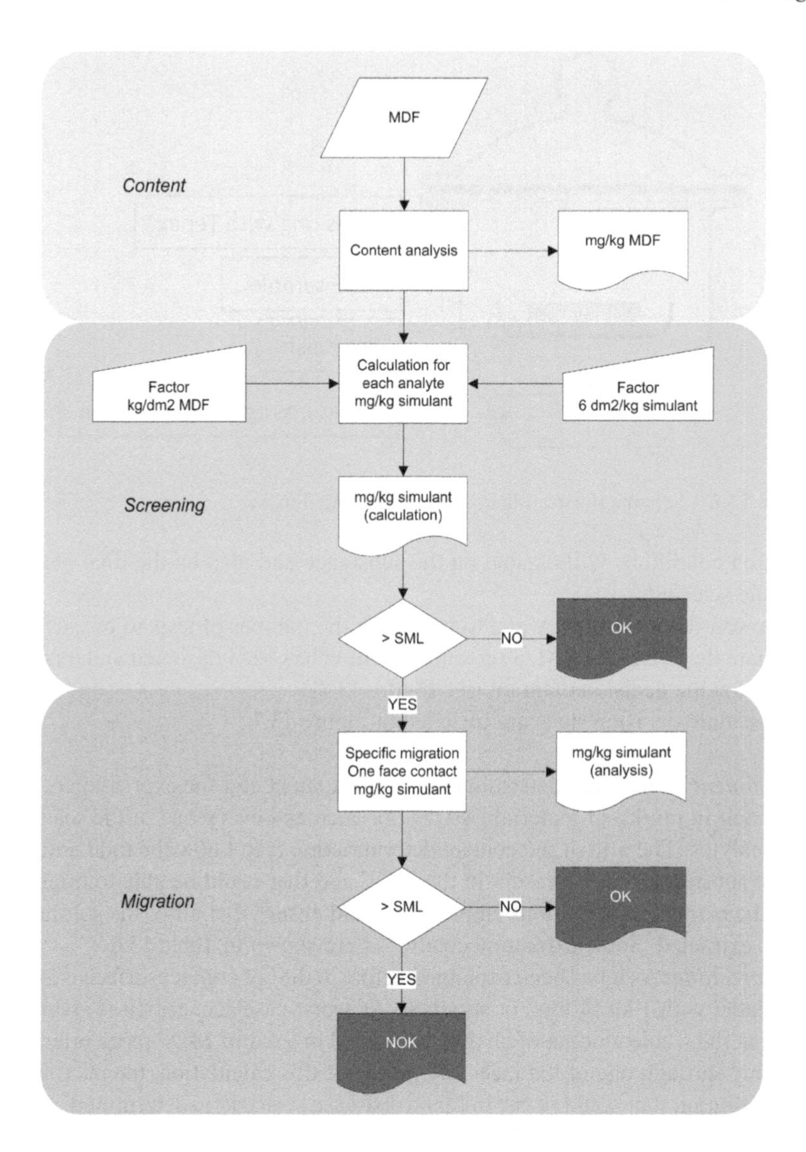

FIGURE 13.7 Proposed methodology for ensuring MDF food safety.

As an example, if in an MDF sample (density: 850 kg/m^3 and thickness: 2.5 mm) a content of melamine of 40 mg/kg MDF is determined, the following calculation should be done: $40 \times 850 \times 2.5 \times 6 \times 10^{-6}$. Thus, the obtained result is 0.5 mg/kg food (corresponds to ×/10) and in this case, as 0.5 is <2.5 (SML of melamine expressed in mg/kg food or food simulant), the specific migration test is not required to be carried out.

Another example could be applied to the content determination of formaldehyde by the perforator method. If the obtained value is 8 mg/100g MDF, this is equivalent

TABLE 13.6
Extraction Conditions Applied to MDF to Determine the Content

Substance	Content
Formaldehyde	ISO 12460-5 perforator
Mineral oils MOAHs	Solvent extraction*
Melamine	Extraction with water at 40°C for 24 h
Bisphenol A	Extraction with acetonitrile at 40°C for 24 h
Pentachlorophenol	Extraction with methanol at 40°C for 24 h
Primary aromatic amines	Extraction with 3% acetic acid at 40°C for 24 h
Metals	Extraction with 3% acetic acid at 40°C for 24 h

* Due to the complexity of this extraction, more information can be found in Bratinova and Hoekstra, 2019.

to 80 mg/kg MDF, and applying the above cited equation (the same MDF sample) the result would be 1.0 mg/kg, that it is less than the SML of formaldehyde (15 mg/kg food or food simulant).

3. *Specific migration*: Only if *X*/10 (calculated maximum specific migration) is > SML, is it necessary to carry out the specific migration test using the test conditions mentioned above.

Following the example of melamine content, for the same MDF sample, if a content of 400 mg/kg MDF is determined, as 5 is > 2.5 it should be necessary to perform the migration test.

13.6.3 Analytical Methodology to Determine Content and SML

Finally, and to evaluate the content of the migration of the cited compounds, analytical methods with suitable performance characteristics should be applied. For example, precision, reproducibility, selectivity and sensitivity of the applied analytical method should allow obtaining confident results at low levels of detection. In a document published by the Joint Research Centre (JRC) more information about all these analytical method characteristics is included, focused on food contact materials. (Bratinova et al., 2009). Regarding the different analytical techniques that could be applied to the different substances, Table 13.7 includes some techniques that could be applied successfully.

Aldehydes and terpenes, small linear chain alcohols and ketones, are identified as predominant compounds in emissions of MDF samples. The presence of small linear chain aldehydes, such as pentanal, hexanal, heptanal, octanal and nonanal is probably due to the degradation of wood or its secondary metabolites (Baumann et al., 2000).

Moreover, the olfactory threshold of the selected compounds in the case of aldehydes and terpenes is important due to the possible impact on the organoleptic

TABLE 13.7
The Main Selected Analytical Techniques

Compound	Analytical technique
Pentanal	GC-MS (P&T) GC-MS
Hexanal	
Octanal	
Nonanal	
α-Pinene	
α-Terpineol	
Caryophyllene	
Bisphenol A	HPLC-FLD
Pentachlorophenol	HPLC-MS/MS
Melamine	HILIC-UV
PAAs	HPLC-MS/MS
Formaldehyde	HPLC-UV
Heavy metals	ICP-MS

characteristic of the fruits and vegetables. These compounds can be analyzed using purge and trap (P&T coupled to gas chromatography–mass spectrometry (GC-MS). Samples are subjected to a slight heating (purge) during which the present volatiles are dynamically dragged by a helium stream being retained in a trap. The trap is then subjected to rapid heating and the compounds are carried to the chromatograph where they are separated. The mass spectrometer in positive electronic impact is typically used and peak identification is determined using mass spectrum libraries (Wiley 8th Mass Spectral Library); standards commercially available should be used for quantification purposes, and if, to achieve low DLs, the acquisition mode selected is ion monitoring (SIM), at least two characteristics masses should be selected for each compound (see Table 13.8). A 6% cyanopropylphenyl, 94% dimethylpolysiloxane

TABLE 13.8
Aldehydes and Terpenes with Its Monitored Ions and Olfactory Threshold, Respectively

Compound	Monitored ion (m/z)	Olfactory threshold ppb
Pentanal (CAS 110-62-3)	44, 58	12
Hexanal (CAS 66-25-1)	56, 44	4.5
Octanal (CAS 124-13-0)	84, 57	0.7
Nonanal (CAS 124-19-6)	57, 98	1
α-Pinene (CAS 80-56-8)	93, 92	6
α-terpineol (CAS 562-74-3)	93, 121	330
Caryophyllene (CAS 87-44-5)	133, 93	64

type GC column can be used for the separation, once it is specially designed for the analysis of volatile organic compounds.

To determine the bisphenol A (BPA) content, a common analytical technique is liquid chromatography with fluorescence spectroscopy (HPLC-FLD). The determination of BPA can be achieved using a reverse phase C18 column and water and acetonitrile as mobile phase. The excitation and emission wavelengths are 225 nm and 305 nm, respectively. The BPA was identified by comparison of the retention time and fluorescence spectra with that obtained with pure standard injected under the same chromatographic conditions.

To determine pentachlorophenol in MDF boards, liquid chromatography coupled to mass spectrometry (HPLC-MS/MS) is a very sensitive technique. Using a gradient elution with water and methanol on a C18 column and a mass spectrometer operating in positive electrospray system ionization (ESI) acquiring data in selected reaction monitoring (SRM) mode, low detection limits can be attained.

Suitable chromatographic conditions to determine the content in MDF boards or migration of aromatic amines primary (PAAs) is described elsewhere (Sendón et al., 2010). Briefly, a determination based on HPLC-MS/MS using a C18 column, ammonium acetate and methanol as mobile phase and acquiring data in selected reaction monitoring (see Table 13.9) enables achieving limits of detection at µg/Kg level as is required.

Melamine content is determined nowadays by hydrophilic interaction liquid chromatography (HILIC) with diode array detector (DAD). The mobile phase to this type of analysis is usually ammonium formate 10 mM at pH = 3 and acetonitrile (5:95 %v/v) in an isocratic mode and a HILIC column is required, and to detect it, a wavelength of 230 nm needs to be selected (Ibarra et al., 2016).

To determine heavy metals at low levels, inductively coupled plasma mass spectrometry (ICP-MS) is required. Li, Al, V, Cr, Mn, Fe, Co, Ni, Cu, Zn, As, Cd, Ba, Hg and Pb can be determined in MDF boards using ^{159}Tb and ^{72}Ge as internal standards. The isotopic masses used for the analyses are shown in Table 13.10.

The formaldehyde in MDF boards and the migration study of it in the selected simulant can be determined by high-performance liquid chromatography (HPLC)

TABLE 13.9
Selected Amines and the Mass Spectrometry MS/MS Conditions Used

Compound	Precursor ion (m/z)	Product ion (m/z)	Collision energy (V)
m-phenylenediamine (CAS 108-45-2)	108.9	91.9, 64.8	15, 15
2,6-toluenediamine CAS (823-40-5)	123.1	107.9, 76.8	18, 40
2,4-toluenediamine CAS (95-80-7)	123.1	107.9, 76.8	18, 40
1,5-diaminonaphthalene CAS (2243-62-1)	158.9	143.1, 114.8	20 ,20
Aniline CAS (62-53-3)	94.1	77.1, 51.2	17, 29
4,4'-diaminodiphenylether CAS (101-80-4)	201.1	107.8, 184.0	20, 20
4,4'-methylenedianiline CAS (101-77-9)	199.1	105.8, 76.8	25, 40
3,3'-dimethylbenzidine CAS (119-93-7)	213.2	107.8, 184.0	20, 20

TABLE 13.10
Isotopic Mass of Each Metal

Element	Isotopic mass
Li	7
Al	27
V	51
Cr	52
Mn	55
Fe	56
Co	59
Ni	60
Cu	63
Zn	66
As	75
Cd	111
Ba	137
Hg	202
Pb	208

coupled to a diode array detector. A previous derivatization step with phosphoric acid and 2, 4-dinitrophenylhydrazine (DNPH) is required followed by the separation on a reverse phase C18 column thermostated at 30°C using water and acetonitrile by a gradient elution. To determine the formaldehyde-2,4-dinitrophenylhydrazone, a wavelength of 360 nm should be selected.

Moreover, to detect and identify any other potential migrant, a general screening of volatile and semi-volatile compounds could be carried out. The above mentioned P&T GC-MS method but in a more drastic condition (purge at 80°C) could be applied for volatiles' determination; for semi-volatiles' determination, a GC-MS method but employing a general purpose column (with a phenyl arylene or (5%-phenyl)-methylpolysiloxane phase) should allow the analysis of MDF extracts (e.g., obtained by solvent extraction at 40°C for 16–24 h with acetonitrile) after injection by split mode. In both cases, the peak identification could be done by employing commercially available mass spectral libraries (e.g., Wiley 8th and Nist Mass Spectral Libraries).

13.7 FINAL REMARKS

MDF is a modern wood-based panel with excellent physical properties and is very easy to mechanize, and because of these characteristics, in recent years it has been widely used as a food contact material for fruits and vegetables.

Unlike for some other FCMs, for MDF there does not exist at present any reference document to use to evaluate its safety when it is used as a packaging for fruits and vegetables. In this chapter, a methodology is proposed to help in the fulfillment of general food contact regulations and that could be followed by industry managers, thus ensuring the suitability of MDF as FCM. Simple, specific and easy to validate analytical methods are proposed to determine the content and migration levels of concerned substances. Moreover, this methodology could be easily updated, if necessary, with other substances, hence warranting the protection of consumer´s health.

ACKNOWLEDGMENTS

This work was financially supported by the MDFasFCM project, which was signed between the USC (University of Santiago de Compostela) and a consortium of MDF manufacturers and wooden food packaging sector companies. Authors are grateful to the mentioned Consortium for the agreement in the publication of the work made within this project's framework.

REFERENCES

Baumann, M. G. D., Lorenz, L. F., Batterman, Stuart A., Zhang, G., 2000. Aldehyde emissions from particleboard and medium density fiberboard products. *Forest Products Journal* 50, 75–82.

BfR, 4th draft elaborated by German Federal Ministry of Food and Agriculture (BfR, 2007) for recycled cardboard for food contact. BMEL: German Federal Ministry of Food and Agriculture.

Bratinova, S., Hoekstra, E., 2019. (Ed) Guidance on sampling, analysis and data reporting for the monitoring of mineral oil hydrocarbons in food and food contact materials. Luxembourg: Publications Office of the European Union. ISBN 978-92-76-00172-0. doi:10.2760/208879, JRC115694.

Bratinova, S., Raffael, B., Simoneau, C., 2009. Guidelines for performance criteria and validation procedures of analytical methods used in controls of food contact materials. *JRC Scientific and Technical Reports*, EUR, 24105.

Cassiano, M. P., Carlos de Francisco, A., Leila, M. L.., Kovaleski, J. L., 2017. Life cycle assessment of medium-density fiberboard (MDF) manufacturing process in Brazil. *The Science of the Total Environment* 575, 103–111. doi:10.1016/j.scitotenv.2016.10.007.

Council of Europe (CoE), 2013. Resolution CM/Res (2013)9 on metals and alloys used in food contact materials and articles. *Adopted by the Committee of Ministers on 11 June 2013 at the 1173rd Meeting of the Ministers' Deputies*. Strasbourg, France. ISBN: 978-92-871-7703-2.

Commission Regulation (EU) 2016/1416 of August 24, 2016 amending and correcting Regulation (EU) No 10/2011 on plastic materials and articles intended to come into contact with food (Text with EEA relevance).

Commission Regulation (EU) No 10/2011 of January 14, 2011 on plastic materials and articles intended to come into contact with food.

European Directorate for the Quality of Medicines & HealthCare of the Council of Europe (EDQM), 2013a. *Metals and Alloys Used in Food Contact Materials and Articles: A Practical Guide for Manufacturers and Regulators*. 1st ed. Strasbourg: Council of Europe (CoE).

European Norm EN 717-1: 2004, Wood-based panels. Determination of Formaldehyde Release. Formaldehyde emission by the chamber method.

FAO. Food and Agriculture Organization of the United Nations, http://www.fao.org/forestry/statistics/80938@180723/en/ (accessed January 29, 2020).

FEDEMCO, n.d. Spanish federation of wood packaging and its components (FEDEMCO). (accessed January 29, 2020).

FEPEX, Federaci6n Española de Asociaciones de Productores Exportadores de Frutas, Hortalizas, Flores y Plantas vivas. https://www.fepex.es/noticias/detalle/frutas-hortalizas-agosto (accessed January 24, 2020).

García Ibarra, V., Rodríguez Bernaldo de Quirós, A., Sendón, R., 2016. Study of melamine and formaldehyde migration from melamine tableware. *European Food Research and Technology* 242(8), 1187–1199. doi:10.1007/s00217-015-2623-7.

International Organization for Standardization, ISO 12460-5:2015, Wood-based panels—Determination of formaldehyde release. Part 5: Extraction method (Perforator method).

ITENE, Technological institute of packaging, transport and logistics (ITENE). Valencia (Spain). (accessed January 29, 2020).

Kartal, S. N., Green, F., 2003. Decay and termite resistance of medium density fiberboard (MDF) made from different wood species. *International Biodeterioration and Biodegradation* 51(1), 29–35.

Kawouras, K. P., Koniditsiotis, D., Petinarakis, J., 1998. Resistance of cured urea-formaldehyde resins to hydrolysis: A method of evaluation. *Holzforschung* 52(1), 105–110. doi:10.1016/j.biortech.2004.12.003.

Kim, T., 2019. Production planning to reduce production cost and formaldehyde emission in furniture production process using medium-density fiberboard. *Processes* 8(8), 7–529. doi:10.3390/pr7080529.

MAPA, Spanish ministry of agriculture, fisheries and food (MAPA). https://www.mapa.gob.es/es/agricultura/temas/producciones-agricolas/frutas-y-hortalizas/ (accessed January 24, 2020).

Standard EN 310:1993, Wood-based panels—Determination of modulus of elasticity in bending and of bending strength.

Standard EN 317:1994, Particleboards and fibreboards—Determination of swelling in thickness after immersion in water.

Standard EN 318:2002, Wood based panels—Determination of dimensional changes associated with changes in relative humidity.

Standard EN 319:1994, Particleboards and fibreboards. Determination of tensile strength perpendicular to the plane of the board.

Standard EN 322:1994, Wood-based panels—Determination of moisture content.

Standard EN 323:1994, Wood-based panels—Determination of density.

Standard EN 324-1:1994, Wood-based panels—Determination of dimensions of boards—Part 1: Determination of thickness, width and length.

Standard EN 324-2:1994, Wood-based panels—Determination of dimensions of boards—Part 2: Determination of squareness and edge straightness.

Standard UNE-EN 622-5:2010, Fiberboards. Specifications. Part 5: Requirements for dry process boards (MDF).

Standard ISO 13302:2008, Sensory analysis. Methods for assessing modifications to the flavour of foodstuffs due to packaging.

Plastics Europe, https://www.plasticseurope.org/en/about-plastics/packaging (Plastics Europe) (accessed January 29, 2020).

Regulation (EC) No. 1935/2004 of the European Parliament and of the Council of 27 October 2004 on materials and articles intended to come into contact with food and repealing Directives 80/590/EEC and 89/109/EEC.

Robertson, G. L., 2012. *Food Packaging: Principles and Practice.* CRC Press. Boca Raton, FL.

Sendón, R., Bustos, J., Sánchez, J. J., Paseiro, P., Cirugeda, M. E., 2010. Validation of a liquid chromatography–mass spectrometry method for determining the migration of primary aromatic amines from cooking utensils and its application to actual samples. *Food Additives and Contaminants: Part A* 27(1), 107–117. doi:10.1080/02652030903225781.

Simoneau, C., Raffael, B., Garbin, S., Hoekstra, E., Mieth, A., Lopes, J. A., Reina, V., 2016. Non-harmonised food contact materials in the EU: Regulatory and market situation, EUR 28357 EN. doi:10.2788/234276 .

Wilson, B., 2010. Wood and fiber science. *Society of Wood Science and Technology* 42, 107–124.

14 Applications of Nanotechnology in Food Packaging

Aswathy Jayakumar, Sabarish Radoor, Jasila Karayil, Radhakrishnan E.K, Sanjay M.R, Suchart Siengchin, and Jyotishkumar Parameswaranpillai

CONTENTS

14.1 INTRODUCTION

The increasing consumer needs for quality foods with higher nutritional value without compromising the safety aspects paved the way for searching for new packaging systems. The innovations in nanotechnology have a great impact on various scientific and industrial fields. It uses nanoscale materials to revolutionize the field of food science and food microbiology. And this includes food packaging, food safety, food processing, development of functional food, and extension of shelf life. The application of nanotechnology in food packaging involves the direct incorporation into food, and food packaging material [1]. And also, the emergence of polymer nanotechnology has many advantages over traditional packaging. Nanotechnology offers advanced features like containment, convenience, protection and preservation along with marketing and communication with the end consumer [2, 3].

Nanocomposites may be defined as combinations of nanomaterials with traditional food packaging systems. Nanocomposites offer better packaging properties than other composite materials. They have a major role in imparting enhanced mechanical, barrier as well as broad antimicrobial spectrum [4–5]. The higher surface area and surface energy of nanofillers provide strong interfacial interactions between polymer and nanofiller, thereby enhancing the properties of polymers. However, the success of nanotechnology depends upon the demands of the consumer and its commercialization. Hence, the applications of nanomaterials in food packaging systems with some insights into safety issues and future perspectives have been discussed in this chapter.

14.2 NANOSTRUCTURES IN FOOD PACKAGING

The emergence of nanomaterials in food packaging is gaining more importance in the food sector. For effective food packaging, the incorporation of nanomaterials in polymers can contribute to improved mechanical, barrier and antimicrobial properties. Basically, several nanostructured materials such as spherical nanoparticles, nanorods, nanosheets and nanotubes have been reported on [6]. Depending upon the materials being incorporated, nanotechnology-based innovations are of active or smart packaging and intelligent or improved packaging (see Figure 14.1). Both these packaging systems can offer better properties such as containment, convenience, protection, preservation, marketing and communication without compromising the quality of foodstuffs (see Figure 14.2).

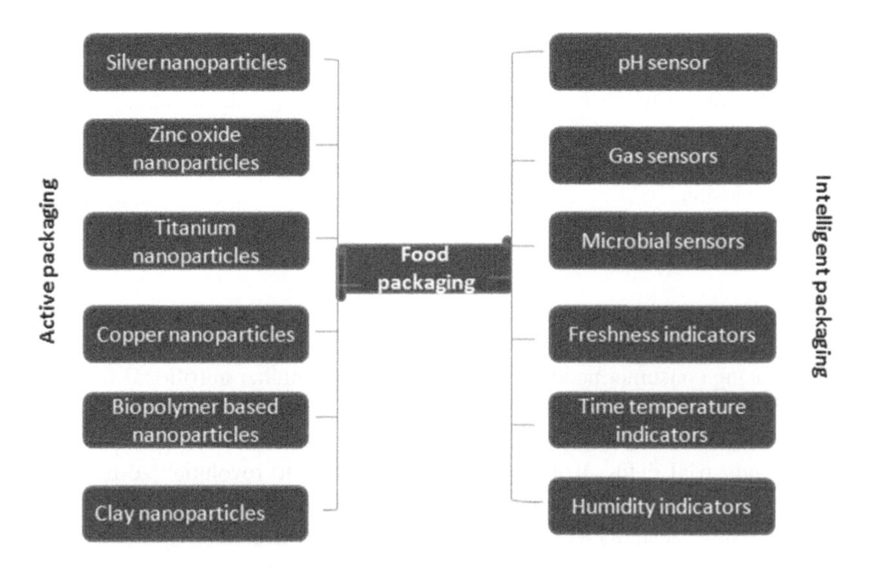

FIGURE 14.1 Nanomaterials involved in food packaging.

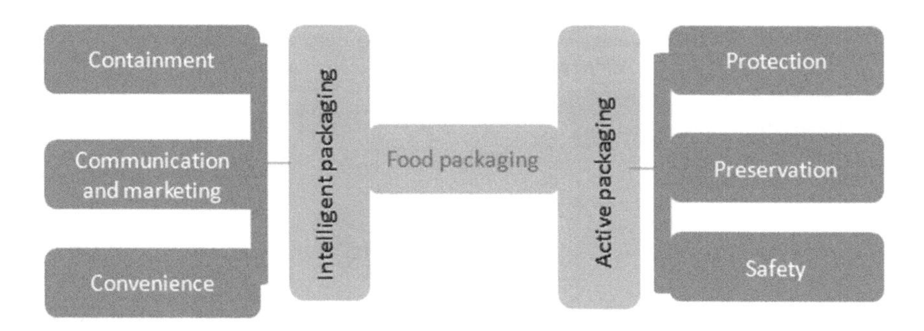

FIGURE 14.2 Features of nanotechnology-based food packaging systems.

14.3 ACTIVE PACKAGING

Active packaging encompasses the incorporation of various components such as antimicrobials, antioxidants, food preservatives, oxygen and water vapor absorbers and ethylene absorbers that extend the shelf life and improves the quality of food products [7–8]. This involves the use of various organic and inorganic nanostructures. Metal-based nanomaterials include the use of silver nanoparticles, zinc-oxide nanoparticles, copper nanoparticles, clay, and silicon-based nanoparticles, titanium-based nanoparticles and magnesium-based nanoparticles. These nanoparticles may directly contact or can slowly migrate from packaging materials to foodstuffs [9]. The mechanistic basis of antimicrobial activity is via the direct interaction with microbes by the disintegration of the bacterial cell membrane, oxidation of cell wall components, production of reactive oxygen species and membrane protein damage ultimately leading to cell death [10].

14.4 MECHANISTIC BASIS OF ANTIMICROBIAL ACTION OF NANOPARTICLES

Nanoparticles can induce bacterial cell death by several mechanisms such as electrostatic interaction, cell membrane penetration, metal ion release and by the induction of reactive oxygen species. And this bacterial cell death would inhibit translation, cause DNA damage, cause protein damage, disrupt the electron transport chain and alter cell membrane permeability of microbes (see Figure 14.3) [11]. The electrostatic interaction between cationic nanostructured materials and the anionic microbial cell membrane enables the adhesion of these materials to microbial cells. This ultimately leads to the shrinkage and the detachment of the cell membrane and causes the rupturing of the cell wall. Nanoparticles also have the capacity to penetrate into the cell membrane and can efficiently bind the biomolecules such as DNA, proteins or lipids.

The interaction of nanoparticles with sulfur groups present in microbial cell wall proteins leads to irreversible changes in the structure of the cell wall and disturbs the integrity of the lipid bilayer, thereby increasing the permeability of the cell membrane [12]. The other mechanisms involve the ability of these nanoparticles to induce the formation of reactive oxygen species that cause oxidative stress in

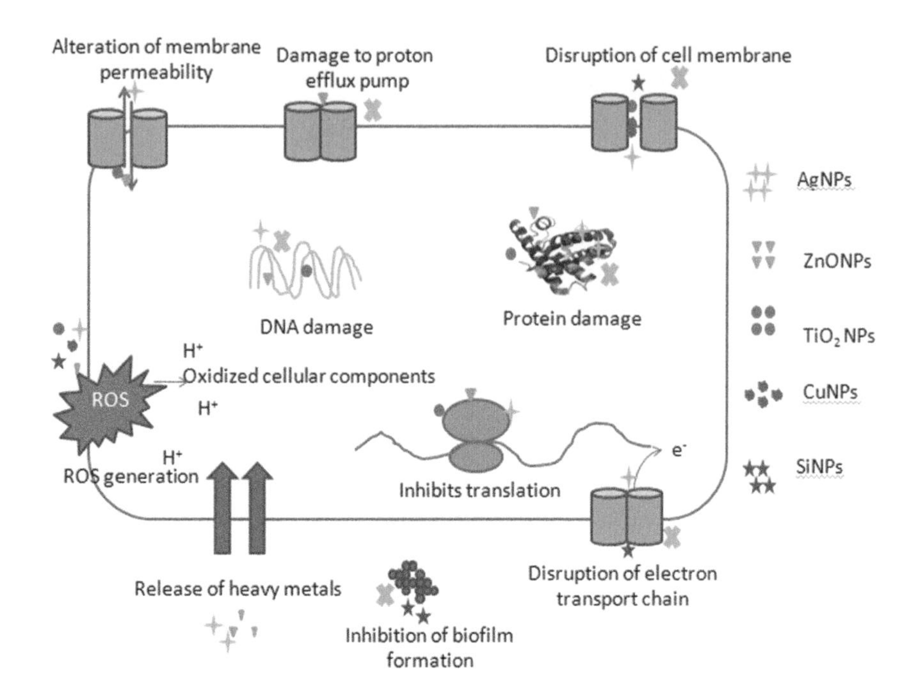

FIGURE 14.3 Antimicrobial mechanisms of nanoparticles [11].

microbial cells that ultimately leads to cell death. The increase in reactive oxygen species and other free radicals causes the dysfunction of endoplasmic reticulum and biomolecule damage that leads to genome instability. Nanostructures are also involved in altering the signal transmission pathways of the microbial cell and inhibit biofilm formation [13].

14.5 SILVER NANOPARTICLES

Silver (Ag) nanoparticles are one of the most commonly used nanoparticles, and are gaining potential interest in the research field due to their excellent antimicrobial potential against multidrug-resistant pathogens. They are gaining more attention due to their low toxicity and antimicrobial and thermally stable nature. The incorporation of these nanoparticles with many polymers such as starch, chitosan, cellulose and gelatin for packaging applications has already been reported. For example, the reinforcement of tragacanth/hydroxypropyl methylcellulose/bee wax edible films with silver nanoparticles exhibited antimicrobial activity against Gram-positive *B. cereus*, *S. aureus*, *S. pneumonia* and *L. monocytogenes* and Gram-negative, *E. coli*, *S. typhimurium*, *P. aeruginosa* and *K. pneumonia* in a dose-dependent manner [14]. Mathew et al. [15] developed bionanocomposite films with silver nanoparticles with improved mechanical and optical properties along with antimicrobial activity against *S. aureus* and *S. typhimurium*. Similarly, tetramethyl-1-piperidinyloxy radical (TEMPO)-oxidized nanocellulose (TNC), grape seed extract and TNC immobilized with silver nanoparticles-based films exhibit improved mechanical properties, good thermal stability, and low water vapor and oxygen permeability along with strong antimicrobial activity against *E. coli* and *S. aureus* [16].

Roy and his co-workers [17] reported that the addition of silver nanoparticles to carrageenan-based films resulted in the enhancement of thermal stability, mechanical strength and UV barrier along with activity against foodborne *E. coli* and *L. monocytogenes.* Wang et al. [18] developed eco-friendly films by incorporating silver nanoparticles into bacterial nanocellulose. These films were reported to be antimicrobial against *E. coli* and exhibit excellent oxygen barrier capacity along with the increase in the shelf life of raw beef. The incorporation of banana peel powder and silver nanoparticles into cellulose matrix resulted in enhanced tensile properties, thermal stability and activity against *E. coli, P. aeruginosa, S. aureus* and *B. licheniformis.* Polyvinyl acetate (PVA) films incorporated with greenly synthesized silver nanoparticles were reported to have excellent mechanical, light barrier, water barrier and antimicrobial activity against *S. typhi and S. aureus* [19]. Kumar et al. [20] reported about how incorporating silver nanoparticles with chitosan/gelatin films enhanced the mechanical property and extended the shelf life of red grapes. The reinforcement of silver nanoparticles into poly(lactide) resulted in the increased mechanical and water vapor barrier of films with prominent antimicrobial activity against *E. coli* and *L. monocytogenes* [21].

14.6 TITANIUM NANOPARTICLES

Titanium dioxide (TiO_2) nanoparticles are highly efficient UV absorbers that exhibit higher photocatalytic activity and thermal stability. As the size of the TiO_2 nanoparticles decreases, the surface energy increases which can have a great impact on polymer performance. The addition of TiO_2 nanoparticles has immense promise in elevating the antimicrobial, mechanical and UV barrier properties of food packaging films. Ecovio® incorporated with titanium dioxide nanoparticles have been reported to have the ability to block UV radiation [22]. The reports by Kaewklin et al. [23] suggest that chitosan titanium dioxide nanocomposite films have ethylene photodegradation activity which delays the ripening in cherry tomatoes. Ahmadi et al. [24] reported the fabrication and characterization of carboxymethyl cellulose (CMC)-reinforced with titanium dioxide nanoparticles for packaging applications. The prepared bionanocomposite films were found to have higher thermal stability, UV barrier and possess antimicrobial activity against *E. coli* and *S. aureus.*

The improvement of mechanical and barrier properties of TiO_2 nanoparticles incorporated with wheat gluten films was demonstrated by El-Wakil et al. [25]. Their reports suggested that these films have potential antimicrobial activity against both gram-negative *E. coli* and gram-positive *S. aureus.* Zhu et al. [26] developed polyacrylonitrile titanium dioxide nanofibers (PAN@TiO_2) with potent ethylene scavenging activity for the storage of fresh fruits and vegetables. Salicylimine-chitosan/TiO_2 nanocomposite showed a higher thermal property, improved mechanical strength along with activity against *S. aureus* and *P. aeruginosa* [27]. The dispersion of titanium dioxide nanoparticles in cellulose nanofiber coating enhances the UV absorption without affecting gas barrier properties [28]. Similarly, the studies conducted by Siripatrawan et al. [29] reported that chitosan-titanium dioxide nanoparticles-based films have excellent ethylene photodegradation capacity. These films were also reported to be active against bacteria such as E. *coli*, *P. aeruginosa*, *S. aureus*, *S. typhimurium* and fungi (*Aspergillus* sp. and *Penicillium* sp.).

14.7 ZINC OXIDE NANOPARTICLES

Zinc oxide (ZnO) nanoparticles have applications in food, pharmaceutical and cosmetic industries [30]. Due to its antibacterial property, it has been used in food packaging materials to preserve color and quality of food. Interestingly, the antibacterial activity of ZnO nanoparticles is retained even when its particle size is reduced to nano dimensions. ZnO nanoparticles are a cheap, non-toxic material which possesses good physio-chemical stabilities, mechanical property, photocatalytic activity and UV filtering ability [31–34]. It shows pronounced antibacterial activities toward foodborne pathogens and high-temperature-resistant spores. The antibacterial activity of ZnO nanoparticles is attributed to its greater interaction with the bacterial cells, thereby inhibiting their growth. The antiseptic, anti-cancerous and anti-inflammatory properties of ZnO nanoparticles are also well documented in literature [35–36].

According to the Food and Drug Administration (FDA), ZnO nanoparticle is a safe material to be used as a food additive and food packaging material [37]. There are several reports on the usage of ZnO nanoparticles for food packaging applications [38–40]. Sarojini et al. [41] observed that ZnO nanoparticles improve the thermo-mechanical and barrier property of polyurethane/chitosan film. The antibacterial activities of the composite were threefold greater than commercial film and the film extends the shelf life of carrots. Recently, Kumar et al. [42] implemented a green approach to synthesizing ZnO nanoparticles and studying its food preservative abilities. They observed that the incorporation of ZnO nanoparticles into the agar matrix improved its mechanical, thermal and preservative properties. The food preservative ability of the composite was found to be the function of ZnO nanoparticles concentration and the composite loaded with high percentage of ZnO nanoparticles preserved the green grapes for 21 days. Similarly, Happy et al. [43] reported a green protocol for the synthesis of ZnO nanoparticles. The mechanism of the antibacterial activity of ZnO nanoparticles was also discussed by Happy et al. Zinc oxide nanoparticle-epoxy resin composite (ZnO NPs-ER) was found to be a promising candidate for storing marine fish species. It was reported that the ZnO NPs-ER-treated silvery pomfret fish showed negligible growth of *Aeromonas* and *Shewanella* microbes which are generally present in spoiled seafood [44]. Laila et al. [45] coated chitosan and ZnO nanoparticles on polyethylene film. Besides improving the water affinity of the film, the zinc oxide nanoparticle also contributes to the antibacterial activity of the film. The reduction in the growth of pathogens is ascribed to the combined effect of chitosan and ZnO nanoparticles. The zinc nanoparticles loaded to poly(butylene adipate-co-terephthalate) (PBAT)-PLA composite displayed good tensile strength, stiffness, UV barrier properties, water barrier properties and antibacterial activity of the composite films indicate its ability to extend the shelf life of food items. In another work, Shankar and his team [46] studied the utility of PLA/ZnO nanoparticles for storing cooked minced fish paste. Their experimental results revealed that food wrapped in PLA/ZnO nanoparticles is free from pathogens like *Escherichia coli* and *Listeria monocytogenes*. Shankar et al. claim that ZnO nanoparticles and Zn^{2+} ions are responsible for the bacteriostatic effect of the film and inhibit the pathogens by destroying the bacterial cell wall and also interfere with the DNA replication and protein synthesis of the bacteria.

14.8 CLAY NANOPARTICLES

Nanoclay belongs to the phyllosilicate family and contains tetrahedrally coordinated silicon atoms bounded to octahedral sheets of aluminum or magnesium hydroxide. The thickness and width of nanoclay are reported to be in the order of 1 nm and 70–150 nm, respectively [47]. Montmorillonite, bentonite, kaolinite, hectorite, and halloysite are the most widely used nanoclays [48]. Recently, the surface of nanoclay has been modified with organic moiety to generate organic clay such as cloisite 30B and cloisite 20A. The attractive features of nanoclay are high aspect ratio, good thermal and chemical stability, good barrier property, ability to improve the tensile strength of composite and good compatibility with thermoplastics. These superior features along with its inexpensive nature make nanoclay one of the widely explored nanoparticles [49]. Nanoclay has been used for specific applications such as filler in polymer composites, as adsorbent and as rheological modifiers. The antibacterial and sterilizing property of nanoclay has been exploited in food packaging industries. The global nanoclay market for food packaging is accelerating [50].

There are several reports on nanoclay-based film for food packaging applications [51–53]. Davachi et al. [54] developed an edible film from salvia macro siphon seed mucilage (SSM) and nanoclay and evaluated its properties. The incorporation of nanoclay improves the mechanical, physical, thermal and barrier performance of composite. The good antibacterial and hydrophobicity of the composites make it a potential food packaging material. In another study, Abreu et al. [55] utilized quaternary ammonium modified montmorillonite clay to prepare starch-based nanocomposite. They observed that film with nanoclay and silver nanoparticles has low permeability for oxygen and water vapor, which is desirable for an ideal food packing material. Further, the food contact and antibacterial test results revealed that the prepared film is safe to be used as food packing material. Bodaghi et al. [56] reported on the fabrication of low-density polyethylene (LDPE)/nanoparticle film for food packaging applications. They report that the simultaneous presence of nanoclay and TiO_2 enhance the mechanical and barrier property of the film. The film also possesses excellent antibacterial activity against *Pseudomonas spp.* and *R. mucilaginosa* bacteria which is responsible for the staling of food items. Meanwhile, Tornuk et al. [57] loaded essential oil into LDPE/nanoclay composite and tested its food packaging ability. Their studies show that the prepared films were able to extend the shelf life of fresh and processed meat by inhibiting the growth of *E.coli*. It also retards the meat discoloration and the meat stored in this film retains its color for up to four days. Water vapour permeability (WVP) and water solubility (WS) are the two important parameters which are usually monitored for food packaging films. A low WVP and WS show the resistance of packaging films to water, which is an important factor for food packages due to the probability of contamination in the presence of water. Rostamzad and co-workers [58] observed a significant decrease in the WS and WVP of fish myofibrillar protein (FMP) while incorporating a small amount of nanoclay and microbial transglutaminase (MTGase) into it. Rostamzad et al. thus suggest that the film could be a suitable candidate for food preservation.

14.9 COPPER NANOPARTICLES

The antibacterial property of copper was exploited by humankind to treat bacterial infections and to disinfect water [59]. It is reported that the antibacterial activity of copper could be tuned by varying the concentration of copper, type of pathogens and environmental conditions [60]. The antibacterial activity of copper can be further improved by conversion into copper nanoparticles. As copper nanoparticles are cheap and possess good physico-mechanical properties, it has been increasingly used for the fabrication of nanocomposite for food packaging applications [61]. Nouri et al. [62] introduced montmorillonite (MMT)-copper oxide nanofiller to enhance the antibacterial activity of chitosan films. Nouri et al. report that even a small amount of MMT-CuO nanoparticle enhances the mortality rate against food pathogens like *E.coli*, *P.aeruginosa*, *S. aureus* and *B.cereus*. The high UV protection of the nanocomposite indicates its potential applications as food packaging material. In a separate study, Bruna et al. [63] systematically studied the effect of montmorillonite-modified copper nanoparticles on the antibacterial activity of LDPE nanocomposite. Biological and mechanical studies revealed that the nanocomposite possesses significant antibacterial and mechanical properties, which are some of the required criteria for food packing material. The effect of concentration of copper nanoparticles on the properties of kefiran-carboxymethyl cellulose film shows that on increasing the concentration of copper nanoparticles, the properties of film, namely contact angle, WVP, UV barrier and mechanical properties along with its antibacterial property, significantly improve. Thus, it is a promising material for the storage of food items [64]. Longano and co-workers [65] adapted a simple and effective laser ablation technique to generate copper nanoparticles. Longano et al. utilized these copper nanoparticles to generate PLA nanocomposite, with good antibacterial activity. This novel biodegradable nanocomposite was proposed to be a suitable material for preservation of food items.

Recently, Shankar and his team conducted a series of studies on the development of novel bionanocomposite for food packaging applications [66–67]. In one particular study, Shankar et al. fabricated agar-based nanocomposite by blending copper nanoparticles and demonstrated its antibacterial effect against gram-positive and gram-negative bacteria [66]. Shankar et al. also report the preparation of a novel Cu nanoparticles-embedded polysaccharide-based composite film for packaging applications [67]. Here Shankar et al. adapted green strategy for the development of Cu nanoparticles. Cu nanoparticles-loaded bio composites displayed high thermal stability, high strength, a desirable water barrier property and UV protection behavior. The antibacterial activity of the composite film is attributed to the destruction of bacterial cell walls and DNA by Cu^{2+} ions. Vasile et al. [68] investigated the effect of ZnO Cu/Ag nanoparticles incorporated into PLA matrix for food packaging application. It was found that the incorporation of nanoparticles content into PLA leads to a reduction in the water vapor and gas permeability transmittance rate. The other highlights of this composite are good mechanical properties, high resistance to bacteria, desirable UV protection and low migration of nanoparticles into food. Zhong et al. [69] developed copper nanoparticles using TEMPO, TNFC and CMC. The incorporation of copper nanoparticles into the PVA matrix results in a biocomposite with enhanced thermal and mechanical stability. In addition to this, the composite also exhibits good antimicrobial properties which make it a potent food packaging material.

14.10 BIOPOLYMER-BASED NANOSTRUCTURES

The environment-related issues of nondegradable materials have paved the way for searching for biopolymer-based innovations [70]. However, low mechanical and barrier properties limit its use in the preparation of food packaging materials. The reinforcement of nanostructures in biopolymer-based packaging systems resulted in the enhanced mechanical, barrier and antimicrobial properties of films [71]. For example, Lin et al. [72] developed moringa oil-loaded nanoparticles-reinforced gelatin nanofibers for food packaging applications. The developed nanofiber films were found to be active against *L. monocytogenes* and *S. aureus* in cheese without affecting the quality. These films were also reported to have enhanced water barrier properties as well as mechanical properties. Zein films incorporated with chitosan nanoparticles and cinnamon essential oil were also reported to possess antimicrobial properties against *S. aureus* and *E. coli* [73]. Wu et al. [74] have reported the development of chitosan/ε-polylysine bionanocomposite films. The prepared films were found to be active against *E. coli* and *S. aureus* with enhanced water barrier and mechanical properties.

Glaser et al. [75] functionalized the polyethylene and polypropylene using chitosan nanoparticles loaded with resveratrol for the development of active food packaging materials. The developed films were found to be active against *S. aureus* and *E. coli* with enhanced gas barrier properties. Cellulose nanocrystals blended with silver/alginate solution were reported to have excellent UV barrier properties along with water barrier and mechanical properties [76]. Shahbazi et al. [77] have reported the development of *Mentha picata* essential oil and cellulose nanoparticles incorporated coatings that extend the shelf life of silver carp fillets. The developed coatings were active against *Pseudomonas* spp. and *Entero bacteriaceae* without affecting the odor, color and acceptability. Similarly, nanocellulose with activated carbon-based films was reported to have biosensing properties with enhanced mechanical and thermal properties [78]. Maliha et al. [79] developed nanocellulose sheets incorporated with bismuth phosphinate for active food packaging. The developed nanosheets inhibited the growth of bacteria and fungi with enhanced mechanical and barrier properties.

14.11 CARBON NANOSTRUCTURES

Liu et al. [80] examined the ability of electrospun polylactic acid/carbon nanotube/chitosan composite films for the preservation of strawberries. Their reports suggest that the developed materials have enhanced mechanical properties and are active against *E. coli* and *S. aureus* and thus can be used for the preservation of fruits and vegetables.

14.12 INTELLIGENT PACKAGING

Intelligent packaging not only offers safety, shelf life and quality but also aids in detection, sensing, recording, tracking and communication with the end consumer about the current status of foods [81]. These systems are applied in features like indicators, sensors and radio frequency identification tags (RFID). The intelligent packaging systems involve pH sensors, O_2 sensors, CO_2 sensors, microbial sensors, freshness indicators, humidity indicators, time and temperature indicators, electronic tongue etc. Nanosensors are used to detect the changes in environmental conditions of packaged foods like microbial contamination, humidity, temperature, pH, gas barriers and product degradation. They are applied as coatings or labels which ensures good packaging integrity as a function of intelligent packaging. Generally, it can be used for the detection of leakage of food stuffs packed in vacuum or inert atmosphere, variation in temperature and pH and microbial safety.

14.13 PH SENSORS

Chitosan, purple corn extract, and silver nanoparticles-based films have been reported to have better mechanical strength, light and water barrier properties, antioxidant activity and antimicrobial activity against *E. coli*, *S. typhi*, *S. aureus* and *L. monocytogenes* along with PH-sensitive property [82]. Ma et al. [83] reported the development of cellulose nanocrystals blended with tara gum and grape skin. The developed films were found to be pH sensing with the ability to monitor the quality of food. Starch-polyvinyl alcohol-roselle anthocyanins-based films developed by Zhang et al. [84] have excellent antioxidant activity along with pH sensing activity. The developed films were found to have the ability to monitor the freshness of pork. Recently, Jayakumar et al. [85] developed pH sensing antimicrobial wraps by incorporating starch, zinc oxide nanoparticles and the extract from jamun fruit. The developed films were found to have enhanced water barrier, mechanical properties and showed activity against *S. typhimurium*. Poly vinyl alcohol-chitosan nanoparticles incorporated with mulberry extract-based films were reported to monitor pH variations. The observed color change from red to green indicated the spoilage of fish and the films thus can be used a label to detect the spoilage of food [86]. Guterrez and his team [87] demonstrate the pH response behavior of nanocomposite prepared from montmorillonite and blueberry extract. This nanocomposite has purple color in neutral pH, however at acidic pH the color switches to red while in the basic medium it turns into green. This ability could be utilized to check the quality of fish and fruit items, as these foods on standing create an alkaline and acidic environment, respectively.

14.14 NANOSENSORS FOR PATHOGEN DETECTION

Nanosensor functions as an indicator that collects and correlates the alterations in environmental conditions such as humidity, temperature, chemical changes and microbial contaminations. For example, biosensors based on carbon nanotubes are gaining increasing attention due to their sensitivity, cost effectiveness, accuracy and simplicity. They can be used to detect microbial pathogens, toxins and other chemical alterations by the degradation of products in food. The change in conductivity can be observed by the interaction of these nanotubes with toxins. The electronic tongue consists of an array of nanosensors that detect the alteration by providing signals on gases that are released by the food items. In the case of quartz crystal microbalance (QCM)-based electric nose, the surface of QCM has various functional groups of biological molecules and various polymers that detect the various odorants and chemicals [88]. Joung et al. [89] reported the development of alumina nanoporous membrane incorporated with hyaluronic acid-based impedimetric immunosensor. The membrane was able to detect bacterial pathogens in milk without pre-treatment. Such a kind of sensor can have immense promise as it can be used directly.

14.15 TEMPERATURE AND TIME INDICATORS

Time-temperature indicators monitor the quality of food and accurately translate the quality of food products to consumers. It can be incorporated as self-adhesive labels in which the variation in temperature can be observed by the change in color. And this may be due to the various physical, biological and chemical changes. Nanoparticles based on methyl cellulose and poly ethylene glycol core with anthocyanidin have been reported to possess antibacterial and temperature sensitivity. These films also have the capacity to efficiently absorb UV light along with antioxidant activity [90]. Nanocomposites based on chitosan and gold nanoparticles have been reported to indicate the thermal history of the food packaging. They reported that these sensors can be integrated with the existing food packaging to ensure food quality and safety [91]. Singh et al. [92] have reported the development of temperature sensitive packaging by using soybean oil and tetradecane. The developed thermoregulating material was found to control the temperature of fresh beef with an acceptable range of quality parameters [93].

14.16 SAFETY CONCERNS AND FUTURE PERSPECTIVES

The recent innovations in nanotechnology-based food packaging have revolutionized the existing traditional packaging systems. In addition to the advantages, the safety aspects of these nanoparticles cannot be ignored. Nowadays, most of the studies are based on the functional aspects of nanoparticles-incorporated packaging systems rather than the safety issues. And also, several nanoparticles are considered as being in the generally regarded as safe (GRAS) category and depending on this, most of the studies are carried out without further examining of the safety issues. The continuous use of nanoparticles may cause acute or chronic toxicity [94]. And these ingested nanoparticles can induce cellular or organ damage in the gastro intestinal tract and also reduce the beneficial microbes in the tract which, in turn, affects human health. The major factor that contributes to the toxicity of nanoparticles is their ability to induce reactive oxygen species production and thereby damage the cell membrane of organelles [95]. Several studies have reported the concentration-dependent effects of various nanoparticles on human health and the environment [96–100]. And also, studies suggested that these ingested nanoparticles get absorbed by the gastro intestinal tract. The absorbed nanoparticles may get transferred out, metabolized or get accumulated depending upon the characteristics of nanoparticles [101].

The toxicological assessment of nanoparticles was analyzed by several *in vivo* and *in vitro* approaches [102]. These include the development of certain predictive models for assessment, but many others have focused on histological changes and pharmacokinetic parameters. But studies on the impact of nanoparticles on cardiovascular, nervous, immune systems, hepatic, renal, pulmonary, digestive and hematological health along with molecular studies can provide deeper understandings about its toxicity. However, the area around nanoparticles-induced cytotoxicity is still not clear. Hence, regulatory authorities should monitor the safety aspects of nanoparticles in food packaging and must ensure food safety before releasing it into the market.

14.17 CONCLUSION

The innovations in nanotechnology have transformed the various aspects of food packaging industries. They hold immense promise over the traditional packaging systems as they can assure containment, convenience, protection and preservation along with marketing and communication advantages. Several reports have suggested that the incorporation of these nanoparticles enhanced the mechanical, antimicrobial, antioxidant and barrier properties of packaging systems through active and intelligent packaging systems. However, safety issues and environmental impact of these innovations should be strictly monitored before reaching the market.

REFERENCES

1. Ravichandran, R. 2010. Nanotechnology applications in food and food processing: Innovative green approaches, opportunities and uncertainties for global market. *International Journal of Green Nanotechnology: Physics and Chemistry* 1(2):P72–P96.
2. Sharma, C., R. Dhiman, N. Rokana, and H. Panwar. 2017. Nanotechnology: An untapped resource for food packaging. *Frontiers in Microbiology* 8:1735.
3. Vanderroost, M., P. Ragaert, F. Devlieghere, and B. De Meulenaer. 2014. Intelligent food packaging: The next generation. *Trends in Food Science and Technology* 39(1):47–62.
4. Jaiswal, L., S. Shankar, and J.-W. Rhim. 2019. Applications of nanotechnology in food microbiology. Methods Microbiol 46:43–60.
5. Othman, S.H. 2014. Bio-nanocomposite materials for food packaging applications: Types of biopolymer and nano-sized filler. *Agriculture and Agricultural Science Procedia* 2:296–303.
6. Bratovcic, A., A. Odobasic, S. Catic, and I. Sestan. 2015. Application of polymer nanocomposite materials in food packaging. *Croatian Journal of Food Science and Technology* 7(2):86–94.
7. Majid, I., G. Ahmad Nayik, S. Mohammad Dar, and V. Nanda. 2018. Novel food packaging technologies: Innovations and future prospective. *Journal of the Saudi Society of Agricultural Sciences* 17(4):454–462.
8. Pereira de Abreu, D.A., J.M. Cruz, and P. Paseiro Losada. 2012. Active and intelligent packaging for the food industry. *Food Reviews International* 28(2):146–187.

9. Störmer, A., J. Bott, D. Kemmer, and R. Franz. 2017. Critical review of the migration potential of nanoparticles in food contact plastics. *Trends in Food Science and Technology* 63:39–50.
10. Slavin, Y.N., J. Asnis, U.O. Häfeli, and H. Bach. 2017. Metal nanoparticles: Understanding the mechanisms behind antibacterial activity. *Journal of Nanobiotechnology* 15(1):65.
11. Baptista, P.V., M.P. McCusker, A. Carvalho, D.A. Ferreira, N.M. Mohan, M. Martins, and A.R. Fernandes. 2018. Nano-strategies to fight multidrug resistant bacteria—"A battle of the titans." *Frontiers in Microbiology* 9:1441.
12. Wang, L., C. Hu, and L. Shao. 2017. The antimicrobial activity of nanoparticles: Present situation and prospects for the future. *International Journal of Nanomedicine* 12:1227–1249.
13. Qureshi, N., et al. 2014. Innovative biofilm inhibition and anti-microbial behavior of molybdenum sulfide nanostructures generated by microwave-assisted solvothermal route. *Applied Nanoscience* 5(3):331–341.
14. Bahrami, A., R. Rezaei Mokarram, M. Sowti Khiabani, B. Ghanbarzadeh, and R. Salehi. 2019. Physico-mechanical and antimicrobial properties of tragacanth/hydroxypropyl methylcellulose/beeswax edible films reinforced with silver nanoparticles. *International Journal of Biological Macromolecules* 129:1103–1112.
15. Mathew, S., A. Jayakumar, V.P. Kumar, J. Mathew, and E.K. Radhakrishnan. 2019. One-step synthesis of eco-friendly boiled rice starch blended polyvinyl alcohol bionanocomposite films decorated with in situ generated silver nanoparticles for food packaging purpose. *International Journal of Biological Macromolecules* 139:475–485.
16. Wu, Z., W. Deng, J. Luo, and D. Deng. 2019. Multifunctional nano-cellulose composite films with grape seed extracts and immobilized silver nanoparticles. *Carbohydrate Polymers* 205:447–455.
17. Roy, S., S. Shankar, and J.-W. Rhim. 2019. Melanin-mediated synthesis of silver nanoparticle and its use for the preparation of carrageenan-based antibacterial films. *Food Hydrocolloids* 88:237–246.
18. Wang, W., Z. Yu, F.K. Alsammarraie, F. Kong, M. Lin, and A. Mustapha. 2020. Properties and antimicrobial activity of polyvinyl alcohol-modified bacterial nanocellulose packaging films incorporated with silver nanoparticles. *Food Hydrocolloids* 100:105411.
19. Thiagamani, S.M.K., N. Rajini, S. Siengchin, A. Varada Rajulu, N. Hariram, and N. Ayrilmis. 2019. Influence of silver nanoparticles on the mechanical, thermal and antimicrobial properties of cellulose-based hybrid nanocomposites. *Composites Part B: Engineering* 165:516–525.
20. Kumar, S., A. Shukla, P.P. Baul, A. Mitra, and D. Halder. 2018. Biodegradable hybrid nanocomposites of chitosan/gelatin and silver nanoparticles for active food packaging applications. *Food Packaging and Shelf Life* 16:178–184.
21. Shankar, S., J.-W. Rhim, and K. Won. 2018. Preparation of poly(lactide)/lignin/silver nanoparticles composite films with UV light barrier and antibacterial properties. *International Journal of Biological Macromolecules* 107(B):1724–1731.
22. Mohr, L.C., A.P. Capelezzo, C.R.D.M. Baretta, M.A.P.M. Martins, M.A. Fiori, and J.M.M. Mello. 2019. Titanium dioxide nanoparticles applied as ultraviolet radiation blocker in the polylactic acid bidegradable polymer. *Polymer Testing* 77:105867.
23. Kaewklin, P., U. Siripatrawan, A. Suwanagul, and Y.S. Lee. 2018. Active packaging from chitosan-titanium dioxide nanocomposite film for prolonging storage life of tomato fruit. *International Journal of Biological Macromolecules* 112:523–529.
24. Ahmadi, R., et al. 2019. Fabrication and characterization of a titanium dioxide (TiO2) nanoparticles reinforced bio-nanocomposite containing *Miswak* (*Salvadora persica* L.) extract—The antimicrobial, thermo-physical and barrier properties. *International Journal of Nanomedicine* 14:3439–3454.

25. El-Wakil, N.A., E.A. Hassan, R.E. Abou-Zeid, and A. Dufresne. 2015. Development of wheat gluten/nanocellulose/titanium dioxide nanocomposites for active food packaging. *Carbohydrate Polymers* 124:337–346.
26. Zhu, Z., Y. Zhang, Y. Zhang, Y. Shang, X. Zhang, and Y. Wen. 2019. Preparation of PAN@TiO2 nanofibers for fruit packaging materials with efficient photocatalytic degradation of ethylene. *Materials* 12(6):896.
27. Montaser, A.S., A.R. Wassel, and O.N. Al-Shaye'a. 2019. Synthesis, characterization and antimicrobial activity of Schiff bases from chitosan and salicylaldehyde/TiO2 nanocomposite membrane. *International Journal of Biological Macromolecules* 124:802–809.
28. Roilo, D., C.A. Maestri, M. Scarpa, P. Bettotti, and R. Checchetto. 2018. Gas barrier and optical properties of cellulose nanofiber coatings with dispersed TiO 2 nanoparticles. *Surface and Coatings Technology* 343:131–137.
29. Siripatrawan, U. and P. Kaewklin. 2018. Fabrication and characterization of chitosan-titanium dioxide nanocomposite film as ethylene scavenging and antimicrobial active food packaging. *Food Hydrocolloids* 84:125–134.
30. Buschow, K.H.J. 2001. *Encyclopedia of Materials: Science and Technology*. Amsterdam; New York: Elsevier.
31. Al-Naamani, L., J. Dutta, and S. Dobretsov. 2018. Nanocomposite zinc oxide-chitosan coatings on polyethylene films for extending storage life of okra (Abelmoschus esculentus). *Nanomaterials* 8(7):479.
32. Hatamie, A., et al. 2015. Zinc oxide nanostructure-modified textile and its application to biosensing, photocatalysis, and as antibacterial material. *Langmuir* 31(39): 10913–10921.
33. Newman, M.D., M. Stotland, and J.I. Ellis. 2009. The safety of nanosized particles in titanium dioxide– and zinc oxide–based sunscreens. *Journal of the American Academy of Dermatology* 61(4):685–692.
34. Sirelkhatim, A., S. Mahmud, A. Seeni, N.H.M. Kaus, L.C. Ann, S.K.M. Bakhori, H. Hasan, and D. Mohamad. 2015. Review on zinc oxide nanoparticles: Antibacterial activity and toxicity mechanism. *Nano-Micro Letters* 7(3):219–242.
35. Jiang, J., J. Pi, and J. Cai. 2018. The advancing of zinc oxide nanoparticles for biomedical applications. *Bioinorganic Chemistry and Applications* 2018:1–18.
36. Espitia, P.J.P., C.G. Otoni, and N.F.F. Soares. 2016. Zinc oxide nanoparticles for food packaging applications. In: Barros-Velázquez, J. (ed), *Antimicrobial Food Packaging*, pp. 425–431. San Diego: Academic Press.
37. Espitia, P.J.P., N.d.F.F. Soares, J.S.d.R. Coimbra, N.J. de Andrade, R.S. Cruz, and E.A.A. Medeiros. 2012. Zinc oxide nanoparticles: Synthesis, antimicrobial activity and food packaging applications. *Food and Bioprocess Technology* 5(5):1447–1464.
38. Król, A., P. Pomastowski, K. Rafińska, V. Railean-Plugaru, and B. Buszewski. 2017. Zinc oxide nanoparticles: Synthesis, antiseptic activity and toxicity mechanism. *Advances in Colloid and Interface Science* 249:37–52.
39. Kołodziejczak-Radzimska, A. and T. Jesionowski. 2014. Zinc oxide—From synthesis to application: A review. *Materials* 7(4):2833–2881.
40. Soren, S., S. Kumar, S. Mishra, P.K. Jena, S.K. Verma, and P. Parhi. 2018. Evaluation of antibacterial and antioxidant potential of the zinc oxide nanoparticles synthesized by aqueous and polyol method. *Microbial Pathogenesis* 119:145–151.
41. Indumathi, M.P. and G.R. Rajarajeswari 2019. Mahua oil-based polyurethane/chitosan/nano ZnO composite films for biodegradable food packaging applications. *International Journal of Biological Macromolecules* 124:163–174.
42. Kumar, S., J.C. Boro, D. Ray, A. Mukherjee, and J. Dutta. 2019. Bionanocomposite films of agar incorporated with ZnO nanoparticles as an active packaging material for shelf life extension of green grape. *Heliyon* 5(6):e01867.

43. Happy, A., M. Soumya, S. Venkat Kumar, and S. Rajeshkumar. 2018. Mechanistic study on antibacterial action of zinc oxide nanoparticles synthesized using green route. *Chemico-Biological Interactions* 286:60–70.
44. Xu, J., R. Song, Y. Dai, S. Yang, J. Li, and R. Wei. 2019. Characterization of zinc oxide nanoparticles-epoxy resin composite and its antibacterial effects on spoilage bacteria derived from silvery pomfret (Pampus argenteus). *Food Packaging and Shelf Life* 22:100418.
45. Al-Naamani, L., S. Dobretsov, and J. Dutta. 2016. Chitosan-zinc oxide nanoparticle composite coating for active food packaging applications. *Innovative Food Science and Emerging Technologies* 38:231–237.
46. Shankar, S., L.-F. Wang, and J.-W. Rhim. 2018. Incorporation of zinc oxide nanoparticles improved the mechanical, water vapor barrier, UV-light barrier, and antibacterial properties of PLA-based nanocomposite films. *Materials Science and Engineering: Part C* 93:289–298.
47. Sadegh-Hassani, F. and A. Mohammadi Nafchi. 2014. Preparation and characterization of bionanocomposite films based on potato starch/halloysite nanoclay. *International Journal of Biological Macromolecules* 67:458–462.
48. Xia, Y., M. Rubino, and R. Auras. 2019. Interaction of nanoclay-reinforced packaging nanocomposites with food simulants and compost environments. Advances in Food and Nutrition Research 88:275–298.
49. Busolo, M.A., P. Fernandez, M.J. Ocio, and J.M. Lagaron. 2010. Novel silver-based nanoclay as an antimicrobial in polylactic acid food packaging coatings. *Food Additives and Contaminants: Part A* 27(11):1617–1626.
50. Bumbudsanpharoke, N. and S. Ko. 2019. Nanoclays in food and beverage packaging. *Journal of Nanomaterials* 2019:1–13.
51. Majeed, K., M. Jawaid, A. Hassan, A. Abu Bakar, H.P.S. Abdul Khalil, A.A. Salema, and I. Inuwa. 2013. Potential materials for food packaging from nanoclay/natural fibres filled hybrid composites. *Materials and Design* 46:391–410.
52. Sothornvit, R. 2019. Nanostructured materials for food packaging systems: New functional properties. *Current Opinion in Food Science* 25:82–87.
53. Abreu, A.S., M. Oliveira, A. de Sá, R.M. Rodrigues, M.A. Cerqueira, A.A. Vicente, and A.V. Machado. 2015. Antimicrobial nanostructured starch based films for packaging. *Carbohydrate Polymers* 129:127–134.
54. Davachi, S.M. and A.S. Shekarabi. 2018. Preparation and characterization of antibacterial, eco-friendly edible nanocomposite films containing Salvia Macrosiphon and nanoclay. *International Journal of Biological Macromolecules* 113:66–72.
55. Pereira de Abreu, D.A., P. Paseiro Losada, I. Angulo, and J.M. Cruz. 2007. Development of new polyolefin films with nanoclays for application in food packaging. *European Polymer Journal* 43(6):2229–2243.
56. Bodaghi, H., Y. Mostofi, A. Oromiehie, B. Ghanbarzadeh, and Z.G. Hagh. 2015. Synthesis of clay-TiO_2 nanocomposite thin films with barrier and photocatalytic properties for food packaging application. *Journal of Applied Polymer Science* 132:41764.
57. Tornuk, F., M. Hancer, O. Sagdic, and H. Yetim. 2015. LLDPE based food packaging incorporated with nanoclays grafted with bioactive compounds to extend shelf life of some meat products. *LWT—Food Science and Technology* 64(2):540–546.
58. Rostamzad, H., S.Y. Paighambari, B. Shabanpour, S.M. Ojagh, and S.M. Mousavi. 2016. Improvement of fish protein film with nanoclay and transglutaminase for food packaging. *Food Packaging and Shelf Life* 7:1–7.
59. Różańska, A., A. Chmielarczyk, D. Romaniszyn, A. Sroka-Oleksiak, M. Bulanda, M. Walkowicz, P. Osuch, and T. Knych. 2017. Antimicrobial properties of selected copper alloys on staphylococcus aureus and escherichia coli in different simulations of environmental conditions: With vs. without organic contamination. *International Journal of Environmental Research and Public Health* 14(7):813.

60. Ruparelia, J.P., A.K. Chatterjee, S.P. Duttagupta, and S. Mukherji. 2008. Strain specificity in antimicrobial activity of silver and copper nanoparticles. *Acta Biomaterialia* 4(3):707–716.
61. Tamilvanan, A., K. Balamurugan, K. Ponappa, and B.M. Kumar. 2014. Copper nanoparticles: Synthetic strategies, properties and multifunctional application. *International Journal of Nanoscience* 13(02):1430001.
62. Nouri, A., M.T. Yaraki, M. Ghorbanpour, S. Agarwal, and V.K. Gupta. 2018. Enhanced antibacterial effect of chitosan film using montmorillonite/CuO nanocomposite. *International Journal of Biological Macromolecules* 109:1219–1231.
63. Bruna, J.E., A. Peñaloza, A. Guarda, F. Rodríguez, and M.J. Galotto. 2012. Development of MtCu2+/LDPE nanocomposites with antimicrobial activity for potential use in food packaging. *Applied Clay Science* 58:79–87.
64. Hasheminya, S.-M., R. Rezaei Mokarram, B. Ghanbarzadeh, H. Hamishekar, and H.S. Kafil. 2018. Physicochemical, mechanical, optical, microstructural and antimicrobial properties of novel kefiran-carboxymethyl cellulose biocomposite films as influenced by copper oxide nanoparticles (CuONPs). *Food Packaging and Shelf Life* 17:196–204.
65. Longano, D., et al. 2012. Analytical characterization of laser-generated copper nanoparticles for antibacterial composite food packaging. *Analytical and Bioanalytical Chemistry* 403(4):1179–1186.
66. Shankar, S., X. Teng, and J.-W. Rhim. 2014. Properties and characterization of agar/CuNP bionanocomposite films prepared with different copper salts and reducing agents. *Carbohydrate Polymers* 114:484–492.
67. Shankar, S., L.-F. Wang, and J.-W. Rhim. 2017. Preparation and properties of carbohydrate-based composite films incorporated with CuO nanoparticles. *Carbohydrate Polymers* 169:264–271.
68. Vasile, C., et al. 2017. New PLA/ZnO:Cu/Ag bionanocomposites for food packaging. *Express Polymer Letters* 11(7):531–544.
69. Zhong, T., G.S. Oporto, and J. Jaczynski. 2017. Antimicrobial food packaging with cellulose-copper nanoparticles embedded in thermoplastic resins. In: Grumezescu, A.M. (ed), *Food Preservation*, pp. 671–702. San Diego: Academic Press.
70. Muthulakshmi, L., A. Varada Rajalu, G.S. Kaliaraj, S. Siengchin, J. Parameswaranpillai, and R. Saraswathi. 2019. Preparation of cellulose/copper nanoparticles bionanocomposite films using a bioflocculant polymer as reducing agent for antibacterial and anticorrosion applications. *Composites Part B: Engineering* 175:107177.
71. Ghanbarzadeh, B., S.A. Oleyaei, and H. Almasi. 2014. Nanostructured materials utilized in biopolymer-based plastics for food packaging applications. *Critical Reviews in Food Science and Nutrition* 55(12):1699–1723.
72. Lin, L., Y. Gu, and H. Cui. 2019. Moringa oil/chitosan nanoparticles embedded gelatin nanofibers for food packaging against Listeria monocytogenes and Staphylococcus aureus on cheese. *Food Packaging and Shelf Life* 19:86–93.
73. Vahedikia, N., F. Garavand, B. Tajeddin, I. Cacciotti, S.M. Jafari, T. Omidi, and Z. Zahedi. 2019. Biodegradable zein film composites reinforced with chitosan nanoparticles and cinnamon essential oil: Physical, mechanical, structural and antimicrobial attributes. *Colloids and Surfaces, Part B: Biointerfaces* 177:25–32.
74. Wu, C., J. Sun, Y. Lu, T. Wu, J. Pang, and Y. Hu. 2019. In situ self-assembly chitosan/ε-polylysine bionanocomposite film with enhanced antimicrobial properties for food packaging. *International Journal of Biological Macromolecules* 132:385–392.
75. Glaser, T.K., O. Plohl, A. Vesel, U. Ajdnik, N.P. Ulrih, M.K. Hrnčič, U. Bren, and L. Fras Zemljič. 2019. Functionalization of polyethylene (PE) and polypropylene (PP) material using chitosan nanoparticles with incorporated resveratrol as potential active packaging. *Materials* 12(13):2118.

76. Yadav, M., Y.-K. Liu, and F.-C. Chiu. 2019. Fabrication of cellulose nanocrystal/silver/alginate bionanocomposite films with enhanced mechanical and barrier properties for food packaging application. *Nanomaterials* 9(11):1523.
77. Shahbazi, Y. and N. Shavisi. 2018. Effects of sodium alginate coating containing Mentha spicata essential oil and cellulose nanoparticles on extending the shelf life of raw silver carp (Hypophthalmichthys molitrix) fillets. *Food Science and Biotechnology* 28(2):433–440.
78. Sobhan, A., K. Muthukumarappan, Z. Cen, and L. Wei. 2019. Characterization of nanocellulose and activated carbon nanocomposite films' biosensing properties for smart packaging. *Carbohydrate Polymers* 225:115189.
79. Maliha, M., M. Herdman, R. Brammananth, M. McDonald, R. Coppel, M. Werrett, P. Andrews, and W. Batchelor. 2019. Bismuth phosphinate incorporated nanocellulose sheets with antimicrobial and barrier properties for packaging applications. *Journal of Cleaner Production* 24:119016.
80. Liu, Y., S. Wang, W. Lan, and W. Qin. 2019. Fabrication of polylactic acid/carbon nanotubes/chitosan composite fibers by electrospinning for strawberry preservation. *International Journal of Biological Macromolecules* 121:1329–1336.
81. Pereda, M., N.E. Marcovich, and M.R. Ansorena. 2019. Nanotechnology in food packaging applications: Barrier materials, antimicrobial agents, sensors, and safety assessment. In: Martínez, L., Kharissova, O., Kharisov, B. (eds), *Handbook of Ecomaterials*, pp. 2035–2056. Springer, Cham.
82. Qin, Y., Y. Liu, L. Yuan, H. Yong, and J. Liu. 2019. Preparation and characterization of antioxidant, antimicrobial and pH-sensitive films based on chitosan, silver nanoparticles and purple corn extract. *Food Hydrocolloids* 96:102–111.
83. Ma, Q. and L. Wang. 2016. Preparation of a visual pH-sensing film based on tara gum incorporating cellulose and extracts from grape skins. *Sensors and Actuators. Part B: Chemical* 235:401–407.
84. Zhang, J., X. Zou, X. Zhai, X. Huang, C. Jiang, and M. Holmes. 2019. Preparation of an intelligent pH film based on biodegradable polymers and roselle anthocyanins for monitoring pork freshness. *Food Chemistry* 272:306–312.
85. Jayakumar, A., K.V. Heera, T.S. Sumi, M. Joseph, S. Mathew, P.G., I.C. Nair, and R.E.K. 2019. Starch-PVA composite films with zinc-oxide nanoparticles and phytochemicals as intelligent pH sensing wraps for food packaging application. *International Journal of Biological Macromolecules* 136:395–403.
86. Ma, Q., T. Liang, L. Cao, and L. Wang. 2018. Intelligent poly (vinyl alcohol)-chitosan nanoparticles-mulberry extracts films capable of monitoring pH variations. *International Journal of Biological Macromolecules* 108:576–584.
87. Gutiérrez, T.J., A.G. Ponce, and V.A. Alvarez. 2017. Nano-clays from natural and modified montmorillonite with and without added blueberry extract for active and intelligent food nanopackaging materials. *Materials Chemistry and Physics* 194: 283–292.
88. Singh, T., S. Shukla, P. Kumar, V. Wahla, V.K. Bajpai, and I.A. Rather. 2017. Application of nanotechnology in food science: Perception and overview. *Frontiers in Microbiology* 8:1501.
89. Joung, C.-K., H.-N. Kim, M.-C. Lim, T.-J. Jeon, H.-Y. Kim, and Y.-R. Kim. 2013. A nanoporous membrane-based impedimetric immunosensor for label-free detection of pathogenic bacteria in whole milk. *Biosensors and Bioelectronics* 44:210–215.
90. Kritchenkov, A.S., et al. 2019. Natural polysaccharide-based smart (temperature sensing) and active (antibacterial, antioxidant and photoprotective) nanoparticles with potential application in biocompatible food coatings. *International Journal of Biological Macromolecules* 134:480–486.

91. Wang, Y.-C., C.O. Mohan, J. Guan, C.N. Ravishankar, and S. Gunasekaran. 2018. Chitosan and gold nanoparticles-based thermal history indicators and frozen indicators for perishable and temperature-sensitive products. *Food Control* 85:186–193.
92. Singh, S., K.K. Gaikwad, M. Lee, and Y.S. Lee. 2018. Temperature sensitive smart packaging for monitoring the shelf life of fresh beef. *Journal of Food Engineering* 234:41–49.
93. Zhang, C., et al. 2013. Time–temperature indicator for perishable products based on kinetically programmable ag overgrowth on Au nanorods. *ACS Nano* 7(5):4561–4568.
94. Fröhlich, E. and E. Fröhlich. 2016. Cytotoxicity of nanoparticles contained in food on intestinal cells and the gut microbiota. *International Journal of Molecular Sciences* 17(4):509.
95. Wu, H., J.-J. Yin, W.G. Wamer, M. Zeng, and Y.M. Lo. 2014. Reactive oxygen species-related activities of nano-iron metal and nano-iron oxides. *Journal of Food and Drug Analysis* 22(1):86–94.
96. Buzea, C. and I. Pacheco. 2019. Toxicity of nanoparticles. In: Pacheco-Torgal, F., Diamanti, M.V., Nazari, A., Goran-Granqvist, C., Pruna, A., Amirkhanian, S. (eds), *Nanotechnology in Eco-Efficient Construction*, Elsevier, pp. 705–754. Cambridge: Woodhead Publishing.
97. De Jong, W.H., et al. 2018. Toxicity of copper oxide and basic copper carbonate nanoparticles after short-term oral exposure in rats. *Nanotoxicology* 13(1):50–72.
98. Patel, S., S. Jana, R. Chetty, S. Thakore, M. Singh, and R. Devkar. 2017. Toxicity evaluation of magnetic iron oxide nanoparticles reveals neuronal loss in chicken embryo. *Drug and Chemical Toxicology* 42(1):1–8.
99. Liu, J., Y. Kang, S. Yin, B. Song, L. Wei, L. Chen, and L. Shao. 2017. Zinc oxide nanoparticles induce toxic responses in human neuroblastoma SHSY5Y cells in a size-dependent manner. *International Journal of Nanomedicine* 12:8085–8099.
100. Singh, S. 2019. Zinc oxide nanoparticles impacts: Cytotoxicity, genotoxicity, developmental toxicity, and neurotoxicity. *Toxicology Mechanisms and Methods* 29(4):300–311.
101. Gaillet, S. and J.-M. Rouanet. 2015. Silver nanoparticles: Their potential toxic effects after oral exposure and underlying mechanisms—A review. *Food and Chemical Toxicology* 77:58–63.
102. Yang, Y., Z. Qin, W. Zeng, T. Yang, Y. Cao, C. Mei, and Y. Kuang. 2017. Toxicity assessment of nanoparticles in various systems and organs. *Nanotechnology Reviews* 6(3):279–289.

Index

F

G

H

I